Studies in Energy, Resource and Environmental Economics

This book series offers an outlet for cutting-edge research on all areas of energy, environmental and resource economics. The series welcomes theoretically sound and empirically robust monographs and edited volumes, as well as textbooks and handbooks from various disciplines and approaches on topics such as energy and resource markets, the economics of climate change, environmental evaluation, policy issues, and related fields. All titles in the series are peer-reviewed.

Alexander Wimmers · Fanny Böse ·
Claudia Kemfert · Christian von Hirschhausen
Editors

Nuclear Power

Technology, Geopolitics, and Economics

Springer

Editors
Alexander Wimmers
Workgroup for Infrastructure Policy (WIP)
Technical University (TU) Berlin
Berlin, Germany

German Institute for Economic Research
(DIW Berlin)
Berlin, Germany

Claudia Kemfert
Professorship for Energy Economics
and Energy Policy
Leuphana University Lüneburg
Lüneburg, Germany

German Institute for Economic Research
(DIW Berlin)
Berlin, Germany

Fanny Böse
Federal Office for the Safety of Nuclear
Waste Management (BASE) in cooperation
with TU Berlin
Berlin, Germany

Christian von Hirschhausen
Workgroup for Infrastructure Policy (WIP)
Technical University (TU) Berlin
Berlin, Germany

German Institute for Economic Research
(DIW Berlin)
Berlin, Germany

ISSN 2731-3409 ISSN 2731-3417 (electronic)
Studies in Energy, Resource and Environmental Economics
ISBN 978-3-031-99893-5 ISBN 978-3-031-99894-2 (eBook)
https://doi.org/10.1007/978-3-031-99894-2

This work was supported by Technische Universität (TU) Berlin and DIW Berlin.

This Springer imprint is published by the registered company Springer Nature Switzerland AG
The registered company address is: Gewerbestrasse 11, 6330 Cham, Switzerland

Contents

Contributors

Jochen Ahlswede Federal Office for the Safety of Nuclear Waste Management (BASE), Berlin, Germany

Mahdi Awawda TU Berlin, Berlin, Germany

Stefania Böhnlein Workgroup for Infrastructure Policy (WIP), Technical University (TU) Berlin, Berlin, Germany

Fanny Böse Federal Office for the Safety of Nuclear Waste Management (BASE) in cooperation with TU Berlin, Berlin, Germany

Matthias Englert Öko-Institut e.V., Darmstadt, Germany

Youssef Said Farag Workgroup for Infrastructure Policy (WIP), Technical University (TU) Berlin, Berlin, Germany

Leonard Göke Reliability and Risk Engineering, ETH Zürich, Zürich, Switzerland

Jan Graefje Federal Office for the Safety of Nuclear Waste Management (BASE), Berlin, Germany

Benedikt Heyl TU Berlin, Berlin, Germany

Claudia Kemfert German Institute for Economic Research (DIW Berlin), Berlin, Germany;
Professorship for Energy Economics and Energy Policy, Leuphana University Lüneburg, Lüneburg, Germany

Doug Koplow Earth Track, Cambridge, MA, USA

Johanna Krauß Workgroup for Infrastructure Policy (WIP), Technical University (TU) Berlin, Berlin, Germany

Sophie Kretzschmar Nuclear Verification and Disarmament, RWTH Aachen University, Aachen, Germany

Christoph Pistner Öko-Institut e.V., Darmstadt, Germany

Fabian Präger Workgroup for Infrastructure Policy (WIP), Technical University (TU) Berlin, Berlin, Germany

Elisabeth Saar Department of Earth Systems Sciences, University of Hamburg, Hamburg, Germany

Gero Scheck Workgroup for Infrastructure Policy (WIP), Technical University (TU) Berlin, Berlin, Germany

Hendrik Schopmans Federal Office for the Safety of Nuclear Waste Management (BASE), Berlin, Germany

Björn Steigerwald Workgroup for Infrastructure Policy (WIP), Technical University (TU) Berlin, Berlin, Germany;
German Institute for Economic Research (DIW Berlin), Berlin, Germany

Maren Vieluf Institute for Political Science, University of Innsbruck, Austria / German Federal Foreign Office, Berlin, Germany

Christian von Hirschhausen Workgroup for Infrastructure Policy (WIP), Technical University (TU) Berlin, Berlin, Germany;
German Institute for Economic Research (DIW Berlin), Berlin, Germany

Niklas Wagner TU Berlin, Berlin, Germany

Ben Wealer Workgroup for Infrastructure Policy (WIP), Technical University (TU) Berlin, Berlin, Germany

Jens Weibezahn Copenhagen School of Energy Infrastructure (CSEI), Copenhagen Business School, Frederiksberg, Denmark

Alexander Wimmers Workgroup for Infrastructure Policy (WIP), Technical University (TU) Berlin, Berlin, Germany;
German Institute for Economic Research (DIW Berlin), Berlin, Germany

Insa Zillmann TU Berlin, Berlin, Germany

Abbreviations

ABWR	Advanced Boiling Water Reactor
ACR	Advanced CANDU Reactor
ADP	Accelerated Decommissioning Partners
ADS	Accelerator Driven System
AEC	US Atomic Energy Commission
AGR	Advanced Gas Cooled Reactor
ANL	Argonne National Laboratory
APIS	Accelerated Program for Implementation of Secure VVER Fuel Supply
ARDP	Advanced Reactor Demonstration Program
ARPA-E	Advanced Research Projects Agency-Energy
ASN	French Nuclear Safety Authority
BAEC	Bangladeshi Atomic Energy Commission
BASE	German Federal Office for the Safety of Nuclear Waste Management
BEIS	UK Department for Business, Energy & Industrial Strategy
BfS	German Federal Office for Radiation Protection
BGE	German Federal Company for Final Storage of Radioactive Waste
BGZ	German Federal Company for Interim Storage of Radioactive Waste
BOO	Build-Own-Operate
BOOT	Build-Own-Operate-Transfer
BOT	Build-Operate-Transfer
BWR	Boiling Water Reactor
CANDU	Canada Deuterium Uranium (reactor)
CEFR	China Experimental Fast Reactor
CFS	Commonwealth Fusion Systems
CGN	China General Nuclear Power Group
CISF	Consolidated Interim Waste Storage Facilities
CNNC	China National Nuclear Corporation
CTBT	Comprehensive Nuclear-Test-Ban Treaty
DA	Domestic Agency
DBFO(M)	Design-Build-Finance-Operate-(Maintain)

DCE	Site-specific Decommissioning Cost Estimates
DESNZ	UK Department for Energy Security & Net Zero
DoD	US Department of Defense
DOE	US Department of Energy
DTF	Nuclear Decommissioning Trust Fund
DU	Depleted Uranium
EDF	Électicité de France
EPAct	Energy Policy Act
ESA	Euratom Supply Agency
ESG	Environment, Sustainability, Governance
ESK	German Commission for Nuclear Waste Management
EU	European Union
EW	Exempt Waste
FBR	Fast Breeder Reactor
FDSA	Final Storage and Disposal Activities
FDSP	Final Disposal Selection Procedure
FIA	Fusion Industry Association
FOAK	First-of-a-kind
FR	Reactors with a Fast Neutron Spectrum (Fast Reactors, see also FBR)
FSSR	Final Status Survey Report
FusDIS	International Atomic Energy Agency's (IAEA) Fusion Device Information System
GCR	Gas-Cooled Reactor
GDF	Geological Disposal Facility
GE	General Electric
GFR	Gas-cooled Fast Reactor
GIF	Generation IV International Forum
GNF	Global Nuclear Fuels
GNS	Gesellschaft für Nuklear-Service
GTCC	"Greater than Class C" Waste
HALEU	High-Assay Low Enriched Uranium
HAW	Higher Activity (Radioactive) Waste
HEU	Highly Enriched Uranium
HHI	Herfindahl-Hirschman Index
HLW	High-Level Radioactive Waste
HTR	High-Cooled Fast Reactor
HTS	High-Temperature Superconducting Magnets
HWR	Heavy Water Reactor
IAEA	International Atomic Energy Agency
ICBM	Intercontinental Ballistic Missiles
ICF	Inertial Confinement Fusion
ILW	Intermediate-Level Radioactive Waste
IMF	Inert Matrix Fuels
INFUSE	Innovation Network for Fusion Energy
IO	ITER Organization

IR	Inflation Rate
ISFSI	Independent Spent Fuel Storage Installation
ITER	International Thermonuclear Experimental Reactor
JAEC	Japan Atomic Energy Commission
JET	Joint European Torus
KENFO	German Nuclear Waste Disposal Fund
KEPCO	Korea Electric Power Corporation
KWU	Kraftwerk Union
LANL	Los Alamos National Laboratory
LCOE	Levelized Cost of Electricity
LEU	Low-Enriched Uranium
LFR	Lead-Cooled Fast Reactor
LFTR	Liquid Flouride Thorium Reactor
LILW	Low-and Intermediate Level Radioactive Waste
LLC	Limited Liabiltiy Company
LLW	Low-Level Radioactive Waste
LLWR	Low-Level Waste Repository
LP&S	Legacy Ponds and Silos
LTE	Long-Term Enclosure
LTP	License Termination Plan
LTS	Low-Temperature Superconducting Magnets
LWGR	Light-Water Cooled and Graphite Moderated Reactor
LWR	Light-Water Reactors
MCF	Magnetic Confinement Fusion
MCFR	Molten Chloride Fast Reactor
MIF	Magnetic Inertial Fusion
MIT	Massachusetts Institute of Technology
MLP	Multi-Level Perspective
MOX	Mixed Oxide
MSR	Molten Salt Reactor
MSRE	Molten Salt Reactor Experiment
MSZ	Machine-Building Plant Electrostal (Russian nuclear fuel manufacturing plant)
MTHM	Metric Tons of Heavy Metal
MWh	Megawatt-hour
NCCP	Novosibirsk Chemical Concentrates Plant (Russian nuclear fuel manufacturing plant)
NDA	Nuclear Decommissioning Authority
NEA	Nuclear Energy Agency of the OECD
NFI	Nuclear Fuel Industries
NIF	National Lgnition Facility
NLF	Nuclear Liabilities Fund
NOAK	N-th-of-a-kind
NORM	Naturally Occurring Radioactive Materials
NPP	Nuclear Power Plant

NPT	Non-Proliferation Treaty
NRC	US Nuclear Regulatory Commission
NRS	Nuclear Restoration Services
NSCI	Nuclear-Specific Cost Increase
NTS	Nuclear Transport Solutions
NWM	Nuclear Waste Management
NWS	Nuclear Waste Services
OCC	Overnight Construction Costs
ORNL	Oak Ridge National Laboratory
P&T	Partitioning and Transmutation
PBO	Parent Body Organization
PCAST	Committee of Advisory on Science and Technology
PHWR	Pressurized Heavy Water Reactor
PPA	Power Purchase Agreement
PPP	Public-Private Partnerships
PRDP	Power Reactor Demonstration Program
PSDAR	Post-Shutdown Decommissioning Activities Report
PSFC	Plasma Science and Fusion Center
PUREX	Plutonium Uranium Redox EXtraction
PWR	Pressurized Water Reactor
RAR	Reasonably Assured Resources
RBMK	Reaktor Bolschoi Moschtschnosti Kanalny
REBCO	Rare-Earth Barium Copper Oxide
REIT	Real Estate Invetment Trusts
ROI	Return On Investment
RPV	Reactor Pressure Vessel
RSK	German Reactor Safety Commission
SCWR	Ssupercritical Water Reactor
SFR	Sodium-cooled Fast Reactor
SGR	Sodium-cooled Graphite-moderated Reactor
SIPRI	Stockholm International Peace Research Institute
SLC	Site Licensing Company
SMR	Small Modular Reactor
SNF	Spent Nuclear Fuel
SNR	So-called New Reactors
SORR	Statement of Response Requirements
SSK	German Commission on Radiological Protection
StandAG	Standortauswahlgesetz
ST-system	Socio-Technical System
TFTR	Tokamak Fusion Test Reactor
THTR	Thorium-High Termperature Reactor
TRU	Transuranic (elements)
TVEL	Russian Nuclear Fuel Cycle Company, Part of Rosatom
TWR	Travelling Wave Reactor
UK	United Kingdom

UKAEA	UK Atomic Energy Authority
UKGI	UK Government Investments
UN	United Nations
US	United States (of America)
VHTR	Very-High Temperature Reactor
VLLW	Very Low-Level Radioactive Waste
VLSW	Very Short-Lived Radioactive Waste
VVER	Water-Water Power Reactor (Russian-designed PWR)
WACC	Weighted Average Cost of Capital
WBS	Work Breakdown Structure
WCS	Waste Control Specialists
WKGT	Warth & Klein Grant Thornton

Chapter 1
Introduction

Fanny Böse, Claudia Kemfert, Christian von Hirschhausen, and Alexander Wimmers

1.1 Introduction

Nuclear power is complex, and it is certainly the world's most complex energy generation technology:

i. It is a technology that can generate huge amounts of energy, through nuclear fission or fusion, within parts of a second, but also leads to ionizing radiation. Nuclear waste has to be shielded from nature and people for hundreds of thousands of years. Nuclear fusion, too, involves radioactive substances, such as tritium, with the danger of proliferation.
ii. Since its very beginnings, geopolitical aspects dominate the sector. This ranges from research and development (R&D) efforts on the nuclear bomb in the early 1940s to efforts of nuclear diplomacy nowadays. This implies geopolitically motivated attempts by the nuclear superpowers (the United States (US), Russia, and China) to diffuse nuclear technology to other countries to foster long-term geopolitical relations.
iii. Nuclear power is economically complex, too, due to its dual-use for military and commercial applications, its long time frames of R&D, energy production, and decommissioning and waste management, and to its lack of competitiveness

F. Böse (✉)
Federal Office for the Safety of Nuclear Waste Management (BASE) in cooperation with TU Berlin, Berlin, Germany
e-mail: fab@wip.tu-berlin.de

C. Kemfert · C. von Hirschhausen · A. Wimmers
German Institute for Economic Research (DIW Berlin), Berlin, Germany

C. von Hirschhausen · A. Wimmers
Workgroup for Infrastructure Policy (WIP), Technical University (TU) Berlin, Berlin, Germany

C. Kemfert
Professorship for Energy Economics and Energy Policy, Leuphana University Lüneburg, Lüneburg, Germany

A. Wimmers et al. (eds.), *Nuclear Power*, Studies in Energy, Resource and Environmental Economics, https://doi.org/10.1007/978-3-031-99894-2_1

in energy markets, requiring state support and/or cost pass-through for private operators to consumers.

iv. Last but not least, the "back-end," i.e., decommissioning and nuclear waste treatment, have been under-researched until recently, but now turn to be major challenges with respect to financing, technical challenges, and long time frames, ranging over a century hence.

These four topics make up the four parts of the book.

Why a new book? Technology, geopolitics, and economics are treated in the literature, but they are rarely dealt with jointly, with an integrated approach. Our colleague at the Paris School of Mines, Professor Francois Lévêque (2014, 202), has suggested that "nuclear power is the daughter of science and warfare," implying the interdependence between technology, both fundamental and applied aspects, and geopolitics. This complexity is often underrepresented in more economically focused research and publications. Neither can geopolitical aspects be "economized," nor is a cost–benefit analysis possible and useful. Economic analysis tends to abstract from the complexity of dual use. Therefore, the allocation of fixed costs to certain applications, e.g. bombs, or electricity, becomes quite random. Overall, we feel a need for this approach, while acknowledging more specialized books from which we have benefited. Examples are Paul Bracken (2012) on geopolitics and "The Second Nuclear Age"; Yim (2022) on "Nuclear Waste Management"; Steven Thomas' (1988) "Realities of Nuclear Power"; Geoffrey Rothwell's (2016) "Economics of Nuclear Power"; François Lévêque (2014) "Uncertainties of Nuclear Power" on technology, politics and economics; and last, but certainly not least, William Nuttall's (2022) recent book "Nuclear Renaissance: Technologies and Policies for the Future of Nuclear Power."

For this book, we present the three aspects that make up the complexity of nuclear power, without which neither the historical nor the current dynamics of the industry can be fully understood: technology—geopolitics—economics. The hypothesis is that the issues and controversies of nuclear power cannot be understood by a one-dimensional, disciplinary approach only, such as physics and chemistry, political science, or business and economics. On the contrary, an interdisciplinary approach is required that combines technical aspects, geopolitical motivations, and the business and economics perspective. This interdependence is quite clear in the early years of nuclear fission and fusion, i.e., the 1930 and '40 s, where, based on fundamental science, geopolitical motivation was the main driver of the fission and the fusion (hydrogen) bombs, respectively. After World War II, economic considerations entered the sector, such as the choice of "the best" reactor technology to generate heat and electricity, or the lowest subsidies to incite private companies to engage into a sector dominated by uncertainties. Today, questions remain interdependent, such as technology foci of R&D policies (e.g., fission against fusion research), the need to maintain a knowledge base in the face of geopolitical conflicts, among others, between the US, Russia, and China, or the role of nuclear power in a decarbonized, sustainable energy system. We add a fourth part to the book dedicated to a particularly complex, "wicked" problem—nuclear waste management. In the following, we sketch out the main topics of the book and provide an overview of the chapters.

1.2 Main Topics of the Book and Overview of Chapters

This book navigates between the past and the future and addresses the most urgent issues of the present. The different parts of this book share a conviction that disciplinary analysis of technology, geopolitics, and economics, is useful and necessary to highlight certain aspects of the nuclear industry, but that interdependencies are an essential element of understanding "the big picture" of the sector and, therefore, cross-references between different aspects are an integral part of the analysis. A minimum understanding of all three is required to understand the historical development of nuclear power in the twentieth century, and to be able to participate in the current and future debates of the 21st. Therefore, after this introductory part, the book is structured into three main parts: technology, geopolitics, and economics. To this, we add an additional part on the back end of the sector, to fill a niche in the existing literature on decommissioning nuclear power plants, and long-term disposal of nuclear waste.

1.2.1 Part I: Technology

When Henry Becquerel, Marie Curie, Albert Einstein, Ernest Rutherford, and many more researchers explored the functioning of atoms and their nuclei in the early twentieth century, they were far from realizing that only half a century later, these nuclear forces would be available in almost unlimited amounts, for both military and commercial uses, through nuclear fission and fusion. What is nuclear power, how do we distinguish between nuclear fission and nuclear fusion, and what developments are currently under way?

Part I of the book summarizes the main technological aspects that facilitate the understanding of this complex sector. In Chapter 2, "Nuclear Physics and Technologies," Alexander Wimmers, Fanny Böse and Christian von Hirschhausen provide a laymen's overview of nuclear physics (and some chemistry) and reactor technologies, including a section on radiation, which makes this industry so specific. The authors cover both nuclear fission, i.e., the splitting of atoms, until now the dominating form of nuclear energy, as well as nuclear fusion, the "underdog" that is now challenging fission. In Chapter 3, Ben Wealer, Alexander Wimmers, and Christian von Hirschhausen explain the fission industry structure of "Nuclear Fission as a System Good." This system is comprised of a complex chain of production, from the front-end including resources and fuel production, to the generation of nuclear power itself, and the back end of waste disposable; across these activities runs the dual-use character of military and commercial applications. In Chapter 4, Stefania Böhnlein, et al. analyze "The System Good Nuclear Fusion," to work out technological and organizational challenges of the commercial usage of nuclear fusion, likewise a complex system good, and very likely more complex than nuclear fission.

1.2.2 Part II: Geopolitics

Since July 16, 1945, when the first atomic bomb was detonated at 5:29 a.m. local time near Alamogordo in the New Mexico desert (US), it has been clear that nuclear fission can release enormous amounts of energy and radiation in fractions of a second. The first military application was very soon after, with the bombing and extensive destruction of the Japanese cities of Hiroshima (August 6, 1945) and Nagasaki (August 9, 1945), which resulted in hundreds of thousands of deaths and injuries. Even afterwards, during the budding Cold War between the US and the Soviet Union, considerable efforts were made to further develop nuclear weapons, in particular the hydrogen bomb. It is quite clear that nuclear energy was, is, and will be a geopolitical issue for the time being, in which secondary strands such as the energy industry, innovation policy, etc., can be interwoven, but at a subordinate level. The characterization of nuclear energy as a "daughter of science and warfare" described above, and the understanding of synergy effects between military and commercial activities, will accompany us throughout the text. It also shapes the understanding of geopolitical lines of conflict.

Part II therefore brings together different perspectives of geopolitical analyses. In Chapter 5, Christian von Hirschhausen, and Claudia Kemfert provide a "Historical Overview" about the role of nuclear power in geopolitical conflicts, since the 1940s until 2022, when discussions about the use of nuclear weapons resurged in the context of the Russian invasion of Ukraine. Chapter 6, by Maren Vieluf et al., provides an introduction into the "Nuclear Disarmament and Non-Proliferation" by showing historical motivations and current developments of the nuclear arms race in a multi-polar world. Chapter 7, by Gero Scheck et al., provides a survey of "Game Theoretical Considerations of Nuclear Armament," especially for nuclear deterrence. After giving a short introduction to the status quo of nuclear weapon states, it analyzes the effect that nuclear weapons have on strategic decision-making and conflict. In Chapter 8, Benedikt Heyl and Insa Zillmann are "Assessing Market Concentrations in the Nuclear Fuel Supply Chain" to find potential bottlenecks which could cause Western nuclear countries to depend on Russian nuclear fuel production. In Chapter 9, Alexander Wimmers, Youssef Farag, et al. analyze "Politics of New Build and Consequential Dependencies." This includes the use of nuclear power as an instrument of geopolitics, in this case by Russia, but also China and the United States, and shows the supremacy of Russian reactor exports.

1.2.3 Part III: Economics

After 1945, it took about a decade to launch different reactor types to attempt the production of electricity and heat at scale. This post-war period also includes attempts to utilize nuclear energy for commercial applications, e.g., in the transport sector, medical technology, materials research, and agriculture. Between 1954 and the end of the 1970s, construction began on 267 of the overall 483 commercial nuclear power plants connected to the grid by the end of 2021, i.e., over 55%. By the mid-1970s, however, hopes that nuclear energy would become cost-effective and— through the development of a plutonium economy—could contribute significantly to the energy supply were over. Today, the production of nuclear power is at about the same level as in the 1980s (2,700 terawatt hours), and its share of total electricity production worldwide is less than 10% (after a peak of 17.5% in 2013). This even corresponds to only 3% of global energy production, as other energies besides electricity are also used for heating, transport, etc., and the trend is downwards (Schneider et al. 2024). What does this peculiar development tell us about the economics of nuclear power?

Part III of the book offers a range of perspectives on the economics of commercial nuclear power. Among others, it develops metrics comparable to other energy technologies, but also sheds light on nuclear special characteristics in terms of the costs, financing, and long-term development. Chapter 10, "Economics of Nuclear Fission Power Plants—An LCOE (Levelized Costs of Electricity) Approach," provides an overview of the costs of different reactor types based on the LCOE approach and analyzes challenges that occur when nuclear energy is integrated into an energy system that is largely based on renewable energy sources (Wimmers et al.). In Chapter 11, the development of US nuclear reactor technology is analyzed through the Multi-Level Perspective framework, tracing its evolution from military applications during World War II through five distinct phases spanning 85 years. The chapter examines how socio-political factors, economic constraints, and public opposition contributed to the industry's ultimate stagnation, despite significant political support: "The First Nuclear Century: A Long-term Analysis of Nuclear Reactor Development in the US (1939–2024)" (Böse et al.). In Chapter 12, Doug Koplow dives into the state involvement in nuclear financing, highlighting increasing subsidies in recent years, as well as the hidden costs of nuclear decommissioning: "Nuclear Economics and Finance: Headwinds and Rising State Support". In Chapter 13, Christian von Hirschhausen analyzes "The Economic Value of Plutonium," a very special topic rarely addressed in energy economics. Chapter 14 offers insights into the emerging nuclear fusion landscape characterized by public, and increasingly private, new ventures. First insights into their development status and funding are given: "New Ventures in Fusion Energy: Analyzing a Dynamically Emerging Market" (Steigerwald, Krauß, and Weibezahn).

1.2.4 Part IV: Nuclear Waste Management

The historian of technology, Joachim Radkau (1983, 301), identified nuclear waste management 40 years ago as the "most embarrassing aspect" of the nuclear industry. He thus named the cognitive dissonance that final disposal was only considered very marginally in the early days of nuclear energy, "although—or precisely because—the secular dimension of this problem was seen early on,—earlier than most other problems of nuclear technology." It is not surprising that techno-economic research has largely ignored this important topic over the last decades. However, today we need to ensure that the legacy of nuclear energy, especially radioactive waste, is disposed of as safely as possible. More than 80 years after the first experimental developments at the interface between basic research and military applications, not a single gram of highly radioactive nuclear waste from the operation of commercial nuclear power plants has been disposed of in deep geological repositories so that people and the environment can be shielded from the risks of radiation release as safely and long-term as possible.

Part IV of the book on the back end and nuclear waste management brings together techno-economic considerations regarding the interim, as well as the final long-term storage of radioactive waste. It combines qualitative as well as quantitative analysis and offers various insights into pathways of different countries. In Chapter 15 "Nuclear Waste Disposal: Steps in the Process and Overview of International Experiences," Jochen Ahlswede et al. provide an overview of the different nuclear waste management pathways in multiple country case studies. By showing its socio-technical complexities, a focus is put on the final long-term storage in a deep geological repository and its limited progress. Chapter 16 conducts a deep dive comparison of different organizational models that can be observed in the nuclear decommissioning process in the US and the UK. It focuses on the involvement of the state and its potential strengths and weaknesses: "Organizational Models [and State Involvement] in Nuclear Decommissioning" (Wagner et al.). Awada et al. assess the "Very Long-Term Financing in Nuclear Waste Management" in Chapter 17 by looking at the performance of the German Nuclear Waste Disposal Fund (KENFO). It assesses the adequacy of the fund and provides a first assessment of the fund volumes in the context of a delay of the site selection process. In Chapter 18, Pistner and Englert offer a technological deep dive into alternative disposal pathways that have been discussed for decades: "Partitioning and Transmutation—Technical Overview and Scenarios for Waste Disposal". The chapter assesses the feasibility of proposed advantages by looking at its effects of long-term storage and waste scenarios.

1.3 Conclusions and Outlook

Nuclear power is complex, and this book is complex, too, with many topics, many authors, and the need to go from history into the future. In Part I, we observe technological advances over the past 80 years, but not sufficient to inverse the declining trend of commercial electricity production, essentially commensurate with the end of nuclear euphoria back in the 1970s. We also suggest that recent emphasis on nuclear fusion research might be triggered by difficulties of its "sister sector," nuclear fission. Given the tightening geopolitical tensions around the globe, Part II is particularly timely. However, it was also the most difficult to conceive, as there is no uniform model how to analyze nuclear diplomacy. On the contrary, we have identified "simple" metrics of economic questions in Part III. Given the dynamic development of other energies, and storage and network technology to deal with intermittency, the share of nuclear power in global energy is negligible in the mid-2020s. Chances are high that this will remain the case, given current technological dynamics globally.

Perhaps the biggest surprise for readers is the complexity of nuclear waste management, and the vast technical and economic challenges ahead of us. Part IV highlights approaches from the last 80 years, and sketches out trajectories for the next century, when the problem will still be far from finished. This extremely demanding task merits more attention.

As usual, the book leaves of couple of questions open, e.g., on innovation, security policy, and environmental policy. Do we need an overhaul of the security strategy, and higher expenditures on nuclear weapons? Where do we allocate scare R&D funds, in fission, and/or fusion, and how much? In addition, somewhat parallel to an important climate policy issue, how do we price plutonium and other nuclear radiants, to protect our and the next 30,000 generations against radioactivity? Alternatively, in other terms, how do we achieve climate and plutonium neutrality?

Acknowledgements Many people and organizations have contributed to the development of this book, so thanks are due in particular to many colleagues at the Technical University (TU) of Berlin, the German Institute for Economic Research (DIW Berlin), and other organizations who have contributed to our understanding of various controversies. Analyzing nuclear energy requires interdisciplinary exchanges, e.g., between the natural sciences, engineering, economics, and social sciences, from which we have benefited greatly. It goes without saying that we are deeply grateful to all the authors who contributed to this book—thank you very much!

We thank the participants in the annual AT-OM Research Workshop, held since 2022, which brings together senior and junior researchers in an informal setting. In the context of the World Nuclear Industry Status Report (WNISR) and by participating in the German chapter of the Scientists for Future (S4F), we have been able to gain many insights and would like to thank all those involved. We also had intensive exchanges with members of the Transens research network (https://www.transens.de) on transdisciplinary nuclear waste governance research. Additional interview partners were Arnulf Grübler and Nebojsa Nakicenovic.

Special thanks go to the other members of the Nuclear Energy Research Group at WIP, which was set up by Ben Wealer and is now coordinated by Fanny Böse and Alexander Wimmers, where we research, teach, and advise together with Alexander Buschner, Charlotte Dering, Youssef Said Farag, Fabian Präger, Gero Scheck, Björn Steigerwald, and Paul Ummethun. We have used this material in previous undergraduate and graduate courses, and thank the participants in the AT-TEG seminar, as well as in the sustainability lab. Our team assistant, Petra Haase, has supported the permanent fission and equally dynamic fusion of junior research groups over the last few years, without her there would only be decay products in the department. Special thanks go to Zoe Felder, Friederike Frieß, Annika Froese, Sarah Glück, and Carolin Grüter for their helpful comments on several chapters. We would also like to thank Springer Verlag, in particular Mr. Glaeser, for his editorial support, and other members of the staff for completing the publication. We extend our sincere gratitude to our external lecturer, Amy Kagan from St. Petersburg, Florida. Last but not least, many thanks also to our families, friends, and acquaintances for the good and controversial discussions on the subject!

References

Bracken, Paul. 2012. *The Second Nuclear Age—Strategy, Danger, and the New Power Politics*. New York, USA: Times Books—Henry Holt and Company, LLC.

Lévêque, François. 2014. *The Economics and Uncertainties of Nuclear Power*. Cambridge, United Kingdom: Cambridge University Press.

Nuttal, William J. 2022. *Nuclear Renaissance: Technologies and Policies for the Future of Nuclear Power*. Boca Raton: CRC Press. https://www.taylorfrancis.com/books/9781003038733

Radkau, Joachim. 1983. *Aufstieg und Krise der deutschen Atomwirtschaft 1945–1975: Verdrängte Alternativen in der Kerntechnik und der Ursprung der nuklearen Kontroverse*. Reinbek bei Hamburg: Rowohlt.

Rothwell, Geoffrey. 2016. *Economics of Nuclear Power*. London, UK: Routledge.

Schneider, Mycle, Antony Froggatt, Julie Hazemann, Özgür Gürbüz, Paul Jobin, Phil Johnstone, Timothy Judson, et al. 2024. 'The World Nuclear Industry Status Report 2024'. Paris, France: Mycle Schneider Consulting. https://www.worldnuclearreport.org/IMG/pdf/wnisr2024-v2.pdf

Thomas, Stephen. 1988. *The Realities of Nuclear Power: International Economic and Regulatory Experience*. Cambridge University Press.

Yim, Man-Sung. 2022. *Nuclear Waste Management: Science, Technology, and Policy*. Lecture Notes in Energy. Springer Dordrecht. https://doi.org/10.1007/978-94-024-2106-4

Part I
Technology

Chapter 2
Nuclear Physics and Reactor Technologies (Fission and Fusion)

Alexander Wimmers, Fanny Böse, and Christian von Hirschhausen

2.1 Introduction

Both nuclear weapons and reactors utilize the same form of energy stemming from the power of the splitting or the fusion of atoms. However, with this form of energy come challenges related to the release of radiation that can have potentially hazardous effects on humans and the environment alike, requiring an array of safety and security measures to minimize the release of radiation itself, and radioactive materials in general. A basic understanding of the physics and the chemistry of nuclear power and as well as a systematization of various technologies and their functioning is necessary to understand the energy sector—including the military sector.

This chapter provides a simplified introduction, first to the physics and chemistry of the atom, radiation as well as the different types of chain reactions (Sect. 2.2), and then introduces various types of reactors (Sect. 2.3) that have been tried, tested, operated, and sometimes were also abandoned. In addition to nuclear fission, i.e., the splitting of atoms, we also present the physics of nuclear fusion and the most important research activities in that field Sect. 2.4 concludes.

A. Wimmers (✉) · C. von Hirschhausen
Workgroup for Infrastructure Policy (WIP), Technical University (TU) Berlin, Berlin, Germany
e-mail: awi@wip.tu-berlin.de

German Institute for Economic Research (DIW Berlin), Berlin, Germany

F. Böse
Federal Office for the Safety of Nuclear Waste Management (BASE) in cooperation with TU Berlin, Berlin, Germany

A. Wimmers et al. (eds.), *Nuclear Power*, Studies in Energy, Resource and Environmental Economics, https://doi.org/10.1007/978-3-031-99894-2_2

2.2 Nuclear Physics and Chemistry

2.2.1 Nuclear Forces and the Atom

The following section introduces the terminology of and around the atom to provide a rudimentary understanding of the underlying processes allowing the operation of fission and fusion devices for energy production. The information presented here is based on Neles and Pistner (2012) and Yim (2022).

2.2.1.1 Four Forces

Every performed action can be traced back to one of the four fundamental forces or interactions. While the gravitational and electromagnetic forces are observable in everyday actions, the weak and strong nuclear forces apply only on a sub-atomic level.

- Gravitation is the force of attraction between all particles that have mass. Although the force itself is relatively weak, it has an unlimited range, making it responsible for forming large structures such as galaxies, stars, and planets.
- The electromagnetic force arises from the electric charge of particles. It can be either attractive or repulsive due to different electric charges. This force is responsible for the chemical behavior of atoms and molecules. It also has an infinite range.
- The strong nuclear force acts on the subatomic level and is responsible for binding nuclei consisting of positively charged particles that would—if the electromagnetic force were stronger—repulse each other.
- The weak nuclear force applies only to the subatomic level and is responsible for radioactive decay.

2.2.1.2 The Atom

Atoms are the building blocks of material. In general, they consist of three types of particles:

- Electrons are negatively charged.
- Protons are positively charged.
- Neutrons have no charge.

Non-ionized atoms have the same number of electrons and protons and, therefore, are neutrally charged. Protons and neutrons, collectively known as nucleons, are concentrated in the center of the atom, the nucleus. These particles make up most of the atomic mass and weigh approximately one amu (atomic mass unit, $\approx 1.6605 \times 10^{-27}$ kg). The electrons "circulate" around the nucleus and are practically weightless (Fig. 2.1). In reality, however, they are better described as electron clouds.

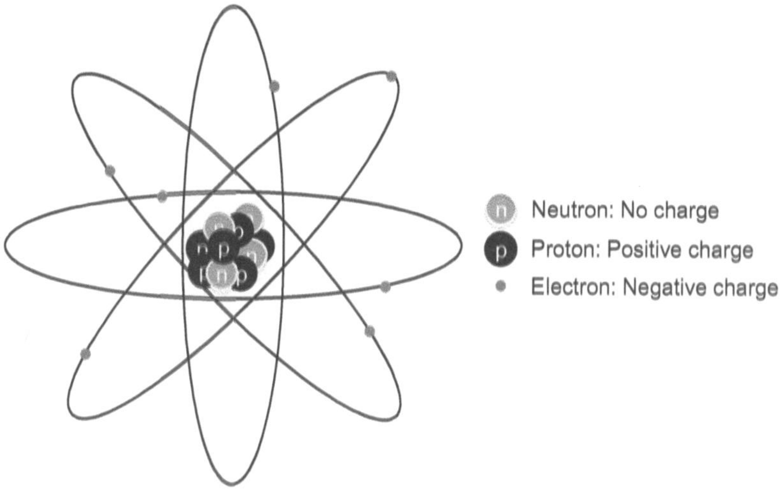

Fig. 2.1 The structure of the atom (*Source* Own depiction derived from Yim [2022, 53])

2.2.1.3 The Periodic Table and the Actinides

The elements of the periodic table are differentiated by their number of protons (Fig. 2.2). Hydrogen (H) has one proton and one electron, helium has two of each, while uranium (U) has 92 protons and electrons. The number of protons is referred to as the atomic number Z. The mass of an atom is determined by the number of protons and neutrons in the nucleus. The sum of both is given in the mass number A. For a given element, the number of neutrons can vary between atoms.

Versions of elements with different mass numbers are referred to as isotopes. Thus, isotopes are distinguished by the mass number. For example, the nucleus of a regular hydrogen atom consists of only one proton; it is therefore called H-1. The hydrogen isotope deuterium (H-2) has an additional neutron, doubling the mass. Another isotope is tritium (H-3) that has two extra neutrons. Other examples are U-235 (uranium isotope with 143 neutrons) or U-238 (uranium isotope with 146 neutrons).

In the context of nuclear fission energy, the actinides, i.e., elements with 89–102 protons, play an important role (Fig. 2.2). Naturally occurring actinides are uranium and thorium that can be used for large-scale nuclear fission. There are two groups of actinides that can be distinguished in reactor physics:

- Transuranic elements: actinides with atomic numbers greater than 92 (uranium) are called transuranic elements. These elements are a result of the neutron capture process during the operation of nuclear fission reactors, where a neutron hits a U-238 atom, which then transforms into transuranic elements such as Neptunium-239 and Plutonium-239 via two beta decays (see Eq. (2.1)).
- Minor actinides: elements that are typically produced in small amounts during the operation of nuclear reactors. These include isotopes such as neptunium,

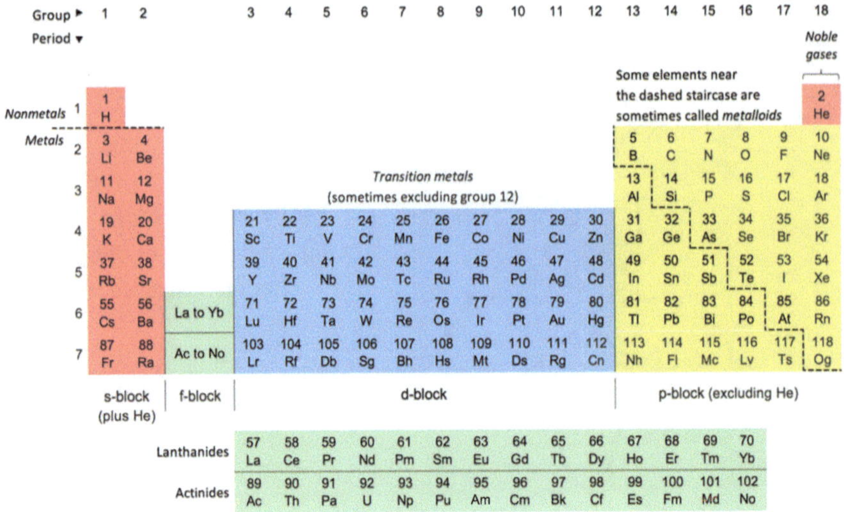

Fig. 2.2 Periodic table of the elements (*Source* Taken from Wikimeadia[1])

americium, and curium.

$$ {}^{238}_{92}U + {}^{1}_{0}n \rightarrow {}^{239}_{92}U \rightarrow {}^{239}_{93}Np + \beta^- \rightarrow {}^{239}_{94}Pu + 2\beta^- \quad (2.1) $$

2.2.1.4 Masses and Binding Energy

Since the strong nuclear force has a relatively short range, it primarily causes attraction between nucleons in close proximity. This means that the strength of this bond is largely independent of the overall size of the nucleus. However, the binding energy of nucleons within the nucleus is also influenced by electromagnetic interaction. As the number of protons in the nucleus increases, their mutual repulsion becomes larger due to their positive charge. The combination of both the strong and the electromagnetic force within the nucleus has two implications:

- The first effect is the resulting binding energy per nucleon. Initially, as the number of nucleons increases, the strength of the binding also increases (Fig. 2.3). This leads to a maximum in binding energy per nucleon at the nucleus of Fe-56 (iron). Consequently, Fe-56 is the most tightly bound nucleus with the optimal binding energy. The binding energy per nucleon is lower for both lighter and heavier nuclei.
- The second effect is a shift in the neutron-to-proton ratio in the nucleus. As nuclei become heavier, electromagnetic repulsion rises, and the binding energy

[1] https://commons.wikimedia.org/wiki/File:Colour_18-col_PT_with_labels.png, accessed on 08–10-2024.

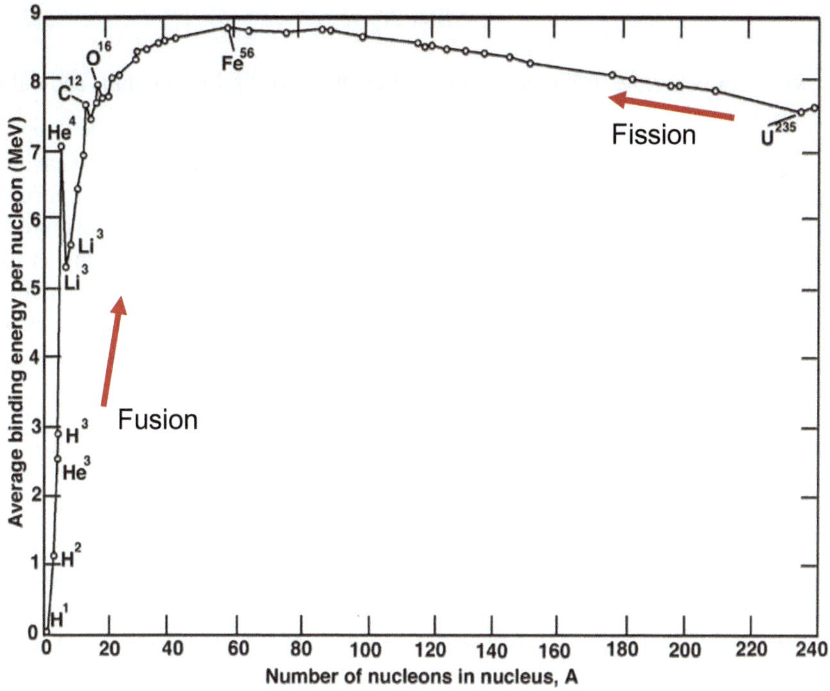

Fig. 2.3 Nuclear binding energy (*Source* Adapted from Wikimedia[2])

per nucleon decreases again. In light nuclei, optimal binding occurs when there is an equal number of protons and neutrons, resulting in stable nuclei. In nuclei with many protons, the repulsion between nucleons increases. For instance, the neutron-to-proton ratio in C-12 (carbon) is exactly one, while for the isotope U-235, the ratio is 1.55, meaning that for this uranium isotope, there are 55% more neutrons than protons in its nucleus. There are two types of nuclear reactions, described by Nuttall (2022, 239) as follows: "Fission obtains energy through the splitting of large heavy atoms [into smaller ones], while nuclear fusion produces far greater levels of energy through the joining together of small light atoms." The difference of energy released from either reaction is related to the nuclear binding energy that is the energy of the electric force subtracted from the value of the strong force. This results in the phenomenon known as the mass defect. For fission, the combined mass of the large atom and the neutron is slightly larger than the combined mass of the new nuclei and fission neutron. For fusion, the combined mass of smaller atoms is slightly larger than the combined mass of the fusion products. The released energy can thus be calculated using Albert Einstein's most famous formula (Eq. (2.2)):

---

[2] https://commons.wikimedia.org/wiki/File:Binding_energy_curve_-_common_isotopes.jpg, accessed on 13-11-2024.

$$\Delta E = \Delta m * c^2 \quad (2.2)$$

with Δm being the difference in mass after fusion or fission (in amu) and c the speed of light (≈298,000 km/s)

Figure 2.3 also provides an intuitive differentiation between nuclear fission and nuclear fusion. The difference between the reactions is shown by the difference in binding energy when moving along the elements, either from left to right (fusion), or right to left (fission). The delta of the binding energy per reaction is a lot higher for fission (≈185.33 MeV for a U-235 fission reaction) than for fusion (≈17.6 MeV for a deuterium–tritium fusion reaction). However, the specific energy release per nucleon is higher for fusion compared to fission as lighter atoms have a smaller number of nucleons involved in one nuclear reaction.

2.2.2 *Ionizing Radiation*

This section explores the topic of radiation by introducing the different types, the dangers arising from exposure to radiation, and the processes of radioactive decay and decay chains. The information presented here is based, among others, on Neles and Pistner (2012) and Yim (2022).

2.2.2.1 Types of Radiation

There are both stable and unstable isotopes. Stable isotopes remain in their current form indefinitely without outside interference. Unstable isotopes, however, undergo a spontaneous disintegration of a nucleus to form a more stable nucleus, a process known as radioactive decay. This decay results in the emittance of ionizing radiation, by which alpha and beta particles as well as gamma rays ionize atoms within (living) cells and thus alter the chemical properties thereof. In high doses, this can cause tissue to die, while in lower doses, it can result in mutations of the genetic programming of cells, which also results in cell death or—if the cell survives—cancer.

There are four types of radiation (Fig. 2.4):

- Alpha radiation: Alpha particles are large in mass (≈nucleus of a helium atom) and highly ionizing. Alpha radiation can cause serious damage, but the large mass of the particles allows alpha radiation to be easily blocked, e.g., by piece of paper, see Fig. 2.4.
- Beta radiation consists of very fast-moving electrons (beta minus) or positrons (beta plus). Due to their lower mass, the potential for biological harm is less than that of alpha radiation, but the speed of beta radiation greatly increases its penetration power, which can cause severe tissue damage. It can be blocked by metal foil.

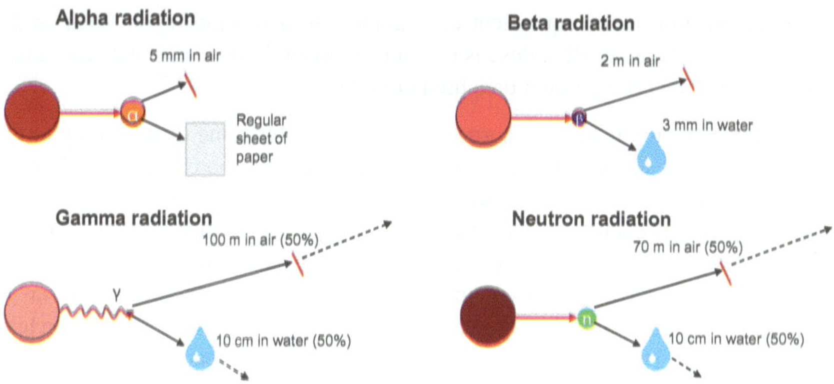

Fig. 2.4 Different types of radiation (*Source* Own depiction following [Neles and Pistner 2012, 41 s.q.])

- Gamma radiation is a very high-energy form of light that is not electrically charged or has mass but is extremely penetrating. It can only be stopped by several centimeters of lead, several meters of concrete or water.
- Neutron radiation: Free neutrons are produced during nuclear fission and fusion. They have a particularly high energy and penetration capability. They are well moderated by water, for example. Fast neutrons can also cause tissue damage.

Additionally, a handful of large nuclides, i.e., with a mass number of 230 or higher, can undergo so-called spontaneous fission in which the nucleus is split into two fragments or fission products. This process also releases gamma radiation and several neutrons. For example, in 3.1% of the cases, Californium-252 can decay spontaneously, and in all the other cases, it emits an alpha particle and is transformed to Curium-248 (see Sect. 2.2.3).

2.2.2.2 Doses and Biological Effects

Ionizing radiation is characterized by:

- The emittance of radiation by an object is measured in Bequerel (Bq) (disintegration per second).
- The energy dose, i.e., the energy that was absorbed by an object, is measured in Gray (Gy) (joules per kilogram).
- Different types of radiation can have different biological effects. Thus, the equivalent dose measures the energy that was absorbed by an object while accounting for the difference in effects of the types of radiation in terms of their biological effects by applying weight factors (i.e., the damage caused proportional to the mass charge or the energy of a given particle). It is measured in Sievert (Sv) (also expressed as joules per kilogram).

- The organ dose is the equivalent dose applied to a specific organ, such as the thyroid, and the equivalent dose is the sum of organ doses to calculate the actual biological effect of radiation on a human body.

Ionizing radiation is highly energetic and can transfer some of its energy to the skin or internal organs. This leads to physical processes in the body's atoms, primarily ionization, where electrons are ejected from an atom, leaving a positively charged ion. Another process is excitation, where energy is absorbed and then released, causing further reactions. These physical effects are not always harmful. For example, they may have no impact if they occur in insensitive tissue, like the outer skin layer. However, if vital molecules like proteins, enzymes, or DNA are damaged, it can lead to serious health issues, such as leukemia or cancer, even years later. Damage to reproductive cells may result in birth defects in offspring. Although repair mechanisms in the body can fix radiation-induced damage, this is not always successful, leaving a risk of long-term effects. Refer to Yim (2022, 155–215) for further details on the biological effects of radiation.

Leukemia, cancer, and genetic damage are examples of stochastic radiation effects. These occur with a certain probability that increases with higher radiation doses. The severity of the damage, e.g., the aggression of a cancer, is independent of the initial dose. In contrast, deterministic effects, like nausea, vomiting, diarrhea, fever, and anemia, occur only after a high radiation dose is received in a short time. These effects have a threshold dose above which the likelihood of fatal outcomes increases. At around two Gy, death is possible, and survival is unlikely above six Gy. The effects of radiation on non-human species remain understudied (Wimmers et al. 2024).

2.2.2.3 Radioactive Decay and Decay Chains

Oftentimes, decaying nuclei result in the formation of other radioactive nuclei. Radioactive decay can continue in a series also known as a decay chain. The decay process of radioactive elements occurs stochastically, meaning that many elements decay with a variation of decay forms, e.g., alpha, beta-plus, beta-minus, or gamma, and with a variation of energy levels, to transform into different (radioactive) elements. The decay occurs spontaneously but with a certain probability, but can also be artificially induced. For example, 64.06% of Bismuth-212 atoms decay to Polonium-212 via beta-minus decays, and the remaining ones (35.94%) decay to Thallium-208 via alpha decays. Both isotopes decay to the stable isotope Pb-208 (lead).

Actinides decay via one of four decay chains. The thorium, uranium, and actinide series begin with naturally occurring elements, also referred to as primordial isotopes,[3] and each end in a stable isotope of lead. The neptunium series begins with Np-237, contains only two naturally occurring isotopes, and was first discovered in

[3] Primordial isotopes were created during the Earth's formation and still remain present today due to their long half-lives.

1947 (Thoennessen 2016). The series terminates with the stable isotope Thallium-205. Figure 2.5 shows the decay processes of the uranium (dashed arrows) and thorium (solid arrows) decay chains.

The speed of the decay is measured as the so-called half-life. This is the time it takes for half of the nuclei of an isotope to decay. For example, Cobalt-60 decays to Nickel-60 with a half-life of about 5.3 years. This means that every 5.3 years, half of the Co-60 of a given quantity will decay, leaving only 25% after 10.6 years, 12.5% after 15.9 years, and so on. Half-lives can range from a few yoctoseconds (10^{-24} s), e.g., Lithium-4, over thousands (Pu-239), to billions of years (Tellurium-128).

Note that for simplicity, Fig. 2.5 does not depict all of the shown elements' isotopes and all of the information provided in the original reference (Magill et al. 2022). Furthermore, not every possible decay process is depicted. As indicated by the different colors in individual isotopes, they can decay in various modes. A small triangle indicates a probability of less than 5%, and diagonal splits indicate higher probabilities of up to 50%. Some isotopes can also decay in three ways, e.g., Ac-226. A white background indicates a so-called isomeric transition, in which a metastable state decays to a lower energy level of the same isotope. This process emits gamma radiation.

2.2.3 *Nuclear Fission*

2.2.3.1 Overview

The energy that is used in nuclear reactors to generate (mostly) electricity is created by the continuous operation of chain reactions.

Moreover, the cross-section describes how likely a certain reaction (in this case, fission) is, depending on the energy of the incoming neutron. As a result, different reactions tend to occur more frequently in either the fast or thermal spectrum. With a fast neutron spectrum, it is more likely to trigger a neutron capture process compared to a thermal spectrum (Fig. 2.6). Furthermore, different neutron spectra lead to different requirements for input materials and to different technological challenges (von Hippel et al. 2019). Both are triggered by the process of external neutron capture, in which a nucleus absorbs a free neutron and is transformed to a different isotope in a so-called excited state to rapidly split into two, very seldomly three, fission products.

2.2.3.2 Fission with a Thermal Neutron Spectrum

The theory of splitting atoms was first described by Irène Curie and Frédéric Joliot (1934) in Paris. It was then performed for the first time by Otto Hahn and Fritz Strassmann (1939) in Berlin (on 17 December 1938), and then explained theoretically by Lise Meitner and Otto Frisch (1939) in January 1939 in Stockholm. In this process, U-235 atoms are hit with a neutron. This results in the (very short-term) existence of

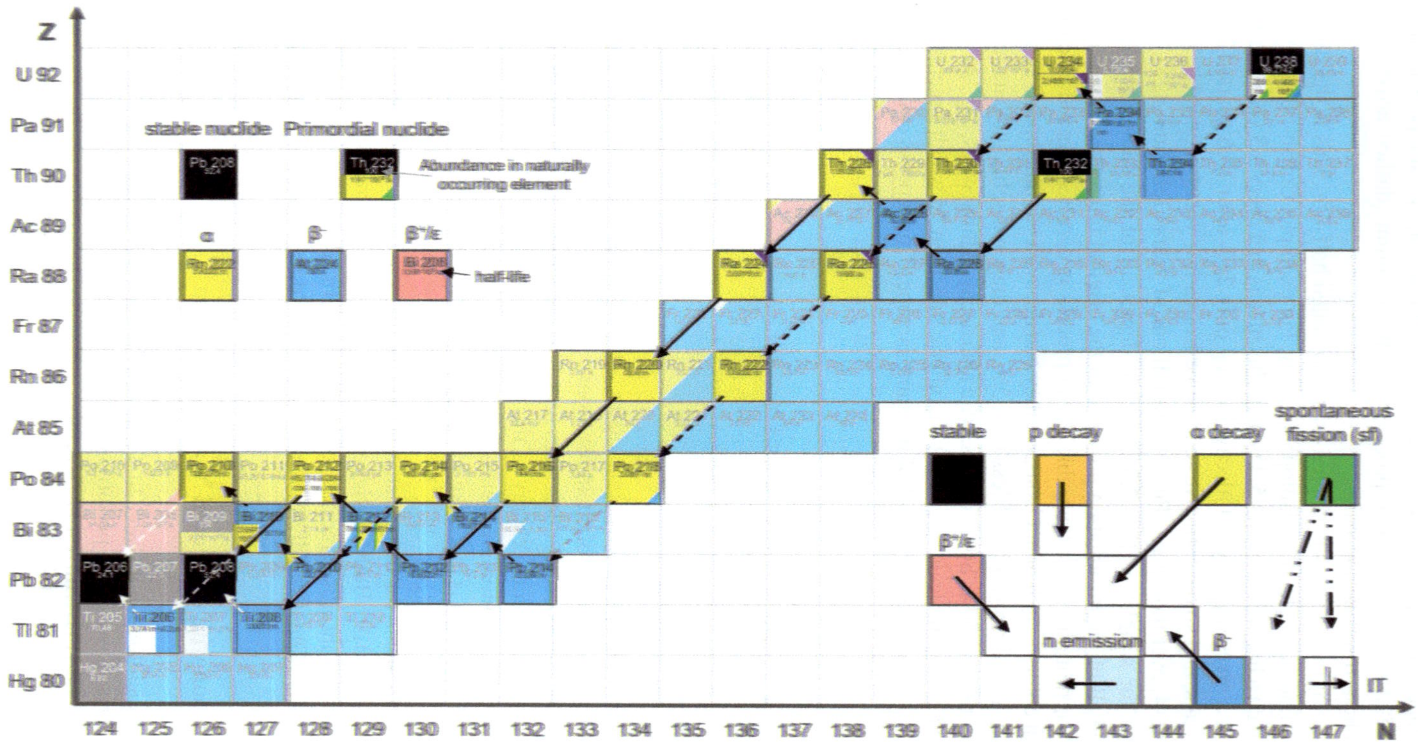

Fig. 2.5 A nuclide map: Thorium (solid arrows) and uranium (dashed) decay chains (*Source* Own depiction following the Karlsruhe nuclide map (Magill et al. 2022)

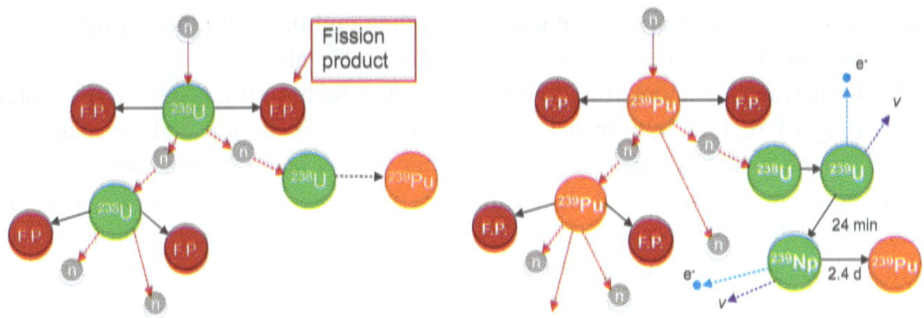

Fig. 2.6 Difference between fission with a thermal (left) or fast neutron spectrum (right) (*Source* Own depiction following von Hippel et al. [2019, 17–20])

a U-236 atom that is then split into two fission products (F.P.). Depending on the F.P.s that are created, a different number of neutrons is once again released. This release of further neutrons results in the chain reaction necessary to operate a nuclear reactor, as these neutrons continue on to hit other atoms. U-235 can fission into different F.P., releasing different amounts of neutrons (2.43 on average). The probability of the creation of certain F.P. combinations is referred to as the fission yield. To ensure an ongoing chain reaction, but at the same time avoid exponential neutron release, the so-called criticality of the reactor must be controlled through moderation, cooling, absorption, and geometry, see below and Sect. 3.1.1.2 for more detail.

A typical fission process is the splitting of U-235 to Krypton-92 and Barium-141 and three neutrons with large amounts of kinetic energy, as shown in Eq. 2.3. Here, fission is induced through neutron absorption, triggering the fission of U-236 into the F.P. Both F.P. are also radioactive and decay to other elements with beta decays until they turn into stable nuclei. Another—albeit much less probable—outcome is the neutron induced fission of U-235 to Zirconium-95 and Tellurium-137. In a given reaction, the number of neutrons stays constant. Thus, the sum of neutrons on the left side (143 neutrons in the U-235 nucleus plus one additional neutron for the induction of the reaction) must be equal to that of the right side (85 in the Ba-141 nucleus, 56 in the Kr-92 nucleus, and three released neutrons to continue the chain reaction in the reactor core).

$$^{235}_{92}U + ^{1}_{0}n \rightarrow ^{236}_{92}U \rightarrow ^{141}_{56}Ba + ^{92}_{36}Kr + 3^{1}_{0}n \quad (2.3)$$

To maintain a chain reaction, it must be ensured that at least one neutron hits another U-235 atom. The criticality k of a system describes whether it can maintain a chain reaction. If k is larger or equal to one, the system is critical and is able to ignite and maintain a chain reaction. To ensure that this is achieved, the probability that neutrons are absorbed by U-235 atoms is increased via the process of moderation. Hereby, the speed of the neutrons is reduced (moderated). The energy is reduced through the collision with other particles of similar weight, resulting in the application of hydrogen or carbon atoms as moderators. At these reduced speeds, moderated (or thermal) neutrons lose no more energy when they collide with other particles, and this substantially increases the chance of hitting U-235 atoms.

In light-water reactors, the dominant reactor technology today, light water (consisting of regular hydrogen isotopes with only one proton and no neutrons) is used as the moderator and for cooling. Other reactor technologies apply graphite or heavy water (consisting of deuterium, i.e., hydrogen with one proton and one neutron) for these purposes, see Sect. 3.1. Light-water reactors require enriched fuel with an increased concentration of U-235 of around 3 to $5\%_{\mathrm{wt}}$. In contrast, natural uranium ore contains only around $0.72\%_{\mathrm{wt}}$ U-235, while U-238 constitutes most of the weight with $99.27\%_{\mathrm{wt}}$. U-234 also occurs naturally but is seldom mentioned because its share in natural ore is marginal. During slow fission, it is inevitable that a neutron hits a U-238 atom. Via the process described on the right side of Fig. 2.6, about 1% of the original uranium is converted to plutonium, mainly Pu-239, plus some other isotopes such as Pu-238, Pu-240, Pu-242, and Pu-244. Pu-239 has a half-life of roughly 24,000 years and is the main source of hazardous radiation in radioactive waste. However, for some, this generation of plutonium is also a desirable process for the establishment of a so-called "plutonium economy," in which the supposed "breeding" of plutonium would solve all future energy challenges (Seaborg 1970).

2.2.3.3 Fission with a Fast Neutron Spectrum

Fast fission occurs when moderation measures are removed from the process. Here, so-called "fast" neutrons fission U-238 to U-239 to Np-239 to, finally, mostly Pu-239, and a small share of other plutonium isotopes, see the right side of Fig. 2.6. The process was performed for the first time in 1941 by Emilio Segrè and Glenn T. Seaborg (Segrè and Seaborg 1941), the latter of whom would later envision the establishment of a plutonium economy (Seaborg 1970). The process of fast fission runs on natural uranium and does not rely on complex enrichment processes. Further, as it seemingly "breeds" plutonium that then in turn further fuels the process, visions of "infinite" energy became the driver for the planned expansion of nuclear power, driven mostly by so-called breeder reactors. The maintenance of this fast fission reaction is technically complex and dangerous, as described in Sect. 3.1.3. The process of fast fission with neutron capture is shown in Eq. (2.4):

$$ {}^{238}_{92}U + {}^{1}_{0}n \rightarrow {}^{239}_{92}U \rightarrow {}^{239}_{93}Np + \beta^{-} \rightarrow {}^{239}_{94}Pu + \beta^{-} \tag{2.4} $$

Here, neutron-induced fission of U-238 results in the creation of a U-239 isotope that decays to Np-239 with a half-life of 24 min with a beta-minus decay. Then, with a half-life of 2.3 days, Np-239 decays to Pu-239 with another beta-minus decay. In the reactor core, Pu-239 is then transformed to Pu-240, Pu-241, etc. via further neutron capture. These processes can go on for millennia and are the main reason for the long-term radiation in spent nuclear fuel (Yim 2022).

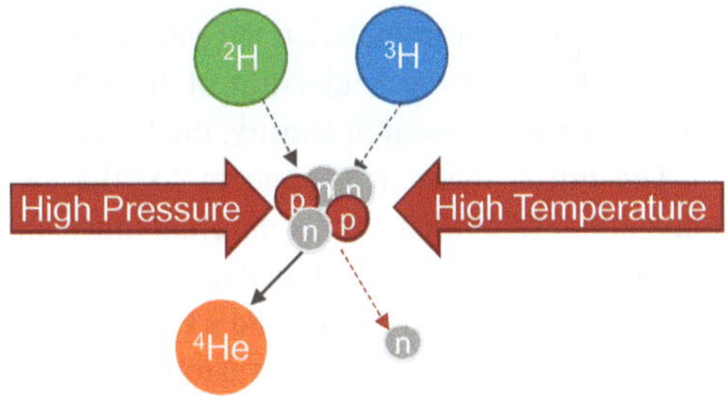

Fig. 2.7 Nuclear fusion reaction (*Source* Own depiction)

2.2.4 *Nuclear Fusion*

2.2.4.1 Overview

The joining of smaller atoms to a single larger atom is called nuclear fusion. This is how the sun produces energy to maintain life on earth. The most common fusion process is the joining of hydrogen isotopes deuterium (H-2) and (radioactive) tritium (H-3) to form a helium atom (He-4) and one free neutron (Eq. 2.5 and Fig. 2.7). This process emits alpha radiation (with the helium atom being the heavy alpha particle) and neutron radiation. The artificial fusion process is technologically complex. Immense pressures and temperatures, ranging up to, depending on the fusion concept, 10^{11} bar and roughly $5*10^7$ K, respectively, are required to fuse the particles. For comparison, the pressure and temperature in the core of the sun are around $2*10^{10}$ bar, and $1.5*10^7$ K. Once achieved, as per a given mass of input material, fusion releases more energy than fission (see Fig. 2.3).

$$ {}^2_1H + {}^3_1H \rightarrow {}^4_2He + {}^1_0n \quad (2.5) $$

2.2.4.2 The Origins of Nuclear Fusion

The theoretical foundations for understanding fusion as an energy process were laid by Arthur Eddington (1926) in the 1920s through his exploration of the mechanisms powering stars and the sun. The first successful nuclear fusion reaction between hydrogen isotopes was achieved in 1934 by researchers Oliphant et al. (1934), who utilized a 100 keV gas discharge, leading to the discovery of tritium, a new hydrogen isotope. The initial application of fusion, however, was directed toward military objectives, specifically the development of a hydrogen bomb (H-bomb). During the Manhattan Project, Edward Teller proposed an early fusion bomb design, known as the "Runaway Super." Prominent scientists like Oppenheimer, Fermi, and Rabi expressed opposition to the development of the H-bomb (Wellerstein 2019). Nevertheless, in 1949, following the first confirmed nuclear weapon test by the Soviet Union, the United States committed to pursuing H-bomb development as a strategic response. This effort included the establishment of Project Matterhorn B at Princeton University under the leadership of John Wheeler, and the foundation of a second nuclear weapon research facility, the Lawrence Livermore Laboratory in California.

The initial theoretical concept for the "Runaway Super" fusion bomb design encountered significant practical challenges. In 1951, with the assistance of Stanislaw Ulam, Edward Teller developed an alternative approach, known as the "Teller-Ulam Design." This design introduced a two-stage mechanism in which the energy from a primary fission bomb initiates fusion by creating intense radiation pressure. The first successful test of an H-bomb using this design, codenamed "Ivy Mike," occurred in 1952 at the Eniwetak Atolls in the Pacific, producing an explosive yield exceeding ten megatons. For comparison, the bombs dropped on Hiroshima and Nagasaki yielded about 15 and 20 kilotons, respectively. Following this breakthrough, other nuclear-armed nations conducted their hydrogen bomb tests, with the Soviet Union's first test occurring in 1955 (Bethe 1995; McCracken and Stott 2005). Refer to Chapter 6 in this book for further technical details, as well as insights on non-proliferation and disarmament efforts.

Although a clear priority was given to the development of weapons, the development of a potential reactor design for energy generation also originated in the 1950s. Next to the military weapons research, project Matterhorn had a second branch under the lead of Lyman Spitzer, who focused on fusion reactor designs for energy production with magnetic confinement and proposed a first design of a so-called "stellarator" in 1951 at Princeton University (Post 1956; Bickerton et al. 1999).

2.2.4.3 Technological Requirements

Nuclear fusion releases immense amounts of energy through the joining of atoms. However, a critical threshold known as the "Lawson criterion" must be surpassed for a fusion reaction to ignite (Eq. 2.6). This criterion is also known as the triple product of plasma density n in ions per cubic meter, the temperature T of those ions in keV ($\approx 11.6 \times 10^6$ Kelvin), and the energy confinement time τ_E in seconds (Lawson 1957).

$$n * T * \tau_E > 6 \times 10^{21} keVsm^{-3} \tag{2.6}$$

Passing this threshold marks the short moment in which fusion is self-sustained. As of today, this has only been achieved by a handful of experiments, and then only for very short time spans in the range of (milli-)seconds (Fig. 2.8). In December 2022, the National Ignition Facility (NIF) at the Lawrence Livermore Lab was the first fusion reaction to achieve a so-called scientific breakeven (see below) when 3.15 MJ were released from a fusion reaction that had been ignited using 2.05 MJ of laser

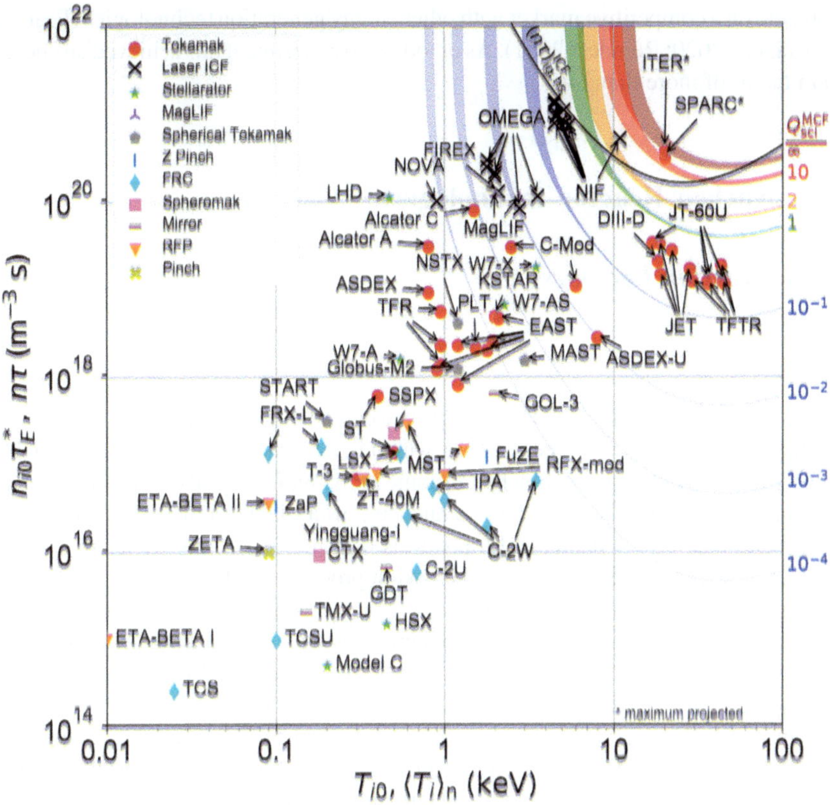

Fig. 2.8 Nuclear fusion developments (with respect to Lawson threshold) (*Source* Wurzel and Hsu [2022, Fig. 2.2])

energy. While this was a significant step in fusion research, the energy required to power the lasers was not included in this calculation.[4]

The goal of a commercial fusion reactor is to produce excess energy to be used for, e.g., electricity generation. Consequently, the energy released by the fusion reaction should be greater than the energy that was required to generate it. To measure this, the fusion energy gain factor Q was introduced as the ratio of power produced by fusion P_{fus} and the power required to keep the plasma heated P_{heat}. While a "scientific breakeven" is defined as Q = 1, a minimum of Q = 5 is expected to be required for a self-sustaining reaction in which the alpha particles released from the fusion reaction itself provide sufficient energy. This is also referred to as "engineering breakeven." "Economic breakeven," estimated at approximately Q = 20, is theoretically reached when the fusion plant covers its own operating costs. "Commercial breakeven" is when the plant covers its own operating and construction costs, and it is able to

[4] Refer to https://www.llnl.gov/article/49301/shot-ages-fusion-ignition-breakthrough-hailed-one-most-impressive-scientific-feats-21st, accessed on 08-10-2024.

compete on a competitive market with other energy generation technologies (Pearson and Takeda 2020; Reinders 2021). As of today, no experiment has achieved an energy gain factor of more than two.

2.3 Nuclear Reactor Technologies

2.3.1 Fission Reactors

2.3.1.1 Overview

A key application of reactor technology is to use the fission process with a "controlled chain reaction" to generate heat, which is then used for various applications, mostly for electricity production but, in some individual cases, also for propulsion in submarines or district heating. Thus, a nuclear power plant is a thermal plant similar to gas and coal power plants and transforms heat into electricity by using a steam cycle. Consequently, the heat from the fission process can then be used to produce steam that drives a turbine, which in turn operates a generator to produce electricity and residual heat. Therefore, it does not emit CO_2 emissions during the process but does generate long-lived and hazardous radioactive waste. See Wimmers et al. (2024) for an overview of externalities induced by nuclear facilities.

Moreover, the fission process can also be used in nuclear weapons to create a large amount of explosive energy and to produce fissile materials such as Plutonium, which are used in nuclear weapons. Thus, nuclear power plants have a dual-use character; the very first application was for military purposes in bombs (Wealer and von Hirschhausen 2020). In the following, we will provide an overview of different technology lines and examples of reactor concepts.

Technology Lines and Reactor Concepts

There are many different reactor designs, which share main characteristics but also can differ in reactor-specific details. Thus, a classification can enhance the technological understanding of different reactor technologies. In the following, we will refer to the classification of Pistner et al. (2021, 2024). Hereby, different technologies are grouped into broader technology lines that have similar main characteristics, such as the same cooling, e.g., light-water reactors (LWRs) that use light water as a coolant. Within these technology lines, different reactor concepts can be distinguished via several characteristics, such as their technical designs (temperature or pressure of the cooling fluid). For example, light water reactors (LWRs) can be split into pressurized water reactors (PWR) and boiling water reactors (BWR). Hence, essential features of nuclear reactors characterize a specific reactor concept or different reactor technology lines. In particular, the following key components of reactors are described

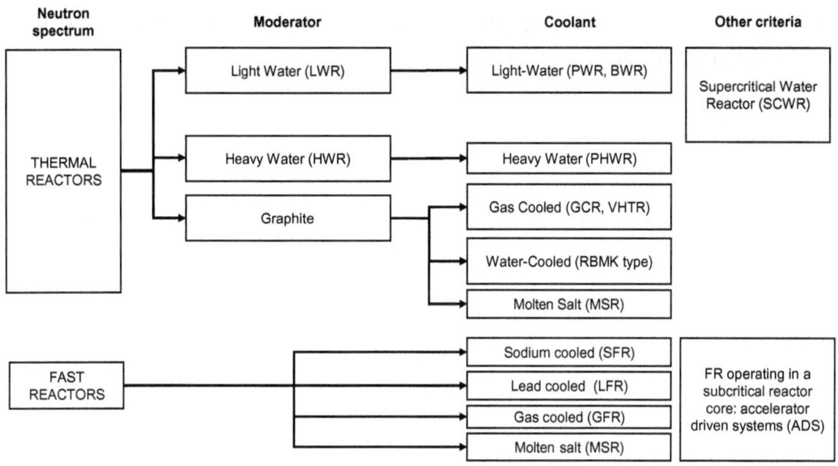

Fig. 2.9 Systematization of fission reactor technologies (*Source* Own depiction based on Pistner et al. [2024])

in further detail including the neutron spectrum, the moderator, the type of primary coolant, the pressure and temperature of the primary coolant, the fuel, and the operational mode (criticality). Each configuration of a reactor concept has to fulfill three main safety goals; control criticality, remove heat, and contain radioactive material to ensure the safe operation of a nuclear reactor. For example, the high energy density demands continuous cooling, as any interruption could damage the core.

Components and Main Characteristics

There are different options to classify reactor technologies. Figure 2.9 illustrates a simplified systematization of reactor technologies based on Pistner et al. (2024).

However, a different classification scheme could also be possible, e.g., to the used fuel or the criticality of the reactor core. Thus, we will shortly describe the five main characteristics of reactor technologies which help to understand their technological similarities and differences.

Neutron Spectrum

The distribution of the energy of neutrons within the reactor core is referred to as the neutron spectrum. Reactor technologies can be distinguished based on whether they use a thermal or a fast neutron spectrum. If a reactor concept uses a fast neutron spectrum, it uses the neutrons with high speed. During the fission process, neutrons are produced that contain high levels of kinetic energy. They are referred to as "fast neutrons." In contrast, a thermal neutron spectrum means that the fast neutrons are slowed down with a moderator. Reactors that operate on a thermal neutron spectrum are called thermal reactors. Thus, thermal reactors use a moderator, whereas reactors with a fast neutron spectrum do not.

Moderator

Reactors operating on a thermal spectrum can be further distinguished by which moderator is used. A moderator is the material that slows down fast neutrons via collision. Light chemical elements that have a similar weight to the neutrons are most suitable as they can slow down neutrons efficiently. Thus, hydrogen isotopes such as deuterium, as found in light and heavy water, are most suitable. In a few cases, graphite is also used as a moderator. The choice of the moderator is also coupled with the fuel that is used. Heavy-water and graphite-moderated reactors can be fueled with natural uranium, whereas light-water reactors require enriched uranium.

Coolant

The primary coolant is used for the direct cooling of the fuel in the reactor core. A coolant serves two purposes. First, it dissipates heat away from the reactor to avoid a potential overheating and subsequent meltdown of the reactor, and second, the medium is used to transport the energy through the plant for further conversion processes. However, often a secondary cooling circuit is added to reduce the risk of radioactive contamination. Thus, different reactor concepts can vary in their choice primary coolant as well as the number of cooling circuits. Almost all operational commercial reactors use light water. In these cases, light-water fulfills not only serves as a coolant but also as a moderator for neutrons.

In addition, heavy water or gases, such as helium or carbon dioxide, are also used for cooling purposes. The coolant can be further distinguished by the operation mode. For example, water cooling can be separated by temperature and pressure in the primary cooling system. In most PWRs, the cooling water is compressed to about 155 bar, while BWRs operate at around 70 bar.

In contrast, fast reactors use coolants, which do not contribute to the moderation of neutrons. They either use a liquid metal, such as sodium (sodium-cooled fast reactor) or lead (lead-cooled fast reactor), or a gas (gas-cooled fast reactor). However, the SFR reactor design has three cooling circuits; two sodium cooling systems and one water circuit for the steam cycle process to reduce sodium-water reactions and radioactive contamination. A special case is molten salts, which can be employed as a coolant in molten salt reactors on fast or thermal spectra.

Fuel

Reactor concepts use different forms of fuel, which contain or produce new fissile materials. This can be natural uranium, enriched uranium, thorium, and plutonium, or a mixture. Natural uranium, which contains mainly U-238 and U-235 (0.72%), of which only U-235 is fissile. A nuclear reactor based on natural uranium cannot be operated with light water but needs another coolant/moderator such as heavy water.

Light-water reactors require a fuel where the proportion of U-235 is increased to enable a sustained chain reaction (typical enrichment levels are 3–5%). In contrast, reactors that use a fast neutron spectrum need a higher proportion of fissile material to sustain a chain reaction since the isotopes like U-235 or Pu-239 are less likely to fission with fast neutrons than with thermal neutrons. Fuels thus either contain higher enrichment levels (<20% of U-235), also referred to as "high assay low enriched uranium" (HALEU), or consist of a mixture of plutonium and uranium oxides, so-called MOX-fuels.

The plutonium is produced from U-238 via neutron capture processes in fast reactors but also in thermal reactors. However, the plutonium production is much more likely in a fast neutron spectrum. If in a reactor more fissile material is produced by neutron capture than it is fissioned, the reactor is also called "breeder." The feature of producing fissile plutonium, desired in a so-called closed nuclear fuel cycle, is enhanced when extra blankets containing U-238 are positioned around the reactor core. Very few countries reprocess and extract plutonium from the spent fuel to be used as mixed-uranium oxide fuels (MOX) or for weapons production. Very few countries reprocess and extract plutonium from the spent fuel to be used as mixed-uranium oxide fuels (MOX) or for weapons production.

Criticality

Lastly, reactors can be distinguished by whether they are operated in critical or subcritical operations. Criticality refers to a condition in which a chain reaction is self-sustaining. In most cases, reactors are used in critical mode in which a continuous chain reaction occurs because each neutron produced during fission leads to a new fission reaction. To ensure that the reaction does not continue uncontrolled, various neutron management measures, such as moderation, cooling, neutron poisoning (e.g., boric acid), or specific geometrical reactor designs, are applied (Yim 2022, 110–11). In contrast, in a subcritical reactor system, neutrons are supplied by an external source, e.g., an accelerator, which expedites protons to high energies directed to the target. In this target, many neutrons are produced, which leads to fission in the reactor core. Thus, these reactor concepts are called accelerator-driven reactors (ADS).

Global Reactor Fleet

Since the first experimental reactors became operational in the Soviet Union at Obninsk and in the US (EBR-I) in 1951, and the subsequent launch of the "Atoms for Peace" program in 1953 (Lévêque 2015), many reactor concepts and technology lines have been tried and tested. In contrast, the global reactor fleet is dominated by light-water reactors (Cowan 1990; Wealer et al. 2018) (Fig. 2.10).

At year-end 2024, the global nuclear fleet comprised 411 reactors in operation with an installed capacity of 369.5 GW (Fig. 2.10). The dominant reactor technology line is the light-water reactor (307 PWRs and 39 BWRs), followed by 45 heavy-water reactors (HWR), and six gas-cooled, graphite-moderated reactors (GCR). The

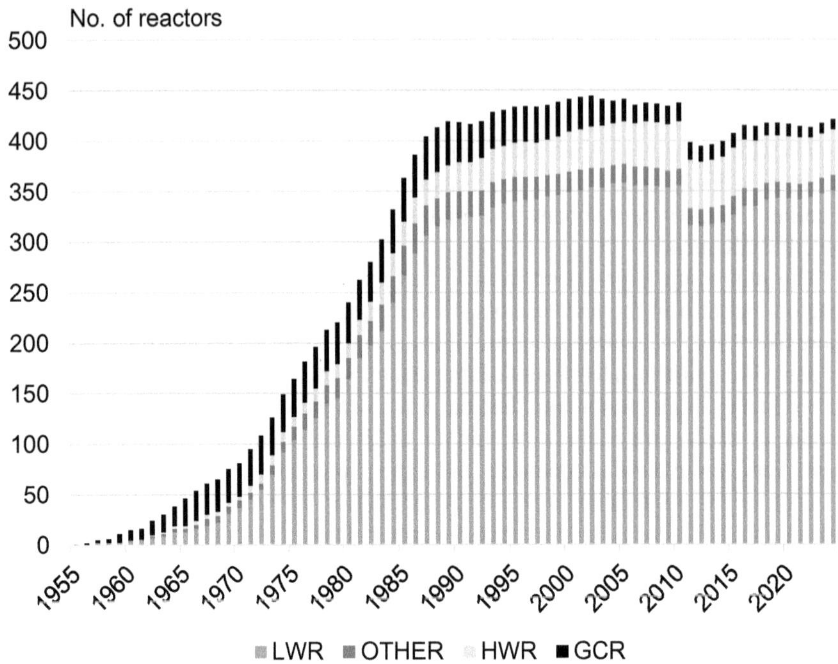

Fig. 2.10 The development of the global reactor fleet from 1954 to 2024 (*Source* Own depiction based on Schneider et al. (2024) and the IAEA PRIS Database.[5])

category "Others" includes 3 reactors with fast neutron spectrum (SFRs), 10 light-water cooled and graphite moderated reactors (LWGR, or RBMK), and 1 gas-cooled high-temperature reactor (HTR). Furthermore, installed capacity increased during the 1970s and has started to stagnate since 1985. However, the construction starts to decline already during the 1970s. The average age of the global nuclear fleet was 31 years.

As of mid-2024, 59 reactors were under construction, of which around 40% were behind schedule. The most recent Western construction projects to be completed are the reactors 3 and 4 of the Vogtle plant in the US, Olikuoto-3 in Finland, and Flamanville in France. These projects were all delayed for several years and were well over budget. Another large-scale project in the United Kingdom (UK, Hinkley Point C) is behind schedule. Others include those planned and sometimes funded by Russia, and many new builds in China by domestic vendors. Here, data and project information are ubiquitous, making project assessments challenging.

[5] https://pris.iaea.org/pris/home.aspx, last accessed on 08–01-2025.

Technology-Oriented Reactor Generations

The evolution of LWR technology from the 1950s to the present reflects significant technological advancements. For example, early LWRs from the 1950s are quite different from those of the 1980s, and again from later developed concepts, such as the European Pressurized Reactor (EPR) and the AP1000. The progression of nuclear reactor technology is classified into generations, each representing a new level of capability and focus. Generation 0 includes the first experimental reactors, designed primarily to demonstrate technical feasibility. Generation I reactors were the first commercial reactors aimed at electricity production as demonstration projects. Generation II reactors represent more advanced, large-scale commercial reactors intended for wider deployment, while Generation III reactors aim to improve safety and economic performance.

As shown in Table 2.1, the development of various reactor technology lines over the past 80 years reflects these generational advancements. Most GEN 0 reactors date back to the 1950s and 1960s, though some, such as Russia's LFR Brest-OD300 and Belgium's European ADS Myrrha, are still being developed. While many reactor types have advanced to Generation I, only a few, primarily LWRs, have progressed to Generation II. All current Generation III reactors are light-water PWRs and BWRs, although some pressurized heavy water reactors (PHWR) are also under development.

Many reactor technologies other than LWRs are often called "new" reactor concepts (SNRs) or "GenIV" and differ in essential characteristics from the current dominant LWR reactor technology, e.g., neutron spectra. This refers to the technology lines very-high temperature reactor (VHTR), the sodium-cooled fast reactor (SFR), the lead-cooled fast reactor (LFR), gas-cooled fast reactor (GFR), molten-salt reactors (MSR), super-critical water reactors (SCWR), or accelerator-driven systems (ADS). These reactor concepts can only build on operational experiences from LWR to a limited extent, and instead require newly developed materials, components, and verification methods. Furthermore, they are characterized with long development histories, which have resulted in some experimental or demonstration projects, but also in many project cancellations. Thus, the SNR reactor concepts have yet to go beyond Generation I. After the 1950s, extensive experiments with VHTR and FBRs took place, but no commercial diffusion transpired (Böse et al. 2024). The Russian BN-1200 is currently under development and might become a commercial Generation II reactor.[6]

[6] Alternative approaches to define "generations" of technology lines exist. Thus, Radkau (2008, 364) reports that in the early days of commercial nuclear power, the logical pathway would be from experimental reactors (generation 1) to light-water reactors (generation 2), on to breeding reactors (generation 3), and then … on to nuclear fusion reactors (generation 4). Baade (1958, 140) also states this explicitly: "Experts consider as the "fourth generation" of nuclear power plants devices relying not on the principle of nuclear fission, but on nuclear fusion." (own translation). The "Generation IV International Forum (GIF) distinguishes reactor technologies between four generations: a first generation of early prototype reactors from the period 1950–1960, a second generation of large power reactors built between 1970 and 1990, some of which are still in operation today, a third generation of power reactors constructed from around 1990 onwards or currently under

Table 2.1 Reactor generations within technology lines

Technology line	Initial experimental reactors	Generation I	Generation II	Generation III
PWR	MTR, S1W, S2W, MZFR	Shippingport, Obninsk, Obrigheim	Konvoi (Isar-2)	AP-1000, VVER-1200, EPR
BWR	BORAX-I, II, III, IV, Kahl	Dresden-I, Grundremmingen-A	SWR-72	(KERENA), ABWR
PHWR	ZEEP, NRX, NRU	Rolphton	CANDU 500, CANDU 6	(EC 6, ACR-1000)
GCR	CP-1	Windscale, Calder Hall, Marcoule	AGR	
VHTR	Dragon, AVR, HTR-10	Peach Bottom, THTR, HTR-PM, VHTR		
SFR	Fermi I, Br-10, CEFR, KNK I and II, Rapsodie, TWR	BN-800, Monju, Superphénix	(BN-1200)	
LFR	(BREST-OD 300)			
GFR	(GFR)			
MSR	ARE, MSRE	(LFTR, MCFR)		
SCWR	HDR	(CSR 1000)		
ADS	(MYRRHA)			

Source (Pistner et al. 2024, 44)

2.3.1.2 Current Reactor Technology Lines

Light-Water Reactors

In the mid-2020s, over 75% of all globally operational reactors were light-water reactors (LWR). Of these, once again, 75% were pressurized water reactors (PWR), and the remainder were boiling water reactors (BWR). Both reactor types use light water as moderator and coolant and thus require enriched uranium as fuel. Over the years, PWRs have emerged as the dominant reactor technology, and out of 58 new build projects ongoing in 2024, only four were not PWRs, nor BWRs.

construction, and a fourth generation of so-called "GEN IV" or "advanced" reactor concepts that could become available on the market after 2030 as proposed by the GIF. This generation concept assumes that generations are defined by the continuous time periods, and period IV (for non-light water reactors) follows "logically" from period III (for light water reactors). However, this concept does not correspond to technological developments and is therefore not suited to understand the industrial and technological dynamics of the sector.

In both LWR designs, fuel is composed of cylindrical pellets, each about one centimeter in diameter and height, encased in metal tubes approximately four meters long. These tubes, made of zirconium alloy, are sealed to contain radioactive fission products. Multiple tubes form a fuel assembly, which includes positions for control rods made from neutron-absorbing material. These rods regulate the reactor operations by controlling the nuclear chain reaction. The reactor core, in a high-capacity PWR (~one GWe), consisting of about 200 fuel assemblies with a total of about 100 tons of uranium, generates heat. The energy density is approximately 38 MW per ton of uranium, making the reactor's compact design economically attractive.

Pressurized Water Reactors

During the operation of a PWR, uranium fuel is consumed, with each fuel assembly remaining in the reactor for about 3–5 years, depending on its initial enrichment. As fuel is used, boron is added to the coolant to balance the reactor's power output. The heat from the reactor is removed by the primary cooling loop, which uses boron-doped, demineralized water under high pressure (about 155 bar) to prevent boiling, even at 330 °C. The heat is then transferred to a secondary loop through steam generators, converting water into steam to drive turbines and generate electricity. Residual heat is released to the environment via cooling systems. The reactor is housed within a containment structure to prevent the release of radioactivity, while spent fuel is stored and cooled on-site before being moved to interim storage (not shown in Fig. 2.11).

Boiling Water Reactors

BWRs differ from PWRs mainly in that steam driving the turbine is generated directly within the reactor pressure vessel, rather than in a separate steam generator. Therefore, a BWR has a single water-steam loop, rather than separate primary and secondary loops. The pressure in this loop is around 70 bar, causing part of the coolant to evaporate as it heats to about 290 °C while passing through the fuel assemblies. In the reactor core of a BWR with an electrical output of 1.3 GWe, approximately two tons of water are vaporized per second. Above the core, steam separators remove water from the steam. In BWRs, control rods are inserted into the core from below, and for rapid shutdown, they are held up using compressed air. Recirculation pumps mix heated water from above the core with newly added water below, which influences the temperature and steam content around the fuel assemblies. Since steam has a lower density than water, increasing the steam content reduces moderation, allowing the pumps to control the chain reaction and reactor power.

The turbine, generator, and condenser setup is similar in both BWRs and PWRs, see Fig. 2.12. However, in a BWR, radioactive steam directly drives the turbine, leading to stricter radiation protection measures outside the containment, particularly in the turbine building. Unlike PWRs, the containment of BWRs is smaller and

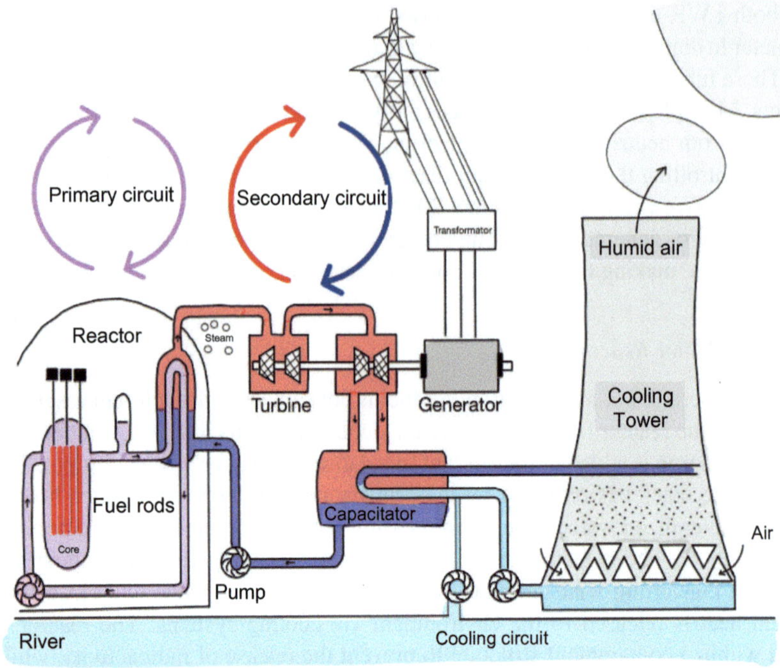

Fig. 2.11 Schematic scheme of a pressurized water reactor (*Source* Own translation taken from Camurdas et al. [2023, 21])

not spherical, as it does not house steam generators. Due to this smaller size, overpressure from a major pipe break may not necessarily be fully contained, requiring special pressure relief systems. Additionally, the spent fuel pool is located outside the containment within the reactor building, which, like in PWRs, is made of reinforced concrete for external protection, but has a rectangular shape rather than a dome.

Light-Water Cooled and Graphite-Moderated Reactors

RBMKs (Russian original: reaktor bolschoi moschtschnosti kanalny) use enriched uranium fuel, with graphite moderating neutrons, to sustain the fission chain reaction, while water serves as the coolant to absorb heat and generate steam for electricity production. Control rods, typically made of boron carbide, are inserted into the reactor core to absorb excess neutrons. By adjusting the depth of the control rods in the reactor, operators can control the rate of the nuclear reaction and maintain safe operation. A key flaw of this design is its positive void coefficient, where steam formation increases reactivity, potentially causing a dangerous feedback loop. Additionally, early control rod designs could temporarily increase reactivity upon insertion. These design flaws, combined with operational errors, contributed to the Chornobyl disaster in 1986. Since then, safety measures were implemented to other operational

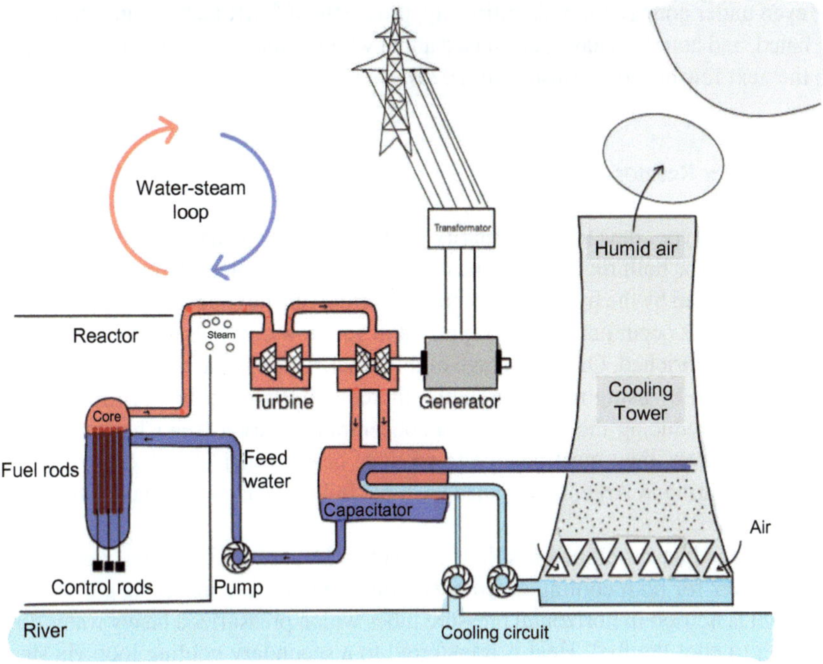

Fig. 2.12 Schematic scheme of a boiling water reactor (*Source* Own translation taken from Camurdas et al. [2023, 22])

RBMKs to address and improve these design issues. However, all RBMKs that were in operation outside of Russia have been shut down.

So-Called Small Modular Reactor Concepts ("SMRs")

Following the US DOE (Chu 2010) and Pistner et al. (2021), so-called "small modular reactors" (SMRs) can be classified as light-water reactors with a net capacity of less than 300 MWe.[7] These reactors are supposed to rid current high-capacity LWRs from their "lumpiness"[8] by providing factory-built modules faster and cheaper (Markard et al. 2020). Because of their lower capacity, many designs also incorporate so-called passive safety systems. Current SMR projects include Russia's two floating KLT-40 s reactors, which have been in operation since 2020, and the Chinese ACP-100, currently under construction. Meanwhile, the US SMR design VOYGR, developed by NuScale, was canceled in 2023. As of 2025, no commercial project was operational

[7] The term is used differently among studies and sometimes can also include reactor concepts with different coolants and a net capacity of less than 300 MWe. However, it is useful to focus on LWRs to focus on a proven reactor concept but which is aimed to build "smaller.".

[8] "A technology is 'lumpy' when it is characterized by high-capacity units, high investment costs, and an increased likelihood of upscaling in unit size" (Böse et al. 2024, 4).

or even under construction. Additionally, prospects of SMR technologies have been debated, and concerns have been raised about whether these technologies can deliver in the next few decades (Böse et al. 2024).

Heavy Water Reactors

This reactor type, developed primarily by Canada for electricity generation, uses heavy water for both moderation and cooling. In heavy water, the regular hydrogen atom is replaced by the hydrogen isotope deuterium that has one extra neutron. While deuterium does occur naturally, it is present in very low concentrations in light water and must be enriched. One advantage of this reactor type is the ability to sustain the chain reaction by using natural uranium, thereby eliminating the need for uranium enrichment, reducing the reliance on a concentrated market (see Chapter 8 of this book). However, they must be able to produce heavy water. As of 2024, there are HWRs operating in Argentina, Canada, China, India, Pakistan, Romania, and South Korea.

The Canadian-developed CANDU (Canadian Deuterium Uranium) reactor uses heavy water for both cooling and moderation, though thcsc functions are separate. The fuel is housed in horizontal pressure tubes which pressurized heavy water flows through to cool the fuel. Heat is transferred to a secondary cooling loop via steam generators, similar to a PWR, where light water drives a turbine in a water-steam cycle. The pressure tubes are situated in a moderator tank filled with non-pressurized, low-temperature (about 70 °C) heavy water, which acts as an effective neutron moderator due to its high density. Because the fuel elements are installed in separate pressure tubes, the fuel can be replaced without shutting down the reactor, ensuring continuous operation. The option to remove single fuel elements during operation allows for rather easy production of high-purity Pu-239, which is of high military value. The plutonium, for e.g., India's first nuclear explosions, was produced in CANDU reactors.

Gas-Cooled, Graphite Moderated Reactors

GCRs use graphite as moderator and a gas, e.g., carbon dioxide, for cooling. Since graphite absorbs few neutrons, these reactors can run on natural uranium and can thus be used to generate plutonium. The UK pioneered this reactor type with the first commercial Magnox (magnesium non-oxidizing) reactor, Calder Hall, beginning its operation in 1956. Magnox reactors use special metallic uranium fuel with magnesium alloy cladding. The first French reactors were also GCRs (the UNGG reactors; French: uranium naturel graphite gaz). They are all closed since 1994. Due to the increased reactivity of magnesium at high temperatures, the operating gas temperature was limited to 360 °C, drastically reducing efficiency. GCRs are larger than their water-based counterparts due to the low thermal conductivity of the gases used (e.g., carbon dioxide), necessitating large volumes and high flow rates. A more advanced version of the Magnox design, the Advanced Gas-cooled Reactor (AGR), uses uranium dioxide fuel, allowing higher power densities but requiring enriched uranium. Today, eight AGRs remain operational in the UK. They are scheduled to close in the coming years.

2.3.1.3 So-Called New or Advanced Reactors ("SNRs") and Technology Lines

The following shows that other reactor technologies have often been under development for many decades. These other technology lines reactor category is sometimes collectively referred to as "new" or "advanced." In the context of this work, we shall refer to these reactors as "so-called new reactor" (SNR) concepts that primarily incorporate non-light-water reactor designs from the 1940s and 1950s. This includes sodium-cooled fast reactors (SFR), lead-cooled fast reactors (LFR), gas-cooled fast reactors (GFR), high-temperature gas-cooled reactors (HTR), supercritical water-cooled reactors (SCWR), molten-salt reactors (MSR), and accelerator-driven systems (ADS). Except for ADS concepts, research and development (R&D) efforts are coordinated by an international forum established in 2001 (called GenIV). Its members include the US, China, Russia, and the EURATOM states, and it aims to renew R&D efforts. Some SNR concepts, in particular fast reactors or ADS, are also envisaged to be used for alternative disposal pathways such as Partitioning and Transmutation. (see Chap. 18 of this book).

Fast Critical Reactors (SFR, LFR, GFR)

Reactors with a fast neutron spectrum, so-called fast reactors (FR) (sometimes also referred to as "fast breeders") can be distinguished by their coolants: sodium, lead, or gas. There have been several attempts to establish FRs, but none have succeeded on a commercial scale. The following briefly addresses the three reactor technology lines.

Sodium-cooled fast reactors (SFR) gained increased interest in response to concerns about uranium depletion. A fast neutron spectrum could reduce reliance on natural uranium because of the breeding of additional fissile material, in particular, plutonium. In such a fuel cycle, plutonium from spent fuel would be reused via reprocessing and MOX-fuel fabrication to decrease the demand for natural uranium (closed fuel cycle). Proponents envisaged the shift to a "plutonium economy" (Seaborg 1970), in which fast-neutron reactors would replace LWRs and energy would become "too cheap to meter" (Strauss 1954, 9). However, due to new uranium discoveries and the reactors' economic and technological challenges, SFRs have not been commercially successful and faced many project cancellations in the US, as well as France, Germany, Japan, and the UK. Operational challenges call high utilization rates into question as long-lasting interruptions have been observed in past projects, such as Superphènix in France or Monju in Japan. Some projects are still in development, especially in Russia, but also in China and India. Furthermore, the US rejoined the development of SFRs, e.g., through a reactor concept being developed by TerraPower. One major challenge is reactivity of sodium with air and water, leading to frequent technical failures during operation and making repair and maintenance complex and lengthy. The use of separated uranium and plutonium from reprocessing would reduce the use of natural uranium, but the separation processes are also complex.

Lead-cooled reactors on fast neutron spectra (LFR) operate similarly to SFRs, but use lead as a coolant, which is less reactive and poses fewer operational risks. It also induces other challenges, such as corrosion. The only operational experience with LFRs comes from Russian submarine reactors, and only one project is currently under construction (BREST-OD-300). A demonstration LFR project in Belgium (MYRRHA) is also under development. Challenges for further development include the high melting point of lead, polonium production in lead–bismuth, the opacity of the coolant making inspections more difficult, increased protection against seismic activity due to the high density of the coolant, and the potential for corrosion and erosion if the coolant comes into contact with materials and components.

Gas-cooled fast reactors (GFR) mainly use helium as a coolant. They are similar to HTRs but operate on a fast neutron spectrum. Developments of GFR technology occurred in the US and Germany in the 1960s and 1970s, and later in the UK and Japan. However, no GFR was ever constructed or operated. LFR and GFR could have cost advantages compared to SFR because their cooling properties do not require a secondary cooling circuit. Detailed cost estimations for construction and operation are unavailable today.

High Temperature Reactors (HTR)

Similar to GCRs, high-temperature reactors (HTRs) use a thermal neutron spectrum moderated by graphite and cooled by a gas. They were developed in the 1960s and have limited operational experience due to significant technological challenges faced by early prototypes like the Dragon reactor in the UK and the THTR-300 in Germany, which were eventually abandoned. So far, no R&D program has been able to operate a reactor at a high utilization rate over an extended period of time. The most intensive development programs (Germany, US, and South Africa) have so far been discontinued at the demonstration power plant phase due to a lack of interest from operators, which was in turn based on concerns about the economic and reliable operation of the plants. Currently, operational reactors are confined to China with two prototypes, and the US company X-energy is developing an 80 MW HTR, the Xe-100, which is still in the design phase. HTRs are designed to provide high-temperature heat, up to 1000 °C. Cost information is mostly speculative. The economics of HTR will essentially depend on the outlet temperature because higher temperatures enable additional applications, e.g., for industrial process heat. Historically, nuclear process heat had to compete with fossil fuels, and interest in HTR technology increased during the oil crisis in the 1970s. Major oil companies, such as Gulf and Shell, made significant financial commitments to HTR. However, due to various unresolved technical problems, orders were cancelled and the HTR technology was abandoned.

Supercritical Water Reactors (SCWR)

The supercritical water-cooled reactor (SCWR) aims to achieve higher steam temperature (along with a higher steam pressure) compared to today's light- and heavy-water reactors, thereby increasing the thermal efficiency of the overall reactor system. When water reaches its so-called critical point at pressures of at least 221 bar and temperatures of at least 374.12 °C, it becomes supercritical. This means that the aggregate state cannot be defined. Here, water has the same density as the fluid state, but the same viscosity[9] as the gaseous state. Furthermore, supercritical water has no surface tension. This means that separation processes in the operational fluid (steam in "regular" reactors that also contains liquid water) are not necessary anymore, and dilution capabilities also change. Furthermore, in theory, a supercritical reactor would achieve higher thermal efficiencies than its "regular" counterparts. Various reactor concepts are in discussion for which distinctions can be made between thermal or fast neutron spectra. Moreover, there can be further separation between pressure vessels as used in LWRs or pressure tubes as used for HWR. Otherwise, the SCWR is very similar to today's LWR concepts. For the first experimental reactor, extensive R&D is regarding for high-temperature resistant materials.

The idea intertwines with improvements of modern coal-fired power stations in recent decades, in which the steam temperature has been raised above the critical point to enable operations at higher temperatures and raise thermal efficiency. Early reactor concepts were discussed as early as the 1950s and 1960s. For example, Westinghouse designed a light-water moderated reactor cooled by supercritical water. However, no SCWR has been built to date. Cost advantages compared to LWRs could be realized due to potential savings in capital costs and higher thermal efficiency.

Molten Salt Reactors (MSR)

Molten salt reactors (MSR) dissolve fissile and fertile[10] materials in liquid molten salt (fluoride or chloride salt). Thus, fission and potential breeding processes occur within the salt, which acts as a coolant at the same time in most designs. MSR technologies could enable fuel handling and chemical processing during operations, eliminating usually necessary periodical reactor shutdowns. Furthermore, the molten salt serves as a coolant used in a primary and secondary circuit. Via heat exchangers, a third cooling circuit operating with water drives the turbines. However, due to

---

[9] Viscosity defines the dynamic resistance of a fluid. For liquids, this can be translated into the term of "thickness." For example, honey or syrup has a higher viscosity than water.

[10] Fissile isotopes such as U-235 and PU-239 can sustain fission reactions, where fertile materials such as U-238 cannot sustain fission reactions but can produce fissile materials via neutron capture.

the technological challenges of liquid MSR fuels, solid fuels are also discussed. In this configuration, the molten salt is only used as a coolant, but the fuel is solid, e.g., TRISO fuel pellets. MSRs can be designed to operate on a fast or thermal spectrum. For the latter, graphite can be used as a moderator. MSRs offer high heat output, with salt types melting at 500 °C and boiling at 1400 °C, allowing for outlet temperatures between 600 and 700 °C, making them potentially useful for heat applications in decarbonized energy systems. MSRs were originally developed for aircraft propulsion in the US at the end of the 1940s, resulting in the first experimental plant in 1954, which was followed by the "Molten Salt Reactor Experiment" (MSRE). After the closure of the experimental reactor in 1960, smaller research programs continued at Oak Ridge National Laboratory. Additionally, countries such as the UK, France, China, and Russia carried out R&D programs. Today, no commercial MSR is operational, but several concepts are under development in China, Switzerland, Russia, and Germany, among others. Cost information of MSR designs is sporadic and not reliable (see Chapter 10).

Accelerator Driven Systems (ADS)

The principle of an Accelerator-Driven System (ADS) is to operate a fast neutron reactor with a subcritical reactor core. Such an operational mode cannot maintain a stable fission chain reaction, therefore, the fission process relies on an external neutron source. The external neutron source is typically a proton accelerator whose particle beam produces high-energy neutrons via a spallation process.[11] ADS concepts are envisaged to produce energy and transmute radionuclides. The original idea of using large accelerators and neutrons from the spallation process for transmutation emerged in the 1950s. Since the 1990s, plans to generate energy with ADS concepts have experienced a resurgence. To date, there is no prototype ADS in operation anywhere. Development efforts remain in the conceptual and planning stages. Current efforts encompass not only the development of a reactor with a fast neutron spectrum (mostly LFRs are planned) but also that of a reliable spallation neutron source. For energy production, the accelerator must operate for months without interruption, but current accelerators are still expensive and large. There are costs associated with the construction of the accelerator and spallation neutron source and their operation; an ADS would therefore be more expensive than a comparable LFR. Nevertheless, ADSs could have significant safety advantages due to the subcritical arrangement.

[11] Spallation occurs when fragments of an object are ejected after an impact or stress. In nuclear physics, this refers to the ejection of neutrons from a nucleus after being hit by another high-energy particle.

2.3.2 *Fusion Reactors*

2.3.2.1 Overview

Fusion devices have been under research since the 1950s. However, in opposition to their fission counterparts, actual research and demonstration reactor experience is limited to a handful of sites. Currently, there are an array of new ventures aiming to establish themselves in the realm of fusion devices for energy production, even though most projects remain in early development stages and are unlikely to achieve commercial applicability in the next decades (Meschini et al. 2023).[12]

Within the rapidly growing nuclear fusion sector, a range of technologies—some innovative and others established for decades—have emerged in the pursuit of fusion energy. These concepts can be divided into two main technology lines: MCF, inertial confinement fusion (ICF); and the magneto inertial fusion (MIF) approach (Pearson and Takeda 2020). In addition, many other reactor types exist. Figure 2.2 shows a variety of fundamental and operational reactor designs (see for details Moynihan and Bortz (2023)).

One of the main challenges of operating a fusion device is to ensure that the extremely high temperatures of the plasma[13] are contained within the reactor without causing the materials of the outer walls to evaporate. Two dominant approaches are under development to control the plasma. The first approach is the magnetic confinement fusion (MCF) that takes advantage of the ionized nature of the plasma. Hereby, the high-temperature-low-density plasma can be directed and confined by applying magnetic fields. This raises the energy confinement time in Eq. (2.6), see Sect. 2.3.2.3. The second approach originates back to military applications and aims to confine the plasma by its own inertia. Here, intense energy beams heat and compress the fuel to fusion so fast that the plasma has no time to reach the walls. This is achieved by increasing the density of the plasma. The approach is inertial confinement fusion (ICF). Both approaches induce extreme temperatures and shall be introduced in the following section (McCracken and Stott 2013; Nuttall 2022). We also mention the broad field of "other" devices and concepts (Fig. 2.13).

2.3.2.2 Current Technology Lines

Magnetic Confinement

MCF takes advantage of the fact that plasmas have an electrical charge and can be directed using magnetic and electrical fields. MCF reactors use magnets to confine the

[12] Find more information on nuclear fusion devices in Chapters (4) and 14 of this book.

[13] The term plasma refers to a mixture of ions, electrons, and neutral atoms or molecules at high temperatures. It describes an aggregate state, such as solid, liquid, or gaseous. The main difference to these, however, is the electromagnetic charge and the resulting electric conductivity of materials in this state.

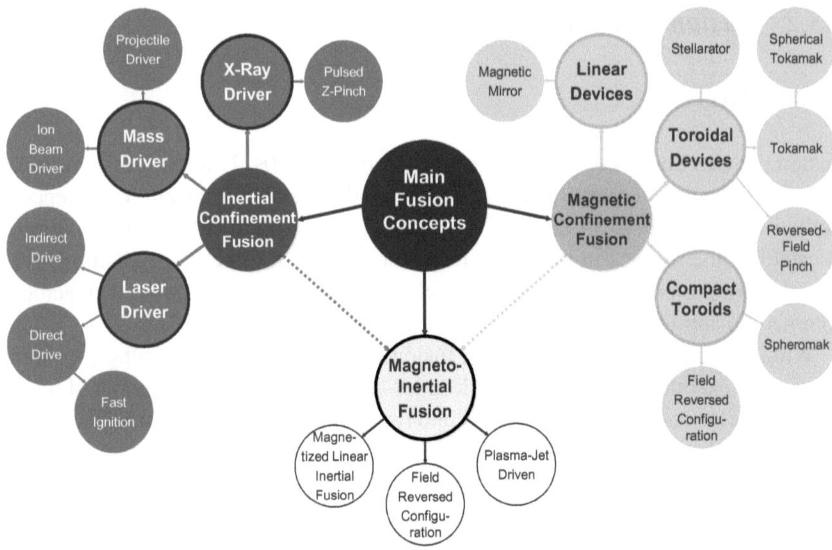

Fig. 2.13 Main fusion concepts and their configurations. (*Source* Own depiction)

plasma, allowing longer energy confinement times. An MCF can be realized through different reactor designs, such as the tokamak or the stellarator, among others. In tokamak devices, a set of magnetic coils arranged in a circular design confines the plasma in a ring form, while an orthogonally oriented central magnetic solenoid[14] at the center of the ring plasma guides it in one direction, resulting in a sort of donut-shaped plasma. Many research devices based on the tokamak design exist, e.g., the Joint European Torus (JET) in the UK or the Tokamak Fusion Test Reactor (TFTR) in the US, see Sect. 2.3.2.3. These reactors have proven the scientific feasibility of the MCF approach. As of today, they have not yet achieved an energy gain factor $Q > 1$. The International Thermonuclear Experimental Reactor (ITER) project, so-far the largest tokamak reactor, aims to demonstrate this. ITER is scheduled to generate its first plasma in 2034 after decade-long delays [15] (see Chapter 4 of this book for details on the ITER project). With the knowledge obtained from ITER, and once several technological, economic, and resource-related obstacles are overcome, a demonstration reactor may follow in the second half of the twenty-first century, after which commercial fusion reactors for energy generation are envisioned (Takeda et al. 2023). A major challenge that tokamak designs must overcome is the disruptive instability of the plasma. Here, a sudden plasma energy release to the wall accompanied by rapid collapse of the electrical and magnetic currents can result in high heat

[14] A solenoid is an electromagnet created by a wire wound into a helical coil, with its length being significantly greater than its diameter.

[15] Michael Banks, "ITER fusion reactor hit by massive decade-long delay and €5bn price hike", 03 July 2024, Physics World, https://physicsworld.com/a/iter-fusion-reactor-hit-by-massive-decade-long-delay-and-e5bn-price-hike/.

loads and electromagnetic forces stressing both the reactor vessel and other structural components (Bickerton et al. 1999). Other MCF designs include the stellarator, in which the magnetic coils are shaped in a complex way as to both confine the plasma and drive the current. While constructing such a reactor is more complex than a tokamak, the main advantage of inherent plasma stability makes external systems driving the current unnecessary. There are a handful of operating stellarators, such as the Wendelstein 7-X in Germany or the Heliotron-J in Japan. Here too, several technological challenges still lie ahead.

Historical Development of MCF Devices

Early efforts to develop nuclear reactors for energy generation began in the 1950s and 1960s, though military applications remained the priority. The second branch of Project Matterhorn, led by Lyman Spitzer at Princeton University, introduced the concept of the stellarator in 1951,[16] while Soviet scientists, including Andrei Sakharov, Igor Tamm, and Oleg Lavrentev, focused on the development of a tokamak concept. This culminated in the construction of the T-3 tokamak, which began operating in 1962. The declassification of atomic research by the US, the UK, and the Soviet Union at the 1958 UN Geneva Conference facilitated international collaboration (Bickerton et al. 1999; Dean 2005).

The late 1960s and 1980s saw a surge in enthusiasm for tokamak designs following the T-3's record achievement in magnetic confinement at the Kurchatov Institute in 1968 (Fig. 2.8). British scientists from Culham validated these results (Peacock et al. 1969). This shift in global focus toward tokamak development spurred the construction of several tokamak devices worldwide. Notable examples include the Tokamak Fusion Reactor (TFR), built in France in 1973; the US-based devices Princeton Large Torus (PLT) and TFTR, constructed in 1975 and 1982, respectively; JET, built in the UK in 1983; and finally, the Japanese JT-60, that became operational in 1985.

The 1985 Geneva Summit marked the beginning of the ITER project, a collaborative international initiative proposed by Mikhail Gorbachev to symbolize West–East scientific cooperation. The project involved the Soviet Union, the US, the European Community (predecessor of the European Union), and Japan. The goal was to advance fusion technology by testing superconducting magnets, developing tritium breeding, and exploring ignited plasma physics. The 1990s saw a reduction in fusion funding due to the collapse of the Soviet Union and economic crises in Russia and Japan. The US also reduced its fusion budget, shutting down the TFTR and temporarily withdrawing from the ITER project (McCracken and Stott 2005).

However, in the 2000s, the US rejoined the ITER project, joined by new members China and South Korea. In 2005, the ITER construction site was confirmed in France, with the ITER organization established in 2007 and construction commencing in 2010. The project has faced significant delays and budgetary challenges, with plasma operations now projected to start sometime in the 2030s. At the same time, other magnetic confinement fusion approaches, such as stellarator designs, regained

[16] The Latin word for "star" is "stella."

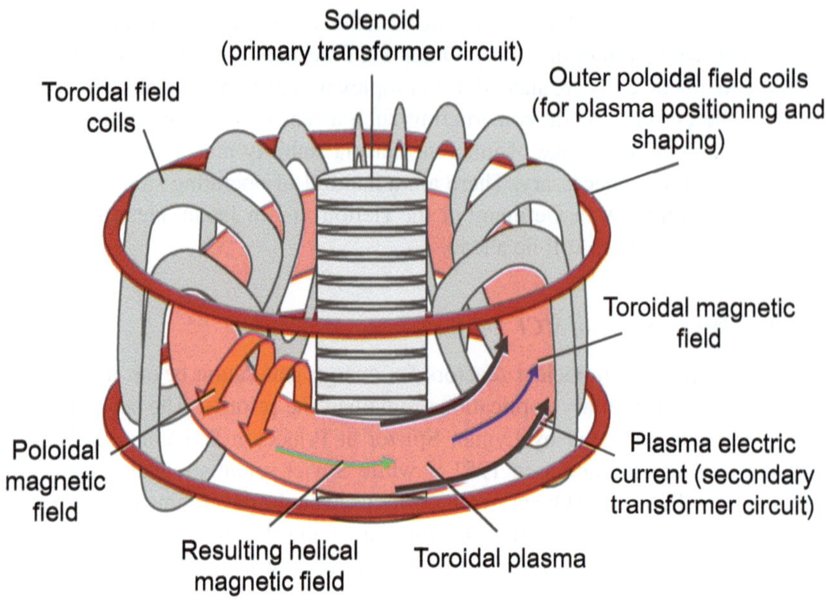

Fig. 2.14 Schematic of a tokamak chamber and magnetic profile (*Source* Own depiction following Li et al. (2014))

interest. Notably, Germany's Wendelstein 7-X stellarator began operation in 2015, continuing parallel research into alternative fusion configurations. In the following, we provide a brief overview of the two main MCF devices.

Tokamak

The term "tokamak" originates from the Russian phrase for "toroidal chamber and magnetic coil" (original: toroidalnaja kamera s magnitnymi katuschkami). This design uses strong magnetic fields to confine hot plasma in a toroidal (doughnut-shaped) geometry. While the temperature of the plasma can reach several million Kelvin, the temperature at the outer walls of the reactors is far lower (in the range of thousands of K) thanks to a very steep temperature gradient. The torus of a tokamak must be made of heat-resistant materials, such as beryllium, tungsten, and carbon. Unique to the tokamak is its function as a large electromagnetic transformer, where the plasma itself acts as the secondary winding. A current induced in the plasma by changes in the primary coil generates a toroidal plasma current, creating helical flows of electrons and nuclei in opposite directions, stabilized by a set of toroidal field coils known as the torus. Figure 2.14 shows a schematic design of a tokamak device. In theory, this configuration should enable effective plasma confinement and stability.

One advantage of the tokamak design is "Ohmic heating" or resistive heating. As the plasma initially resists smooth flow, energy losses contribute to plasma heating.

Over time, more energy is directed toward establishing the helical plasma flow, thus achieving both motion and heating essential for fusion. This heating mechanism, combined with strong toroidal confinement, makes the tokamak a promising approach. The ITER project aims to demonstrate the tokamak's potential for power generation, creating conceptual layouts of future fusion power plants. In ITER, the fusion reaction is intended to be self-sustaining through the internal generation of tritium through lithium blanket breeding in the reactor walls (Nuttall 2022).

In May 2024, the WEST tokamak, operated by the French nuclear agency CEA, achieved a new record by sustaining a plasma field for 6 min. In terms of energy released, the JET in the UK achieved a record of 69.26 MJ (the equivalent of burning 2 kg of coal) in 2023. However, the plasma was only sustained for 6 s. The JET was closed in December 2023.[17]

Stellarator

Unlike tokamaks, stellarators do not rely on transformer action to generate the helical toroidal motion of plasma ions and electrons. Instead, they use externally applied, geometrically complex magnetic fields to maintain plasma confinement (Fig. 2.14). This design allows stellarators to operate continuously in a steady-state mode, which may become an advantage for potential power station applications. However, the lack of transformer action also means that stellarators cannot rely on resistive (Ohmic) heating, historically limiting their ability to reach the high temperatures necessary for fusion. Advances in plasma heating technologies, such as microwave and neutral beam injections, have addressed some of these challenges, replacing earlier heating methods that often caused plasma instabilities. The recent feasibility of stellarators is largely due to improvements in computational power, which have enabled the modeling of complex magnetic configurations and have facilitated their design and construction. The Wendelstein 7-X in Greifswald, Germany, is currently the leading stellarator project, operational since 2015. This device has achieved significant milestones, including setting a stellarator record for the fusion triple product at $6*10^{19}$ keVm^{-3} s.[18] However, despite these advancements, stellarators still fall short of the confinement performance seen in top-performing tokamaks (Nuttall 2022).

While stellarators remain far fewer in number and less dominant than tokamaks in magnetic confinement fusion research, they continue to provide valuable insights into continuous operational modes and magnetic confinement physics. Although further development is needed before stellarators can compete as the leading fusion technology, their potential should not be dismissed (Fig. 2.15).

---

[17] https://euro-fusion.org/eurofusion-news/dte3record/.

[18] Following Lawson's criterion, the required threshold for self-sustaining fusion is $6*10^{21}$ keVm^{-3} s, see Sect. 2.4.3.

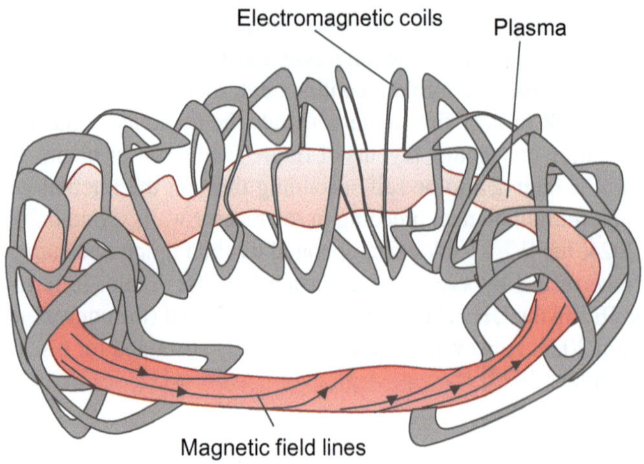

Fig. 2.15 Schematic of a stellarator design (*Source* Own depiction)

Inertial Confinement

ICF reactors enable fusion reactions by increasing the density of the plasma by pulsed devices, rather than by long energy confinement times. This is done by rapidly compressing and heating a small fuel target with one or several laser beams. There are different concepts to realize ICF. Here, we focus on laser-driven concepts.

Historical Development of ICF Devices

In the 1960s, following H-bomb testing, Edward Teller and colleagues at Lawrence Livermore National Laboratory explored fusion applications beyond weaponry, including potential commercial power generation. Unlike the bomb design, where a fission bomb serves as the primary, this new approach aimed for ignition through radiation implosion within a small so-called "hohlraum" (German, meaning a cavity) energized by an external non-nuclear primary. Early fusion research in this area received limited funding from the Department of Defense. By the late 1960s, laser development had advanced sufficiently to become central to the ICF approach. However, initial device completion estimates were optimistic, and lasers of the time lacked sufficient power. In the 1970s and 1980s, secret experiments, including the Halite series at Lawrence Livermore and Centurion at Los Alamos National Laboratory, were conducted at a nuclear test site in Nevada. These tests confirmed the feasibility of the ICF concept, but revealed that achieving fusion would require laser energy between 20 and 100 MJ, far exceeding initial estimates. With the nuclear test ban treaty coming into effect in the 1990s and halting underground testing, concerns arose over maintaining nuclear weapon safety and capabilities, leading to the establishment of a national stewardship program under the National Defense Authorization Act. This program drove the construction of significant laser facilities, such as the

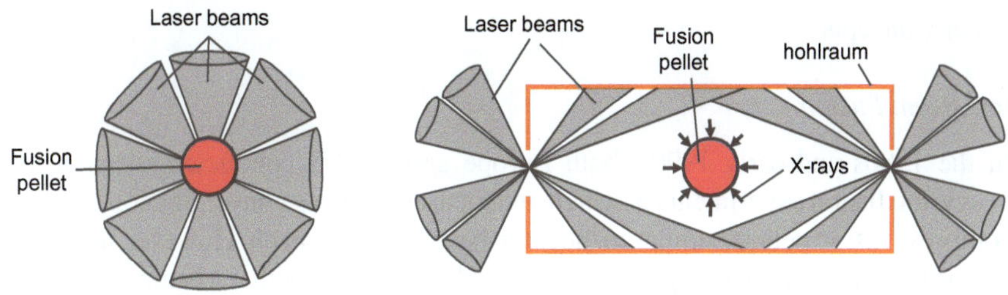

Fig. 2.16 ICF concepts. Direct drive (left) and indirect drive (right) (*Source* Own depiction based on Badziak (2012, Fig. 2.2))

National Ignition Facility, to support laser fusion research and stockpile stewardship (McCracken and Stott 2005, 2013).

Technical Details

Laser-driven ICF devices can be implemented through either direct-drive or indirect-drive methods (Fig. 2.16). In the direct-drive method, multiple ultraviolet laser beams directly and symmetrically irradiate the fusion pellet, compressing it directly. The indirect-drive method places the pellet inside a small cylinder called a "hohlraum," usually made of materials like gold or uranium. In this approach, laser beams heat the inner walls of the hohlraum, producing hot plasma that emits thermal X-rays. These X-rays are then absorbed by the target surface, causing an implosion similar to direct laser irradiation. Although X-ray absorption is more efficient than direct laser absorption for the fusion pellet itself, a significant portion of the input energy is used to heat the hohlraum, reducing the overall efficiency of the process. For both schemes, approximately 1 MJ of input energy is needed for fusion ignition (Badziak 2012).

Both direct- and indirect-drive methods face similar challenges, including achieving high laser-to-fuel energy transfer efficiency, controlling the implosion symmetry, preventing fuel pre-heating by hot electrons and X-rays, avoiding premature mixing of hot and cool fuel caused by hydrodynamic instabilities, and ensuring the formation of a strong shockwave at the fuel center. To address these challenges, short-wavelength radiation is essential, along with extremely precise pellet fabrication, allowing deviations of only a few micrometers on both the inner and outer surfaces. Laser or X-ray beams must be highly accurate and synchronized to arrive at all points on the pellet simultaneously. Furthermore, the pellets must be positioned accurately to be hit by a laser pulsing at several Hertz, for which large amounts of power[19] must be provided. (Badziak 2012, Meier et al. 2014, Reinders 2021).

[19] The NIF Powering Conditioning System (PCS) delivers 330 MJ of electrical energy to power a single laser shot; peak power is one TW; more than the entire US uses in the same fraction of a second, see: https://lasers.llnl.gov/about/how-nif-works/power-conditioning; The laser itself delivers 1.8 MJ with a peak power of 500 TW, see: https://pubmed.ncbi.nlm.nih.gov/17514286/.

Other Concepts

Ion-Beam Fusion

In the 1980s and early 1990s, both Europe and the US conducted research into heavy-ion fusion experiments in which one focused on using converging ion beams to compress fusion pellets, similar to the laser-based ICF approach. However, achieving uniform irradiation and compression remained challenging with ion beams, leading to decreased interest over time. Most experiments were closed down by the mid-1990s. While heavy-ion research did continue, for example, at the Relativistic Heavy Ion Collider (RHIC) at Brookhaven National Laboratory, these efforts primarily served basic science goals, such as investigating fundamental nuclear physics, rather than focusing on fusion energy applications.

In the twenty-first century, however, ion-beam fusion research has seen renewed interest, albeit with a different focus. Rather than using heavy ions to compress fuel pellets, the current approach involves accelerating ions of the fusion fuel itself. A prominent example is the aneutronic fusion strategy, where proton beams are fired onto Boron-11. This approach is pursued by US-based TAE Technologies. Here, proton beams are directed tangentially to a boron-rich plasma, causing it to spin within a magnetic confinement. This approach is more similar to magnetic confinement fusion (MCF) than traditional ICF due to the use of a hot, low-density plasma in a magnetic environment (Nuttall 2022).

Z-Pinch

In Z-pinch experiments, a precisely constructed metal cylinder acts as a "hohlraum," producing symmetric X-ray radiation to implosively compress the fusion pellet, similarly to indirect laser-based ICF. The pinch occurs along the Z-axis in a three-dimensional space. This is achieved by passing a massive electrical current (up to 20 MA in the Z-machine at Sandia National Laboratories, New Mexico, US) through a cylindrical plasma, generating a magnetic field that drives the plasma radially inward, creating an intense compression or "pinch" effect. A significant portion of this kinetic energy is converted into X-rays, which heat the hohlraum to high temperatures, allowing for symmetric irradiation of the pellet. The Z-machine produces brief, high-power pulses (up to 50 TW), generating X-rays that can cause significant implosive compression. Innovations in Z-pinch include the wire array method, where a fine wire cage is vaporized to create plasma during the discharge, and the dynamic Z-pinch, which utilizes a moving hohlraum for increased efficiency.

While promising, the Z-pinch approach remains inefficient, as it requires considerable X-ray output to achieve uniform irradiation. The concept has seen limited funding compared to tokamak, and laser-based fusion projects. However, the Z-pinch has one major advantage: its fuel pellet can be larger than in laser-driven systems, potentially yielding higher fusion energy per shot (up to 1.2 GJ, approximately 80 times that of laser systems). If fuel elements could be mass-produced, an industrial-scale Z-pinch power station with repeated fueling might be feasible in the future (Nuttall 2022).

2.3.2.3 Overview of Reactor Projects and Main Technological Challenges

FusDIS Statistics on Reactor Projects

The IAEA provides an overview of operating and planned nuclear fusion projects in its database titled "Fusion Device Information System" (FusDIS).[20] As of April 2024, the database lists a total of 153 fusion devices in 29 countries. The US (39), Japan (25), China (13), Russia (13), and the UK (10) lead the list in terms of entries. The devices are differentiated regarding their type, status, design, and ownership. Three quarters of projects are publicly owned and funded. Tokamak concepts make up 53% of total devices, followed by alternative concepts (28%), including Z-Pinch approaches, stellarators (11%) and laser-based ICF concepts (8%). Notably, each of the 101 operating devices is a so-called "experimental" reactor, while demonstration reactors, designed for concept proof for commercialization, are all still in the planning stage (14 in total). Given the vast amount of remaining technological challenges, briefly discussed below, it is uncertain when and how these demonstration projects will be designed, licensed, built, and operated.

Remaining Technological Challenges

Despite ongoing developments, fusion concepts face enormous technological challenges that must be addressed before experimental reactors may become available for commercial energy generation. The following section will give a brief overview of the main challenges based on Abdou et al. (2021); Meschini et al. (2023); Reinders (2021); and others referenced above.[21]

Firstly, the challenge of fuel availability must be overcome. Most fusion concepts rely on deuterium–tritium (DT-) fusion and thus require tritium to function reliably. An initial stockpile of tritium is required to ignite a first reaction, either once the reactor is constructed and ready to operate or after maintenance work or plasma disruptions. A 1000 MW fusion reactor would require roughly 56 kg of tritium per year. While tritium shall be "bred" during fusion reactor operations, the start-up inventory challenge remains. The current global tritium stockpile is less than 27 kg, and the half-life of only 12.32 years means it cannot be produced in advance for

[20] https://public.tableau.com/app/profile/mattepev/viz/ATF/FusDIS, last accessed on 29–04-2024.

[21] Two critical assessments (in German), written at a time interval of over two decades, are Grunwald et al. (2002), and Grünwald (2024).

long-term storage. The only reliable production facilities for tritium today are fission heavy water reactors ("CANDUs") in Canada and India.

Second, the challenge of tritium scarcity is hoped to be overcome by fuel breeding via Lithium-6. While this has yet to be tested on a large scale, Li-6 is, in theory, an abundant material. However, other industries, such as battery production, also require lithium, creating potential competition for the material. Other materials, such as beryllium, are required for the manufacture of magnets. Further, specially designed materials are yet to be developed to withstand the extremely high temperatures and steep temperature gradients over many hours of the year.

Thirdly, proponents of fusion technologies refer to a potential fusion plant as a so-called "baseload" power capable of producing electricity "around the clock." This has never been demonstrated. For MCF devices, the main challenge lies with plasma stability; current tokamaks usually uphold the necessary conditions for only a few seconds. Laser-based ICF devices must ensure the precise positioning of pellets, the consistent provision of sufficient laser power, and a very high repetition rate of the pellet provision and laser pulses to generate consistent heat for electricity production. Individual fusion reactions have been triggered by these devices, but never several in a row. If a fusion reactor must be restarted, it would require a very high amount of power (from the grid)—up to several hundred megawatts—to ignite the initial fusion reaction.

The fourth challenge relates to the induced neutron radiation during the fusion process. The high-energy neutrons released from DT-fusion are planned to be used for energy generation by transferring heat into the blankets embedded in the walls of the fusion device (and in parallel breed tritium). This process inevitably leads to neutron activation of the materials, making them radioactive and generating non-negligible amounts of low- and intermediate-level radioactive wastes. Furthermore, this activation mechanism can alter the physical and chemical properties of the walls' materials and potentially lead to unexpected material behavior that remains unstudied today.

Finally, to become commercially available in the future, fusion reactors should ideally provide cheap energy to compete in an energy system dominated by very low-marginal-cost renewables. This leads to flexible operations challenges (referring to reliability) and raises cost concerns. As discussed in Chapter of this book, a vast number of fusion cost studies exist, but none of these concepts have ever been operated. Consequently, economic challenges will need to be tackled, either through extensive state subsidies; alternative use cases, such as non-electrical applications; or yet-to-be-proven cost reductions.

2.4 Conclusion

The chapter introduced nuclear science and various fission and fusion technologies. All of these processes rely on nuclear physics and chemistry, and all produce radiation and radioactive waste. While nuclear fission is well understood and "works" in demonstration reactors, nuclear fusion is a more complex animal and has achieved lower technological readiness levels thus far.

Fission and fusion can be realized in a wide range of reactor configurations. Therefore, various designs were under development in parallel. Regarding fission, light water technology (LWR) became the dominant reactor type and was deployed in several countries. However, as orders stagnated and new construction projects became increasingly costly and delayed, renewed interest in other fission technologies can be observed. However, these, too, have different challenges to overcome. Compared to LWRs, so-called novel reactors (SNRs), are unavailable today and require further R&D. A system change, away from the high-capacity LWR route, is not to be expected in the coming decades.

Different approaches are under development regarding fusion technologies, too, particularly MCF and ICF. While the research focused mainly on MCF tokamaks, laser fusion has recently gained increasing importance. However, both approaches require significant further development.

References

Abdou, Mohamed, Marco Riva, Alice Ying, Christian Day, Alberto Loarte, L.R. Baylor, Paul Humrickhouse, Thomas F. Fuerst, and Seungyon Cho. 2021. 'Physics and Technology Considerations for the Deuterium–Tritium Fuel Cycle and Conditions for Tritium Fuel Self Sufficiency'. *Nuclear Fusion* 61 (1): 013001. https://doi.org/10.1088/1741-4326/abbf35

Baade, Fritz. 1958. *Welt Energiewirtschaft: Atomenergie—Sofortprogramm oder Zukunftsplanung.* Hamburg: Rowohlt.

Badziak, J. 2012. 'Laser Nuclear Fusion: Current Status, Challenges and Prospect'. *Bulletin of the Polish Academy of Sciences: Technical Sciences* 60 (4): 729–38. https://doi.org/10.2478/v10175-012-0084-8

Bethe, Hans A. 1995. 'Bombs After Hiroshima:Dark Sun. The Making of the Hydrogen Bomb.' *Science* 269 (5229): 1455–57. https://doi.org/10.1126/science.269.5229.1455

Bickerton, R. J., R. S. Pease, R. Hawryluk, J. E. Allen, K. Lackner, and D. C. Robinson. 1999. 'History of the Approach to Ignition [and Discussion]'. *Philosophical Transactions: Mathematical, Physical and Engineering Sciences* 357 (1752): 397–413. http://www.jstor.org/stable/55120

Böse, Fanny, Alexander Wimmers, Björn Steigerwald, and Christian von Hirschhausen. 2024. 'Questioning Nuclear Scale-up Propositions: Availability and Economic Prospects of Light Water, Small Modular and Advanced Reactor Technologies'. *Energy Research & Social Science* 110 (April 2024): 103448. https://doi.org/10.1016/j.erss.2024.103448

Camurdas, Lena-Jülide, Reni Safitri, Insa Zillmann, Fabian Präger, Alexander J. Wimmers, Christian von Hirschhausen, Alexander Tetsch, and Sibylle Tetsch. 2023. *Einfach mal abschalten—und dann? Die Geschichte der deutschen Atomkernkraft und ihr radioaktives Erbe.* München: oekom verlag.

Chu, Steven. 2010. 'America's New Nuclear Option: Small Modular Reactors Will Expand the Ways We Use Atomic Power'. *Wall Street Journal*, 23 March 2010. https://www.wsj.com/articles/SB10001424052748704231304575092130239999278

Cowan, Robin. 1990. 'Nuclear Power Reactors: A Study in Technological Lock-In'. *The Journal of Economic History* 50 (3): 541–67. https://doi.org/10.1017/S0022050700037153

Curie, Irène, and Frédéric Joliot. 1934. 'I. Production Artificielle d'éléments Radioactifs—II. Preuve Chimique de La Transmutation Des Éléments'. *Journal de Physique et Le Radium* 5 (4): 153–56. https://doi.org/10.1051/jphysrad:0193400504015300

Dean, Stephen O. 2005. 'Historical Perspective on the United States Fusion Program'. *Fusion Science and Technology* 47 (3): 291–99. https://doi.org/10.13182/FST05-A708

Eddington, A. S. 1926. 'The Source of Stellar Energy'. *Nature* 117 (2948): 25–32. https://doi.org/10.1038/117025a0

Grunwald, Armin, Reinhard Grünwald, Dagmar Oertel, and Herbert Paschen. 2002. 'Kernfusion'. TAB-Sachstandsbericht 75. Büro für Technikfolgen-Abschätzung beim Deutschen Bundestag. https://www.itas.kit.edu/pub/v/2002/grua02b.pdf

Grünwald, Reinhard. 2024. 'Auf dem Weg zu einem möglichen Kernfusionskraftwerk—Wissenslücken und Forschungsbedarfe aus Sicht der Technikfolgenabschätzung'. Nr. 1. TA-Kompakt. Berlin: Büro für Technikfolgen-Abschätzung beim Deutschen Bundestag (TAB). https://doi.org/10.5445/IR/1000177720

Hahn, Otto, and Fritz Strassmann. 1939. 'Uber den Nachweis und das Verhalten der bei der Bestrahlung des Urans mittels Neutronen entstehenden Erdalkalimetalle'. *Naturwissenschaften* 27 (11). https://www.nssp.uni-saarland.de/lehre/Vorlesung/Kernphysik_SS19/History/Papers/Hahn_Strassmann.pdf

Hippel, Frank von, Masafumi Takubo, and Jungmin Kang. 2019. *Plutonium: How Nuclear Power's Dream Fuel Became a Nightmare*. Singapore: Springer. http://ezproxy.st-andrews.ac.uk/login?url= http://link.springer.com/10.1007/978-981-13-9901-5

Lawson, J D. 1957. 'Some Criteria for a Power Producing Thermonuclear Reactor'. *Proceedings of the Physical Society. Section B* 70 (1): 6–10. https://doi.org/10.1088/0370-1301/70/1/303

Lévêque, François. 2015. *The Economics and Uncertainties of Nuclear Power*. Cambridge, United Kingdom: Cambridge University Press.

Li, S., H. Jiang, Z. Ren, and C. Xu. 2014. 'Optimal Tracking for a Divergent-Type Parabolic PDE System in Current Profile Control'. *Abstract and Applied Analysis* 2014:1–8. https://doi.org/10.1155/2014/940965

Magill, Joseph, Raymond Dreher, and Zsolt Sóti. 2022. 'Karlsruhe Nuclide Chart'. 11th Edition. Nucleonica. https://www.nucleonica.com/wiki/index.php?title=Category%3AKNC

Markard, Jochen, Nuno Bento, Noah Kittner, and Alejandro Nuñez-Jimenez. 2020. 'Destined for Decline? Examining Nuclear Energy from a Technological Innovation Systems Perspective'. *Energy Research & Social Science* 67 (September):101512. https://doi.org/10.1016/j.erss.2020.101512

McCracken, G. M., and P. E. Stott. 2005. *Fusion: The Energy of the Universe*. Complementary Science Series. Burlington, MA: Elsevier Academic Press.

McCracken, G. M., and P. E. Stott. 2013. *Fusion: The Energy of the Universe*. 2nd ed. Amsterdam ; Boston: Elsevier/Academic Press.

Meier, W.R., A.M. Dunne, K.J. Kramer, S. Reyes, and T.M. Anklam. 2014. 'Fusion Technology Aspects of Laser Inertial Fusion Energy (LIFE)'. *Fusion Engineering and Design* 89 (9–10): 2489–92. https://doi.org/10.1016/j.fusengdes.2013.12.021

Meitner, Lise, and O. R. Frisch. 1939. 'Disintegration of Uranium by Neutrons: A New Type of Nuclear Reaction'. *Nature* 143 (3615): 239–40. https://doi.org/10.1038/143239a0

Meschini, Samuele, Francesco Laviano, Federico Ledda, Davide Pettinari, Raffella Testoni, Daniele Torsello, and Bruno Panella. 2023. 'Review of Commercial Nuclear Fusion Projects'. *Frontiers in Energy Research* 11 (June):1157394. https://doi.org/10.3389/fenrg.2023.1157394

Moynihan, Matthew, and Alfred B. Bortz. 2023. *Fusion's Promise—How Technological Breakthroughs in Nuclear Fusion Can Conquer Climate Change on Earth (And Carry Humans To Mars, Too)*. Switzerland: Springer.

Neles, Julia Mareike, and Christoph Pistner, eds. 2012. *Kernenergie: eine Technik für die Zukunft?* Technik im Fokus. Berlin: Springer Vieweg.

Nuttall, William J. 2022. *Nuclear Renaissance: Technologies and Policies for the Future of Nuclear Power*. Boca Raton: CRC Press. https://www.taylorfrancis.com/books/9781003038733

Oliphant, Marcus L. E., Paul Harteck, and Ernest Rutherford. 1934. 'Transmutation Effects Observed with Heavy Hydrogen'. *Proceedings of the Royal Society of London. Series A, Containing Papers of a Mathematical and Physical Character* 144 (853): 692–703. https://doi.org/10.1098/rspa.1934.0077

Peacock, N. J., D. C. Robinson, M. J. Forrest, P. D. Wilcock, and V. V. Sannikov. 1969. 'Measurement of the Electron Temperature by Thomson Scattering in Tokamak T3'. *Nature* 224 (5218): 488–90. https://doi.org/10.1038/224488a0

Pearson, Richard J, and Shutaro Takeda. 2020. 'Review of Approaches to Fusion Energy'. In *Commercialising Fusion Energy*, edited by William J Nuttall, Satoshi Konishi, Shutaro Takeda, and David Webbe-Wood, 2–1–2–31. IOP Publishing. https://doi.org/10.1088/978-0-7503-2719-0ch2

Pistner, Christoph, Matthias Englert, Christian von Hirschhausen, Fanny Böse, Björn Steigerwald, Lukas Gast, and Richard Donderer. 2024. 'Analysis and Evaluation of the Development Status, Safety and Regulatory Framework for So-Called Novel Reactor Concepts'. BASE—Research reports on the safety of nuclear waste disposal BASE-031/24. Berlin: Federal Office for the Safety of Nuclear Waste Management. https://www.base.bund.de/shareddocs/downloads/en/expert-info/f/final-report-novel-reactor-concepts.pdf?__blob=publicationFile&v=1

Pistner, Christoph, Matthias Englert, Christian Küppers, Ben Wealer, Björn Steigerwald, Christian von Hirschhausen, and Richard Donderer. 2021. 'Sicherheitstechnische Analyse und Risikobewertung einer Anwendung von SMR-Konzepten (Small Modular Reactors)'. Vorhaben 4720F50500. Forschungsberichte zur Sicherheit der nuklearen Entsorgung. Berlin: Bundesamt für die Sicherheit der nuklearen Entsorgung. https://www.base.bund.de/SharedDocs/Downloads/BASE/DE/berichte/kt/gutachten-small-modular-reactors.pdf?__blob=publicationFile&v=6

Post, Richard F. 1956. 'Controlled Fusion Research—An Application of the Physics of High Temperature Plasmas'. *Reviews of Modern Physics* 28 (3): 338–62. https://doi.org/10.1103/RevModPhys.28.338

Radkau, Joachim. 2008. *Technik in Deutschland: vom 18. Jahrhundert bis heute.* Umfassend überarb. und aktualisierte Neuausg. Frankfurt/Main: Campus.

Reinders, L. J. 2021. *The Fairy Tale of Nuclear Fusion.* Cham: Springer International Publishing. https://doi.org/10.1007/978-3-030-64344-7

Schneider, Mycle, Antony Froggatt, Julie Hazemann, Özgür Gürbüz, Paul Jobin, Phil Johnstone, Timothy Judson, et al. 2024. 'The World Nuclear Industry Status Report 2024'. Paris, France: Mycle Schneider Consulting. https://www.worldnuclearreport.org/IMG/pdf/wnisr2024-v2.pdf

Seaborg, Glenn T. 1970. 'The Plutonium Economy of the Future'. Presented at the Fourth International Conference on Plutonium and Other Actinides (5 October 1970), Santa Fe, New Mexico. http://fissilematerials.org/library/aec70.pdf

Segrè, Emilio, and Glenn T. Seaborg. 1941. 'Fission Products of Uranium and Thorium Produced by High Energy Neutrons'. *Physical Review* 59 (2): 212–13. https://doi.org/10.1103/PhysRev.59.212.2

Strauss, Lewis L. 1954. 'Remarks Prepared by Lewis L. Straus, Chairman, United States Atomic Energy Commission, For Delivery At The Founders' Day Dinner, National Association of Science Writers, On Thusday, September 16, 1954, New York, New York.' Speech presented at the Founders' Day Dinner, National Association of Science Writers, New York, September 16. https://www.nrc.gov/docs/ML1613/ML16131A120.pdf

Takeda, Shutaro, Alexander Ryota Keeley, and Shunsuke Managi. 2023. 'How Many Years Away Is Fusion Energy? A Review'. *Journal of Fusion Energy* 42 (1): 16, s10894–023–00361-z. https://doi.org/10.1007/s10894-023-00361-z

Thoennessen, Michael. 2016. *The Discovery of Isotopes.* Cham: Springer International Publishing. https://doi.org/10.1007/978-3-319-31763-2

Wealer, Ben, Simon Bauer, Nicolas Landry, Hannah Seiß, and Christian von Hirschhausen. 2018. 'Nuclear Power Reactors Worldwide—Technology Developments, Diffusion Patterns, and Country-by-Country Analysis of Implementation (1951–2017)'. Data Documentation 93.

Berlin: DIW Berlin, TU Berlin. https://www.diw.de/de/diw_01.c.583380.de/publikationen/data_documentation/2018_0093/nuclear_power_reactors_worldwide_technology_developments_nd_country-by-country_analysis_of_implementation__1951-2017.html

Wealer, Ben, and Christian R. von von Hirschhausen. 2020. 'Nuclear Power as a System Good: Organizational Models for Production along the Value-Added Chain'. DIW Discussion Papers 1883. Berlin: Deutsches Institut für Wirtschaftsforschung (DIW). http://hdl.handle.net/10419/222865

Wellerstein, Alex. 2019. 'John Wheeler's H-Bomb Blues'. *Physics Today* 72 (12): 42–51. https://doi.org/10.1063/PT.3.4364

Wimmers, Alexander, Fanny Böse, Jasmin Beppler, Pauline Morawe, Maximilian Weber, and Christian Von Hirschhausen. 2024. '(Re)Integrating Radioactive Materials and Waste into a Global Sustainable Development Context'. *Radiation and Environmental Biophysics*. https://doi.org/10.1007/s00411-024-01088-x

Wurzel, Samuel E., and Scott C. Hsu. 2022. 'Progress toward Fusion Energy Breakeven and Gain as Measured against the Lawson Criterion'. *Physics of Plasmas* 29 (6): 062103. https://doi.org/10.1063/5.0083990

Yim, Man-Sung. 2022. *Nuclear Waste Management: Science, Technology, and Policy*. Lecture Notes in Energy. Springer Dordrecht. https://doi.org/10.1007/978-94-024-2106-4

Chapter 3
Nuclear Fission: A System Good Analysis

Ben Wealer, Christian von Hirschhausen, and Alexander Wimmers

3.1 Introduction

As the title of this book suggests, nuclear power is generally considered to be one of the most complex forms of energy generation since it combines sophisticated chemo-physical processes with a very sophisticated engineering production structure and constitutes multiple socio-technical challenges, such as very long-term storage of waste or the international control of fissile materials (proliferation). As such, technological issues cannot be strictly separated from geopolitical strategies. Furthermore, economic and institutional aspects also shape the dynamics of the sector. The production and consumption of nuclear power do not obey any technical or economic "law" but result from several socio-technical pathways along a chain that builds the "system" of nuclear fission power (for nuclear fusion, a similar approach can be applied, refer to Chapter 4 of this book).[1]

The complexity of nuclear power is seldom part of public debates, and even among scientists, it is often oversimplified. Naïve questions sometimes suggest simple answers, where reality is complex and non-linear: What is the best way to generate nuclear power? What is the cheapest technology? How much does one megawatt-hour (MWh) of firm nuclear electricity cost? However, a deep dive into the system is merited to understand the complexity of the socio-technical system of nuclear fission. This facilitates the reader's orientation in the following chapters of this book and allows nuclear power to be put into the context.

[1] For the remainder of this chapter, "nuclear power" and other related terms refer to fission processes.

B. Wealer · C. von Hirschhausen (✉) · A. Wimmers
Workgroup for Infrastructure Policy (WIP), Technical University (TU) Berlin, Berlin, Germany
e-mail: cvh@wip.tu-berlin.de

C. von Hirschhausen · A. Wimmers
German Institute for Economic Research (DIW Berlin), Berlin, Germany

A. Wimmers et al. (eds.), *Nuclear Power*, Studies in Energy, Resource and Environmental Economics, https://doi.org/10.1007/978-3-031-99894-2_3

Therefore, in this chapter, we sketch out the main elements of the system good nuclear power and provide a socio-technical description of the system to assess different organizational models for private or public action structured along the production chain of the nuclear industry. Organizational models in the sense of this contribution are institutional solutions for the production or provision of the complex system good "nuclear power," or of individual subsystems along the production chain. The chapter adopts a positive perspective of describing the system parts as they are and not how they "could" or "should" be (which would be a normative perspective). Where possible, the chapter also describes the governance structure of the involved companies (state, private, semi-private); their degree of vertical integration, and the market shares; as well as the form of transaction, which can be done, e.g., through the market or hierarchies, i.e., long-term contracts or vertical integration.

This work is organized in the following way: Sect. 3.2 introduces the applied methodology. Section 3.3 describes the system good nuclear power in general and thus provides a stylized representation of the sector. It also highlights the interfaces between the different stages, including the military sector. The subsequent sections then provide the analysis of the production stages. Section 3.4 covers the "front end," consisting of mining and processing, uranium enrichment, and fuel fabrication. Section 3.5 deals with the construction, operation, and decommissioning of nuclear power plants (NPP). Section 3.6 covers the "back end," mainly interim storage and final disposal. Section 3.7 concludes.[2]

3.2 Methodological Background: System Good Analysis[3]

Providing goods and services requires multiple activities conducted successively or in parallel. Consequently, coordination between involved actors becomes imminent. Following Beckers et al. (2012), we refer to these complex activities as system goods and introduce a framework for the analysis thereof.

This framework is structured around the analysis of three main components. The goal of the first, the technical system, is the analysis of necessary goods, assets, processes, and interdependencies along the value chain to supply the system goods. In the second component, one can identify roles in the technical system. These roles "contain a structured bundle of tasks relating to the supply of a defined good" (Beckers et al. 2012, 4). Coordination needs to arise because roles are interrelated to varying degrees, and decisions in one area affect tasks in others. From these roles emerges the final component of the system good analysis: that of actors and institutions.

---

[2] This chapter builds on earlier research by the authors, mainly (Wealer and Hirschhausen 2020) (with additional references), as well as sectoral updates on individual segments of the system, such as Wimmers and von Hirschhausen (2024) for decommissioning. We thank Fanny Böse and participants of the graduate course on "Nuclear Power" for constructive comments. The usual disclaimer applies.

[3] This section is largely taken from (Wimmers and Von Hirschhausen 2024, 2–3) with ample references to the cited work on institutional system good analysis.

In this framework, actors take on certain roles corresponding to their specific assets and resources. Institutions define the field of action for actors. For example, coordination is guided by contracts or laws, while certain actors might be restricted from taking on some roles by licensing requirements. The complete description of all activities to supply a system good comprises an organizational model. The optimal design of such an organizational model, consisting of actors and institutions, is a primary consideration of normative analyses, for example, following the concept of New Institutional Economics, which aims to "[get] the institutions right" to achieve functional and efficient markets (Radulovic 2005, 1886).

Gizzi (2016) suggests a three-step iterative approach to define an organizational model. In the first step, the system good itself must be described. This approach includes the central properties of assets, goods, and processes, as well as technical and value chain requirements. The second step constitutes the three-stage mapping of the system good in the analysis framework, for which three different model types are created (see Fig. 3.1). First, in the "sector model," a stylized description of the system good is mapped onto the technical system by identifying all relevant assets, processes, and goods and their interdependencies. An arrow indicates demand-driven pressure. Following this, the decision model can be defined. Here, roles are described in a way that minimizes the required number thereof, meaning that some roles may be able to take on several processes or tasks, as indicated by the surrounding shapes. This framework is the basis for the later analysis of organizational models. Next, the needs for coordination and associated coordination areas can be defined, and relations must be analyzed, as indicated by the arrows. In the third and final step, organizational models are created. These organizational models include assumptions about which actors are involved in various roles. Actor models, such as monopolistic structures, can be derived from these assumptions. Further assumptions are made on the effect's institutions have on actors' behavior.

3.3 The System Good Nuclear Power: A Stylized Description

Nuclear power is one of the most complex system goods imaginable, with many overlapping and interdependent production stages. The description relies on understanding nuclear power as having emerged as "the daughter of scientific research and the military" (Lévêque 2013, 170). This understanding follows a logic of economies-of-scope, i.e., the joint production of military and commercial products and services. However, for clarity, we do omit a detailed description of the system good "nuclear weapons complex" from this analysis and instead illuminate the interfaces between the commercial and military sectors. Nuclear power cannot be provided without a complex web of physical, legal, and institutional infrastructure. Figure 3.2 shows the technical system of nuclear power, its subsystems (which are also system goods),

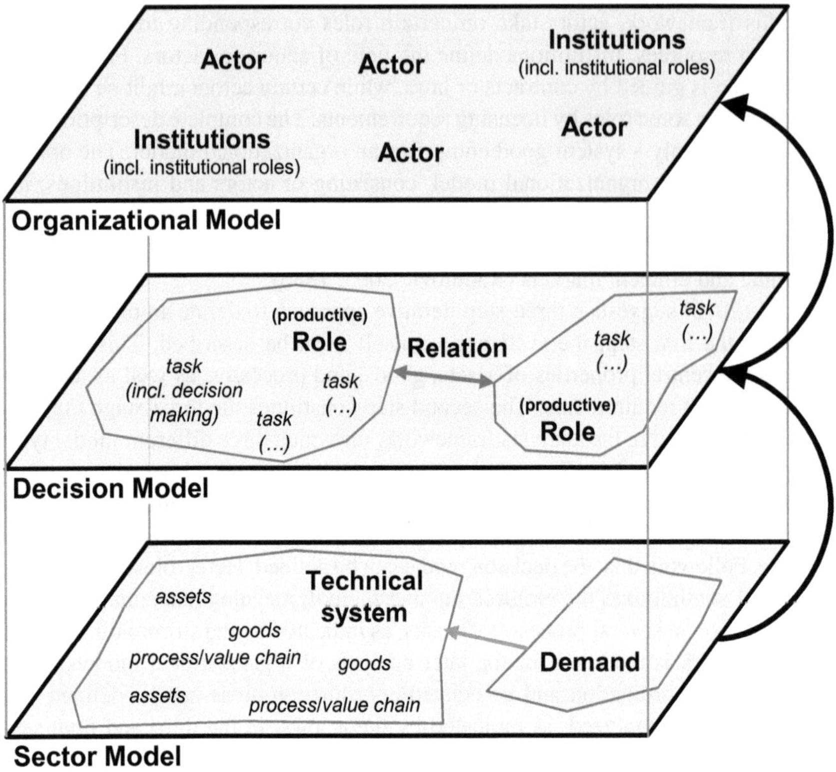

Fig. 3.1 Key elements of the system good analysis (*Source* Own translation adapted from Gizzi [2016, 10]) (*Note* The grey shapes surrounding individual components of the models indicate that an aggregate comprises several aspects that should be considered, such as the technical system or a role (shown here in bold). Arrows indicate the direction of influence)

and the interactions. In the subsequent sections, the individual stages of the value chain are examined in more detail.

3.3.1 The Front-End: Mining and Processing (A), Enrichment (B), and Fuel Fabrication (C)

For uranium ore (or thorium) to be used as fuel for electricity generation (or for military applications), the naturally occurring ore must undergo several processes.[4] The chain begins with the mining process, i.e., the extraction of ore from the ground. The mined natural uranium (consisting mainly of U-238) is grounded and washed,

[4] This section focuses on the production of fuel from uranium ore. For an assessment of the potential use of thorium, refer to Nelson (2012).

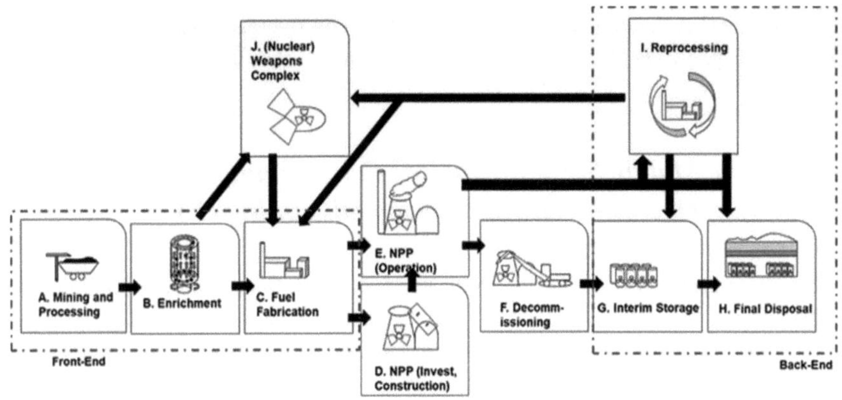

Fig. 3.2 The technical system and subsystems of the system good nuclear power (*Source* Own depiction)

leaving this production stage as a so-called "yellow cake." The commercial fleet of NPPs consists mainly of light water reactors (LWR), which require low-enriched uranium (LEU). Today, only a few commercial reactors operate with natural uranium (e.g., the Canadian-designed CANDU reactors in operation in Canada, India, and Romania). LWRs were not originally part of the technical portfolio in the late 1940s and early 1950s, as uranium enrichment was complicated and expensive. During the Manhattan Project, the United States (US) developed the first large enrichment plant in Oak Ridge, Tennessee.[5] Enrichment technology is a so-called "dual-use asset," i.e., the technology is, in principle, applicable to both commercial and military assets. For example, uranium for fuel used in LWR is generally enriched to 3 to 5% of U-235. But the technology can also be used to enrich uranium to a higher degree and thus produce fissile material that can be used in weapons. A by-product of this enrichment is depleted uranium (DU), which can also be used for military purposes.[6]

Once the uranium has been enriched, the production of fuel elements for subsequent use in a power plant follows. The production of fuel elements has a high degree of specificity because these high-tech products must be built according to the individual specifications of the reactor types. This requires complex coordination processes, which is why some fuel element manufacturers are also reactor manufacturers. This stage also has an interface to the nuclear weapons sector. In the US, for example, plutonium from the weapons arsenal was used to manufacture so-called

[5] Refer to Chapter 2 (Technologies) of this book for details on the development of nuclear weapons and reactors.

[6] The military services of the US and UK use depleted uranium in ammunition and projectiles as well as in the armor of vehicles. One of the main reasons is that this "waste product" is cheap and available in large quantities, and uranium is a very dense and hard metal with a high penetrating power, i.e., it is armor-piercing. DU ammunition was first used on a large scale in the 1991 Gulf War, in Bosnia and Herzegovina in 1995, and in Kosovo in 1999. For more information regarding the harmful consequences and risks, see Fairlie (2009) Bleise et al. (2003), and Giannardi and Dominici (2003).

mixed oxide (MOX) fuel elements (Hyatt 2017). This overall process (sub-processes A to C in Fig. 3.2) is also known as the "front end" of the fuel chain.

3.3.2 The Middle Section: Nuclear Power Plant Construction (D), Operation (E), and Decommissioning (F)

The middle section comprises the NPP itself: construction, operation, and decommissioning. Construction consists of various upstream and simultaneous processes. Particularly noteworthy are the upstream processes of research and development. This may relate to determining the feasibility of the reactor technology or simply understanding the design logic if the technology is not domestically sourced. This is followed by designing and manufacturing (or procuring) the nuclear steam supply system, as well as the design and procurement of the remaining equipment and its integration with the nuclear steam supply system into the overall station design. The actual process of building the plant and monitoring its progress completes the construction phase.

Once the power plant is completed, the reactor core is loaded, and the correct license has been granted, the actual operating phase of the NPP begins. Initially, the current power plant fleet was generally designed to run for 30–40 years. Some plants might operate for up to 60 years. This state includes maintenance and repair of technical equipment, loading and unloading of fuel elements, disposal of operational wastes, and (interim) storage of spent fuel. Interfaces to other production stages are imminent: Fuel elements must be procured, and high-level radioactive waste must be disposed of or temporarily stored or transported to reprocessing.

Decommissioning has an essential interface to "operation," since the closure of the reactor is a necessary condition for decommissioning the plant. During operations, the power plant licensee must prepare and plan the decommissioning process, which induce lengthy regulatory processes. The so-called post-operational phase of an NPP, in which the reactor is defueled, often occurs under the operating license. The operator or the organization responsible for dismantling must also be able to access information, personnel, or operating facilities during dismantling. During dismantling, large quantities of low- and intermediate-level radioactive waste and high-level radioactive waste are produced, which must be disposed of or temporarily stored. With only 23 decommissioned nuclear power reactors worldwide and increasing grid decoupling of NPPs, nuclear decommissioning is increasingly focused upon.[7]

[7] Refer to Chap. 15 of this book for a detailed assessment of organizational models in nuclear decommissioning.

3.3.3 *The Back-End: Interim Storage (G), Final Disposal (H), and Reprocessing (I)*

Nearly all processes in the different value-added stages generate radioactive waste. The first types of waste that emerge are the tailings (excavated rocks to access the uranium ore) at the uranium mine, and all the subsequent processes in the front end also create radioactive waste. Excluding Slovakia and Russia, European commercial NPPs have, for instance, generated more than 2.5 million m^3 of low intermediate-level waste (LILW), of which 20% is currently in interim storage. Over 60,000 tons of spent nuclear fuel are in interim storage, of which 81% are in wet storage, which comes with particular safety risks. The process of decommissioning may produce another 1.4 million m^3 of LILW. Over its (currently estimated) lifetime, European nuclear reactors may produce around 6.6 million m^3 of nuclear waste, which must be disposed of below ground (Besnard et al. 2019).

The most challenging task is the final disposal of high-level waste (HLW). What exactly constitutes HLW depends on the national policy and on the nature of the fuel chain. Most HLW consists of spent nuclear fuel (SNF). The most widely used strategy for HLW is to store the spent fuel in interim storage facilities while constructing a deep geological disposal repository for final disposal. However, more than seven decades after the construction of the first nuclear reactors, no deep geological repository for HLW is operating.[8]

Another option is the reprocessing of SNF with the idea of extracting uranium and plutonium, of which only a small part will reenter the fuel chain as fresh fuel. Only some of the separated plutonium is reintroduced into the fuel cycle through the fabrication of MOX fuel rods, while the depleted uranium is mainly stored, constituting vast stockpiles. In addition, reprocessing spent nuclear fuel creates large quantities of LILW, resulting in a greater volume of waste to be managed compared to when SNF is treated directly as waste. After reprocessing, the residual fission products and actinides in liquid form are evaporated and converted to solids by a vitrification process before being disposed of. Reprocessing also has an interface to the military sector and raises proliferation concerns. If extracted by reprocessing, the plutonium contained in SNF becomes a proliferation risk.

3.3.4 *Nuclear Weapons (J)*

Within the system good framework, nuclear weapons play a specific role, and are localized orthogonally to the production chain (Fig. 3.2). A central theme of the geopolitics of nuclear power is that nuclear weapons were at the very origins of the sector, and they continue to play a central role in many nuclear programs today. In the days of World War II, both the US and the Soviet Union were focusing on

[8] Refer to Chapter 16 of this book for a detailed assessment of global HLW storage activities.

fundamental and applied R&D to develop nuclear weapons. These activities have proliferated to a point where today, anybody with decent technical knowledge and access to fissile material is able to build some form of a nuclear bomb.

The two routes to the bomb involve either the enrichment of uranium or the separation and enrichment of plutonium. Both were pursued on a large industrial scale in Project Manhattan, where a large number of research groups and the military were involved (Groves 1983; Rhodes 2012)[9]:

- Originally, the enrichment of uranium was instigated for weapons development. For this purpose, a huge industrial complex was built in Oak Ridge, Tennessee (US) from 1942 onwards, where the gaseous diffusion process was used for the first time on a large industrial scale. This corresponds to the arrow from the enrichment to the weapons complex (Fig. 3.2).
- Later, however, the second possibility of a plutonium bomb was also pursued. For this purpose, the world's first large-scale industrial plutonium factory was commissioned in Hanford, Washington (US) in 1943. The Soviet Union had obtained the technical foundations through espionage from Los Alamos, the main site of the Manhattan Project, and also developed a bomb in the post-war period, which was fed with plutonium from an experimental reactor. This corresponds to the arrow from reprocessing to the weapons complex.

In contrast to electricity generation, which today mainly uses material enriched to 3–5% U-235, the construction of a uranium bomb requires highly enriched uranium (HEU). An enrichment level of less than 20% is referred to as LEU, which cannot be used for nuclear weapons. For nuclear weapons, uranium is typically enriched to 90%. HEU is used according to the "gun-type" principle by combining two subcritical U-235 masses to start the chain reaction. This type of bomb was used in Hiroshima and developed an explosive force equivalent to around 15,000 tons of the conventional explosive TNT (Englert 2012).

Plutonium is produced in a nuclear reactor via neutron capture from more abundant U-238. This avoids the enrichment processes. The dual-use principle also applies here: The plutonium produced in a commercial nuclear reactor can (but does not necessarily have to) be used for military purposes. Similar to uranium, a distinction is made between reactor plutonium with a proportion of around 60% Pu-239 and weapons-grade plutonium with more than 93% Pu-239. In the "implosion" bomb design, a sphere of fissile plutonium is coated with conventional explosives that compress the fissile material to become supercritical. The bomb that destroyed Nagasaki was an implosion bomb with plutonium as the explosive; it reached 21,000 tons of TNT-equivalent explosive force (Englert 2012).

Since the 1940s, there have been further developments in nuclear weapons. The next generation was created in the early 1950s by combining nuclear fission and nuclear fusion, the so-called thermonuclear or hydrogen bomb. In the first stage,

[9] This section draws on the chapter by Matthias Englert (2012) on nuclear weapons, summarized (in German) by von Hirschhausen (2023, chap. 2.3). Further information is provided in Chapter 6 and 7 of this book.

nuclear fission is primarily used to initiate the fusion process of different hydrogen isotopes, releasing very large amounts of energy. These weapons can achieve an explosive force thousands of times greater than that of the first fission bombs, i.e., millions of tons of TNT equivalents. By increasing the explosive force, it also became possible to use nuclear weapons on intercontinental missiles.

Modern nuclear weapons can be built with as little as four kilograms of plutonium or twelve kilograms of uranium. Much less fissile material is needed for so-called "dirty" bombs, in which radioactive substances are added to bombs with conventional explosives to increase the destructive effect of the bomb on site through radioactive contamination. Constructing simple fission bombs, such as those used in Hiroshima and Nagasaki, is no longer a technical challenge, even for emerging countries. Today, there are 10 countries with the theoretical capability to build nuclear weapons. In addition, Ukraine, Belarus, and Kazakhstan, three other now independent countries, were formerly involved in nuclear weapons programs as former parts of the Soviet Union. The biggest hurdle today, therefore, is not technology, but getting hold of HEU and weapons-grade plutonium.

At present, due to increasing geopolitical tensions, the nuclear weapons industry is thriving in all nuclear weapon countries. The three nuclear superpowers, China, Russia, and the US, are putting significant resources towards the development of new weapon systems. The US and Russia are not only modernizing their warheads but also the total system, including production facilities, missile aircraft, and delivery systems. As a dynamic newcomer, China is pursuing an active expansion and has a growing number of warheads (approx. 410 in 2023) (SIPRI Yearbook 2024: Armaments, Disarmament and International Security 2024). The yearbook of the Stockholm International Peace Research Institute (SIPRI) (2023, 12) suggests that China's "nuclear stockpile is expected to continue growing over the coming decade and some projections suggest that it will deploy at least as many intercontinental ballistic missiles (ICBMs) as either Russia or the US in that period."[10]

In the following sections, the production stages "nuclear front-end" (A, B, C in Fig. 3.2), the "nuclear middle section" (construction, operation, and decommissioning, D, E, F), and the "nuclear back-end" (G, H, I in Fig. 3.2) are mapped in the analysis framework. First, the technical system is described in detail, which allows deriving possible tasks, roles, and actors, as well as the relationships between these roles and actors. This allows for identifying coordination needs as well as essential interfaces. In the third step, organizational models to produce the good of this value-added stage are described with respect to the companies involved, their governance structure, and competition.

---

[10] The SIPRI yearbooks provide an overview of world nuclear forces, though they do not cover industry structures in much detail. For further details on weapons, see Bondansky (2004, chap. 17), on fissile materials see the reports by the "International Panel on Fissile Materials" (https://fissilematerials.org/), and on interlinkages with the commercial sector see Stirling and Johnstone (2018; 2024). Also, refer to Chapter 6 of this book for further insights on nuclear non-proliferation.

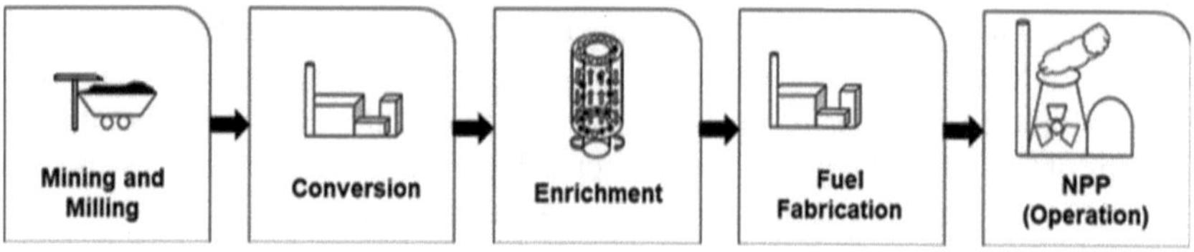

Fig. 3.3 Technical system representation of the "front-end" for light water reactors (*Source* Own depiction)

3.4 Nuclear Front-End: From Mining to Fuel Fabrication

3.4.1 Mapping of the System Good "Front-End" in the Analysis Framework

3.4.1.1 Description of the Technical System

To be used as a fuel for electricity generation (or for military applications), uranium (or thorium) must undergo several procedures. In this section, we describe the necessary processes to supply nuclear fuel to LWRs. These are generally attributed to the so-called "front end" of the nuclear fuel chain. Different production steps must be carried out. In some cases, and these require highly specific assets. In addition, some of these processes are interlinked, which leads to the identification of interfaces between corresponding assets. Figure 3.3 presents this segment of the technical system, which shall be described in more detail below.

Mining and Milling

The first step in the "front-end" is the mining of the ore. This section focuses on uranium, though other ores, e.g., thorium, can also be used.[11] To explore and mine uranium ore, deposits must first be discovered and classified; a specific technical asset for this are, for instance, air gamma spectrometry tools for resource exploration. Depending on the specific deposit, different mining processes can be used: Open pit and underground mining are conventional mining processes, while in some cases, leaching (in-situ, in-place, or heap leaching) is performed. Depending on the exploitation technique, required assets are, e.g., chemicals or pumps for leaching or trucks and conveyors for open pit or underground mining. The most common production method in 2021 is in-situ leaching (63.3%), followed by open-pit (17.1%) and

[11] Thus far, no other ores than uranium have been used on a commercial scale. The use of thorium would, amongst others, require a reshuffling of the entire system (i.e., conversion, enrichment, fuels, reactors, etc.), both the front- and middle-section, which is likely to prevent an industrial-scale use of thorium in the next decades (Nelson 2012). See David (1985) on the role of path dependencies in technical systems.

Table 3.1 Shares of global production of uranium ore by production method

Production method	2018	2019	2020	2021 (Preliminary)
Open-pit mining	17.0	16.6	18.7	17.1
Underground mining	20.8	19.9	16.1	15.1
In-situ leaching	55.5	56.9	58.3	63.3
Co-product/by-product	6.6	6.5	6.6	4.5
Other	0.1	0	0	0

Source NEA and IAEA (2023)

underground mining (15.1%). An overview is provided in Table 3.1 (NEA and IAEA 2023).

The next step is the extraction process of transportable material from the uranium ore by milling and leaching with acid. The solution is then cleaned, and the uranium is precipitated and dried. The end product is concentrated triuranium octoxide (U_3O_8), also called "yellow cake" (Bodansky 2004, 197).[12] Mining and milling are both transformation and transfer processes. In most cases, conventional mines operate a mill on their production site, so the transfer is mainly between the intermediary good "yellow cake" and the transport to the conversion service provider. There is a very low degree of substitution for uranium.

Conversion and Enrichment

Mined uranium consists of about 99.28% U-238 and only 0.72% U-235 (relevant for nuclear fission), the remainder being a minimal share of U-234. The to-be-used uranium must be enriched to 3–5% U-235 for LWR operation. The majority (~90%) of operational commercial power plants are LWRs and thus use enriched fuel. Only a fraction of power plants, e.g., the Canadian CANDU reactors, use natural uranium as a fuel and do not need this step of the nuclear fuel chain. The intermediate process before uranium can be enriched is converting the yellow cake into uranium hexafluoride (UF_6). This extremely corrosive gas converts to hydrofluoric acid when it encounters water, so the handling is technically complex.

Historically, uranium enrichment, i.e., the separation of isotopes U-238 and U-235, was carried out using the gaseous diffusion process, a centerpiece of Manhattan Project in Oak Ridge (Tennessee, US). Today, however, the centrifuge process has become established: The gaseous UF_6 is fed into a centrifuge, and the heavier U-238 atoms are separated from the U-235 atoms.[13] This step must be repeated about ten times or more to achieve the desired enrichment level of U-235. What remains is depleted uranium. The technology used in uranium enrichment is highly specific and

---

[12] Mining produces a lot of external effects, like radon exposure of the workers and environmental damages in the form of mining residues, the so-called tailings (Bodansky 2004).

[13] The performance of a uranium enrichment plant is measured in separative work units (SWU). See Bodansky (2004, 200–201) for more details and a representation of further enrichment technologies.

Table 3.2 Processes and assets of the technical system "front-end"

Elementary process	Processes	Assets
Mining and Milling	• Discovery and classification of uranium deposits • Mining of uranium ore • Transportation of ore to mill • Milling and leaching of ore to produce concentrated U_3O_8	• Uranium mines (or ore) • Mining infrastructure • Transport logistic
Conversion	• Conversion of U_3O_8 into UF_6	• Conversion plant
Enrichment	• Enrichment of UF_6	• Enrichment plant (e.g., centrifuge)
Fuel Fabrication	• Chemical conversion of UF_6 into uranium dioxide (UO_2), • Transformation into ceramic UO_2 pellets • Fabrication of the metal framework for the fuel assembly • Finalization of final fuel assembly	• Fuel fabrication plant • Decontamination tools

Source Own depiction

very sensitive and, therefore, constantly under international control because of the proliferation risks of weapons-grade uranium. Both conversion and enrichment are transformation processes. The enrichment stage also requires highly specific assets.

Fuel Fabrication

The final process of the front-end is fuel fabrication. First, the enriched UF_6 is chemically converted into uranium dioxide (UO_2), from which ceramic pellets are produced. Then, the fabrication of the metal framework for the fuel assembly consisting of fuel rods follows, into which the fuel pellets are loaded. The rods are then sealed and finally assembled into the final fuel assembly structure. The resulting fuel elements are high-tech products built according to each reactor type's individual specifications (Wimmers et al. 2025). Fuel fabrication is a transformation process that has interdependencies to the production stage of the "middle section" of the system good, more specifically to the power utility. Thus, a strong need for coordination arises between the supplier of fuel elements and the utilities. This interface needs to be designed especially for compatibility.

The analysis of the technical system leads to an array of interdependent processes, which are characterized mainly by a high degree of specificity, a low degree of substitutivity, and some compatibility. Table 3.2 overviews the processes and assets needed in the nuclear front end.

3.4.1.2 Tasks, Roles, and Actors

In this step of the analysis, a role model can be set up, in which the major roles required for the analysis and the relationships between these roles are included. We focus here on the role type offered by the elementary process, which primarily comprises the combined use of assets to perform elementary processes. All the tasks needed for the elementary processes of "mining and milling," "conversion," "enrichment," and "fuel fabrication" could each be provided by one specific role. Between each of these roles, an ordering relationship would exist; for instance, the role "conversion" would order from the role "mining and milling," which would provide concentrated U_3O_8. In addition, time-complementary relationships are traceable between the roles, as the different processes must be carried out in a certain order. Possible areas of coordination are between the enrichment plant and the provision of the conversion service, and between the nuclear utility and fuel fabrication, especially towards fuel element structure and reactor specifics.

Actors involved in the nuclear front-end are the countries where the uranium deposits are located. The elementary processes of the "front-end" can be provided by private companies, state companies, or a mixture of both. One actor is the mining company, which owns and exploits the mine. Another actor is the provider of conversion services. The specific asset for this process is a commercial conversion plant. Next, the enrichment services provider has the specific asset of an enrichment conversion plant. The last actor in the nuclear front end, at least from the production point of view, is the provider of the conversion, pelletizing, and final fuel assembly construction services.

3.4.2 *Organizational Models for the Provision of ...*[14]

3.4.2.1 ... Uranium

For 2021, the "Red Book" of the NEA and IAEA (2023) estimated recoverable uranium deposits[15] to around 6 million tons (recoverable at costs of less than $130 per kgU), only a slight decrease from the previous estimates for 2019. Depending on future nuclear reactor capacity development, these reserves could last for about 100 years. The Red Book lists identified recoverable uranium resources in 55 countries, with Australia, Kazakhstan, and Canada in the top three, accounting for nearly 47% of global resources. This includes retrievable resources costing up to $260 per

[14] Refer to Chapter 8 of this book for a more detailed market concentration analysis on the nuclear front-end.

[15] "Reasonably assured" and "inferred resources"; the former category refers to uranium reserves for which the quantities and concentrations can already be estimated very precisely and reliably; the second category refers to deposits for which less precise information is available (Neles and Schmidt 2012).

Table 3.3 The 10 biggest uranium mines, as of 2022

Mine	Main owner	Production [Tons]	Global share%
Cigar Lake (Canada)	Cameco/Orano	6,928	14
HUSb (Namibia)	Swakop Uranium (CGN)	3,358	7
Inkai, sites 1–3 (Kazakhstan)	Kazatomprom/Cameco	3,201	7
Olympic Dam (Australia)	BHP Billiton	2,813	6
Karatau (Budenovskoye 2) (Kazakhstan)	Uranium One/Kazatomprom	2,560	5
Rössing (Namibia)	CNNC	2,255	5
SOMAIR (Niger)	Orano	2,020	4
Four Mile (Australia)	Quasar	1,740	3
Central Mynkuduk (Kazakhstan)	Ortalyk	1,650	3
South Inkai 4 (Kazakhstan)	Uranium One/Kazatomprom	1,600	3
Total (top 10)		28,125	57

Source Compiled by authors, based on World Nuclear Association

kgU. The top 10 countries account for 84.9% of total global uranium resources.[16] Most of these countries do not operate any NPPs. France, as one of the largest nuclear-operating countries, does not operate its own mines, but operates mines in Canada, Gabon, Kazakhstan, Mongolia, Namibia, and Niger.

The mining companies are key players in uranium mining.[17] It should be noted that it typically takes many years for a uranium mining project to move from discovery to production (OECD/NEA 2008, 45). In 2022, 15 countries produced uranium, with a total production of about 49,355 tons of uranium. In 2022, Kazakhstan produced the largest share of the world's uranium supply (43% or 21 kt). The ten largest mines cover 57% of global demand. The mine with the largest share is the Cigar Lake Mine in Canada. In 2022, about 7,000 tons were mined there, corresponding to 14% of the world production. A joint venture between Canadian Cameco and French Orano operates the mine. This joint venture approach is followed at other mines, too. In some cases, the mine output is shared between the partners or one of the partners may be responsible for selling all the uranium produced (OECD/NEA 2008, 51). Table 3.3 provides an overview of the ten largest mines, their production, and the main mine owners, ordered by production. Table 3.4 shows the top four mining companies.

[16] Global distribution of the identified resources (as of January 1, 2021): Australia 21.7%, Kazakhstan 17.4%, Canada 10.4%, Russia 7%, Namibia 6%, South Africa 5.9%, Brazil 5.5%, Niger 4.7%, China 4%, Uzbekistan 2.2%, Ukraine 2.2%, Mongolia 1.8%, India 1.3%, United States of America 1.1%, Tanzania 1%, Botswana 1%.

[17] Since 2011, spending on uranium exploration and mine development steadily declined from over US-$2 billion in 2014 to only US-$250 million in 2020. In 2021, the trend began to reverse, when global spending increased by US-$30 million (NEA and IAEA 2023).

Table 3.4 The top four uranium mining companies, as of 2022[18]

Company	Main owner	Production [tons]	Share worldwide%
Kazatomprom	The Kazakh National Wealth Fund Samruk-Kazyna JSC owns 81.28% of the share capital of NAC Kazatomprom JSC, and 18.72% is in free float on the Astana International Exchange and the London Stock Exchange	11,373	23
Orano	The French State directly or indirectly controls 86.52% of the capital	5,519	11
Cameco	Canadian State[19]	5,675	12
CGN	Chinese state[20]	4,385	8

Source Compiled by authors, based on World Nuclear Association, and various websites

The supply side of the uranium market is somewhat concentrated, with the top four producing mining companies serving 56% of the uranium market (Table 3.4). Kazakhstan continuously increased its production year after year, overtaking Canada and Australia in 2009. Since then, Kazatomprom has been the market leader, with a market share of 22% in 2018.[21] The other three major mining companies, Orano, Cameco, and CGN, have roughly half this market share (10–12% each). These four companies are majority-owned by national states, i.e., the Kazakh, French, Canadian, and Russian states. The US is only a small producer: In 2023, about 87% of uranium was imported,[22] down from 93% in 2018. This has led to US companies demanding uranium import quotas. Refer to Wimmers et al. (2025) for an overview of European dependencies in the front-end. The EU imports nearly all its uranium supply, and as of 2023, 96% was sourced from only five countries: Canada, Russia, Kazakhstan, Niger, and Australia.

Vestergaard (2015) reports that uranium deals are made under long-term contracts, executed several years before the first delivery, and purchased through a bidding process. Only about 20% of all uranium was sold on the spot market, and the remaining 80% under long-term contracts, typically lasting for 5 years or, in some cases, up to 10 years or more. Buyers tend to diversify their uranium supply and the corresponding conversion, enrichment, and fuel fabrication services. Utilities can tell mining companies how much uranium to deliver to which converter and then to

[18] Other mining companies are: Uranium One and ARMZ (both Russia), Navoi Mining (Uzbekistan), CNNC (China), BHP (Australia), and General Atomics (US).

[19] The shares of Cameco are subject to restrictions on foreign (non-Canadian) ownership (OECD/NEA 2008, 46).

[20] https://web.archive.org/web/20160303213254/http://www.slideshare.net/UKTI2014/galleon-suite-27th-market-briefings, accessed 21–08-2024. This date holds for all other online references unless otherwise indicated.

[21] In 2018, Kazatomprom acquired 40.5% of the shares of Energy Asia (4% market share).

[22] US Energy Information Administration, as of June 2023, https://www.eia.gov/energyexplained/nuclear/where-our-uranium-comes-from.php.

which enrichment service provider. The prices in long-term contracts can be fixed, or they can vary over the contract duration; in this case, the prices depend on the spot market price. The demand for uranium is inelastic. The fact that there is no real substitute for uranium reinforces the singularity of the uranium market. One of the largest secondary sources of uranium stems from nuclear weapons programs. For instance, under an intergovernmental agreement between Russia and the US (in force between 1993 and 2013), highly enriched uranium from military sources was delivered from Russia to the US, which was then divided between four companies: AREVA (30%), Cameco (30%), TENEX (30%), and NUKEM (10%) (OECD/NEA 2008). This is roughly the equivalent of approximately 9,000 tons per year in the form of low-enriched uranium, which also affected the demand.[23]

3.4.2.2 … Conversion and Enrichment Services

The only commercial conversion plants in 2023 were located in the US, Canada, France, Russia, and China, and thus, except for Canada, exclusively in nuclear weapons states.[24] ConverDyn, the multinational US company, is the only company not in state ownership. However, the US conversion plant halted production in 2017 due to a worldwide oversupply of UF_6 at the time.[25] The French plant operated by Orano increased its output from only 2,500 tons in 2018 to 9,000 tons in 2022 as a supposed response to the Russian invasion of Ukraine.

The conversion services market actors are also involved in the mining business. The largest conversion plants are operated by Orano in France and by CNNC in China; both also operate uranium mines. This vertical integration makes sense, as these companies are solely involved in the nuclear sector. In contrast, uranium mining is not the only mining service provided by other major mining companies like BHP or Rio Tinto. UF_6 conversion services can be bought on the spot market, but most conversion supply is organized via long-term contracts.[26] Table 3.5 provides an overview of the commercial conversion plants, their production capacity, and the rate of capacity utilization.

Until the 1980s, the US monopolized the "Western" market for enrichment services with its diffusion process, while the USSR was the monopolist for the "Eastern" market. However, since the 1980s, companies with gas centrifuge technology, such as the British-Dutch-German company Urenco, have gained increasing market shares. Currently, the only operational large-scale commercial enrichment plants are located in the US, the UK, France, Russia, China, Japan, Germany, and the Netherlands. Because of the inherent high risk of proliferation, only the nuclear weapon

---

[23] For further information on the uranium market, see, e.g., Monnet et al. (2017), as well as Rooney et al. (2015).

[24] US Energy Information Administration, op.cit.

[25] Supposedly, the plant was restarted in 2024. Refer to Chapter 8 of this book for more.

[26] A transaction on the spot market is usually in the form of a purchase of uranium and conversion together (i.e., UF_6) but priced separately (OECD/NEA 2008, 56).

Table 3.5 Conversion plants and their capacity as of 2022

Company	Location (country)	Capacity [Tons]	Capacity utilization [Tons]
Orano	Pierrelatte and Malvesi (France)	15,000	9,000 (60%)
CNNC	Lanzhou und Hengyang (China)	15,000	11,000 (73%)
Cameco	Port Hope (Canada)	12,500	10,600 (85%)
Rosatom	Seversk (Russia)	12,500	12, 000 (96%)
ConverDyn	Metropolis (US)	7,000	0 (0%)
Total		62,000	42,600

Source Own depiction based on Finegold (2023)

states have enrichment technologies on a commercial scale, with one exception being the Urenco Company.[27] The Russian state owns Tenex, which has by far the largest enrichment capacity located at four enrichment plants, followed by the French state-controlled Orano. These two companies have also vertically integrated enrichment services into their portfolio. Japanese JNFL and Chinese CNNC mainly serve their respective domestic markets. Due to the dual-use nature and the complex institutional environment, the majority of supply is conducted under long-term contracts, often with a duration of five or more years (OECD/NEA 2008). Table 3.6 gives an overview of the global enrichment companies, their location, as well as their operational and planned capacity.

3.4.2.3 … Nuclear Fuel

Nuclear fuel assemblies are highly engineered products that must be adapted to the physical and chemical characteristics of the reactor. Due to this high specificity, many fuel fabrication companies are also reactor vendors (or owned by them). The top suppliers, Westinghouse and Framatome, are major reactor vendors, while the Russian company, TVEL, which produces the fuel for the Russian PWRs (VVER), is part of the Rosatom conglomerate. The LWR market comprises three single markets (OECD/NEA 2008): a market for boiling water reactors (BWR, dominated by GE Hitachi); a market for (Western designed) pressurized water reactors (PWR, dominated by Orano/Framatome); and the VVER market, where TVEL dominates. The company has a share of almost 100%, albeit efforts are ongoing to diversify fuel supply, especially for Eastern European power plants after the Russian invasion of Ukraine in 2022. In general, however, the market for LWR fuel has become more competitive lately (see Chapter 8 of this book), as several competing suppliers can now supply different fuel types; Russian TVEL can manufacture Western PWR fuel, and Western companies can produce VVER fuel. In addition, there has been

[27] Urenco is regulated by the 1971 Treaty of Almelo. It is owned one third by the German energy companies PreussenElektra and RWE, one third by the Dutch government and one third by the British government. The company is a specialized company and operates enrichment plants in Germany, the U.K., and the Netherlands.

Table 3.6 Global enrichment capacity, operational, and planned

Company	Location	2013	2015	2020	2025	2030
Orano	Georges Besse I & II (France)	5,500	7,000	7,500	7,500	7,500
Urenco	Gronau (DE); Almelo (NL); Capenhurst (UK)	14,200	14,400	13,700	17,300	16,300
Urenco	Urenco (New Mexico)	3,500	4,700	4,900		
Tenex	(Angarsk, Novo Uralsk, Zelenogorsk, Seversk; Russia)	26,000	26,578	27,700	26,200	24,800
CNNC	CNNC (Hanzhun & Lanzhou; China)	2,200	5,760	6,300	11,000	17,000
JNFL	Japan (Rokkashu)	150	150	150	150	1,500
Others	Argentina, Brazil, India, Pakistan, Iran	75	100	66	375	525
Total separative work units/year (thousands p.a.)		51,700	58,750	55,416	62,525	66,275

Source Own depiction based on World Nuclear Association (2022)[28]

some rationalization in the LWR fuel fabrication industry. The Japanese company Toshiba bought Westinghouse Electric and Nuclear Fuel Industries (NFI) and formed a joint venture with GE Hitachi (Global Nuclear Fuels, GNF). Orano and Mitsubishi Heavy Industries acquired shares of Mitsubishi Nuclear Fuel and have formed a fuel fabrication joint venture in the US. Orano and Kazatomprom agreed to build a fuel fabrication plant (a 1200 t/yr) in Kazakhstan. Kazatomprom plans to vertically integrate another elementary process of the fuel chain into its portfolio. It also plans to provide a third global fuel supply by 2030, with a possible major customer being China.

The supply of nuclear fuel is generally based on long-term contracts. Changing the fuel supplier can have costly and regulatory consequences.[29] Some fuel fabrication companies do not possess the asset "conversion plant." This service is then provided by other actors, who might even sell UO_2 powder. Substitute relations between the different players can be expected to emerge as several companies can supply LWR

[28] World Nuclear Association, 2020. "Uranium enrichment". Accessed June 5, 2020. https://www.world-nuclear.org/information-library/nuclear-fuel-cycle/conversion-enrichment-and-fabrication/uranium-enrichment.aspx and Updated Webpage as of 11 October 2022.

[29] "In order to change supplier, the NPP must operate for 2 or 3 years with a mixed core (i.e., both old and new fuel together), which can be complicated to license and can limit the coverage of fuel warranties from both fuel vendors. The alternative is to reload the whole core with new fuel, which is wasteful and expensive and thus would only be adopted in unusual circumstances. Other complications of changing fuel vendor include: the need to obtain regulatory approvals and make license amendments; differences in the methodologies used for reload safety analyses; and the need to introduce and license new or modified software for reload calculations. For such reasons, switches of fuel supplier are relatively rare, and ten-year contracts with an option for prolongation are not unusual" (OECD/NEA 2008, 73).

Table 3.7 Global LWR fuel fabrication companies and production capacity in tons per year as of 2021

Company	Location (country)	Conversion	Pelletizing	Rod/ assembly
Westinghouse AB/**	AB: Västeras (Sweden) **: Springfields (UK); Columbia (US)	3,337	2,794	3,614
Framatome FBFC/ANF/ Inc	FBFC: Romans (France); ANF: Lingen (Germany); Inc: Richland (US)	3,800	3,250	3,250
TVEL MSZ/ NCCP	MSZ: Elektrostal (Russia); NCCP: Novosibirsk (Russia)	1,950	2,700	2,760
GNF Japan/ Americas	Kurihama (Japan); Wilmington (US)	1,200	1,620	1,630
CJNF Jianzhong	Yibin (China)	800	800	800
KNFC	Daejeon (Korea)	700	700	700
NFI	PWR: Kumatori (Japan); BWR: Tokai-Mura (Japan)	0	633	534
ENUS	Juzbado (Spain)	0	500	500
Mitsubishi Nuclear Fuel	Tokai-Mura (Japan)	450	440	440
INB	Resende (Brazil)	160	120	400
CBNF	Baotou (China)	0	0	400
CNNFC	Baotou (China)	200	200	200
DAE Nuclear Fuel Complex	Hyderabad (India)	48	48	48
Ulba	Ust Kamenogorsk (Kazakhstan)	0	108	200
Total		12,645	13,913	15,476

Source Own depiction based on World Nuclear Association (2021)[30]

fuel. Table 3.7 informs about the LWR fuel fabrication companies and their respective production capacity in tons per year with respect to conversion, pelletizing, and fuel assemblies as well as production sites.

[30] World Nuclear Association, Updated 13 October 2021. https://world-nuclear.org/information-library/nuclear-fuel-cycle/conversion-enrichment-and-fabrication/fuel-fabrication.

3.4.3 Environmental Effects of the Front-End[31]

The effects of uranium mining begin with the tailings, i.e., the unprocessed remains of the uranium ore, which are usually stored on site in the open air. The tailings not only contain residues of the uranium itself but also radioactive decay products and other heavy metals such as arsenic, copper, and cadmium. In addition to the dangers of direct contact or inhalation of the decay product radion-222, dust from tailings disposal sites also leads to a widespread of pollutants. Furthermore, groundwater bodies are at risk due to seepage of pollutants. In poorly stabilized tailings basins, there is a risk of dam failures and the flooding of settlements and agricultural land.

Spoil heaps and pits from unfilled opencast mines can cause further environmental effects. These can also contain heavy metals, generate dust, and contribute to pollutant inputs into groundwater bodies. Potentially similar effects are caused by the in-situ leaching process, which dissolves the uranium ores using a chemical extraction fluid that is only partially disposed of, if at all. Once the process plants of the front-end area are dismantled, they may still contain pollutants. This can result, to a small extent but over a very long period of time, in perpetual costs for which the original operators can no longer be held liable.

The environmental costs and subsequent clean-up costs of uranium mining and processing are not included in the pricing of electricity from nuclear energy but represent negative external effects for the affected regions and the population. Examples include uranium mining in Niger, the Jackpile-Paguate uranium mine in the US, and the in-situ leaching of the Beverley uranium mine in Australia (Neles and Schmidt 2012, 154–58). In Germany, the reappraisal of uranium mining during the GDR era—the third largest producer in the world at the time—is relevant: At Wismut, up to 120,000 people worked at sites in Thuringia and in Saxony, extracting a total of 231,000 tons of uranium in over 1,000 individual projects. Although Wismut GmbH, founded after reunification, has progressed in surface and underground remediation work since 1990, considerable tasks remain in the long-term management of tailings disposal sites and at certain sites, which should be completed as late as 2040.[32]" (von Hirschhausen 2023, 35–36).

---

[31] This section was translated into English from the German original (von Hirschhausen 2023, 35–36). Refer also to Wimmers et al. (2024) for an overview of radioactive releases into the environment.

[32] At just under US-$ 40 per ton of uranium mined, the costs of remediation (originally €6.3 billion, now over €8 billion) are in line with the lower cost group for uranium ore and only slightly below the global market price, which has been valid for a long time. If the total cost of the measure is taken as a lower limit for the environmental effects, the uranium price on the market should almost double.

3.5 The Middle-Section: Nuclear Power Plants (Construction, Operation, and Decommissioning)

3.5.1 Mapping of the System Good "Nuclear Power Plant" in the Analysis Framework

3.5.1.1 Description of the Technical System

The technical system "Nuclear Power Plant" in the middle section can be divided into five subsystems with three input and three output variables (Rothwell 2016). Figure 3.4 gives an overview of the individual subsystems. Subsystem 1 contains the reactor itself, an apparatus used to maintain nuclear fission in a self-sustaining chain reaction. The reactor is part of the nuclear steam supply system (NSSS) and consists of the reactor core, reactor coolant system, and auxiliary systems. The nuclear steam supply system produces steam or hot gases. It can include other technical components, depending on the reactor technology (e.g., heat exchanger and pressurizer in a PWR). The steam is transported to the turbine building (subsystem 3), where the steam is expanded and converted into electrical energy in the generator. The expanded wet steam is transported to the condenser (subsystem 5), where it is cooled, and the water is then transported back to the steam supply system, thus completing the water cycle. The balance of the plant includes the other systems, i.e., components not included in the nuclear steam supply system (e.g., measurement and control systems, electrical engineering). The transformer (subsystem 4) transmits electrical power to the transmission grid connection at the busbar.

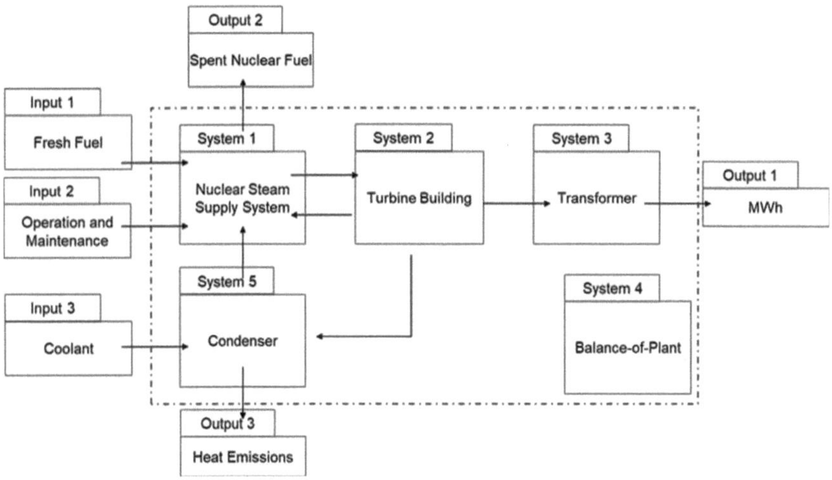

Fig. 3.4 System diagram of a nuclear power plant (*Source* Own depiction based on Rothwell (2016, 3) and NRC 10 CFR §170.3)

This division into five subsystems with in and output variables is also helpful for identifying the needed elementary processes and assets for providing an NPP. In this analysis, we distinguish between the NSSS (or nuclear island) and the rest of the plant. The construction of an NPP consists of several processes: Research and development in the field of technology to determine its viability, or to develop an understanding of the logic of the design if the technology does not originate in the respective country; design and manufacturing (or procurement) of the nuclear steam supply system; design and procurement of the remaining equipment and its integration with the NSSS into the overall station design; and the construction of the plant (Thomas 1988).

Some elements do not require specific assets and are of lower strategic importance, such as the turbine building, transformer, balance-of-plant, and condenser; these are also needed for conventional power plants and are available on a competitive market. However, the nuclear island and the construction and engineering services can usually only be supplied by specialist nuclear vendors. Therefore, from a production point of view, it makes sense to distinguish the "nuclear island" from the rest of the power plant.[33] The nuclear steam supply system is often manufactured specifically for a particular reactor design. Parts of this system are the most critical components, as they require heavy forgings. For example, the containment vessel, reactor pressure vessel, and steam generators of a third-generation LWR are made from nearly 200 forgings comprising over 4,000 tons of steel, with some ingots weighing 500–600 tons each. Only a limited number of forging presses in the world can handle such large blocks.

From a project management view, the supply and construction of the nuclear island is on the critical path of the project as the "heart of the power station." It is thus interlinked to all other construction processes. All these processes are complementary to the overall functioning of the power plant. Figure 3.4 helps to identify some other interfaces: Another important technical interface (Input 1) exists for the fuel fabrication company. Additional interfaces appear towards the production stage, "storage" or "disposal," as spent nuclear fuel (output 2) must be removed from the reactor and then stored.

3.5.1.2 Tasks, Roles, and Actors

Various relationships exist among the roles that need to be fulfilled to construct an NPP. There are ordering relationships, for instance, between the constructor and the reactor vendor or between the reactor vendor and a forging company. Also, various time-complementary relationships exist and must be managed. Here, we focus on the role type of providing the elementary process, which comprises primarily the combined use of necessary assets. The elementary processes NSSS provision and the provision of the RPV are focused upon. While in some cases, both activities can be

[33] It also makes sense from a cost perspective, because the reactor equipment alone can account for between 20 (University of Chicago 2004) and 40% (Rothwell 2016) of direct costs.

fulfilled by one role, i.e., the reactor vendor, the vendor must subcontract RPV manufacturing task to a specialized forging company. Of course, a time-complementary relationship exists between these two roles: Reactor vendors usually provide the design and manufacturing (or procurement) of the NSSS. The remaining equipment and its integration with the NSSS into the overall station design can be carried out, for instance, by the role of the architect-engineer. The constructor supervises the construction.

Of course, different institutional settings influence the roles in an NPP construction project. OECD/NEA (2008) provides an overview of the three main contracting approaches for constructing NPPs: turnkey approach, split-package, and multi-contract approach. In the turnkey approach, one large contract between the reactor vendor (or consortium) and the customer covering the supply of the entire plant is agreed upon. This includes everything from the design and licensing work to when the vendor hands the "key of a working plant" over to the customer, e.g., supply of all equipment and components, all on-site and off-site fabrication, assembly and construction work, testing, and commissioning. The vendor can subcontract work that they cannot supply themselves.[34] The customer can also opt for the split-package approach. Here, the project is divided into a few main systems (like the systems described in Fig. 3.4), each contracted to a different supplier.[35] The multi-contract approach gives the customer maximum control over the design and construction of the plant, but, in this approach, they also have the responsibility for the overall project. As only a few large nuclear utilities have the necessary resources (i.e., nuclear in-house expertise) to carry out this role, an architect-engineer will usually be contracted as the overall project manager. The architect-engineer is responsible for, e.g., the overall design, licensing, and contractor selection for each of the plant's systems, management, and monitoring of the actual construction work, and finally, for plant testing and commissioning.

3.5.2 *Organizational Models for the Construction of Nuclear Power Plants*

The nuclear industry, rooted in the interplay between science and warfare, has historically been governed by collective planning rather than competitive private investment. The early phase of the commercial use of nuclear power in the post-war period (1945 to the 1950s) was marked by the looming Cold War between the US and its partner countries and the Soviet Union, including satellite states. The focus was on

[34] Refer to Chapter 9 of this book for an analysis of specific turnkey contract designs applied by Russia in Bangladesh, Egypt, and Türkiye.

[35] In a split-package approach it is necessary to allocate responsibility for design and licensing for the overall power plant, an important elementary process is the "integration of the various packages to ensure that the power plant's systems work together correctly". This role can be filled by, e.g., the plant's owner.

the further development of nuclear weapons and other military uses and was, of course, firmly in state hands. The first private companies entered the nascent reactor market towards the end of the 1950s.[36]

The first and second-generation reactors were mainly built by companies such as Westinghouse Electric in the US, Kraftwerk Union (KWU) in West Germany, and Toshiba in Japan. This first expansion was also marked by these companies mainly serving their respective domestic markets—they were indigenous or "home suppliers." The large number of companies involved in the early days of nuclear power was first reduced through industrial restructuring in the 1970s, as the nuclear construction market was not large enough. The decline in construction projects in the 1980s encouraged further industry consolidation, leaving only a handful of consolidated nuclear reactor vendors remaining in the 2010s (WNA 2016). Among them are the traditional reactor vendors Westinghouse and Framatome, which both went bankrupt in the 2010s and were "rescued" through state interventions. Since 2000, Russia has become the emerging nuclear vendor and now dominates the reactor market with more technology agreements than the next four largest suppliers combined (France, US, China, and Korea). Other players are the state-owned companies from China (CGN, CNNC) and Korea (KEPCO).

Most of the current new-build projects are situated in Asia and in the former USSR. They are still conducted by home suppliers, e.g. in China by majority-owned Chinese companies, in Korea by the state-owned monopolist KEPCO, or in Russia by state-owned Rosatom. Since the 1980s, there has been a trend of transferring reactor licenses, e.g., the French PWR fleet is based on a Westinghouse license, and the Chinese fleet is partially based on the Western technology transfers, e.g., the AP-1000. The EPR license was also transferred to China. Table 3.8 shows the reactor vendors and the number of their construction projects (as of 2024). This table shows the number of construction projects each vendor is responsible for and their respective market share in percentages. In total, there are 62 construction projects accounted for, including suspended projects, with an overall Hirschman-Herfindahl Index (HHI) of market concentration of 1511, indicating a moderately concentrated market among reactor vendors as of late 2024.[37] Accumulating projects by countries gives an HHI of 2526, indicating a high market concentration skewed towards Chinese actors that build domestically.

For the construction of NPPs, the degree of integration is of interest, especially the degree of horizontal integration. This gives a reactor vendor more control over production capacity and prices, i.e., a higher degree of self-sufficiency, as they would be able to supply a high proportion of the needed components for reactor construction

[36] Refer to Chapter 11 of this book for a detailed assessment of the nuclear industry developments in the US over the last 80 + years.

[37] The Hirschman-Herfindahl Index (HHI) measures the concentration of a market, it is defined as the sum of the squares of the market shares of all firms within the industry (in absolute %-values), and thus it can range between 0 and 10,000. A value of less than 1,500 is considered a competitive market, values between, up to, 2,500 indicate moderate concentration, while values exceeding 2,500 indicate a high market concentration. See Chapter 8 of this book for a detailed calculation of the HHI for the nuclear fuel supply chain.

Table 3.8 Ongoing and suspended NPP construction projects as of October 2024

Vendor	Origin country	Number of projects (of which abroad)
CNNC	China	17 (0)
CGN	China	5 (0)
SPIC	China	5 (0)
NPCIL	India	7 (0)
Rosatom (including AtomStroyExport)	Russia	13 (9)
J-Power	Japan	2 [suspended]
KEPCO	South Korea	2 (0)
Energoatom	Ukraine	2 [suspended]
EDF	France	3 (2)
CNEA	Argentina	1 (0)
Electrobrás	Brazil	1 [suspended]
SE	Slovakia	1 (0)
Total		62 (11)

Source Own compilation, based on World Nuclear Association.[38]

from their own factories (Thomas 1988). A classic horizontally integrated supplier is Framatome (former Areva), which owns the forge site Le Creusot, where RPVs are produced. In comparison, Westinghouse had already begun to outsource forging tasks in the 1960. For example, both RPVs for the Prairie station in Minnesota were forged by Le Creusot. Today, the production of large components will generally be subcontracted to specialist companies and built on a one-off basis, presumably at higher costs in countries such as Japan and China. A concept similar to horizontal integration, but from a national and not a corporate point of view, is the degree of localization, i.e., the existence of a self-reliant domestic nuclear supply chain. A high degree of localization can be observed in France, Japan, Korea, China, and Russia. The US, however, has abandoned localization and depends on imports for many components, most notably RPVs. The degree of localization can also influence production costs and give companies a competitive edge.

The most critical components of an NPP are those attributed to the NSSS, as they require heavy forgings. Worldwide, only a limited number of forging presses can accommodate such large ingots. The WNA estimates the annual worldwide production capacity of RPVs to be sufficient for 30 large reactors (WNA 2016, 98). Table 3.9 shows the large forging companies in operation (as of 2021), their heavy forging press capacities, and the number of RPVs (or other large components) they can produce in one year (no data provided on Le Creusot).

The main very heavy forging capacities are located in Asia, the primary actor being Japan Steel Works (JSW), which accounts for 80% of the world market for

[38] https://world-nuclear.org/information-library/current-and-future-generation/plans-for-new-reactors-worldwide.

Table 3.9 Operational forging companies for reactor pressure vessel production and their production capacity as of 2021

Company	Country	Heavy forging presses [Tons]	RPVs per year
Japan Steel Works	Japan	14,000 × 2	12
China First Heavy Industry	China	15,000 and 12,500	5
China Erzhong & Dongfang	China	16,000 & 12,700	5
Shanghai Electric Group	China	16,500 and 12,000	6
OMZ Izhora	Russia	15,000	4
Le Creusot, Areva	France	11,300 and 9,000	–

Source World Nuclear Association[39]

large, forged components for NPPs and is also contracted to supply Framatome. In 2009, Westinghouse was already constrained as the RPV covers and steam generator parts for the AP1000 could only be supplied by JSW. Another leading RPV supplier is Mitsubishi Heavy Industries (MHI), which together with JSW supplied the reactor pressure vessel to Olkiluoto-3. Other companies are China First Heavy Industry in China, OMZ Izora in Russia, and Le Creusot in France. According to the World Nuclear Association, new capacity is being built and planned in Japan, China, South Korea, Czechia, Russia, the UK, and India. But there has been no data update since 2021.

All nuclear power construction is carried out by state companies, or by private companies benefitting from cost plus regulation or other incentives. Examples are the state-owned companies Framatome and EDF in France, KEPCO in Korea, or equity interests in joint venture agreements, e.g., the joint venture between China General Nuclear (CGN) and Electricité de France (EDF) for the construction of the Taishan EPR in China or EDF/CGN for Hinkley Point C in the UK. Other forms of government financing mechanisms include additional cost recovery rates or surcharges on electricity sales (e.g., Vogtle project in Georgia, US), the use of national funds, the creation of a government-led private bank, banks to finance infrastructure, and asset pooling. Support and incentives for nuclear projects can also be provided by governments in the form of loan guarantees (e.g., Vogtle project), guaranteed long-term electricity contract agreements (e.g., Hinkley Point C), or with export credit agencies (e.g., the France Coface loan to the Finnish energy utility TVO for Olkiluto-3 in Finland).[40]

In general, the complex nature of an NPP requires a considerable degree of "after sales" service from the vendor. In most cases, the vendor also supplies fuel services, as the majority of the major vendors (e.g., Westinghouse, Rosatom, Framatome) have vertically integrated fuel companies into their firm structure. Most reactor vendors also provide maintenance and replacement services during the reactor's lifetime.

[39] WNA 2021, https://wna.origindigital.co/information-library/nuclear-fuel-cycle/nuclear-power-reactors/heavy-manufacturing-of-power-plants.

[40] Refer to Weibezahn and Steigerwald (2024) for an overview of NPP project funding mechanisms.

3.5.3 *Organizational Models for the Operation of Nuclear Power Plants*

3.5.3.1 Overview of Tasks

NPPs are designed to generate heat from a nuclear reactor to produce heat that is, analogously to other thermal power plants, used to generate steam, the energy of which is then transformed to kinetic energy in a turbine and, finally, to electrical energy in a generator. The final product of all operational NPPs is electricity. There are examples of district heat provisions from NPPs in CHP operation, mostly in Eastern Europe, but these are limited to low-demand centers in close proximity to the plants (IAEA 2019).

The main task of NPP operations is the control of reactor output through the manipulation of control rods. When the control rods are inserted further into the reactor, more neutrons are absorbed, reducing the reactor's power output. Conversely, withdrawing the control rods results in an increase in the reactor's power output. This management is also referred to as criticality control, with criticality being the state of neutron flow under which the fission chain reaction is self-sustaining. Furthermore, fuel changes must be conducted every 10 to 12 months, during which PWRs are typically taken off the grid, while some other reactor types can continue to operate. Fuel procurement from specialized vendors is usually based on long-term contracts, as discussed in Sect. 3.2.3. Once removed from the reactor, spent fuel is transferred to wet storage where it remains for around five years until the waste is typically transferred to on-site interim storage facilities (Kopáčik et al. 2020).

An NPP is operated to produce electricity and generate revenue through the sale thereof. In regulated markets, nuclear power is fed into the grid at a near constant rate, and operators are compensated on fixed-price terms. In liberalized markets, NPPs compete with other electricity generation technologies, especially renewables and, in the US, with comparatively cheap natural gas. Prices are usually set pay-as-clear, meaning that in times of high renewables penetration, low electricity prices reduce NPP competitiveness, and the low flexibility of NPPs hinders the exploitation of peak demand prices. Refer to Chapter 10 for a detailed analysis of NPP economics. Consequently, extensive governmental support has been introduced to keep NPPs operational, see Chapter 12.

What differentiates NPPs from conventional thermal power plants is the risk of exposure to potentially hazardous radiation. Consequently, safety measures must be in place to minimize this risk. The continuous monitoring of the integrity of safety barriers is integral to NPP operations. Annual revisions are usually accompanied by more complex decadal revisions necessary for operating license extensions and regular IAEA inspections. This can include preparations for lifetime extensions. Furthermore, NPPs must be protected from external threats, necessitating an array of security measures that must also be managed. These measures consist of monitoring, cyber-security, and physical barriers, as well as the implementation of stringent processes.

3.5.3.2 Actors, Roles

The organization of NPP operations can be intertwined with construction and also relates to oversight. To construct, operate, and later decommission an NPP, one or more licenses are required. Usually, regulatory oversight is conducted by one or several agencies involved directly or indirectly with the NPP licensee. They can, in some cases, even be owners. A given NPP has one licensee. This licensee may be owned by one or several entities who can themselves be subject to scrutiny by agencies or be state actors themselves. Traditionally, construction and operation tasks were conducted by separated actors, but recently, vertically integrated actors, most notably Rosatom, have been emerging, see Chapter 9. Similar developments have occurred on the other end of reactor lifetimes, decommissioning, see Sect. 5.4.

In Europe and the US, reactor vendors such as Westinghouse or KWU supplied the nuclear plant ("construction") and potentially fuel-related services, while specially created subsidies of the utilities carried out the operation. An exception was the attempt of French utility EDF, also to provide reactor construction services abroad. This attempt has failed most notably in Finland and the UK, where projects have been over budget, delayed by several years, and, as a consequence, unprofitable.

This vertical integration attempt is more successfully applied by Russian state-owned supplier Rosatom that both builds and operates NPPs, both domestically and abroad. Notably, the Akkuyu plant in Türkiye shall be built and operated by Rosatom, and the electricity will be sold to Turkish customers at a fixed price until the project breaks even. Chinese CNNC is also vertically integrated, albeit solely on the domestic market; it builds and operates its own NPPs, and is also involved in uranium mining, see Sect. 4.2.1.

3.5.4 Decommissioning

Nuclear decommissioning is an important step in the lifetime of an NPP that merits the provision of various complex tasks and, oftentimes, the interaction of many individual actors. The complexity of this system good "nuclear decommissioning" shall be derived in the following section.[41] Notably, of the roughly 200 closed reactors as of mid-2024, only 23 have been fully decommissioned, and only 145 are undergoing some form of active decommissioning. About 400 reactors are currently operational worldwide. Consequently, decommissioning is a historically neglected and currently overlooked challenge in the nuclear system (Schneider et al. 2024).

---

[41] This section is based on Wimmers and von Hirschhausen (2024).

3.5.4.1 Overview of Tasks

Approaches to Nuclear Decommissioning

Nuclear decommissioning must be completed safely and swiftly to minimize radiological risks to the environment and the public. Further, the removal and safe storage of radioactive material increase security by reducing the risks of proliferation and intentional targeting of the sites. Decommissioned nuclear facilities can be released from regulatory oversight and then reused or sold.

There are three strategies to decommission an NPP. The first, immediate or direct dismantling, is the fastest option. Here, decommissioning begins directly after plant closure. While radiation is still high, remaining institutional knowledge allows for more efficient decommissioning, potentially reducing overall cost. Delayed or deferred dismantling involves placing the reactor into a long-term enclosure (LTE) for years to decades to allow short-lived radioisotopes to decay. This reduces health risks for workers and allows externally organized decommissioning funds to accumulate. This strategy might be chosen at sites with gradual reactor closure to allow for parallel decommissioning. Disadvantages include the loss of institutional knowledge and possible loss of structural integrity of buildings. The third strategy, entombment, can be applied to sites that have experienced an accident. Here, the site is sealed since dismantling poses too large a risk to workers.

Regardless of the chosen strategy, operations tend to follow a similar pattern once physical dismantling begins. In the first stage, the "warm-up" stage, which includes the post-operational phase, the reactor is defueled, and the first components are removed to allow access to the reactor core. In the subsequent "hot-zone" stage, the dismantling of highly contaminated parts, such as the reactor pressure vessel, is conducted. In the final "ease-off" stage, the remaining internal components and buildings are decontaminated. Depending on the target state of the site, buildings are demolished ("greenfield") or decontaminated to be eligible for other (non-nuclear) purposes ("brownfield"). Table 3.10 shows the current status of global decommissioning efforts.

Processes and Assets

Nuclear decommissioning involves various interconnected and interdependent processes. While some fleets are somewhat homogeneous, standardized techniques may only apply to some plants, requiring the development of specialized tools with limited reusability. The availability of skilled personnel, particularly with access to institutional knowledge, is crucial for successful decommissioning. Site-specific conditions and opposition from local populations can further affect progress. Uncertainty regarding radiological contamination and activation arises during the dismantling of remaining buildings and components. Consequently, the need for coordination among actors arises to ensure safety, security, and efficiency to decommissioning projects. Further, access to waste disposal routes and facilities is crucial for successful

Table 3.10 Overview of global decommissioning status as of July 2024

Country	Closed reactors	Warm-up (of which defueled)	Hot-zone	Ease-off	LTE	Completed (of which greenfield)
US	41	4 (4)	6	4	10	17 (5)
UK	36	21 (14)	9	0	6	0
Germany	36	11 (5)	11	9	1	4 (3)
Japan	27	26 (5)	0	0	0	1 (1)
France	14	3 (0)	3	0	8	0
Russia	11	4 (3)	0	0	7	0
Sweden	7	2 (0)	5	0	0	0
Canada	6	1 (1)	0	0	5	0
Bulgaria	4	4 (0)	0	0	0	0
Italy	4	2 (1)	2	0	0	0
Taiwan	4	4 (0)	0	0	0	0
Ukraine	4	0	0	0	4	0
Belgium	3	2 (2)	0	1	0	0
Slovakia	3	1 (1)	0	2	0	0
Spain	3	1 (0)	0	0	1	1
Lithuania	2	2 (2)	0	0	0	0
South Korea	2	2 (0)	0	0	0	0
Switzerland	1	1 (1)	0	0	0	0
India	1	1 (1)	0	0	0	0
Armenia	1	0	1	1	1	0
Kazakhstan	1	0 (0)	0	0	1	0
Netherlands	1	0 (0)	0	0	1	0
Pakistan	1	1 (0)	0	0	0	0
Total	213	93 (38)	36	16	45	23 (9)

Source Taken from Schneider et al. (2024)

decommissioning. Nuclear waste must be disposed of at specialized facilities, the availability of which depends on the type of waste. Interim storage facilities for high-level waste, mostly spent fuel, are often located on-site, and remain operational after technical decommissioning of the plant itself is completed. The release of a site from regulatory oversight requires the removal of all waste from the premises, either to a consolidated interim storage facility or to (currently unavailable) final repositories (Bärenbold et al. 2024).

Production

Two stylized organizational models could occur in different countries, depending on their regulatory setting and domestic nuclear decommissioning industry. The first model is vertical integration, where a single state-owned utility owns and operates all NPPs. Decommissioning is funded by state funds, and there is limited private industry involvement. Possible benefits may include the utility's incentivization to provide a better service to reduce radiation risks and minimize public expenditure resulting from secure funding and organizational proximity to the overseeing state. Efficiency gains are possible due to knowledge concentration, reduced risks from external agents, and limited unforeseen events. However, a publicly owned utility may not prioritize efficiency. The second model is market competition, with multiple entities owning or operating NPPs. The state acts solely as an overseeing agency, and decommissioning is financed via external segregated funds. There is a market consisting of multiple profit-maximizing decommissioning companies. Learning effects and fund owners keeping surplus cash could lead to faster and cheaper decommissioning if asset specificity and uncertainty remain limited. For highly asset-specific plants, the risk of stranding in final state liability remains. In reality, hybrid solutions dominate, in which individual tasks are contracted to different agents, and public entities carry out some activities (Bärenbold et al. 2024). Two models and their efficiency are discussed in Chapter 15 of this book.

Funding

Five options exist to finance nuclear decommissioning projects. Firstly, for some decommissioning projects, funds come directly from the state budget. Secondly, the option of internal segregated funds allows for licensees to self-administer funds, but these are earmarked in balance sheets and may not be used otherwise. Thirdly, internal non-segregated funds must not necessarily be separated from other assets or business interests. Fourthly, if decommissioning is organized via external segregated funds, licensees or electricity consumers make payments to funds that are managed by independent fund operators. These funds are completely separated from other assets. Finally, although only practiced in the US, licensees may use a combination of financial instruments such as surety bonds, letters of credit, or other guarantees to provide surety methods to the regulator.

3.5.4.2 Actors and Institutions

For most countries, the organization of nuclear decommissioning follows a pattern: The government, represented by responsible ministries, defines the relevant laws and regulations, e.g., mandates regarding decommissioning strategies, and provides agency and funds to the overseeing departmental agencies. The number of agencies involved differs among countries as responsibilities for oversight can be split up, e.g., regarding radiation safety or financial oversight. The nuclear licensee can outsource individual decommissioning tasks or the whole process to third-party contractors who in turn interact with on-site staff at the NPP itself. Generated waste is potentially transferred to a waste disposal, on-site. The organization thereof is also country-specific. Refer to Bärenbold et al. (2024) for an overview of organizational models for nuclear decommissioning in six countries.

The operational stage of the value-added chain has interfaces with the military and security sectors. For example, nuclear weapon states tend to stick to nuclear power and invest in lifetime extensions (or new power plants), such as the US, France, and the UK. Edward Moniz, former US Secretary of Energy under US President Obama (2013–2017), for example, emphasizes that it is essential that policymakers recognize that the civil nuclear sector and the associated supply chains are closely linked and are crucial factors for US national security (Energy Futures Initiative 2017). A study by the Atlantic Council estimated the value that the civilian nuclear power sector contributes to the US national security apparatus to be around US$42.4 billion. In other words, the authors argue that the lack of a civilian nuclear sector would present an immediate and significant economic shock (and impact on the labor force) (Ichord and Oosterveld 2019). At present, the UK governments also seem to be committed to nuclear power, and one possible reason for continuing to operate NPPs is to maintain national capabilities to build and operate nuclear-powered submarines (Stirling and Johnstone 2024).

3.6 The Back-End: Nuclear Waste Management

3.6.1 Mapping of the System Good "Waste Management" in the Analysis Framework

3.6.1.1 Description of the Technical System

Most processes in the different value-added stages generate nuclear waste. The first kind of waste that emerges is the tailings (excavated rocks to access the uranium ore) at the uranium mine; all the subsequent processes in the front-end create nuclear waste. Figure 3.5 gives an overview of the nuclear fuel chain and the different generated waste forms along this chain. In this section, we focus on operational, decommissioning, and high-level wastes in the form of spent nuclear fuel (SNF) and vitrified waste from reprocessing.[42]

---

[42] This section is largely based on own publications, amongst other the "Quantities of Nuclear Waste" published in the World Nuclear Waste Report (Besnard et al. 2019), and Wealer et al. (2019). For further (technical) details on nuclear waste management, refer to Yim (2022).

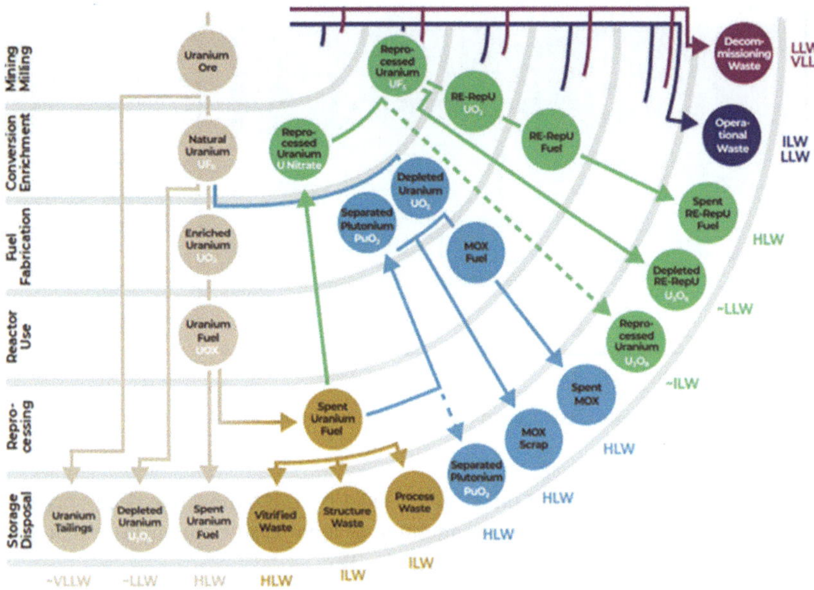

Fig. 3.5 The nuclear fuel chain and generated wastes (*Source* WISE-Paris in Besnard et al. [2019])

Waste Classification and Management Approaches

The IAEA (2009) classifies nuclear waste according to its activity concentration into exempt waste, very short-lived waste (VSLW), very low-level waste (VLLW,) low-level waste (LLW,) intermediate-level waste, and high-level waste (HLW), see Fig. 3.6.[43] LLW, ILW, and HLW require the most sophisticated waste management infrastructure. LLW is above clearance levels but with limited amounts of long-lived radionuclides that require robust isolation and containment for up to a few hundred years. ILW consists mainly of long-lived radionuclides, particularly alpha-emitting radionuclides, and requires more sophisticated containment. ILW needs little to no provision for heat dissipation during storage and disposal. Because ILW consists mainly of long-lived radionuclides that will not decay to an activity concentration acceptable for near-surface disposal, ILW must be disposed of at greater depths, in the order of tens to a few hundred meters. HLW has levels of activity concentration high

[43] Exempt waste is waste that meets the criteria for clearance, exemption, or exclusion from regulatory control. Very short-lived Waste (VSLW) has radionuclides with very short half-lives; VSLW can be stored up to a few years to profit from the radiological decay. After this period, it can either be cleared from regulatory control for uncontrolled disposal, use, or discharge, depending on the arrangements with the regulatory body. VSLW consists primarily often used for research and medical purposes. Very low-level Waste (VLLW) does not need a high level of containment and isolation and is suitable for disposal in near surface landfill type facilities—that may contain other hazardous wastes—with limited regulatory control. Typical VLLW is for example soil and rubble with low levels of activity concentration; concentrations of longer-lived radionuclides are generally very limited.

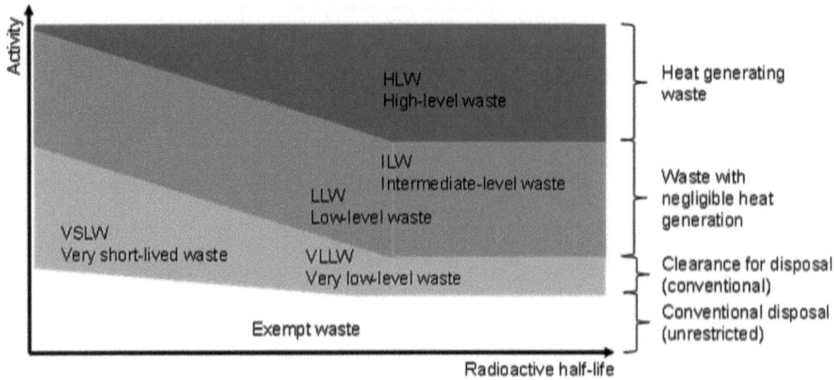

Fig. 3.6 Radioactive waste classification (*Source* Bärenbold et al. [2024, 10])

enough to generate significant quantities of heat by the radioactive decay process. HLW also includes waste with large amounts of long-lived radionuclides that need to be considered in the design of a disposal facility.

For each of these waste classes, the IAEA recommends disposal routes. VLLW can be disposed of in a specific landfill, similar to conventional landfills. LLW is most suited to be disposed of in near-surface disposal facilities, engineered trenches, or vaults on the ground or tens of meters below ground level. ILW is to be disposed of "deeper in the ground" in underground facilities, which consist of constructed caverns, vaults, or former mines in tens of meters up to hundreds of meters below ground. HLW is supposed to be disposed of in geological disposal facilities; this requires the siting of a facility deep underground in a stable geological setting (e.g., clay, salt, granite). In addition, the waste shall be surrounded by engineered barriers that protect the waste and limit the movement of radionuclides if they are released from the waste packages.

Operational and Decommissioning Waste

The operation of an NPP produces different types of nuclear waste in different physical states, most of which are classified as LLW and ILW (collectively referred to as LILW). The generation rate of operational waste depends on many factors, such as the deployed reactor technology and the age of the reactor. A typical LWR of one GW_e produces 250–600 m^3 of LILW over its operational lifetime (IAEA 2007). The state of matter is relevant for unconditioned operational waste, e.g., whether the waste is liquid or solid. Solid waste includes protective clothing, replaced plant components, or insulation material. Contaminated cooling water, oils, vaporizer concentrates, filter substances, or sludge, which forms when solid matter collects as sediment at the bottom of pumps, are examples of liquid operational wastes. Waste management involves various transformation processes (e.g., pre-treatment, compaction, and incineration). Among the main elementary processes is conditioning, in which waste is brought into a stable and immobilized form, which makes it suitable for transportation, storage, and, eventually, disposal. After conditioning, the waste is either stored in drums, or storage, transport, and disposal casks (storage process), for which a necessary asset is an interim storage facility. Alternatively, the waste is disposed of, for which an indispensable asset is a disposal facility.

During the decommissioning of a nuclear facility, significant quantities of mainly LILW are produced. Implementing a well-functioning radioactive waste management program during the decommissioning of the facility depends primarily on the availability of sufficient storage space for the required drums, casks, and other containers. A typical LWR of one GW_e capacity produces 5,000 to 6,000 tons of short-lived LILW and 1,000 tons of long-lived LILW and HLW (IAEA 2007). This estimate must be taken cautiously as decommissioning experience of high-capacity reactors is limited. Like operational waste, the quantity of decommissioning waste again depends on the reactor technology, the clearance level of waste, the decommissioning strategy, and the operating time, among other factors.

High-Level Waste: Reprocessing and Spent Nuclear Fuel

What constitutes HLW depends on the national policy and the nature of the fuel chain. In general, there are three different strategies for dealing with nuclear waste: the "Open Cycle," the "Closed Cycle," and the "Wait and See option" (IAEA 2002, 34). However, the term "cycle" is misleading. Instead, the term "nuclear fuel chain" is more accurate as only a tiny part of the reprocessed material will reenter the fuel chain as fresh fuel. Only some of the separated plutonium is reintroduced into the fuel chain through the fabrication of MOX fuel rods, while depleted uranium is mostly stored and constitutes significant stockpiles. In addition, reprocessing SNF creates large quantities of LILW, meaning that the total volume of waste to be managed is much greater than if the spent fuel is treated directly as waste.[44]

After reprocessing, the residual fission products and actinides in liquid form are evaporated and converted to solids by a vitrification process before intended further disposal. When SNF is not reprocessed, it must be stored for about 40 years to allow for activity levels to decline. SNF is stored first in the spent fuel pools and then in interim storage facilities, usually close to the facility where it was created. These interim storage facilities can either be "wet" (analogous to spent fuel pools) or "dry," i.e., consisting of casks made of concrete or reinforced steel. Depending on the waste management strategy, SNF might be again transferred into the final disposal casks for final disposal. This process is usually done underwater. Table 3.11 gives an overview of the elementary processes and needed assets.

The analysis indicates several interdependent processes, which can be of a high degree of specificity, especially the disposal of ILW and HLW. The technical system of nuclear waste management consists of many transformation, storage, and transfer processes, some of which are interlinked. An important technical interface requiring

[44] Refer to Chapter 18 of this book for an overview of transmutation and partitioning.

Table 3.11 Processes and assets of the technical system "storage" and "disposal"

Elementary process	Processes	Asset
Treatment and conditioning of LILW	Depending on the waste: • Compaction • Solidification • Incineration • Conditioning	• Waste treatment tools and infrastructure • Casks, drums
Storage of LILW	• Transport to interim storage facility • Storage	• Nuclear transport logistics • Interim storage facility
Treatment of SNF	• Loading fuel into spent fuel pools (immediate cooling) • Loading fuel into storage/transport casks • Transport of casks to interim storage facility or reprocessing facility	• Spent fuel pool • Storage/transport casks • Interim storage facility • Nuclear transport logistics
Interim storage of SNF/HLW	• Wet or dry storage of SNF • storage of HLW	• Storage/transport casks • Wet or dry interim storage facility
Reprocessing SNF	• Loading fuel into storage/transport casks • Transporting SNF to reprocessing facility • Extracting plutonium and uranium • Vitrifying waste • Conditioning waste • Transporting waste back to waste producer	• Reprocessing facility • Storage/transport casks • Nuclear transport logistic • Vitrification plant
Disposing of LLW	• Conditioning waste • Transporting waste to facility • Disposing of waste	• Near-surface disposal facility • Transport and disposal casks • Nuclear transport logistic
Disposing of ILW	• Conditioning waste • Transporting waste to facility • Disposing of waste	• Belowground disposal facility • Transport and disposal casks • Nuclear transport logistic
Disposing of HLW	• Conditioning waste • Transporting waste to facility • Disposing of waste	• Deep geological disposal facility • Transport and disposal casks • Nuclear transport logistic

Source Own depiction

coordination is between the treatment, storage, and disposal processes, especially regarding containers or casks. Their design depends on the storage and disposal methods. Interdependencies between waste management and operations exist, too. With the start-up of a new reactor, operational waste and spent fuel are generated, which must be treated and stored. Another noteworthy interface is that between decommissioning and storage. When decommissioning starts, large quantities of waste are generated in a short time that must be either stored (interim) or disposed of. A lack of disposal facilities (or storage) can also hinder the decommissioning process or even render the regulatory release of a given site impossible.

3.6.1.2 Tasks, Roles, and Actors

All the tasks needed for the elementary processes of the "treatment" of SNF or LILW can be handled by the nuclear licensee or be taken over by a dedicated waste management organization; both could be public or privately owned. The same goes for the process of "disposal." In some cases, the responsibility is shifted to national governments. An important actor is the provider of containers for transporting, storing, or disposing of nuclear waste. Recently, most notably in the US, private actors have emerged that offer interim waste management in combination with decommissioning services. Consequently, the privatization of the services of "treatment," "storage," and "reprocessing" could also intensify in the foreseeable future.[45]

3.6.2 Organization Models for the Provision of …

3.6.2.1 … Low- and Intermediate-Level Waste Storage

Disposal facilities for LILW have been implemented in some countries for decades; mainly specific landfill disposal and engineered trenches, near-surface disposal and below-ground facilities. After the early US landfills showed signs of leakages, protective functions of the disposal facilities were reinforced with additional safety barriers in the form of concrete components and structures (mainly in France and the US). Nevertheless, small amounts of radioactive material can also be washed out from these facilities to environmental contamination. Underground facilities for LILW were constructed in Sweden and France, while former mines are used for the disposal of LILW in Germany, Spain, and the Czech Republic.

The majority of LILW disposal facilities are constructed and operated by state agencies or public companies (e.g., Andra in France; Enresa in Spain; BGE in Germany; and Radioactive Waste Management Ltd (RWM), which is owned by the public body NDA, in the UK). In some cases, the waste management organization is a private company. In Sweden, for instance, the utilities have created the

[45] See Wimmers and von Hirschhausen (2024) for details on this development.

private company SKB, the Swedish Nuclear Fuel and Waste Management Company, to operate the disposal facilities. The situation in the US is more complex as LILW disposal is managed on a state level; HLW management is a federal issue. Some private companies operate disposal facilities, including EnergySolutions, a major nuclear waste management company.

3.6.2.2 … Reprocessing

As reprocessing means the extraction of plutonium, there is no actual "commercial" or "private" provider of reprocessing services. All reprocessing service providers are state-controlled. Western countries can send their SNF for reprocessing to France (operated by majority state-owned Orano in La Hague), while Russia (formerly, the USSR) historically provided this service to a few central European countries. The other Western facility at Sellafield, UK, closed in 2018. Thus, French Orano has a monopoly position and has thus fully integrated the nuclear front- and back-end into its operation, with the exception of final disposal. The same holds for Atomenergoprom, which is part of Rosatom. A Japanese reprocessing plant has been under construction since 1993. With the completion date being delayed more than 20 times, it remains uncertain if and when the plant will become operational. Other facilities are operated in India for heavy water reactor fuel and in Israel and Pakistan for military purposes.

Reprocessing contracts are long-term. One reason is the long time from unloading SNF from the reactor through reprocessing, MOX fuel fabrication, and return of vitrified waste; another is the high specificity and costs for commissioning a reprocessing plant. In Europe, reprocessing is still practiced by France, the Netherlands, and Russia. However, most countries have suspended or stopped it, mainly for economic reasons (e.g., Belgium, Bulgaria, Germany, Hungary, Sweden, Switzerland, and the UK).

3.6.2.3 … High-Level Waste Storage

The final disposal of HLW is usually the responsibility of a dedicated government agency or other approved bodies established for this purpose. There are exceptions to this; for instance, in Sweden and Finland, the nuclear utilities created private companies to manage SNF (interim storage and final disposal). As no disposal facility is available worldwide,[46] interim storage issues are becoming relevant. Interim storage is often the scope of the utilities, while the disposal is in the hands of the national governments. In most countries, SNF is stored in a decentralized way, mainly at the reactor site, either in casks (dry storage), or in spent fuel pools. As of 2016, 81% of Europe's SNF is still in wet storage; although dry storage might improve the safety (Besnard et al. 2019). In Germany, the utilities have built interim storage facilities on

[46] Finland began trial runs at its final waste disposal site at Onkalo in 2024.

the reactor site to store their SNF. In 2016, with a reorganization of the nuclear waste management, the ownership of the interim storage facilities was transferred to the federally owned company for interim storage (BGZ). In the future, the LILW storage facilities on the reactor sites will also be transferred to the public company. LILW disposal site Konrad and a future HLW disposal site are to be owned and operated by the sister company for final waste disposal BGE.[47]

Important actors in the waste management services are the cask producers. Casks made of metal are in use in most European countries, while in the US, concrete casks are the norm. The main actors are the German utilities-owned GNS (Gesellschaft für Nuklear-Service mbH), which produces the CASTOR casks, French Orano, and US company Holtec International.

As the disposal of nuclear waste is mostly a national matter and done by in the majority of the cases by state-controlled companies or public bodies, there is no real competition in this production stage of the system good nuclear power. There is some competition regarding specialized waste management companies providing casks or conditioning services. The state dominates the forthcoming final disposal process. However, it places special demands on regulators due to multiple principal-agent problems resulting from information asymmetries between the actors (Hirschhausen and Wimmers 2023).

3.7 Conclusion

The analysis conducted in this chapter reveals that nuclear power, as a system good, involves numerous interdependencies and interfaces across its production stages, requiring significant coordination. Major actors like state-controlled Rosatom and Orano/Framatome exhibit high levels of vertical integration, participating across almost all stages. The Chinese nuclear industry is also vertically coordinated through state planning, though it has introduced competition between state-controlled companies.

In uranium supply, the market is dominated by long-term contracts, with about 80% of uranium traded this way. Leading producers like Kazatomprom, Orano, Cameco, and Uranium One control nearly half of the uranium market, with joint ventures common in large projects, such as the Cigar Lake mine. Similarly, the nuclear fuel fabrication market is shaped by the reactor-specific nature of fuel assemblies, with major reactor vendors like Westinghouse, Framatome, and TVEL controlling much of the market. Vertical integration is also prominent in the other front-end processes (milling, conversion, and enrichment) with companies like Kazatomprom expanding into fuel fabrication.

NPP construction, primarily driven by state decisions and centrally planned, is characterized by high degrees of planning and strong government involvement. The

[47] Refer to Chapter 16 of this book for a detailed assessment of global HLW disposal facility developments.

reactor market is dominated by state-owned companies, particularly in Russia, China, and South Korea. The lack of experience in nuclear decommissioning presents ongoing challenges, with competitive approaches emerging, resulting in potential efficiency gains.

Final disposal, the most complex stage, poses numerous principal-agent challenges, with waste management typically handled by state agencies. Competition is minimal, limited to companies providing waste conditioning or casks, while reprocessing remains under state control, with Orano holding a monopoly in Western Europe.

The remainder of this book explores the complex interrelations in and around different production stages of the nuclear system good and analyzes these with an array of economic methods. For example, Chapter 9 reflects on Rosatom's organizational models for its foreign new build projects in Bangladesh, Egypt, and Türkiye, while Chapter 8 analyzes the competition in the market for front-end processes. The whole Part IV of this book is dedicated to the vast array of challenges of the nuclear back-end, such as decommissioning (Chapter 15), long-term funding (Chapter 17), and extended interim-storage.

References

Bärenbold, Rebekka, Muhammad Maladoh Bah, Rebecca Lordan-Perret, Björn Steigerwald, Christian Von Hirschhausen, Ben Wealer, Hannes Weigt, and Alexander Wimmers. 2024. 'Decommissioning of Commercial Nuclear Power Plants: Insights from a Multiple-Case Study'. *Renewable and Sustainable Energy Reviews* 201:114621. https://doi.org/10.1016/j.rser.2024.114621

Beckers, Thorsten, Florian Gizzi, and Klaus Jäkel. 2012. 'An Approach to Analyze "System Goods"—Classification, Presentation, and Application'. 2021–02. WIP-Working Paper. Berlin: Workgroup for Infrastructure Policy (WIP), Technische Universität Berlin. https://www.tu.berlin/fileadmin/www/10002415/WIP_Publikationen_PDF/beckers_gizzi_jaekel_2012-system_good_analysis_wip-wp_2012-02-12-1.pdf

Besnard, Manon, Marcos Buser, Ian Fairlie, Gordon MacKerron, Allison Macfarlane, Eszter Matyas, Yves Marignac, Edvard Sequens, Johan Swahn, and Ben Wealer. 2019. 'The World Nuclear Waste Report. Focus Europe'. Berlin, Brussels. https://worldnuclearwastereport.org/

Bleise, A., P.R. Danesi, and W. Burkart. 2003. 'Properties, Use and Health Effects of Depleted Uranium (DU): A General Overview'. *Journal of Environmental Radioactivity* 64:93–112.

Bodansky, David. 2004. *Nuclear Energy: Principles, Practices, and Prospects.* 2nd ed. New York: Springer.

David, Paul A. 1985. 'Clio and the Economics of QWERTY'. *The American Economic Review* 75 (2):332–37. http://www.jstor.org/stable/1805621

Energy Futures Initiative. 2017. 'The U.S. Nuclear Energy Enterprise: A Key National Security Enabler'. Washington DC, USA: Energy Futures Initiative.

Englert, Matthias. 2012. 'Kernwaffen—Das Zusammenspiel von Kernenergienutzung und Atombombe'. In *Kernenergie—Eine Technik für die Zukunft?* edited by Julia Neles and Christoph Pistner, 189–208. Technik im Fokus. Berlin, Heidelberg: Springer Vieweg.

Fairlie, Ian. 2009. 'Depleted Uranium: Properties, Military Use and Health Risks'. *Medicine, Conflict and Survival* 25 (1): 41–64. https://doi.org/10.1080/13623690802568962

Finegold, Ben. 2023. 'A Nuclear Stranglehold: Russia's Unsanctionable Position in the Nuclear Industry'. Ocean Wall. https://oceanwall.com/wp-content/uploads/2023/08/OWL-Russia-Report-Final.pdf

Giannardi, Cristina, and Daniele Dominici. 2003. 'Military Use of Depleted Uranium: Assessment of Prolonged Population Exposure'. *Journal of Environmental Radioactivity* 64:227–36.

Gizzi, Florian. 2016. 'Implementierung komplexer Systemgüter: Ein methodischer Ansatz für ökonomische Untersuchungen und seine Anwendung auf Verkehrstelematiksysteme für die Straße'. Berlin: Technische Universität Berlin.

Groves, Leslie M. 1983. *Now It Can Be Told: The Story of the Manhattan Project*. New York, USA: Da Capo Press.

Hirschhausen, Christian von. 2023. *Atomenergie: Geschichte und Zukunft einer riskanten Technologie*. 1. Auflage. Wissen. München: C.H.Beck.

Hirschhausen, Christian von, and Alexander Wimmers. 2023. 'Rückbau von Kernkraftwerken und Entsorgung radioaktiver Abfälle in Deutschland: Ordnungspolitischer Handlungsbedarf'. *Perspektiven der Wirtschaftspolitik* 24 (3). https://doi.org/10.1515/pwp-2023-0032

Hyatt, Neil C. 2017. 'Plutonium Management Policy in the United Kingdom: The Need for a Dual Track Strategy'. *Energy Policy* 101 (February):303–9. https://doi.org/10.1016/j.enpol.2016.08.033

IAEA. 2002. 'Factors Determining the Long Term Back End Nuclear Fuel Cycle Strategy and Future Nuclear Systems'. IAEA-TECDOC-1286. Vienna: International Atomic Energy Agency.

IAEA. 2007. 'Estimation of Global Inventories of Radioactive Waste and Other Radioactive Materials'. IAEA-TECDOC 1591. Vienna, Austria: International Atomic Energy Agency. https://www-pub.iaea.org/MTCD/Publications/PDF/te_1591_web.pdf

IAEA. 2009. 'Classification of Radioactive Waste'. IAEA Safety Standards GSG-1. Vienna, Austria: International Atomic Energy Agency. https://www-pub.iaea.org/MTCD/Publications/PDF/Pub1419_web.pdf

IAEA. 2019. 'Guidance on Nuclear Energy Cogeneration'. Technical Report NP-T-1.7. International Atomic Energy Agency. https://www.iaea.org/publications/13385/guidance-on-nuclear-energy-cogeneration

Ichord, Robert F., and Bart Oosterveld. 2019. 'The Value of the US Nuclear Power Complex to US National Security'. Washington DC, USA: Atlantic Council. https://www.atlanticcouncil.org/wp-content/uploads/2019/10/Nuclear-Power-Value-IB-final-web-version.pdf

Kopáčik, Alojz, Ján Erdélyi, and Peter Kyrinovič. 2020. 'Design, Construction, and Operation of Nuclear Power Plants: Geodetic Activities'. In *Engineering Surveys for Industry*, by Alojz Kopáčik, Ján Erdélyi, and Peter Kyrinovič, 197–213. Cham: Springer International Publishing. https://doi.org/10.1007/978-3-030-48309-8_10

Lévêque, François. 2013. *Nucléaire On/Off—Analyse économique d'un pari*. Paris: Dunod.

Monnet, Antoine, Sophie Gabriel, and Jacques Percebois. 2017. 'Analysis of the Long-Term Availability of Uranium: The Influence of Dynamic Constraints and Market Competition'. *Energy Policy* 105 (June):98–107. https://doi.org/10.1016/j.enpol.2017.02.010

NEA, and IAEA. 2023. 'Uranium 2022: Resources, Production and Demand'. NEA No. 7634. Nuclear Energy Agency and International Atomic Energy Agency. https://www.oecd-nea.org/jcms/pl_79960/uranium-2022-resources-production-and-demand?details=true

Neles, Julia Mareike, and Gerhard Schmidt. 2012. 'Urangewinnung—Von der Mine bis ins Kraftwerk'. In *Kernenergie: eine Technik für die Zukunft?*, edited by Julia Mareike Neles and Christoph Pistner, 143–59. Technik im Fokus. Berlin: Springer Vieweg.

Nelson, Andrew T. 2012. 'Thorium: Not a near-Term Commercial Nuclear Fuel'. *Bulletin of the Atomic Scientists* 68 (5): 33–44. https://doi.org/10.1177/0096340212459125

OECD/NEA. 2008. *Market Competition in the Nuclear Industry*. Paris: OECD Publishing.

Radulovic, Verena. 2005. 'Are New Institutional Economics Enough? Promoting Photovoltaics in India's Agricultural Sector'. *Energy Policy* 33 (14): 1883–99. https://doi.org/10.1016/j.enpol.2004.03.004

Rhodes, Richard, ed. 2012. *The Making of the Atomic Bomb.* 25th anniversary ed. New York, USA: Simon & Schuster Paperbacks.

Rooney, Matthew, William J. Nuttall, and Nikolaos Kazantzis. 2015. 'A Dynamic Model of the Global Uranium Market and the Nuclear Fuel Cycle'. *Resources Policy* 43 (March):50–60. https://doi.org/10.1016/j.resourpol.2014.11.003

Rothwell, Geoffrey. 2016. *Economics of Nuclear Power.* London, UK: Routledge.

Schneider, Mycle, Antony Froggatt, Julie Hazemann, Özgür Gürbüz, Paul Jobin, Phil Johnstone, Timothy Judson, et al. 2024. 'The World Nuclear Industry Status Report 2024'. Paris, France: Mycle Schneider Consulting. https://www.worldnuclearreport.org/IMG/pdf/wnisr2024-v2.pdf

SIPRI Yearbook 2024: *Armaments, Disarmament and International Security.* 2024. Oxford: Oxford University Press.

Stirling, Andy, and Phil Johnstone. 2018. 'A Global Picture of Industrial Interdependencies Between Civil and Military Nuclear Infrastructures'. SPRU Working Paper Series 2018–13 (August). Sussex, UK: University of Sussex. https://www.sussex.ac.uk/webteam/gateway/file.php?name=2018-13-swps-stirling-and-johnstone.pdf&site=25

Stirling, Andy, and Phil Johnstone. 2024. 'Illuminating the "UK Nuclear Complex". Implications of Hidden Links between Military and Civil Nuclear Activities for Replacing Negative with Positive Irreversibilities around Nuclear Technologies'. 2. York IND Research Report. University of York. https://www.york.ac.uk/media/politics/documents/yorkindproject/2024%20Report2%20Stirling%20&%20Johnstone%20Civil%20and%20military%20link%20in%20the%20UK%20nuclear%20weapons%20complesx.pdf

Stockholm International Peace Research Institute (SPRI). 2023. *Sipri Yearbook 2023: Armaments, Disarmament and International Security.* Oxford, UK: Oxford University Press.

Thomas, Stephen. 1988. *The Realities of Nuclear Power: International Economic and Regulatory Experience.* Cambridge University Press.

University of Chicago. 2004. 'The Economic Future of Nuclear Power'. Chicago, IL: University of Chicago.

Vestergaard, Cindy. 2015. 'Governing Uranium Globally'. DIIS Report. Copenhagen, Denmark: Danish Institute for International Studies.

Wealer, Ben, and Christian von Hirschhausen. 2020. 'Nuclear Power as a System Good. Organizational Models for Production Along the Value—Added Chain'. DIW Discussion Paper 1883. Berlin, Germany: DIW Berlin.

Wealer, Ben, Jan Paul Seidel, and Christian von Hirschhausen. 2019. 'Decommissioning of Nuclear Power Plants and Storage of Nuclear Waste: Experiences from Germany, France, and the U.K.' In *The Technological and Economic Future of Nuclear Power*, edited by Reinhard Haas, Lutz Mez, and Amela Ajanovic, 261–86. Wiesbaden: Springer VS. https://doi.org/10.1007/978-3-658-25987-7_12

Weibezahn, Jens, and Björn Steigerwald. 2024. 'Fission for Funds: The Financing of Nuclear Power Plants'. *Energy Policy* 195 (December):114382. https://doi.org/10.1016/j.enpol.2024.114382

Wimmers, Alexander, Fanny Böse, Jasmin Beppler, Pauline Morawe, Maximilian Weber, and Christian Von Hirschhausen. 2024. '(Re)Integrating Radioactive Materials and Waste into a Global Sustainable Development Context'. *Radiation and Environmental Biophysics.* https://doi.org/10.1007/s00411-024-01088-x

Wimmers, Alexander, Paul-Jonas Ummethun, Anke Herold, Claudia Kemfert, Björn Steigerwald, and Christian von Hirschhausen. 2025. 'Russische Exporte von Uran und Atomtechnologie in die EU—Optionen für alternative Beschaffung'. Ein Debattenbeitrag 20. Diskussionsbeitragäge der Scientists for Future. https://doi.org/10.5281/zenodo.14728811

Wimmers, Alexander, and Christian Von Hirschhausen. 2024. 'Organizational Models for the Decommissioning of Nuclear Power Plants: Lessons from the United Kingdom and the United States'. *Utilities Policy* 91 (December):101843. https://doi.org/10.1016/j.jup.2024.101843

WNA. 2016. 'The World Nuclear Supply Chain: Outlook 2035'. London: World Nuclear Association.

Yim, Man-Sung. 2022. *Nuclear Waste Management: Science, Technology, and Policy*. Lecture Notes in Energy. Springer Dordrecht. https://doi.org/10.1007/978-94-024-2106-4

Chapter 4
The System Good Nuclear Fusion—Organizational Approaches and Two Case Studies on Magnetic Confinement Fusion

Stefania Böhnlein, Fanny Böse, and Alexander Wimmers

4.1 Introduction

In recent years, research and development (R&D) efforts for nuclear fusion energy technologies have accelerated. A surge in publicly and privately funded projects has emerged, with many private ventures aiming to commercialize fusion technology.[1] Currently, 101 fusion devices are operating worldwide, with 14 under development and 48 planned, each based on different technological approaches. However, these devices are limited to research and prototype reactors.[2] Despite accelerating efforts, no commercial-scale fusion power plant has so far been realized, as the technological and economic feasibility of a large-scale fusion device for commercial production is still unproven (IAEA 2023; Grünwald 2024). Developing and building experimental and commercial fusion power plants involves a series of complex decision processes, including selecting appropriate technologies, sourcing critical components, complying with regulatory frameworks, and securing funding. This complexity necessitates the alignment of the technological approach with suitable organizational models to address the vast array of specific challenges that arise in fusion device development and construction.

[1] Refer to Chapter 14 of this book for a detailed assessment of fusion ventures.

[2] https://nucleus.iaea.org/sites/fusionportal/Pages/FusDIS.aspx, last accessed on 16–12-2024.

S. Böhnlein · F. Böse · A. Wimmers (✉)
Workgroup for Infrastructure Policy (WIP), Technical University (TU) Berlin, Berlin, Germany
e-mail: awi@wip.tu-berlin.de

A. Wimmers
German Institute for Economic Research (DIW Berlin), Berlin, Germany

A. Wimmers et al. (eds.), *Nuclear Power*, Studies in Energy, Resource and Environmental Economics, https://doi.org/10.1007/978-3-031-99894-2_4

Most current development projects are nascent or in very early stages of development. However, magnetic confinement fusion (MCF) devices have been under development for several decades (see Chapter 2 of this book), and recently, several projects have accelerated, organized in public or private ownership structures, or a combination of both. This chapter compares two organizations that plan on constructing a similar reactor type considered the most advanced fusion approach: the tokamak (Moynihan and Bortz 2023). The first venture is the so-called ITER[3] project, a large-scale international public project under development for decades, characterized by cost escalations and substantial delays. The goal is to prove the technological feasibility of fusion energy and pave the way for future electricity-producing demonstration plants (DEMOs) by gathering research and technological knowledge. The second project is that of Commonwealth Fusion Systems (CFS), a private new venture applying a shorter timeline and a more agile, commercially driven approach within its strategy to operate the SPARC[4] reactor within the following decade. This project shall gather information and lessons learned to develop their follow-up commercial energy-producing reactor, the ARC (IAEA 2022; 2023). The comparison of these models provides insights into the complexities of fusion projects and offers lessons learned that can be applied to the design of organizational models and strategies for future applications.

Given the technological and organizational complexity, the comparison conducted here follows the system good analysis framework derived from Beckers et al. (2012). The provision of goods and services involves multiple activities carried out either sequentially or in parallel, necessitating coordination among participating actors. This framework focuses on three main components: The first is the technical system, which examines the required goods, assets, processes, and their interdependencies along the value chain to deliver the system good. Secondly, within this system, roles are identified as structured bundles of tasks tied to supplying specific goods. Coordination becomes essential because these roles are interrelated, with decisions in one area impacting others. The final component of the framework encompasses actors and institutions, which emerge from the defined roles. Actors adopt roles based on their resources and capabilities, while institutions set the parameters for their actions. For instance, coordination may be regulated by contracts or laws, and licensing requirements may limit certain actors from assuming specific roles. The complete set of activities required to supply a "system good" forms an organizational model.[5]

The remainder of this chapter is structured as follows: Sect. 4.2 defines the "system good nuclear fusion power," following the above-mentioned framework, and explains how fusion works in this context. From this, we conclude the main organizational and technological challenges of fusion device development, focusing on magnetic fusion.

[3] The term 'ITER' in this paper refers both to the ITER fusion reactor and the broader ITER project, which involves its organizational, technical, and collaborative dimensions.

[4] The term 'SPARC' in this paper refers both to the SPARC fusion reactor and the broader SPARC project, which involves its organizational, technical, and collaborative dimensions.

[5] Refer to Chapter 2 of this book for more details on the framework.

Then, in Sect. 4.3, we analyze both cases introduced above: ITER and CFS/SPARC. Here, we focus on their respective approaches to overcome identified challenges. In Sect. 4.4, the differences are discussed, and Sect. 4.5 concludes.

4.2 System Good Nuclear Fusion Power

4.2.1 Technical System of Nuclear Fusion

In the following, the full generalized production chain of the "system good nuclear fusion power" is described, for all fusion technologies, including MCF, Inertial Confinement Fusion (ICF), and Magnetic Inertial Fusion (MIF). The stylized description covers the subsystems and processes required at each stage of the production chain, beginning with fuel extraction and preprocessing, followed by fuel fabrication, plant construction, operation, decommissioning, waste management, and the associated proliferation risks. We also highlight the sequence of these stages and their interconnections to show how each step influences the next, and potentially vice versa. As the production chain for nuclear fusion is still evolving, some assumptions are made about its future development for both commercial and experimental fusion applications. Figure 4.1 illustrates the stages of the production chain. The primary focus of the subsequent system good analysis will be on the construction stage, where organizational models play a significant role in shaping the outcome of the provision of nuclear fusion devices. Further, organizational models for the operation of (commercial) fusion plants for energy production cannot be analyzed as of today because they do not exist.

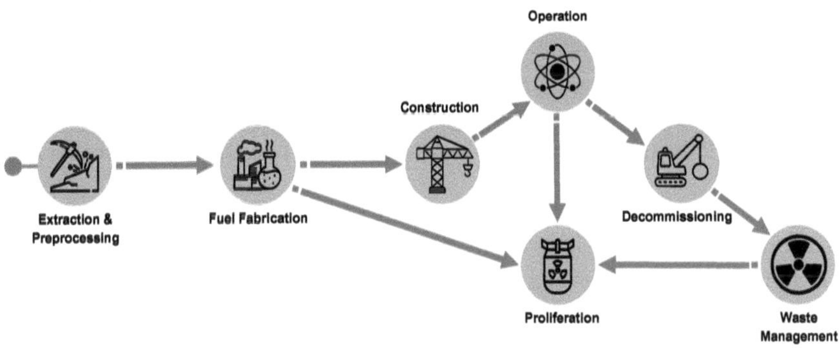

Fig. 4.1 Stylized technical system of nuclear fusion power (*Source* Own depiction)

4.2.1.1 Extraction and Preprocessing

Fusion reactions involve the fusion of atomic nuclei and require specific types of fusion fuel depending on the technology and reaction. The most common fusion reactions under consideration for future technologies are deuterium–tritium (D-T), deuterium-deuterium (D-D), deuterium-helium-3 (D-He-3), and proton-boron-11 (p-B-11). Each of these reactions requires different fuels, which must be extracted and processed from various sources before being used in fusion reactors (IAEA 2023). The energy requirements for each fuel differ substantially. For example, plasma density requirements are higher by several orders of magnitude for D-He-3 and p-B-11 than for D-T fusion.

Deuterium and tritium are both hydrogen isotopes. Deuterium has one extra neutron, is an abundant material, and can be easily extracted from seawater through electrolysis. This makes it one of the most accessible fusion fuels. Tritium has two extra neutrons, roughly tripling its mass compared to a regular hydrogen isotope (H-1). It is not available naturally in quantities necessary for the operation of fusion reactors and must therefore be produced. It is typically created in nuclear fission reactors, mostly heavy water reactors, as a byproduct. Furthermore, tritium is radioactive and has a half-life of 12.4 years; consequently, it cannot be stockpiled to ensure an abundant supply until fusion devices become operational.

To overcome the challenge of tritium supply in an ongoing fusion reaction, some fusion reactor concepts based on D-T fusion envision embedding lithium-6 (Li-6) in so-called breeding blankets to produce more tritium. When high-energy neutrons from the fusion reaction hit the blanket walls containing Li-6, they trigger a reaction that produces tritium which can then be injected into the plasma for additional reactions, allowing to create and maintain a self-sustaining supply of tritium.[6] Since Li-6 is consumed in the process and must to be re-integrated in the blankets, a reliable supply chain for the isotope is required in order to build up a sufficient inventory (Giegerich et al. 2019). Li-6 must first be extracted and enriched from naturally occurring and highly available lithium, via mining or from seawater extraction.[7] Due to its potential use in thermonuclear weapons, the production and enrichment of Li-6 are heavily regulated to prevent its proliferation. Hence, as of today, Li-6 is not commercially available in the industrial quantities needed for large-scale fusion reactors.[8]

B-11 is an isotope of boron. It can be easily mined and refined to meet the purity standards required for use in fusion reactions (McKenzie et al. 2023). Protons used in p-B-11 fusion are abundant and can be extracted from water or hydrocarbons.[9] Thus,

[6] https://www.iter.org/machine/supporting-systems/tritium-breeding. Last accessed on 17–02-2025. This date holds for all other online references unless otherwise indicated.

[7] https://www.iter.org/node/20687/will-fusion-run-out-fuel.

[8] https://news.newenergytimes.net/2022/01/08/lithium-lithium-everywhere-and-none-to-use-for-fusion-reactors/

[9] https://cicenergigune.com/en/blog/hydrogen-production-methods-colours, last accessed on 15–11-2024.

on-site fuel production for fusion devices relying on such fuel can be viable. He-3, while being rare on Earth, can be sourced from other fusion reactions or potentially mined from lunar soil in the future. For instance, reactors using D-D fusion fuel generate He-3 as a byproduct and could theoretically create an internal fuel supply (Entler et al. 2023).

4.2.1.2 Fuel Fabrication

The fuel materials used for fusion reactions, often in gaseous form, need to be further processed to become suitable for fusion, typically by transforming them into fuel pellets. After the raw materials are extracted, they are transported to a specialized fuel fabrication facility, where the fuel is processed into the required form for injection into the reactor. For B-11, refinement may occur either during the fuel fabrication stage or earlier in the extraction process. This fluid transition between "extraction" and "fuel fabrication" shows a possible overlap between the two stages of the production chain. Once the fuel is prepared, it can be transported to the fusion reactor and injected into the core to initiate or sustain the fusion reaction.[10]

4.2.1.3 Construction

The construction phase of a fusion power plant involves several interdependent steps, starting with efforts to comply with nuclear safety regulations and ending with the assembly of the reactor. Hence, critical steps are to prove compliance with legislation, conduct R&D to improve the reactor's design and operation through experimental and modeling studies, and finalize the designs for critical reactor components. Once the designs are completed, procurement and production of components take place, along with the development of the necessary supporting systems. After the components have been produced, they are transported to the construction site, which includes specialized buildings like the reactor and assembly hall. Here, they are integrated with the reactor's plant and auxiliary systems, such as diagnostics, control rooms, and, depending on the specific fusion technology, waste storage and disposal facilities. (Bigot 2016; IAEA 2023).

4.2.1.4 Operation

Once the construction of the fusion power plant is completed, the reactor components are fully assembled, and the necessary operating licenses and authorization have been obtained, the operational phase of the fusion reactor begins. For most reactors, this is still in the future. Primary processes include initiating and sustaining plasma

---

[10] https://hb11.energy/ and https://www.iter.org/fusion-energy/making-it-work

generation; continuous fuel loading and injection, each depending on the fuel; real-time monitoring and control of plasma conditions; and regular maintenance and repair of equipment, while also analyzing the reactor performance (IAEA 2002). A critical part of this stage is the provision of sufficient energy, from the electrical grid or other sources, to initiate a fusion reaction, either for pulses in an ICF device, or when the plasma must be ignited in an MCF device after plasma failure or after maintenance work. ITER alone will require electricity loads of up to 600 MW during plasma operations.[11]

4.2.1.5 Decommissioning

Once the reactor reaches the end of its operational lifetime, it must be dismantled and decommissioned. During this phase, components and materials that were contaminated or activated through neutron absorption during operation must be disposed of in compliance with regulatory and safety requirements. Although fusion reactors do not produce the same long-lived and highly radioactive waste as fission reactors, materials near the plasma can become radioactive and must be managed carefully to comply with safety standards. These materials are decontaminated and processed to reduce risks before disposal or recycling. Efforts are being made to minimize the potential for materials to become highly radioactive during fusion reactor operations by using reduced-activation construction materials (IAEA 2023). Most fusion reactors, especially those for commercial operation, are still under development, and none have yet been decommissioned, so there is no experience to draw from. Therefore, the time and effort to decontaminate and deactivate components cannot be predicted at this stage. However, experience from fission reactor dismantling shows that this vital part of the production chain is often underestimated and induces substantial costs, while potentially requiring several decades to complete.

4.2.1.6 Waste Management

After decommissioning, most of the waste from fusion reactors will be activated materials such as structural components that have been exposed to neutron radiation or radioactively contaminated materials. Fusion does not produce high-level waste like fission reactors, but intermediate level waste may be generated during operations. This waste requires special storage facilities to mitigate radiation risks, especially for tritiated materials, which require dedicated tritium waste management strategies, as handling and disposal of tritiated components is challenging. However, the volume of waste generated during operation can vary significantly depending on the reactor design and materials used. Efforts to use reduced activation materials, such as special steel, further aim to minimize the radioactive footprint of fusion reactors (IAEA 2023). Regardless, low- and intermediate-level wastes must be safely stored for up

[11] https://www.iter.org/machine/supporting-systems/power-supply, accessed on 07–01-2025.

to several thousands of years to minimize risks to the environment and to humans; see Chapters 16 and 18 of this book. Infrastructure for these types of wastes, however, is more readily available than that for high-level wastes, such as spent nuclear fuel.

4.2.1.7 Proliferation

Some components and materials used in fusion technologies present proliferation risks that make careful management necessary. Fusion plants using D-T fusion fuel, for instance, produce high-energy neutrons during operation that could be used to breed plutonium from natural uranium. D-T fusion reactors further need extensive inventories of tritium and Li-6, both necessary for fusion, but also required for advanced nuclear weapons (see Chapter 6 of this book). Moreover, ICF devices add further proliferation risks as they require advanced technology and specialized knowledge, including high-precision simulations that could be repurposed for nuclear weapons development programs. For example, the National Ignition Facility at the Lawrence Livermore Laboratory, California, is pioneering ICF research, but its research is primarily focused on thermonuclear weapons stockpile management rather than (commercial) energy generation.[12] Fusion systems that do not involve fissionable materials are currently not subject to proliferation-related regulations, like the "Treaty on the Non-Proliferation of Nuclear Weapons." It is, therefore, necessary to establish or leverage regulations to address the proliferation risks associated with the development and use of fusion technologies to ensure they are used solely peacefully (DOE 2024).

4.2.2 *Construction of MCF Fusion Devices*

In the following, the technical system for the construction of a MCF device power plant is analyzed. The analysis provides a detailed understanding of the system by describing key assets and processes, as well as interactions and interdependencies with other processes that are necessary to build a fully functional reactor. Table 4.1 offers an overview of these processes and assets. Note that while the technical system is generalized, specific implementations may vary depending on the size and technical approach of the fusion project. For instance, ITER and SPARC are both tokamaks and plan to use D-T fusion fuel, but they differ both in scope and technology: ITER, a large-scale public project, integrates low-temperature superconducting magnets (LTS) and includes blanket modules to test tritium-breeding and fuel-cycle technologies.[13] SPARC, however, is a smaller-scale and privately funded project in which high-temperature superconducting magnets (HTS) will be used, and blankets are

---

[12] https://lasers.llnl.gov/news/ignition-experiment-advances-stockpile-stewardship-mission, accessed on 07–01-2025.

[13] https://www.iter.org/few-lines

omitted (Creely et al. 2020). The following analysis considers experimental devices and elements required for future power plants for (commercial) energy production. This includes, for instance, the ARC reactor, which is envisioned as the large-scale successor of SPARC, and shall include additional systems like blanket modules or supporting systems for electricity production (Sorbom et al. 2015).[14]

Critical steps in building a MCF power plant include the demonstration of compliance with nuclear safety requirements, R&D programs in design and operation via experimental and modeling studies, the design completion for critical components and the tokamak complex, procurement and production of critical components, the development of supporting systems, the construction of the tokamak complex, and the assembly of the tokamak reactor and its supporting systems (Bigot 2016).

The elementary process, "procurement and production of critical components," and the associated assets, as well as all preceding elementary processes, are critical for manufacturing the tokamak components. These include the vacuum vessel, which contains the plasma; a cryostat to provide the ultra-low temperature environment; blanket modules for neutron moderation and heat conversion; and a diverter for heat management and tritium recycling. Furthermore, superconducting magnets create magnetic fields for maintaining plasma stability and allow to sustain fusion reactions. In addition to these interdependent components, the tokamak is supported by many other systems, e.g., diagnostics and control systems. The transformation of these assets into final components, followed by their assembly, results in a fully functional tokamak reactor (IAEA 2002).

The components of the reactor must be tailored to its requirements and specifications, which shows the high level of specificity for such endeavors. For instance, in the case of ITER, the vacuum vessel construction requires 1,000 steel forgings which are custom-made for the project's needs, with the largest weighing ten tons.[15] Many of the heavy and large components of the reactor cannot be manufactured on-site. Instead, production is often conducted elsewhere, followed by the transportation of fabricated components to the project site for assembly. This process requires specialized logistics to handle the transport of oversized components. The construction plan for a fusion reactor further involves building a facility to house the tokamak, along with auxiliary and plant systems, and additional structures, such as tritium provision and diagnostics facilities. This complex usually requires a reinforced concrete wall structure to ensure safety and structural integrity, which is necessary for such a sensitive environment (IAEA 2002).

Table 4.1 provides an overview of all processes and the associated assets that are necessary for the construction of experimental and commercial MCF power plants, especially tokamak reactors. However, these may vary in scale and complexity

[14] Naturally, each of these processes requires the availability of qualified personnel trained in plasma physics, thermodynamics, mechanical, civil and computational engineering, data science, and more. This requirement is omitted from the table below because it is necessary for every process, and we focus on the physical assets.

[15] https://www.iter.org/node/20687/central-solenoid-fabrication-photo-reportage

Table 4.1 Overview of processes and assets for the construction of an MCF device

Elementary process	Detailed processes	Required assets
Demonstration in accordance with nuclear safety requirements	• Conduct safety analysis and risk assessments • Perform regulatory reviews and secure approvals • Implement safety protocols and monitoring systems	• Safety analysis software • Monitoring and diagnostic systems
R&D Programs in Reactor Design and Operation	• Theoretical modelling and simulations • Experimental testing and data analysis • Design optimization based on test results and simulations	• Simulation software • Research laboratories • Data analysis software
Design Completion for Critical Components and the Tokamak Complex	• Complete engineering drawings • Complete architectural drawings • Validate design with physical and computational prototypes • Review and update designs based on R&D feedback	• CAD software • Prototype fabrication tools (e.g., CNC machines) • Simulation software tools
Procurement and Production of Critical Components	• Find and qualify suppliers • Contract suppliers via tender process • Contract management • Manufacture of critical components • Manage supply chain logistics • Oversee fabrication and quality control of components	• Supply chain management systems (e.g., logistics) • Transportation networks • Fabrication facilities and tools • Quality control systems • Special resources and raw materials (e.g., vanadium alloys for magnets or superconducting alloys such as niobium-tin)
Development of Supporting Systems	• Design and test supporting systems (e.g., cooling and electrical system) • Integrate control systems for operational management • Ensure compatibility and safety of all supporting systems	• Electrical infrastructure • Control systems software • Materials for thermal management (e.g., silicon carbide) • Diagnostics and monitoring systems • Safety equipment (e.g., radiation shielding)

(continued)

Table 4.1 (continued)

Elementary process	Detailed processes	Required assets
Construction of the Tokamak Complex	• Site preparation and infrastructure development • Install and secure major components and systems	• Heavy construction equipment • Building materials • (Heavy) lifting equipment (e.g., cranes)
Assembly of the Tokamak Reactor and Supporting Systems	• Sequential and modular assembly of reactor components • Integration testing of assembled components • Final system calibrations and security checks • Correction of alignment deviations	• Manufactured components • Assembly tools and machinery (e.g., laser alignment tools) • Integration testing equipment (e.g., oscilloscope) • (High precision) calibration instruments

Source Own compilation based on (IAEA 2002; Bigot 2016; FIA 2024a). Abbreviations used: CAD: computer-aided design; CNC: computer numerical control; R&D: research and development

depending on the size of the reactor device and its associated power plant. Interfaces exist between the preceding and subsequent elementary processes necessary for constructing a fusion power plant. This shows that the sequential elementary processes are interdependent, as each process relies on the outcomes of the previous one to operate properly. For instance, the process of "design completion for critical components and the tokamak complex" depends on the results and data of the R&D programs. In contrast, the finalized design is required for the process "procurement and production of critical components."

4.2.2.1 Tasks, Roles, and Relations in Fusion Construction Projects

After defining the technical system and key processes for developing a MCF device, a role model can be established to clarify actor relationships and responsibilities.[16] The role of "supply of elementary processes," which takes on tasks such as the "procurement and production of critical components" and the "assembly of the tokamak reactor," requires significant coordination. Many fusion device components are sourced from a single supplier and pose challenges due to their complexity, high cost, long lead time, and custom specifications. Given the current underdevelopment of the fusion production chain, vendors often subcontract tasks to specialized manufacturers, further increasing coordination demands. Main contractors must ensure timely and budget-compliant project completion. Underdeveloped supply chains and monopolistic vendor structures could escalate costs and increase interdependencies.

[16] This assessment is based on insights compiled from the Directorate-General for Energy et al. (2021; 2023), FIA (2024a), WNA (2008), and Chapter 9 of this book.

Roles relating to the "supply of a physical good" often involve subcontracting, logistics, and on-site assembly by technicians, engineers, and inspectors, all requiring close supervision and regulatory oversight. Temporal-complementary relationships exist between roles, such as vendor and subcontracted firms, or between construction workers and machine assembly staff, emphasizing the need for coordinated efforts. Processes relating to research, e.g., regarding projected plasma behavior, depend on nuclear scientists, while the reactor design itself involves engineers who rely on experimental results to design components. This also forms a time-complementary link. Civil engineers and architects contribute by designing and constructing the tokamak infrastructure and supporting systems.

Coordination is primarily managed through contracts that align motivations via incentives or penalties. Fusion projects typically involve public–private partnerships, with governments often acting as principals and private entities handling the construction and operations. Funding comes from public, private, or hybrid sources, with public projects benefiting from government support, and private ventures relying on investments, subsidies, or research grants.

The construction of fusion plants is heavily influenced by regulatory frameworks, which rely on standards designed for nuclear fission. These are often unsuitable for the unique requirements of a fusion device, necessitating tailored national regulations. Until these are available, fusion projects adhere to existing nuclear regulations, with smaller-scale experiments potentially subject to less stringent requirements than comparable fission projects due to their lower risk profiles.

4.2.2.2 Main Challenges in the Development of Fusion Devices

The development and construction of fusion devices and power plants, whether experimental or commercial, comes with technical and organizational hurdles. Addressing them efficiently can reduce risks and increase the likelihood for successful provision. Companies and projects building fusion devices must account for these challenges by pursuing a proper strategy and organizational model to tackle them. The following sections show the hurdles faced by fusion developers. Some have led to the design of individual organizational models, trying to address, and overcome these challenges. In the subsequent section, the models and strategies implemented for ITER and SPARC will be reviewed to identify which model proves to be the most efficient (ex-ante). As the analysis focuses on organizational models for the construction of MCF devices, the technological challenges will be categorized into those arising during the development of fusion devices in general and those specific to the development of MCF devices.

4.2.3 *Technological Challenges*

4.2.3.1 Fusion in General

Burning Plasma

One of the biggest challenges in developing MCF devices is achieving self-sustaining plasma, also referred to as "burning plasma," which is necessary to reach a sufficient energy gain. In this plasma state, most of the heating required to maintain the high plasma temperatures comes from the energy generated by the fusion reactions themselves. To extract energy from the plasma that can be used for electricity generation, the energy used to heat the plasma must be lower than that generated by the fusion reactions. The energy gain factor Q_{fus} is a useful indicator, as it is defined as the ratio of the power produced from the fusion reaction and the power required to ignite and sustain the reaction (refer to Chapter 2 for more details). At $Q_{fus} > 1$, the so-called scientific breakeven is achieved in which the energy yield is greater than the input. However, since not all the energy produced by fusion reactions can be kept within the plasma, a higher Q_{fus} is needed to achieve a burning plasma state which varies depending on the technological approach. For instance, in the case of D-T fusion, a ratio of $Q_{fus} = 5$ is often targeted, as this marks the entry point into the burning plasma regime. This threshold needs to be surpassed since only one-fifth of the energy produced in D-T fusion reactions is retained in the plasma for self-heating via alpha particles; the remainder escapes as neutron energy, potentially to be used for electricity generation and tritium breeding (Meschini et al. 2023). However, a commercial power plant will have to cover additional components and account for thermal losses, necessitating even higher energy gain factor values. Reaching the desired Q_{fus} depends on efficient plasma confinement, which keeps the plasma within the required temperature, and particle density ranges for a sufficient confinement time.

However, these parameters and the overall performance of burning plasma are poorly understood. Existing knowledge is mostly derived from theoretical calculations and simulations, which makes it difficult to predict plasma behavior in fusion devices and subsequently optimize confinement strategies and reliably produce energy (U.S. GAO 2023). Getting a simulated burning plasma into a real-world system is a major milestone for fusion energy. Without achieving this, fusion power plants would remain fully dependent on external energy sources, which would make a commercial power plant economically unviable. To date, only one laboratory has achieved a burning plasma state, albeit through ICF for only one single pellet, which underscores the uncertainties and risks associated with obtaining the desired energy gains in other fusion projects (Zylstra et al. 2022).

Plasma Confinement and Stability

Developing a self-sustained fusion system with sufficient energy gain requires plasma confinement to tune the plasma within a specific temperature and density for a certain time. This keeps the plasma stable and prevents it from expanding or encountering the reactor walls, which would result in temperature losses and instabilities. The three main technological approaches, MCF, ICF, and MIF, address this challenge in different ways (Meschini et al. 2023; Pearson and Takeda 2020).

- MCF devices use magnetic fields to confine and manipulate plasma. MCF devices are characterized by low plasma density, requiring longer confinement times to overcome the necessary thresholds. External beams or electromagnetic waves are used to heat the plasma to the required temperature to achieve a sufficiently high fusion triple product.[17] Although magnetic fields initially heat the plasma by inducing electrical currents, the heating effect diminishes as plasma temperatures increase, necessitating the use of external heating systems to maintain the required conditions.[18] MCF systems are sensitive to small perturbations in the plasma column, which can grow and destabilize the system, potentially damaging the reactor walls.
- In contrast, ICF devices compress the fusion fuel to create plasma at extremely high densities and temperatures by using high-power lasers or ion beams to target the fuel— either directly or indirectly— which creates a symmetrical implosion triggered by shock waves. Although ICF plasmas have noticeably short confinement times, keeping near-perfect symmetry in the shock wave is crucial. Even small disturbances can reduce compression efficiency. This can lower the plasma density and temperature and increase the risk of instabilities.
- MIF reactors combine elements of MCF and ICF as they produce plasmas of medium density, higher than MCF but lower than ICF, with intermediate confinement times. A driver, similar to those used in ICF, compresses the plasma to achieve the required fusion triple product.

Therefore, each confinement approach relies on external sources, such as plasma heating, to achieve the required temperatures, densities, and confinement times. These methods must precisely address the challenges arising from differences in the parameters. In addition, instabilities are still a critical challenge for all fusion technologies, even though their nature and the way to mitigate are different for each approach.

---

[17] The triple product, also referred to as the "Lawson criterion" is calculated as the product of plasma temperature, density and confinement time. In addition to the energy gain factor, it marks a specific threshold for different fusion reactions that must be overcome for plasma to ignite (Lawson 1957; Wurzel and Hsu 2022).

[18] https://www.iter.org/machine/supporting-systems/external-heating-systems, last accessed on 15–12-2024.

Resource and Fuel Availability

Further challenges arise from the choice of fusion fuel, which depends on the respective fusion approach, as some fuels are abundant and others not so much. For instance, most of the existing global tritium stock is expected to be consumed by ITER experiments, which raises concerns about powering future fusion reactors aimed to generate energy with the presently available tritium reserves. To address this limitation, reactors that rely on D-T fusion reactions are usually designed to produce, or "breed" their own tritium using breeding blankets containing Li-6 (Abdou et al. 2021).

For tritium breeding to be sustainable, the system's tritium breeding ratio, i.e., the ratio of produced and consumed tritium, must be greater than one. Achieving this ensures a sufficient tritium supply during reactor operation without relying on external sources. The breeding process with Li-6 is exothermal, thus generating additional energy that can be extracted for energy conversion in the plant. Breeding is also possible with Li-7, which is more abundant, but this reaction is endothermal (Biel 2014).

However, breeding blankets have not yet been developed, tested, or demonstrated in operational environments. For this reason, some experimental devices relying on D-T fusion reactions, such as SPARC, do not include breeding blankets, while others, like the ITER project, incorporate them primarily for testing purposes (Meschini et al. 2023). Therefore, it is still uncertain whether future fusion power plants will be self-sufficient regarding their tritium supply. A consistent reliance on external tritium supply could risk the viability of D-T fusion reactors for large-scale and continuous operation due to the limited availability of tritium.

These concerns do not apply to fusion energy systems that do not rely on tritium. For instance, other fusion fuels, such as He-3, can be produced through an open fuel cycle using D-D reactions. B-11, which is used in aneutronic fusion reactions with protons (i.e., those fusion reactions that do not produce neutrons), is widely available and mineable. This eliminates supply chain issues and makes the installation of on-site breeding systems redundant (Meschini et al. 2023). However, these alternative fuels require significantly higher plasma temperatures to achieve the desired fusion energy gain, which presents an additional technical challenge for reactor design and operation (Gordinier et al. 2003). For instance, among the most studied fusion reactions, the highest energy release occurs during He-3-He-3 reactions. However, this reaction requires much more energy to fuse, which is not practical with current technologies. In contrast, D-T fusion reactions have the highest cross section at comparatively low energy levels, which makes it the most promising and widely used for fusion concepts today (Meschini et al. 2023; Entler et al. 2023). Developers must, therefore, trade off between the higher energy input and reactor complexity for high energy output fusion reactions like p-B-11 versus the lower energy input and simpler reactor design of D-T reactions with lower potential energy gain.

Radiation Damage

Another challenge arises from radiation damage caused by the plasma environment. Plasma-facing components and materials, including the superconducting magnets, are continuously exposed to long-term irradiation from high-energy neutrons produced during fusion reactions. Superconducting materials are highly sensitive to low levels of radiation damage, but are essential for the magnet system in MCF and MIF devices. Using fuels that result in aneutronic reactions, such as p-B-11 or He-3, could mitigate these concerns, thus helping to reduce radiation damage and simplify constraints on reactor design and material durability (Meschini et al. 2023).

Power Generation

For a fusion device to be economically viable and practical as a future energy source, net energy production is required. Hence, a power conversion system must be installed to convert the energy released by fusion reactions into electricity or some form of heat. Considering D-T fusion devices, most of the energy released by fusion reactions is carried by neutrons in the form of kinetic energy. This energy is captured in the blanket where neutrons are slowed down, collide, and interact with the blanket material, converting their kinetic energy into thermal energy. The generated heat is then transferred to a coolant flowing through or around the blanket[19] for further use. The heated coolant later drives turbines via a secondary coolant loop to generate electricity. This implies that D-T fusion devices require an effective power conversion system, including a blanket system, a coolant system, and turbines. The consensus is to use turbines to generate electricity. In contrast, aneutronic fusion reactions have the potential to generate electricity directly. Since no neutrons are produced, the integration of a blanket for capturing neutrons or including turbines for converting thermal energy into electricity is redundant, as the kinetic energy of charged alpha particles, which have been produced by the fusion reaction, is captured directly (Meschini et al. 2023). For instance, the alpha particles can be slowed down within a reactor sphere, where they will be neutralized upon contact, allowing their charge to be captured as a source of electricity (Lobner 2021).

4.2.3.2 Magnetic Confinement Fusion

Magnet Technology

MCF devices create a magnetic field to confine the plasma. Achieving a cost-efficient and strong magnetic field requires the use of superconducting magnets, which can be LTS or HTS. Superconductors are materials that show zero electrical resistance and repel magnetic field lines when cooled to cryogenic temperature levels. This

[19] Note that the blanket coolant may be either liquid or gaseous.

property allows electric currents to flow through these materials without energy loss and to create a powerful magnetic field. Superconducting magnets are wounded wires made from materials such as niobium-tin or rare-earth barium copper oxide (Ou et al. 2023). Magnets wound with non-superconducting materials can also achieve high magnetic fields, but they consume high amounts of energy during operation due to power dissipation, which makes them less efficient for fusion power plants.

LTS are widely used and are made of materials such as niobium-titanium, which become superconducting when cooled to a critical temperature of 10.7 K, enabling the creation of magnetic fields of up to 17 T. Similarly, LTS made of niobium-tin must be cooled to 18.3 K and can generate magnetic fields of up to 22 T (Huang and Laifeng 2018). Cooling the superconducting wires to cryogenic temperatures requires a cooling system which adds operational complexity and energy requirements.[20] In contrast, HTS have the potential to generate magnetic fields of up to 30 T at a higher temperature level than LTS.[21] This reduces the energy needed for cooling which makes HTS magnets more efficient. However, the material costs of HTS cables are higher due to low demand. But as demand for fusion magnets increases, economies of scale could lower the cost of HTS materials (Huang and Laifeng 2018). Research into HTS materials and their industrial applications is relatively young. Therefore, many uncertainties and reliability concerns still exist. For instance, thermo-magnetic instabilities, which can cause the loss of superconductivity (a phenomenon known as magnet quench), pose the risk of damaging the system. Fully developed solutions to address these challenges are not available yet (Meschini et al. 2023).

Material Resistance

Materials in a given MCF device must endure simultaneous stresses such as high temperatures, intense neutron flux, and ion damage. The combination of these factors is beyond the durability and resistance of currently available materials. Hence, today's experimental fusion devices are not designed for continuous or long-term operation. To avoid frequent repairs or component replacement in future demonstration and commercial power plants, further materials development is necessary. For instance, it is estimated that materials in a DEMO device must last approximately one year under extreme fusion conditions to be economically viable (Meschini et al. 2023).

---

[20] https://www.iter.org/mag/5/superconductivity-it-gets-current-flowing

[21] The temperature levels are still cryogenic. For example, the first HTS to be discovered, the alloy yttrium barium copper oxide, requires temperatures of 93 K to become supercondcting (Wu et al. 1987).

4.2.4 *Organizational Challenges*

4.2.4.1 Skilled Workforce

Fusion projects rely on a solid organizational structure in which fusion specialists and a skilled workforce, including scientists, engineers, project managers, operators, and technicians, cooperate. Given the small size of the fusion community, attracting, educating, and training specialists is essential to address the challenges inherent in fusion technology development. Collaborating with the industry can help foster cross-discipline skills, bridge technical disciplines, and ensure that expertise is applied across all project lifecycles (IAEA 2024a). Following this approach could accelerate the path toward commercialization by bridging skill gaps more effectively. However, the current shortage of skilled professionals is a significant challenge. Without sufficient workforce capacity, project timelines will be delayed, and the overall development of fusion energy will be hindered. Therefore, fusion projects often have to recruit foreign workforces, emphasizing the need for global collaboration between research institutions, governments, and private industries (U.S. GAO 2023).

4.2.4.2 Tailored Regulation

Regulatory uncertainty further hinders the development of fusion devices and power plants. Presently, fusion projects mainly operate under legislation and regulations originally designed for nuclear fission. Applying strict fission-based regulations to fusion projects, which often might not require such strict oversight, will lead to delays and increased costs. Overregulation or mismatched regulation complicates project execution and creates barriers for potential investors. Establishing a dedicated regulatory framework for fusion is a challenge for governments and regulators, as the process is likely complex and time-consuming (U.S. GAO 2023).

4.2.4.3 Limited Supply Chain

Due to the high specificity of fusion components, the supply chain for fusion energy systems is underdeveloped, with only a limited number of suppliers and vendors. Several critical components, such as diagnostic systems, are produced by only a few companies worldwide. This limits the ability to source the necessary components in the required quantities and to scale up fusion energy systems. For instance, semiconductors, which are needed for magnets, are also in high demand in other industries, making it difficult to meet the supply of fusion projects (U.S. GAO 2023). To address these challenges, it might be necessary to prove the feasibility and commercial potential of fusion energy to encourage companies to invest in the sector and develop a reliable supply chain for reactor components.

4.2.4.4 Funding

Developing and building a fusion reactor involves uncertainties in investments, cost predictions, and estimations which are influenced by factors such as plant size, material requirements (such as steel or concrete), and the underdeveloped industry for standardized production of highly specific reactor components. Since these uncertainties will persist until the fusion industry matures, the acquisition of sufficient funding will pose an ongoing challenge for the development and commercialization of fusion technology for the next decades. The financial requirements for fusion reactors vary significantly depending on whether the plant is experimental, demonstrative, or commercial. For instance, the expected costs for ARC—the commercialization-oriented reactor after SPARC—are estimated to be around US-$5–6 billion. In contrast, ITER was estimated to cost €5 billion[22] in 2001, but its costs have since escalated to approximately €20 billion in 2024 and might further increase due to repeated project delays, as shown in Sect. 4.4. In addition, uncertainties in investments and the financial risks of building commercial fusion plants might foster financing costs induced by high interest rates (Meschini et al. 2023). Future costs of fusion energy and devices will also depend heavily on the lifetime and robustness of their components, especially their resistance to plasma-induced damage. Frequent component replacement and early plant closures due to material degradation could substantially impact the economic feasibility of fusion devices (US GAO 2023).

4.3 Case Studies

The analysis focuses on the organizational frameworks that govern the construction of ITER and SPARC. It examines the decisions made to address the challenges associated with the development and construction of fusion devices, since such projects are inherently uncertain. We focus on the technological approaches adopted; decisions within the "production" phase, which include procurement, manufacturing, and assembly of essential components; the funding models; and the institutional environments that shape each organizational model.

Following the multiple case study approach proposed by Yin (2014, 105), we aim to "predict contrasting results but for anticipable reasons." The analyzed projects ITER and SPARC are comparable due to their common aim of MCF construction, but differ sufficiently in both organizational, e.g., public versus private ownership, and technological approaches, e.g., regarding reactor sizes. The organizational models are compared regarding their (proposed) effectiveness in achieving the respective technological and organizational challenges. The comparison is conducted relatively, and when no conclusion on a challenge can be made, primarily due to the early development stages of both projects, this is also indicated.

[22] €8.12 billion as of 2024.

4.3.1 *ITER*

ITER, the International Thermonuclear Experimental Reactor, is a large-scale experimental tokamak currently under construction in southern France. It is designed to prove the feasibility of fusion energy by investigating and demonstrating the "burning plasma" state and testing integrated technologies, such as advanced magnets and tritium breeding modules. ITER also serves as a prototype for future devices, such as the planned DEMO reactor. The project is managed through intergovernmental collaboration, with funding and resources provided by seven member entities: China, the European Union (through Euratom), India, Japan, Korea, Russia, and the United States. The idea for ITER originated in 1985, when the Soviet Union proposed building a tokamak through international collaboration. Over the following decades, this vision evolved into a finalized engineering design in the early 2000s. In 2006, the seven key members formally signed the ITER Agreement, through which the ITER Organization (IO) was established to oversee the project. Actual construction work for the facility and components began in 2010, with the first operations originally projected for 2016, and the first experiments validating fusion power planned for 2020. However, a series of delays and quality issues over the years, which were caused by factors such as the COVID-19 pandemic, corroding parts, and inconsistencies between components requiring repairs, have led to multiple postponements. The current schedule estimates construction completion by 2036, with full operation expected to begin in 2039.[23]

4.3.1.1 Technological Approach

ITER is a MCF device based on tokamak technology, designed to rely on D-T fusion reactions. The design draws on the physical and technical knowledge gained from the Joint European Torus (JET) in the UK that operated from 1978 to 2023.[24] Its primary goal is to prove the scientific and technological feasibility of fusion energy production. This means achieving a scientific energy gain of $Q_{fus} \geq 10$ for plasma burn durations of 300–500 s, as well as developing long-pulse, non-inductive scenarios to maintain $Q_{fus} \approx 5$ for up to 3000 s (IAEA 2022). ITER will further test tritium breeding technologies and the concept of "burning plasma." Additionally, ITER material behavior during plasma exposure (IAEA 2002).

To confine the plasma and achieve the required triple product, ITER's magnet system uses LTS which were chosen for their cost effectiveness and availability during the concept design for ITER. The LTS are cooled using liquid helium. At the time of design, HTS technology was not sufficiently advanced to be included in the concept (Whyte et al. 2016; Haack 2024). Approximately 850 metric tons of superconducting materials are required for ITER's LTS (Grunblatt et al. 2005; IAEA

[23] https://www.nature.com/articles/d41586-024-02247-2. Note that the case studies rely on a large number of URLs and other sources, that we do not reference extensively in this chapter.

[24] https://euro-fusion.org/devices/jet/jet-history/

2022). Plasma heating is provided by external systems such as neutral beam injection and high-frequency waves. The reactor will be located on a 180-hectare site, weigh approximately 23,000 metric tons, and be 30 m tall (IAEA 2022).

The device is planned to run on an external tritium supply from CANDU reactors, as this is the only commercial source of tritium. Over its operational lifetime, ITER will require an estimated amount of 18 kg of tritium. While this will be enough for ITER's research goals and meet its demand, supplying future reactors like DEMO or other experimental devices running simultaneously will pose a significant challenge (Pearson et al. 2018; Abdou et al. 2021). To address this, ITER will test tritium breeding blankets to demonstrate the concept of tritium self-sufficiency. During full-power operation, ITER's electrical consumption will range from to 110 to 620 MW for plasma-related operations, including power of 100 MW required for auxiliary systems, such as the cooling water system and the cryoplant. This total energy consumption will translate to an estimated monthly electricity bill of approximately €2.5 million.

4.3.1.2 Provision, Procurement, Production

ITER's organizational model is a decentralized, international, network-like structured public procurement framework with multiple layers. Fabrication duties are distributed according to each member state's technical expertise and industrial capabilities. The size and complexity of some of the reactor's critical components has led to global cooperative production arrangements between all ITER members. For instance, the vacuum vessel sections are jointly produced by Europe and Korea, and the central solenoid, which is a key component of the magnet system, is produced by the United States and Japan. Further, India and the United States are responsible for the fabrication of the cooling water systems, while China, Europe, Korea, and Russia contribute to the blanket system. Except for India, all ITER members are fulfilling production orders for the magnets.

Each ITER member has established a Domestic Agency (DA) that acts on behalf of its government. Under procurement agreements, these legal entities are responsible for acquiring the assigned components, systems, or infrastructure, and signing contracts with industrial partners. However, due to the complexity of the components, other DAs may collaborate by subcontracting some of the production and delivery. For example, the vacuum vessel is produced jointly, with four segments in Korea and five in Europe, and additional features, such as higher ports and in-wall shielding, are supplied by Russia and India, respectively.

The ITER Organization (IO), governed by the ITER Council, is the central authority responsible for the management and coordination of the ITER project. Its role includes overseeing the design, construction, and operation of the facility, ensuring that technical specifications, budgets, and timeframes are met. Throughout the supply chain, the IO coordinates the actions of the DAs and industrial partners. However, the IO does not govern the DAs; instead, they operate under the authority of their respective countries and have autonomy over the procurement of the components assigned to them.

Industrial partners are responsible for manufacturing components and systems for the ITER facility, as well as providing technical knowledge and support to the DAs throughout the supply chain.[25] During the procurement process, each DA appoints companies to be responsible for goods, services, or labor. The companies are selected based on specific criteria, mostly via tenders. Between January 2008 and December 2016, the European DA awarded 839 contracts and grants, totaling approximately €3.8 billion to industrial companies, small and medium-sized enterprises, and research organizations across Europe (Delasnerie 2023). Although the DAs are state-affiliated institutions representing the ITER countries, 50% of the actual suppliers or industrial partners who fabricate components, supply services, or work for the ITER project are private firms with experience gained within the nuclear fission sector. Notable companies involved in the manufacturing of reactor components include General Atomics, Hyundai Heavy Industries, and Mitsubishi Heavy Industries Ltd.

Fusion experts and laboratories are other players involved in the ITER project. They provide their facilities and expertise or conduct studies on scientific issues related to the performance of fusion reactions (Chiocchio 2006). For instance, the DAs involved in procuring the conductors have signed a service contract with the École Polytechnique Fédérale de Lausanne, which guarantees the availability of SULTAN, a modified conductor test facility. The IO has also nominated nine contractors for assembly and installation.

According to Article 14 of the ITER Agreement, the IO must comply with the applicable national laws and regulations of the Host State, particularly in areas of public and occupational health and safety, nuclear safety, radiation protection, licensing, nuclear substances, environmental protection, and protection against acts of malevolence. Since ITER is being built in France, it must comply with French nuclear safety regulations, enforced by the French Nuclear Safety Authority (ASN).[26]

Figure 4.2 shows the interactions among various stakeholders, the governance relationships, and the delivery flows governed by contractual arrangements, in particular procurement agreements. The ITER project is under the supervision of ASN to ensure strict compliance with nuclear regulations.

4.3.1.3 Funding

ITER's funding structure is a multinational collaboration system of in-kind (material, parts or services) and financial contributions from its seven members. It is aimed to finance the design, building, assembling, commissioning, operation, and eventually

[25] http://www.iter.org/newsline/-/3854, last accessed on 02–12-2024.

[26] https://www.iter.org/newsline/-/3385, last accessed on 22–08-2024.

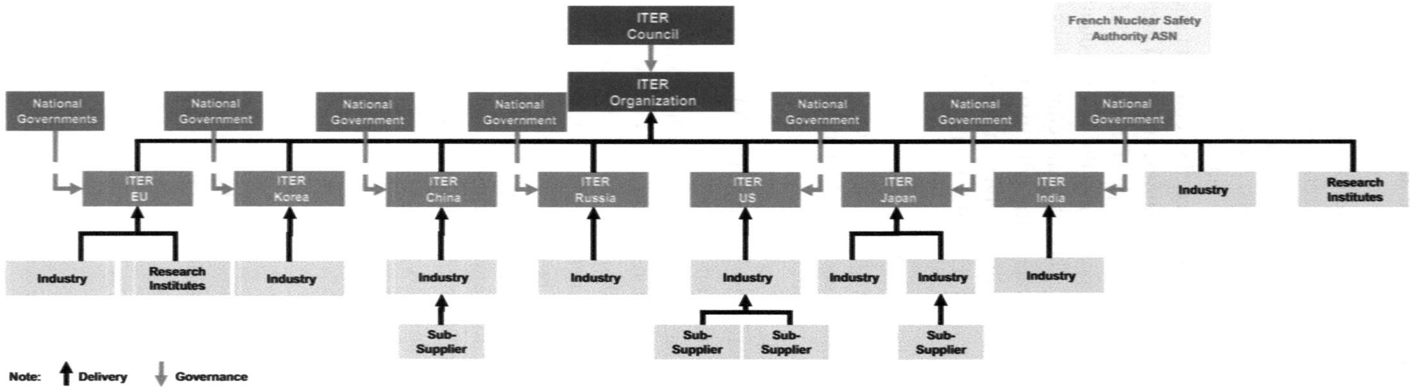

Fig. 4.2 Stylized organizational model of the ITER project (*Source* Own depiction) (*Note* Upward arrows show the delivery of components to the responsible and contracting organization in the upper layer, while downward arrows represent governance flow. Abbreviations used: ASN: Autorité de sûreté nucléaire; EU: European Union; ITER: International Thermonuclear Experimental Reactor; US: United States)

decommissioning of the ITER facilities. Besides being responsible for the procurement of components, each member's DA also acts on behalf of its respective government to provide short- or long-term contributions (ITER Organization 2023). Note that nearly 90% of member contributions to the ITER project are in-kind, meaning they are provided as goods, services, or expertise rather than cash payments.[27]

Each DA's contributions to the project are funded from its respective government budget, hence derived through tax payments and allocated to the project. For instance, between 2014 and 2022, the EU allocated € 6.164 billion from its budget to European companies (Delasnerie 2023). Europe alone is responsible for approximately 45.5% of the construction costs, while each of the other member countries contributes around 9.1%. During operations, the cost burdens shall be redistributed, with Europe paying 34%; Japan and the US 13% each; and China, India, Korea, and Russia each paying 10%.

Due to the global nature of the project, with each participant acquiring in-kind hardware in their own area and currency, it is challenging to convert the total cost of the ITER construction into a single currency. Initially, the ITER project was estimated to cost approximately US-$6 billion.[28] Technical and regulatory challenges, along with supply chain issues, have led to significant delays and cost increases. These have caused milestones and timelines to be postponed and forced the ITER Council to revise the project budget multiple times. There are still concerns about the overall spending, mainly due to the adjusted schedule and the difficulty calculating the precise costs. The costs vary greatly between the seven member states due to their distinct costs for elements like labor, materials and other expenditures related to the development of reactor components. This results in independent assessments ranging from US-$41 to 65 billion, depending on the assumed production location (Kramer 2023). Official estimates assume costs of approximately €17 billion if all manufacturing were done in Europe.[29] Economic factors such as inflation, currency exchange rate fluctuations, and variations in labor and material costs could therefore further increase the overall expense of the ITER project. Note that while the Covid-19 pandemic is blamed as a main reason for project delays,[30] the original target date of achieving first plasma operations in 2016 was already out of scope in June 2010, when the first baseline was defined.[31] Now, first plasma shall be achieved around 2036.

---

[27] http://www.iter.org/faq#How_is_ITER_financed_, last accessed on 07–07-2024.

[28] https://world-nuclear.org/information-library/current-and-future-generation/nuclear-fusion-power. Due to the mentioned uncertainties and the range of the cost estimates, we refrain from providing inflation-adjusted cost values for such values.

[29] http://www.iter.org/faq#Do_we_really_know_how_much_ITER_will_cost, last accessed on 11–11-2024.

[30] https://www.iter.org/node/20687/new-baseline-prioritize-robust-start-exploitation

[31] https://www.iter.org/project/road-iter#event-_6th-iter-council-meeting

4.3.2 CFS / SPARC

SPARC (short for "smallest possible ARC") is a compact demonstration tokamak developed and built in collaboration between CFS and the Plasma Science and Fusion Center (PSFC) of the Massachusetts Institute of Technology (MIT). CFS is a private venture that spun out of the PSFC and was founded in 2018 by former MIT team members. The company is supported by a consortium of institutional partners and venture capital firms. This partnership benefits from the MIT's institutional credibility and the backing of various collaborators.[32] The SPARC reactor aims to demonstrate and exceed scientific energy gain, while also proving the feasibility of high-temperature superconductors which are used to generate the reactor's magnetic field. Following the successful testing of SPARC, a follow-up reactor named ARC ("Affordable, Robust, Compact") is planned to be constructed. ARC will be a commercial fusion device designed to produce and deliver electricity at a commercial scale. The SPARC tokamak, currently under construction in Devens, Massachusetts, with a current delay of two years, is expected to be operational in 2026, and a net energy gain shall be demonstrated by 2027 (Crownhart 2024). The project has undergone several design iterations over two short-term project phases to reach its current design state (Creely et al. 2020).

4.3.2.1 Technological Approach

Like ITER, SPARC is a MCF device based on the tokamak concept using D-T fuel for its fusion reactions. Its goal is to achieve a scientific fusion gain of $Q_{fus} > 2$ with the potential of reaching $Q_{fus} \approx 11$ under nominal conditions. If successful, SPARC would be first MCF device to exceed the net scientific energy threshold. During operation, plasma-facing materials, such as carbon and tungsten, shall be tested, especially by focusing on factors like impurity radiation, erosion, and tritium retention. SPARC combines REBCO-based HTS magnets and will generate a strong magnetic field which allows for a smaller, more compact reactor (Creely et al. 2020). Rare-earth barium copper oxide (REBCO) superconductors, which have become commercially available in recent years, can be cooled with liquid nitrogen, whereas LTS used by ITER require liquid helium, which is more expensive (Haack 2024). Even though costs can be reduced when using less expensive cooling material, HTS are still costly. For instance, in 2021, HTS conductors were at least five times more expensive than LTS materials due to limited production and low demand. However, economies of multiples could make HTS wires cost competitive in the future.[33] SPARC will use

[32] https://news.mit.edu/2018/mit-newly-formed-company-launch-novel-approach-fusion-power-0309, last accessed on 18–08-2024.

[33] https://www.nextbigfuture.com/2024/08/beakthrough-towards-highest-performance-superconducting-wire-beating-copper-on-price-performance.html

10,000 km of HTS tape for its magnet system (compared to 100,000 km of superconducting strands used for ITER's LTS magnets)[34] and will be 40 times smaller in volume than ITER, while achieving a stronger magnetic field, but targeting a smaller energy gain.[35] This compact design could make SPARC an economically competitive reactor, since its smaller size enables faster construction and makes it market-ready in less time while achieving significantly higher magnetic field strength. Furthermore, major cost drivers like steel and concrete are reduced due to the smaller reactor size so that material costs can be potentially minimized (Meschini et al. 2023). Additional heating for SPARC will be provided through an ion cyclotron resonance heating system with 25 MW power input (Creely et al. 2020). Unlike ITER, which requires a large tritium supply, SPARC will only need a small amount of tritium (10 g in total) over its operational lifetime,[36] and might therefore face reduced challenges of tritium supply chain constraints. The follow-up reactor, ARC, will be equipped with breeding blanket modules to achieve potential tritium self-sufficiency. However, SPARC is not designed to integrate and evaluate those modules. Instead, SPARC will gain knowledge essential for the development of its successor by testing high-field magnet technology, plasma physics, and engineering systems, and it will collect data on burning plasmas. The focus of SPARC's R&D is therefore on refining core fusion technologies and physics. This means that there will be no prototype testing for ARC's breeding blanket (Creely et al. 2020).

4.3.2.2 Provision, Procurement, Production

Since the magnets are the most complex and unique part of the SPARC reactor, CFS decided to build them in-house within a dedicated factory while the necessary HTS materials are acquired from commercial vendors. Other critical components such as the vacuum vessel, the cryoplant, power supplies, and plasma-facing parts are procured via the market by signing long-lead procurement agreements with vendors (Brunner 2023). MIT researchers are responsible for designing concepts for the outsourced manufacturing of custom-made components, which are to be manufactured externally.[37]

The construction of the SPARC reactor pursues a project timeline of five years in which several iterations and project phases ending with specific milestones are to be completed. Each iteration serves to build a prototype, test its performance, analyze the results, and implement refinements based on lessons learned which resulted in many design variations over time. Within the first phase of the SPARC project (2018–2021), five design iterations aimed at finalizing a feasible version of the reactor were

[34] https://www.iter.org/facts-figures#:~:text=100%2C000%20kilometres%20of%20niobium%2Dtin,2009%20and%20ended%20in%202014

[35] https://www.technologyreview.com/2023/10/04/1080113/2023-climate-tech-companies-commonwealth-compact-tokamak-carbon-free-electricity-fusion-reactor-electromagents/

[36] https://cfs.energy/devens-campus/devens-info

[37] https://andersondahlen.com/anderson-dahlen-mit-fusion-magnet-development/

to be carried out. This phase focused on constructing and operating an HTS-based magnetic coil as the first milestone, which was followed by completing the SPARC design as the second milestone. The second phase, lasting four years (2021–2025), covers the final stages of design, construction, and commissioning. Compared to ITER, SPARC is scheduled to become operational within a very short timeframe of seven years, of which three years are planned for the design and R&D, and four years for construction and commissioning (Creely et al. 2020). By employing test-driven developments—a key feature of agile methods commonly used in high-risk industries such as space exploration—technological uncertainties can be systematically addressed, risks minimized, and delays prevented (U.S. NRC 2021).

The organizational model is structured around CFS and MIT, with the former managing and governing the project, including attracting investors and industrial allies, and the latter focusing on research and technological development for the tokamak. CFS is also responsible for managing the finances. It also leverages its research relationship by partnering with the MIT Energy Initiative, which provides access to institutional resources and capabilities, while also offering a network to engage with incumbent energy companies, strategic partners, and potential investors.[38] The project is further supported by strategic partnerships and shareholder positions with large energy companies, as well as technological capabilities from governmental institutions. For instance, Italian oil and energy company Eni is a major shareholder of CFS. Through a "Collaboration Framework Agreement" with CFS, Eni increased its share in the project and will use its industrial expertise, resources, and technical support, along with its network of industrial partners, to benefit CFS and the development of SPARC.[39] Another shareholder, Norway's oil and gas group, Equinor, has an observer seat on the board of directors at CFS,[40] thereby exerting influence on the company's decision-making process and strategy. Other key partners include several universities and laboratories, such as the Lawrence Livermore National Lab, the Max Planck Institute for Plasma Physics, the University of Maryland (FIA 2024b), the University of Wisconsin,[41] and the US Department of Energy's Brookhaven National Laboratory. Through these partnerships, CSF can access advanced facilities, resources and the expertise of other scientists,[42] while also becoming part of a horizontal supplier business by building components, especially magnets, on behalf of other institutes. This helps the company gain more experience in the field of fusion and apply this knowledge to the development of the SPARC device.[43]

---

[38] https://news.mit.edu/2018/mit-newly-formed-company-launch-novel-approach-fusion-power-0309, last accessed on 18–08-2024.

[39] https://www.eni.com/en-IT/media/press-release/2023/03/eni-signs-new-collaboration-agreement-with-cfs-development-fusion-energy.html, last accessed on 05–12-2024.

[40] https://www.equinor.com/news/ev/invests-fusion-energy, last accessed on 10–09-2024.

[41] https://magneticsmag.com/commonwealth-fusion-systems-delivers-superconducting-magnets-to-university-of-wisconsins-wham-project/, last accessed on 17–10-2024.

[42] https://www.bnl.gov/newsroom/news.php?a=116853, last accessed on 10–09-2024.

[43] https://magneticsmag.com/commonwealth-fusion-systems-delivers-superconducting-magnets-to-university-of-wisconsins-wham-project/, last accessed on 17–10-2024.

Governmental support has been reinforced by a new regulatory framework, introduced through the US ADVANCE Act and the Fusion Energy Act. These regulations have opened new paths for fusion by separating the regulatory processes for fusion and fission. Since fusion poses fewer risks and causes less environmental harm than fission, it is subject to less stringent regulations. This separation accelerates the development of fusion devices, simplifies the authorization process, and demonstrates the government's support and confidence in fusion technologies. Changes in regulations can minimize investor concerns about the uncertainties and delays caused by strict and time-consuming regulations. This, in turn, simplifies the path to deployment as further investments are encouraged.[44]

4.3.2.3 Funding

The funding model for SPARC consists primarily of private sector investments and energy industry partnerships, with some contributions by the US Federal Government and public–private partnership initiatives. Unlike ITER, which relies mainly on in-kind contributions from member states, SPARC's funding is monetary, with the funds being used to cover research, development, construction, and operational costs through direct investments. CFS is actively searching for investors and raising funds, with a significant portion of those funds going to R&D at MIT.[45] Since its founding in 2018, CFS has raised over US-$2 billion from 54 investors over six funding rounds.[46] A huge part of the funding comes from private investments, venture capital, and hedge funds, with individual contributions ranging from US-$84 million to US-$1.8 billion. Notable investors include big energy companies, venture capital firms, and high-net-worth individuals such as Bill Gates, who invested personally and through his venture firm, Breakthrough Energy Ventures.[47]

One of the early and significant financial support was given from Eni by investing US$50 million in 2018 and thereby becoming a notable shareholder in CFS. This early investment helped build trust and opened the door for other companies to invest in the new venture. Eni further strengthened its partnership with CFS by signing a framework agreement to support projects related to SPARC and ARC. This agreement benefits CFS by allowing it to leverage Eni's network of stakeholders and connect with experts from the energy sector, which potentially helps to attract more

[44] https://heatmap.news/climate/advance-act-fusion

[45] https://news.mit.edu/2018/mit-newly-formed-company-launch-novel-approach-fusion-power-0309

[46] https://pitchbook.com/profiles/company/226641-88#overview, last accessed on 07–09-2024.

[47] https://www.crunchbase.com/organization/commonwealth-fusion-systems/company_financials, last accessed on 07–07-2024.

financial support from the industry.[48] Other contributors include Norwegian Equinor and Google.[49]

The smaller portion of financial contribution comes from public funding. For instance, the US Department of Energy provided a US$15 million grant to promote R&D in fusion energy in 2023 as part of the "Milestone Program." The allocation of the funds were linked to the condition that they be released only when companies achieve their pre-agreed milestones, which helps to minimize the financial burden and risks for taxpayers.[50] Additional investments have been attracted from public–private partnership programs, such as the Advanced Research Projects Agency-Energy (ARPA-E)[51] and the Innovation Network for Fusion Energy (INFUSE), which are both managed by the US Department of Energy. These programs provide smaller grants to fund specific project phases and technological and research endeavors, which are done in collaboration with universities and national laboratories.[52]

4.4 Discussion

The following section compares how ITER and SPARC plan to overcome the technical and organizational challenges.

4.4.1 Technical Challenges

4.4.1.1 Plasma Confinement and Stability & Magnet Technology

Both devices incorporate superconducting magnet technology to stabilize and confine the plasma to fusion conditions because of its strong magnetic field. ITER uses LTS magnets from proven and cost-effective materials. However, due to the limited magnetic field strength, ITER requires a larger reactor size to achieve the conditions necessary for fusion. In contrast, SPARC uses advanced, newly available REBCO-based HTS magnets which produce a much stronger magnetic field, allowing for a smaller, more compact reactor design while still confining the plasma more accurately. HTS magnets, although promising for their ability to produce a stronger

[48] https://www.eni.com/en-IT/media/press-release/2023/03/eni-signs-new-collaboration-agreement-with-cfs-development-fusion-energy.html

[49] https://www.fusionenergybase.com/organization/commonwealth-fusion-systems, last accessed on 19–10-2024.

[50] https://cfs.energy/news-and-media/cfs-signs-15-million-doe-agreement-to-advance-commercial-fusion-energy

[51] https://cfs.energy/news-and-media/commonwealth-fusion-systems-awarded-3.7-million, last accessed on 11–09-2024.

[52] https://cfs.energy/news-and-media/commonwealth-fusion-systems-wins-two-doe-infuse-awards-to-help-commercialize-fusion-energy

magnetic field than LTS, are significantly more expensive and have not been demonstrated in long-term operational environments, thus posing technological uncertainties. For example, unresolved issues, such as magnet quench that could cause system damage, remain a significant challenge. However, CFS is actively researching solutions to these concerns by, for example, optimizing HTS magnets to improve quench resiliency (Hartwig et al. 2024).

4.4.1.2 Resource and Fuel Availability

ITER and SPARC address the possible lack of fuel availability in different ways. ITER relies on an external tritium supply, sourced mainly from CANDU reactors, which should be sufficient for its research needs during operation. Additionally, ITER will test tritium breeding blankets to develop self-sufficiency, thereby addressing a key challenge for future reactors, since ITER will consume most of the globally accessible tritium reserves during its operational lifetime. SPARC, however, avoids tritium supply challenges by requiring only ten grams of tritium over its operational lifetime, which eliminates supply constraints. The follow-up device ARC shall, however, include breeding blankets.

4.4.1.3 Radiation Damage

Since both ITER and SPARC aim to generate fusion reactions using neutronic fuels (D-T), they must address the challenge of radiation damage to reactor components caused by high-energy neutrons. To do so, ITER evaluates plasma-facing materials to mitigate neutron damage under extreme conditions. The data gathered from these evaluations supports the development and advancement of materials for future reactors, such as DEMO (IAEA 2002). SPARC also evaluates plasma-facing materials, particularly carbon and tungsten, focusing on erosion and impurity radiation concerns (Creely et al. 2020). However, since both reactors are not yet operational, predicting the long-term radiation resistance of these materials remains uncertain.

4.4.1.4 Power Generation

ITER and SPARC are both experimental devices designed to prove the feasibility of fusion and advance fusion technologies, rather than focusing on power generation. Consequently, neither device includes a power conversion system. Power generation will be addressed in their respective successor devices, where corresponding systems will be integrated (Meschini et al. 2023). As such, it is not possible to determine whether the challenge of power generation has been addressed efficiently in either project.

4.4.1.5 Material Resistance

Since they are experimental and short-term, neither ITER nor SPARC addresses the issue of material resistance required for future continuous operation fusion reactors. Instead, these projects serve as steppingstones to test materials and gather data, informing the development of advanced materials for future devices like DEMO and ARC. Both reactors are far from reaching the conditions that materials must meet for continuous operation, since their efforts focus on short-term testing and demonstrating fusion feasibility (Creely et al. 2020; Claessens 2023) rather than tackling material durability challenges needed for long-term operation.

4.4.2 Organizational Challenges

4.4.2.1 Skilled Workforce

ITER benefits from its global collaboration with its seven member states, industrial partners and research institutions. In contrast, SPARC relies on the workforce from MIT's PSFC, institutional resources such as MIT Energy Initiative, and partnerships with universities and national labs, while also leveraging industrial expertise from major energy companies.[53]

4.4.2.2 Tailored Regulation

ITER suffers from delays and increased costs, partly due to strict French nuclear safety regulations. These regulations, as previously discussed, are tailored to nuclear fission technologies. They introduce substantial oversight requirements and lengthy approval processes, as fission power requires stricter regulatory control due to its higher associated risks. SPARC, however, benefits from a more favorable regulatory environment in the United States, where fusion is separated from fission under frameworks such as the ADVANCE Act.[54] However, the fusion-specific regulatory framework is still in its early stages of development and must continue to evolve to avoid imposing unnecessary restrictions.

4.4.2.3 Limited Suppliers

ITER relies on a complex, global supply chain for its components, which leads to delays and risks associated with managing multiple suppliers across member

[53] https://news.mit.edu/2018/mit-newly-formed-company-launch-novel-approach-fusion-power-0309, last accessed on 18–08-2024.

[54] https://heatmap.news/climate/advance-act-fusion, last accessed on 14–09-2024.

states, such as supply disruptions or delayed deliveries. Additionally, the reliance on suppliers within an immature sector, where series production for some critical components does not yet exist, combined with the long production timelines required for large-scale components (Meschini et al. 2023), further worsens the potential for delays. SPARC addresses supplier limitations by using in-house manufacturing for critical components like magnets, which reduces its dependency on external suppliers (Brunner 2023). In addition, SPARC benefits from its smaller reactor size, reduced material requirements, and fewer components which makes the reactor less vulnerable to supplier limitations compared to large-scale projects like ITER. SPARC further benefits from the global fusion supply chain established through ITER, which provides access to existing industrial expertise and manufacturing capabilities (IAEA 2024b).

4.4.2.4 Funding

ITER is funded by public money from member states, with contributions mostly in-kind. While this model ensures long-term financial stability through governmental support, it introduces complexities like dependency on global suppliers and exposure to economic or geopolitical instabilities that could disrupt the supply of in-kind contributions. As noted earlier, ITER's funding model is bureaucratic and induces long approval processes due to multiple layers of oversight and regulation which could end in delayed budgetary decisions. This is particularly problematic for large-scale, long-term projects, as prolonged delays can result in component degradation or corrosion, necessitating their replacement, which would further extend the project's timeline. In contrast, SPARC gets private funding from venture capital, energy companies, and government grants (IAEA 2024b), which allows for more flexible decision-making. By using an iterative approach—where technology is built, tested, refined, and improved—SPARC can convince investors and acquire funds in multiple rounds (Creely et al. 2020). Furthermore, as the feasibility of producing sufficient fusion energy is still uncertain, reliance on private venture capital offers a distinct advantage since venture capital investors are more willing to take risks, accept long-term returns on investment, and support the development of innovative materials and technologies.[55]

Overall, the comparison of both projects, ITER and SPARC, presents complementary strategies on how to address the technical and organizational challenges of a MCF device. Each project has its strengths in different areas, but unresolved challenges remain that must be addressed in future applications since both reactors are experimental and, therefore, do not consider all potential hurdles. This evaluation shows how each organization handles these challenges; however, determining which project will succeed technically and organizationally, and thus serve as the most efficient model, remains uncertain since both are yet to be completed.

[55] https://fastercapital.com/content/Risks-of-taking-venture-capital.html, last accessed on 17–10-2024.

4.5 Conclusion

Nuclear fusion is among the most complex of all possible system goods. In this chapter, we provided a general framework for analyzing such systems and applied it to two particularly interesting case studies. We find that ITER and SPARC follow different approaches to overcome similar technological and organizational challenges. While ITER is a large-scale and long-term infrastructure project relying on proven technology, such as LTS, SPARC is based on a more compact and innovative design to be constructed relatively quickly. Additionally, ITER is supported by a global network of countries, each funding and supplying goods, materials, and services. This brings long-term stability, but also induces coordination issues and organizational inertia. In contrast, SPARC is characterized by private funders, agile development, low reliance on external suppliers, and the smaller scale of the project in general.

Despite their differences, both projects face the same challenges, including material durability, radiation damage, and unproven long-term reliability of the reactors. Put together, these projects will likely contribute essential knowledge and technological advancements. This foundation might allow future applications to better address and adapt to technological and organizational challenges.

Beyond energy generation, fusion power offers additional applications, such as isotope production for cancer treatment and hydrogen generation through waste heat utilization (Nuttall 2020). R&D in these fields could foster the economic attractiveness of fusion technology and broaden its appeal to investors and stakeholders.

The inevitable limitations of this work include the reliance on projections and statements made by the respective organizations, as neither SPARC nor ITER are operational as of the time of writing. Furthermore, the general feasibility of a fusion reactor remains uncertain, and whether MCF devices like SPARC and ITER will overcome the technological challenges is questionable. There are many other potential devices that might well overtake the current advantages of tokamak devices regarding operational experience. Consequently, future research should assess other ongoing fusion reactor development projects on their respective approaches to overcome the above-mentioned challenges.

References

Abdou, Mohamed, Marco Riva, Alice Ying, Christian Day, Alberto Loarte, L.R. Baylor, Paul Humrickhouse, Thomas F. Fuerst, and Seungyon Cho. 2021. 'Physics and Technology Considerations for the Deuterium–Tritium Fuel Cycle and Conditions for Tritium Fuel Self Sufficiency'. *Nuclear Fusion* 61 (1): 013001. https://doi.org/10.1088/1741-4326/abbf35

Beckers, Thorsten, Florian Gizzi, and Klaus Jäkel. 2012. 'An Approach to Analyze "System Goods"- Classification, Presentation, and Application'. 2021–02. WIP-Working Paper. Berlin: Workgroup for Infrastructure Policy (WIP), Technische Universität Berlin. https://www.tu.berlin/fileadmin/www/10002415/WIP_Publikationen_PDF/beckers_gizzi_jaekel_2012-system_good_analysis_wip-wp_2012-02-12-1.pdf

Biel, W. 2014. 'Tritium Breeding and Blanket Technology'. Presented at the DPG School 'The Physics of ITER', Bad Honnef, Germany, September 26. https://www.dpg-physik.de/veranstaltungen/2014/biel2.pdf

Bigot, Bernard. 2016. 'Progress in ITER Construction, Manufacturing and R&D'. In *AEA Fusion Energy Conference*. Kyoto, Japan. https://nucleus.iaea.org/sites/fusionportal/Shared%20Documents/FEC%202016/fec2016-preprints/preprint0762.pdf

Brunner, Daniel. 2023. 'Commonwealth Fusion Systems' High-Field Path to Fusion Energy'. Workshop Presentation presented at the Canadian Workshop on Fusion Energy Science and Technology (CWFEST) 2023, October 24. https://www.cns-snc.ca/wp-content/uploads/2024/01/No-07-Daniel-Brunner-Commonwealth-Fusion-Systems-CWFEST-2023.pdf

Chiocchio, S. 2006. 'ITER and International Scientific Collaboration'. In *Proceedings of EPAC 2006*, 3641–45. Edinburgh, United Kingdom. https://accelconf.web.cern.ch/e06/PAPERS/FRYCPA01.PDF

Claessens, Michel. 2023. *ITER: The Giant Fusion Reactor*. 2nd Edition. Vinon-sur-Verdon, France: Springer Nature Switzerland. https://doi.org/10.1007/978-3-031-37762-4

Creely, A. J., M. J. Greenwald, S. B. Ballinger, D. Brunner, J. Canik, J. Doody, T. Fülöp, D. T. Garnier, R. Granetz, and T. K. Gray. 2020. 'Overview of the SPARC Tokamak'. *Journal of Plasma Physics* 86 (5). https://doi.org/10.1017/S0022377820001257

Crownhart, Casey. 2024. 'Inside a Fusion Energy Facility'. *MIT Technology Review*, 31 October 2024. https://www.technologyreview.com/2024/10/31/1106384/inside-a-fusion-energy-facility/

Delasnerie, Alix. 2023. 'ITER Financing by the EU Budget—State-of-Play'. Briefing. European Union: Policy Department for Budgetary Affairs. https://www.europarl.europa.eu/RegData/etudes/BRIE/2023/756469/IPOL_BRI(2023)756469_EN.pdf

Directorate-General for Energy, GRS, and KIT. 2021. 'Study on the Applicability of the Regulatory Framework for Nuclear Facilities to Fusion Facilities—Towards a Specific Regulatory Framework for Fusion Facilities'. Study. Publications Office of the European Union. Luxembourg, Luxembourg: European Union. https://data.europa.eu/doi/https://doi.org/10.2833/787609

Directorate-General for Energy, Matthew Smith, Frank Gérard, Rene Csinszka, Shankar Shashwati, Andrea Finesso, Matthew Moynihan, Simon Woodruff, and Annika Stein. 2023. 'Analysis on a Strategic Public-Private Partnership Approach to Foster Innovation in Fusion Energy'. Study. Rotterdam, Netherlands: European Union. https://data.europa.eu/doi/https://doi.org/10.2833/323326

DOE. 2024. 'Fusion Energy Strategy 2024'. U.S. Department of Energy. https://www.energy.gov/sites/default/files/2024-06/fusion-energy-strategy-2024.pdf

Entler, Slavomir, Jan Horacek, Ondrej Ficker, Karel Kovarik, Michal Kolovratnik, and Vaclav Dostal. 2023. 'Estimation of Fuel Operating Ranges of Fusion Power Plants'. *Nuclear Engineering and Technology* 55 (7): 2687–96. https://doi.org/10.1016/j.net.2023.04.024

FIA. 2024a. 'The Fusion Industry Supply Chain 2024'. Washington, DC, United States: Fusion Industry Association. https://www.fusionindustryassociation.org/wp-content/uploads/2024/07/FIA-Supply-Chain-2024-Report.pdf

FIA. 2024b. 'The Global Fusion Industry in 2024'. Report. Washington, DC, United States: Fusion Industry Association.

Giegerich, T., K. Battes, J. C. Schwenzer, and C. Day. 2019. 'Development of a Viable Route for Lithium-6 Supply of DEMO and Future Fusion Power Plants'. *Fusion Engineering and Design* 149 (111339). https://doi.org/10.1016/j.fusengdes.2019.111339

Gordinier, M. R., J. W. Davis, F. R. Scott, and K. R. Schultz. 2003. 'Encyclopedia of Physical Science and Technology'. In *Nuclear Fusion Power*, edited by Robert A. Meyers, 3rd ed., 671–99. New York, United States: Academic Press. https://www.sciencedirect.com/science/article/pii/B0122274105004919

Grunblatt, G., P. Mocaer, Ch. Verwaerde, and C. Kohler. 2005. 'A Success Story: LHC Cable Production at ALSTOM-MSA'. *Fusion Engineering and Design* 75–79 (November):1–5. https://doi.org/10.1016/j.fusengdes.2005.06.216

Grünwald, Reinhard. 2024. 'Auf dem Weg zu einem möglichen Kernfusionskraftwerk—Wissenslücken und Forschungsbedarfe aus Sicht der Technikfolgenabschätzung'. Nr. 1. TA-Kompakt. Berlin: Büro für Technikfolgen-Abschätzung beim Deutschen Bundestag (TAB). https://doi.org/10.5445/IR/1000177720

Haack, Julia. 2024. 'Superconductivity for Nuclear Fusion: Past, Present, and Future'. *Arabian Journal for Science and Engineering*, February. https://doi.org/10.1007/s13369-024-08720-4

Hartwig, Zachary S., Rui F. Vieira, Darby Dune, Theodore Golfinopoulos, Brian LaBombard, and Christopher J. Lammi. 2024. 'The SPARC Toroidal Field Model Coil Program'. *IEEE Transactions on Applied Superconductivity* 34 (2). https://doi.org/10.1109/TASC.2023.3332613

Huang, Chuanjun, and Li Laifeng. 2018. 'Magnetic Confinement Fusion: A Brief Review'. *Frontiers in Energy* 12 (2): 305–13. https://doi.org/10.1007/s11708-018-0539-1

IAEA. 2002. 'ITER Technical Basis'. 24. ITER EDA Documentation Series. Vienna, Austria: International Atomic Energy Agency. https://www-pub.iaea.org/MTCD/Publications/PDF/ITER-EDA-DS-24.pdf

IAEA. 2022. 'World Survey of Fusion Devices 2022'. World Survey of Fusion Devices. Vienna, Austria: International Atomic Energy Agency. https://www-pub.iaea.org/MTCD/Publications/PDF/CRCP-FUS-001webRev.pdf

IAEA. 2023. 'World Fusion Outlook 2023'. World Fusion Outlook. Vienna, Austria: International Atomic Energy Agency. https://www.iaea.org/publications/15524/iaea-world-fusion-outlook-2023

IAEA. 2024a. 'Fusion Key Elements'. Vienna, Austria: International Atomic Energy Agency. https://www-pub.iaea.org/MTCD/Publications/PDF/p15764-P2099E_web.pdf

IAEA. 2024b. 'IAEA World Fusion Outlook 2024'. World Fusion Outlook. Vienna, Austria: International Atomic Energy Agency. https://www-pub.iaea.org/MTCD/Publications/PDF/p15777-24-02766E_WFO_web.pdf

ITER Organization. 2023. 'Financial Report 2022'. Financial Report. St. Paul-lez-Durance, France: ITER Organization. https://e.issuu.com/embed.html?d=iter_organization_2022-financial-report&u=iterorganization

Kramer, David. 2023. 'ITER Appears Unstoppable despite Recent Setbacks'. *Physics Today* 76 (8): 18–22. https://doi.org/10.1063/PT.3.5287

Lawson, J D. 1957. 'Some Criteria for a Power Producing Thermonuclear Reactor'. *Proceedings of the Physical Society. Section B* 70 (1): 6–10. https://doi.org/10.1088/0370-1301/70/1/303

Lobner, Peter. 2021. 'The Fork in the Road to Electric Power From Fusion'. New South Wales, Australia: HB11 Energy. https://lynceans.org/wp-content/uploads/2021/02/HB11-Energy_Australia-converted.pdf

McKenzie, W., D. Batani, and T.A. Mehlhorn. 2023. 'HB11—Understanding Hydrogen-Boron Fusion as a New Clean Energy Source'. *Journal of Fusion Energy* 42 (17). https://doi.org/10.1007/s10894-023-00349-9

Meschini, Samuele, Francesco Laviano, Federico Ledda, Davide Pettinari, Raffella Testoni, Daniele Torsello, and Bruno Panella. 2023. 'Review of Commercial Nuclear Fusion Projects'. *Frontiers in Energy Research* 11 (June). https://doi.org/10.3389/fenrg.2023.1157394

Moynihan, Matthew, and Alfred B. Bortz. 2023. *Fusion's Promise—How Technological Breakthroughs in Nuclear Fusion Can Conquer Climate Change on Earth (And Carry Humans To Mars, Too)*. Switzerland: Springer.

Nuttall, William J. 2020. 'Commercial Opportunities for Nuclear Fusion'. In *Commercialising Fusion Energy—How Small Businesses Are Transforming Big Science*, 1–21. Bristol, UK: IOP Publishing.

Ou, Y. X., H. Q. Wang, X. Ouyang, Y. Y. Zhao, Q. Zhou, C. W. Luo, Q. S. Hua, X. P. Ouyang, and S. Zhang. 2023. 'Recent Advances and Strategies for High-Performance Coatings'. *Progress in Materials Science* 136 (July):101125. https://doi.org/10.1016/j.pmatsci.2023.101125

Pearson, Richard J., and Shutaro Takeda. 2020. 'Review of Approaches to Fusion Energy'. In *Commercialising Fusion Energy: How Small Businesses Are Transforming Big Science*, edited by William J. Nuttall, 1–3. Bristol, UK: IOP Publishing. https://iopscience.iop.org/book/edit/978-0-7503-2719-0

Pearson, Richard J., Armando B. Antoniazzi, and William J. Nuttall. 2018. 'Tritium Supply and Use: A Key Issue for the Development of Nuclear Fusion Energy'. *Fusion Engineering and Design* 136 (November):1140–48. https://doi.org/10.1016/j.fusengdes.2018.04.090

Sorbom, B. N., J. Ball, T. R. Palmer, F. J. Mangiarotti, J. M. Sierchio, P. Bonoli, C. Kasten, et al. 2015. 'ARC: A Compact, High-Field, Fusion Nuclear Science Facility and Demonstration Power Plant with Demountable Magnets'. *Fusion Engineering and Design* 100 (November):378–405. https://doi.org/10.1016/j.fusengdes.2015.07.008

U.S. GAO. 2023. 'Fusion Energy | Potentially Transformative Technology Still Faces Fundamental Challenges'. Technology Assessment. Washington, DC, United States: United States Government Accountability Office. https://www.gao.gov/assets/gao-23-105813.pdf

U.S. NRC. 2021. 'Developing a Regulatory Framework for Fusion Energy Systems'. https://www.nrc.gov/docs/ML2129/ML21299A313.pdf

Whyte, D. G., J. Minervini, B. LaBombard, E. Marmar, L. Bromberg, and M. Greenwald. 2016. 'Smaller & Sooner: Exploiting High Magnetic Fields from New Superconductors for a More Attractive Fusion Energy Development Path'. *Journal of Fusion Energy* 35 (January):41–53. https://doi.org/10.1007/s10894-015-0050-1

WNA. 2008. 'Structuring Nuclear Projects for Success'. London, UK: World Nuclear Association. https://world-nuclear.org/uploadedFiles/org/WNA/Publications/Working_Group_Reports/Structuring%20Projects%20Report.pdf

Wu, M. K., J. R. Ashburn, C. J. Torng, P. H. Hor, R. L. Meng, L. Gao, Z. J. Huang, Y. Q. Wang, and C. W. Chu. 1987. 'Superconductivity at 93 K in a New Mixed-Phase Y-Ba-Cu-O Compound System at Ambient Pressure'. *Physical Review Letters* 58 (9): 908–10. https://doi.org/10.1103/PhysRevLett.58.908

Wurzel, Samuel E., and Scott C. Hsu. 2022. 'Progress toward Fusion Energy Breakeven and Gain as Measured against the Lawson Criterion'. *Physics of Plasmas* 29 (6): 062103. https://doi.org/10.1063/5.0083990

Yin, Robert K. 2014. *Case Study Research: Design and Methods*. 5. edition. Los Angeles London New Delhi Singapore Washington, DC: SAGE.

Zylstra, A.B., O.A. Hurricane, and D.A. Callahan. 2022. 'Burning Plasma Achieved in Inertial Fusion'. *Nature* 601 (7894): 542–48. https://doi.org/10.1038/s41586-021-04281-w

Chapter 5
Nuclear Power and Geopolitical Conflicts—An Overview and a Classification of Nuclear Countries World-Wide

Christian von Hirschhausen and Claudia Kemfert

5.1 Introduction

From the 1940s and until today, there is a consensus in the literature: The use of nuclear energy is based on a combination of basic scientific research and the militarily motivated search for applications in geopolitical conflicts. The opinion commonly held today that nuclear energy is "what happens in the power plants" has nothing to do with its actual origins and only concerns a small, "commercial" area of electricity generation, where the commercial use for energy production has been marginalized—now at about 3% of global primary energy production.

In this chapter, we provide a summary of the literature on the role of nuclear power in geopolitical conflicts. In addition, we report conceptual attempts to classify nuclear countries according to their geopolitical status. We thus generate a "companion" chapter to the more in-depth analysis in this book" (see Chaps. 6 and 7). The chapter first reviews the dual use character of nuclear power coming out of World War II until today (Sect. 5.2) and then characterizes this period as the "first" and the "second" nuclear age, summarizing the approach by Bracken (2012).The characterization of the about 40 nuclear countries is based on geopolitical assessment, in some cases appended by (more or less pronounced) ambitions for commercial nuclear power; this section reports the political science approach spearheaded by

C. von Hirschhausen (✉)
Workgroup for Infrastructure Policy (WIP), Technical University (TU) Berlin, Berlin, Germany
e-mail: cvh@wip.tu-berlin.de

C. von Hirschhausen · C. Kemfert
German Institute for Economic Research (DIW Berlin), Berlin, Germany

C. Kemfert
Professorship for Energy Economics and Energy Policy, Leuphana University Lüneburg, Lüneburg, Germany

A. Wimmers et al. (eds.), *Nuclear Power*, Studies in Energy, Resource and Environmental Economics, https://doi.org/10.1007/978-3-031-99894-2_5

Stirling and Johnstone (2018). Section 5.3 also describes the current state in the three nuclear superpowers: the US, Russia, and China. Last, in Sect. 5.4 we present ongoing discussions in potential newcomer countries ("newbies") and report on a correlation between nuclear ambition and the level of autocracy in the sample, as suggested by Neumann et al. (2020). Section 5.5 concludes.[1]

5.2 Nuclear Power: Science and Warfare

5.2.1 Science ...

François Lévêque (Lévêque 2013, 170) from the École des Mines de Paris coined the characterization of "atomic energy as a child of science and warfare" in his overview work. He thus describes the development in the first half of the twentieth century, in which atomic energy was 1) scientifically discovered, and 2) developed on a large industrial scale for warfare. Lévêque was by no means the first to come up with this image. On the contrary, analogous interpretations of atomic energy have been handed down since the 1940s: Robert Oppenheimer, head of the Manhattan Project and scientific "father" of the atomic bomb, described his work as "now I have become death, the destroyer of worlds." David Lilienthal, chairman of the US Atomic Energy Commission (AEC) after World War II, wrote in a joint work with co-chair Dean Acheson (the Undersecretary of Defense), Robert Oppenheimer, and others, that the search for commercial applications was only the flip side of military applications (Acheson-Lilienthal Report 1946, 10). The physicist and innovator Amory Lovins, who was awarded the Order of Merit of the Federal Republic of Germany in 2016 as the "inventor" of the energy transition, and his wife Hunter Lovins addressed the ambiguity of nuclear power in detail in their book "Nuclear power and nuclear bombs" (Lovins and Lovins 1980). In the current literature, the topic is explored by Cox et al. (2016) and Stirling and Johnstone (2018).[2]

In 1933, Leo Szilard published the first theoretical work on nuclear fission, closely followed by Enrico Fermi, who published his theory of beta decay ("Fermi interaction") in 1934. Nuclear fission was first carried out experimentally on December 17, 1938 by Otto Hahn and Fritz Strassmann in Berlin-Dahlem (Thielallee 63) by firing neutrons at uranium nuclei and then detecting the fragments, including barium and krypton. The physical interpretation of this was provided by Lise Meitner, Hahn's

[1] This chapter is inspired by the geopolitical literature cited which it summarizes. The chapter refers to sections from the German book on nuclear power (von Hirschhausen 2023, Chaps. 1 and 3). We thank the colleagues in the AT-OM research team for constructive discussions, in particular Alexander Wimmers and Fanny Böse for stringent project management, and William Nuttall, Maren Vielhuf, Björn Steigerwald, and Ben Wealer for earlier discussions on this topic. The usual disclaimer applies.

[2] An analogous line of argument can also be found in the German-language literature, including that of the economist Fritz Baade (1958, 120), the nuclear engineer Friedrich Münzinger (1960, 3), the political scientist Lutz Mez (1979) and technology historian Joachim Radkau (1983, 136).

long-term colleague, who had to emigrate due to her Jewish origins, and her nephew Otto Frisch in Stockholm in January 1939. The theory of maintaining uranium fission as a chain reaction was also described by Joliot-Curie in Paris in 1939 and then later, in 1942, experimentally proven for the first time in Chicago by Leo Szilard and Enrico Fermi with a larger team (Nelson 2014, Chap. 2).

5.2.2 … and Warfare

The potential for military use of nuclear fission was foreseeable to many scientists as early as the 1920s. However, the link between basic science and warfare only emerged in the context of World War II. In the summer of 1939, Hungarian physicists in exile—Leo Szilard, Eugene Wigner, and Edward Teller—urged Albert Einstein, who was also in exile on Long Island, New York, to write a letter to United States (US) President Franklin D. Roosevelt. Einstein did so on August 2, 1939. The physicists warned the political and military leadership of the danger of falling behind Germany's war efforts, soon before the start of World War II on September 1, 1939. They advocated for a rapid acceleration of a nuclear energy research and development program. However, it was not until Japan's attack on Pearl Harbor (December 7, 1941) and the US entry into World War II that the US military committed itself to the development of an atomic bomb. The Manhattan Project, initially based in Manhattan, New York before moving to the New Mexico desert, brought together numerous research institutions and scientists, primarily from the US, Canada, and the United Kingdom. It also integrated a significant number of highly qualified scientists from enemy nations (Germany, Denmark, Austria, Hungary, Italy), such as Hans Bethe, Nils Bohr, Enrico Fermi, Otto Frisch, Leo Szilard, Edward Teller, and Eugene Wigner (Groves 1983; Rhodes 2012; Reed 2013).

From a history of technology perspective, the speed of development between the official launch of the Manhattan Project in 1942 and the creation of the atomic bomb in just three years is both remarkable and terrifying. On December 2, 1942, Fermi, Szilard, and their team in Chicago achieved the first continuous chain reaction of uranium-235 in the so-called Chicago Pile No. 1 (Nelson 2014). After this breakthrough, the challenge that remained for scientists was to scale up this chain reaction to larger energy levels and compress it into bombs. This was achieved through the close collaboration of science, represented by the Manhattan Project's scientific director, Professor Robert Oppenheimer, and the military, represented by General Leslie Groves, the overall project leader. More than 130,000 people worked at over 20 sites on various aspects of bomb development. At Oak Ridge, Tennessee, the focus was on extracting fissile uranium-235, achieving concentrations of over 80% for the gun-type uranium bomb. In Hanford, Washington, plutonium separation was underway for the implosion bomb. In Los Alamos, New Mexico, the various strands came together, where details like ignition and explosives were fine-tuned. On July 16, 1945, the first test of a plutonium implosion bomb took place in Alamogordo, New Mexico. On August 6, the US dropped a gun-type uranium bomb ("Little Boy," which had not been tested beforehand) on Hiroshima, followed by the dropping of a plutonium implosion bomb ("Fat Man") on Nagasaki on August 9, 1945. It is estimated that over 100,000 people died instantly, with over 120,000 people suffering severe injuries, and many more succumbing to radiation effects (Groves 1983, 319, 346). The extent to which these bombings contributed to Japan's eventual surrender remains debated among historians and political scientists.

In other countries, a close tie between basic research and the military also occurred during this period. For a time, the United Kingdom even appeared to be "ahead" of the US, as it boasted a high concentration of physicists and chemists and was at the center of nuclear physics during the 1920s and 1930s. The first bomb design was developed in Birmingham (UK) in 1940 by Rudolph Peirels and Otto Frisch. However, the implementation of larger projects, such as uranium concentration and plutonium production, could not be achieved on the necessary scale. As a result, leading UK physicists like Otto Frisch, James Chadwick, and Rudolph Peirels joined the Manhattan Project in the US. In France, the Joliot-Curies, who deliberately avoided cooperating with the US and UK, continued working on cyclotrons and bomb designs in collaboration with the military. The Soviet Union also worked feverishly on the atomic bomb during the early 1940s, benefitting from the espionage activities of a group led by German-born Klaus Fuchs, who worked for the UK in Los Alamos. This group successfully communicated a variety of bomb plans and designs to Moscow.

The attempts to develop nuclear weapons in Germany remain controversial. In the 1920s and 1930s, cities like Göttingen, Berlin, Leipzig, and Munich were among the global leaders in nuclear physics and chemistry research. Despite the exile of leading researchers, Germany continued its intense efforts in research and development until 1945, which could have potentially led to the construction of a nuclear bomb. The "Uranverein" (Uranium Club), founded in 1939, was tasked with investigating the physical and scientific resource base for such a project, analogous to the US effort. Despite conceptual progress, no experimental breakthroughs were made. The German "pile" (a graphite-based experimental setup for a self-sustaining nuclear reaction) had to be relocated to Haigerloch in Swabia due to the impending Soviet occupation of Berlin. When US troops arrived in April 1945, they found a very rudimentary experimental setup, which surprised the international community, as German scientists were still far from developing a militarily useful application (Goudsmith 1947; Groves 1983; Chap. 17). The failure of the German nuclear program has been attributed to a lack of knowledge, understanding, and interest from the Nazi regime in nuclear technology, a fragmented scientific community, insufficient infrastructure for industrial implementation, and the deteriorating war conditions (Popp 2021).

The close connection between science and warfare persisted in the postwar period, leading to the enhancement of nuclear weapons capabilities during the Cold War. At the center of this was the development of even more powerful weapons, based on the combination of nuclear fission and fusion: The so-called hydrogen bomb ("thermonuclear weapon") uses the implosion of atomic nuclei (fission) as a trigger for the second stage, the fusion of hydrogen. With thermonuclear weapons, it is possible to achieve explosive yields thousands of times greater than the first fission bombs, reaching millions of tons of TNT equivalent. In the US, a group of physicists and chemists led by Edward Teller and Stanislaus Ulam, emerging from the Manhattan Project, worked on this development, which was first tested in 1954. The Soviet Union quickly caught up, with a team including, Andrei Sakharov, testing their own hydrogen bomb by 1955. This spurred further US efforts. Other WW II victors, such as the UK, France, and later China, also pursued nuclear weapons development. These nations formed the five permanent members of the United Nations Security Council, a configuration that remains to this day.

Initially, the development of nuclear energy was exclusively focused on weapons, but the postwar period saw the emergence of various so-called "civilian" applications, primarily electricity generation and medical uses, such as CT scans, X-rays, and radiation therapy. From these dual-use developments, different strategies evolved that combined military and civilian applications. In economics, this is referred to as "economies of scope" (Hirschhausen 2017, 9–13). The nuclear-armed states emphasized the development of commercial uses and, in parallel to weapons development, built extensive nuclear power plant portfolios. Other nations, however, used ostensibly civilian nuclear energy programs to mask their military ambitions, such as Pakistan and later North Korea.

5.3 The First and the Second Nuclear Age and Regional Conflicts

5.3.1 From the First Nuclear Age …

Without geopolitical conflicts, there would be no nuclear industry. The development of the late 1930s and 1940s and the beginning of the first nuclear age can be summarized as simply as that. And it is the central key to understanding national developments in the sector, and thus decisively shapes the design of national nuclear innovation systems. After the end of the Second World War and the race between the US and the German Reich for the atomic bomb, but also implicitly with the Soviet Union, the post-war period was dominated by the geopolitical conflict between the communist Soviet Union and the capitalist US. During the Cold War, most countries in the world aligned themselves with one of the two world powers, except for China and the Non-Aligned Movement. The development of nuclear weapons was an important element of the Cold War, which was waged on all levels, including political-ideological, economic, technical-scientific, cultural-social, but—due to their enormous destructive potential—not with nuclear weapons. In many regional conflicts, however, the Cold War also became a "hot" war, for example in Korea, Vietnam, and Angola.

Military applications in the 1940s and 1950s contributed significantly to the implementation of nuclear reactor technology. Initially by the navy in the form of nuclear submarines, the first of which was deployed by the US Nuclear Navy in 1954 (the "Nautilus") and by the Soviet Union in 1957 (the "Lenin"). To this day, nuclear submarines are part of the DNA of all nuclear weapons states. The spark quickly spread to the air force and army, among other things, through the development of molten salt reactors to power nuclear aircraft and nuclear-powered vehicles on land. The Soviet Union also quickly caught up with the US in the development of nuclear weapons and was soon on par with the US not only in space research but also in nuclear technology. Following the development of the first fission atomic bomb in 1949, which was based on Soviet espionage, the Soviet Union succeeded in advancing its own considerable developments through the intensive combination of science and warfare: It detonated its first hydrogen bomb in 1953, shortly after the US (1952). The nuclear race between the two systems also accelerated in light of the first Berlin Crisis (1948/49) and the Korean War. At the same time, both sides began to lay the foundations for the proliferation of their nuclear technology to allied countries and to control the spread of weapons-grade fissile material (Eisenhower's atoms-for-peace speech in 1954, Khrushchev's speech on peaceful coexistence in 1956).

From today's perspective, it remains unclear why there were no further uses of atomic bombs after 1945 despite the many possibilities and the explicit desire, also from parts of the scientific community, to use nuclear weapons on the respective opponent. There were plenty of reasons: as early as the late 1940s, the mathematician and game theorist John von Neumann suggested dropping the atomic bomb on the Soviet Union in order to eliminate the strategic opponent at an early stage and decide the upcoming conflict in favor of the US (Poundstone 1992). The same was also under discussion at the time of the Berlin Blockade by the Soviet Union in the fall of 1948 (Bracken 2012). In the mid-1950s, atomic bombing was discussed in the context of the Korean War. On the occasion of the secret transfer of medium-range nuclear weapons from the Soviet Union to Cuba in October 1962, US President John F. Kennedy was on the verge of military intervention, possibly including the use of nuclear weapons (Bracken 2012).

To this day, the dominant explanation lies in the game-theoretical figure of "mutual deterrence." According to this, nuclear weapons are not used because even the leading nation cannot make a profit from the first strike, as it will be hit just as hard by the opponent's nuclear weapons in the second stage, the unstoppable response. In other words, the "successful" attacker dies a few minutes later than the "loser." Game theory was developed in the nineteenth century by mathematicians to explain non-competitive market structures (so-called oligopoly markets) (Cournot, Bertrand). Its further development, which was also supported by German-speaking scientists in the 1920s (von Neumann, von Morgenstern, von Stackelberg), led to further cooperation between science and warfare. In the late 1940s, the mathematician John Nash developed non-cooperative game theory as an employee of the Rand Cooperation, a research organization for the United States Armed Forces. Furthermore, mathematician Robert Auman and economist Thomas Schelling developed their theories and applications of war games to explain mutual deterrence and the resulting balance of power. Schelling in particular has also been active as a strategy consultant since the 1950s, including in "war games," in which strategic behavior in anticipated war conflicts is simulated in order to gain insight into options for action (Bracken 2012, 84 ff.). Nash as well as Auman and Schelling received the Alfred Nobel Memorial Prize in Economic Sciences sponsored by the Swedish Riksbank, in 1994 and 2005, respectively.

5.3.2 … to the Second Nuclear Age …

In the fall of 1989, the "Iron Curtain" between Eastern and Western Europe, and the Wall between the GDR and West Germany and between East and West Berlin, came down. In autumn 1991, the Soviet Union dissolved after an attempted military coup failed in response to the decentralization ("perestroika") initiated by state president and party secretary Mikhail Gorbachev. In the euphoria of this period of change, the "Cold War" was also seen as over, and the West as its winner. In the exuberance of the time, the phrase "the end of history" made the rounds (Fukuyama 1992). Even back then, this a-historical view was criticized, as it placed a heavy burden on the transformation process of the post-socialist states.

With regard to the nuclear age, the idea of the end of the Cold War is not only wrong, but also neglects important developments in the period that have been underestimated due to the focus on the bilateral Cold War, with far-reaching consequences. In fact, there was no break between the two nuclear ages as a result of the fall of communism; rather, developments had already taken place in the shadow of the Cold War, which long before 1989/91 had already led to a multipolar world and the regionalization of geopolitical conflicts, the characteristics of the second nuclear age. This suggests a seamless transition to the second nuclear age.

Bracken (2012, Chap. 4) explains in detail the distinction between two perspectives on the atomic ages and explains why the interpretation of a smooth transition better reflects the developments. Figure 5.1 shows the different perspectives on the process:

- In the upper part of the figure, the two ages are separated and there is a period, e.g., the 1990s, without a nuclear threat thanks to the end of the Cold War.
- The lower part of the figure, on the other hand, shows a gradual but overlapping transition from the first, bilateral nuclear age towards the multilateral second nuclear age that prevails today.

There are several "candidates" for the period of transition between the first (bilateral) and the second (multilateral) nuclear age:

- The latter may already begin in 1964 with the entry of China. China had already joined the circle of nuclear powers in 1964 through a nuclear test and even possessed a hydrogen bomb from 1967. In the 1980s, China, for a long time a latecomer in the nuclear business, then set out to become a global nuclear power.
- A second possibility is the year 1974, when India, in permanent conflict with Pakistan, carried out its first nuclear weapons test ("smiling Buddha"). India and

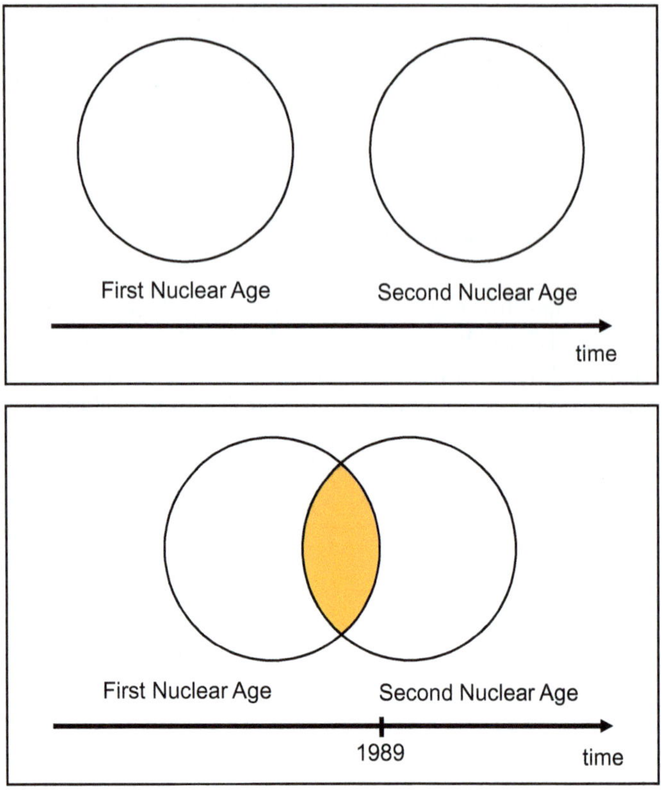

Fig. 5.1 Different concepts of the "Second Nuclear Age." *Source* (Bracken 2012, 96/97)

Pakistan are typical players in the second nuclear age, and both were already active in the nuclear business in the 1970s.

- At the very latest, the second nuclear age began in the 1990s, i.e., still in the twentieth century, which ended with further nuclear weapons tests by India and Pakistan. The engineer Abdul Quadeer Khan from Pakistan, a student of metallurgy at the Technical University of Berlin in the early 1980s, developed the nuclear bomb for Pakistan in the 1990s with help from the West and tried to sell it to Libya, Iran and North Korea through intermediaries, for which he was celebrated as a national hero in Pakistan (Corera 2009).

Bracken (2012, 122) even goes far beyond the horizon of the Cold War in his characterization: while the first nuclear age is still a Europe-centric one (including the US as a natural extension of Europe), the second nuclear age is a post-colonial one, in which not (only) fifty years of Cold War have ended for the majority of the affected countries and the world's population, but also a 500-year period of Western domination, which he calls the "post-Vasco da Gama era."

The second nuclear age is a multi-player game with several decision-making centers and regions, such as South Asia, East Asia, and the Near and Middle East, which exist and act largely independently of each other. Instead of two dominant countries (US, Soviet Union), there is now a third nuclear superpower, China. There are also several other "second order" nuclear powers such as the traditional UN security powers Great Britain and France, the unofficial nuclear powers India and Pakistan, the special cases of South Africa (nuclear weapons program ended in 1989) and Israel (nuclear weapons without a civilian nuclear power program), and the newcomer North Korea. Iran is allegedly working on a nuclear weapons program. Due to easier access to nuclear technology and the fact that the proliferation of fissile materials (highly enriched uranium or plutonium) cannot be fully controlled, the construction of nuclear weapons is currently also possible for other state or non-state actors, e.g., terrorist organizations. Instead of reaping a peace dividend, we are now deeper into the age of nuclear weapons than ever before.

5.4 Characterization of Nuclear Countries

5.4.1 Nuclear Countries, Geopolitics, and Some Commercial Energy

In this section, we have arrived at the present day and provide a characterization of different groups of nuclear states that differ in their military and/or commercial activities. The issue of economies of scope from military and commercial co-production (dual-purpose use) has only recently been discussed publicly again, although it has shaped the development of the nuclear industry since 1945. In a very detailed analysis, Stirling and Johnstone (2018) examine the interdependence between military nuclear activities, specifically the commissioning of new nuclear submarines, and commercial activities, specifically the announcement of an extensive new-build program for nuclear power plants, for the United Kingdom. In this context, they develop a detailed categorization of 45 (out of a total of 195) countries that currently use nuclear energy militarily and/or commercially (Fig. 5.2). Categories of countries are developed, differentiating between commercial and military aspects; of course, each country can appear in different categories.

The following categories are formed with regard to geopolitical-military use:

- Countries with a permanent seat on the United Nations Security Council ("World Security Council")
- Countries that are officially or unofficially nuclear weapon states
- Countries with national capacities to build nuclear submarines
- Countries with the status of "regional power"

In addition, the following categories are formed with regard to commercial use:

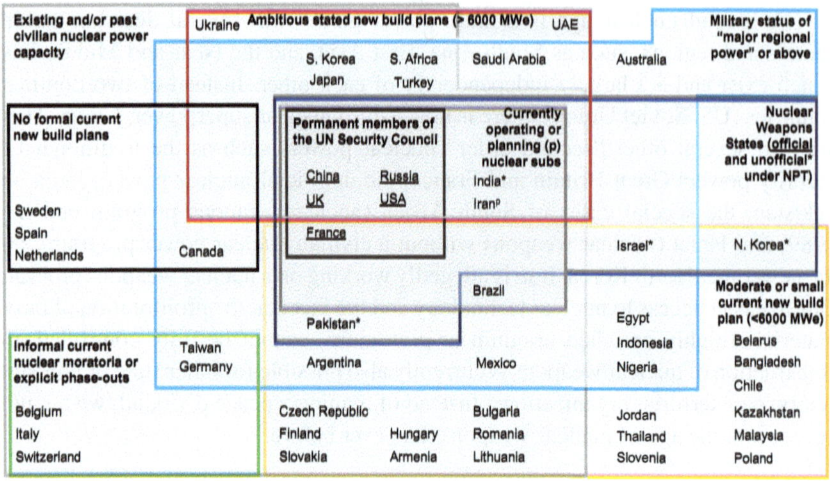

Fig. 5.2 Categorization of nuclear states with military and/or commercial use. *Source* Stirling and Johnstone (2018) with some updates

- Countries with conventional use of nuclear energy
- Countries with extensive expansion plans for commercial nuclear energy (more than 6 GW)
- Countries with minor expansion plans for commercial nuclear energy (less than 6 GW)
- Countries with commercial use but no expansion plans
- Countries with expansion moratoria or planned termination of commercial use

The depiction provides insights into connections and correlations, from which causalities can also be derived, taking historical context into account. The focus here is on the obvious correlation between military and commercial activities. For example, 21 of the 23 countries designated as regional powers have commercial nuclear energy (the exceptions are Australia and Egypt). These 21 countries in turn include all (ten) countries that have developed military nuclear technology to date (with one exception: North Korea). Economic strength alone cannot be used as an explanation for nuclear commitment, as economically strong countries are not always the ones with the most extensive military and/or commercial expansion plans. The clearest correlation is between the highest geopolitical rank, the status as a permanent member of the UN Security Council, and the official status as a nuclear weapons state according to the Non-Proliferation Treaty (NPT). This correlation is perfect, involving the US, the United Kingdom, France, Russia, and China. Furthermore, the correlation to national nuclear submarine capabilities is also strong. Only India and Brazil have these and are not permanently represented in the Security Council; however, both are strong regional powers and have ambitions for a permanent seat, with India as an unofficial nuclear weapons state. Obviously, there is a close nexus between geopolitical status and the pursuit of military nuclear technology.

A look at commercial expansion programs also provides interesting insights. Only eight (out of a total of 195) countries worldwide are currently pursuing expansion programs of over 6 GW: South Korea, Saudi Arabia, South Africa, China, India, the United Kingdom, and the US. At the time of the study, only France was not one of the UN Security Council states. This has since changed with the announcement of a major expansion program in 2021 by President Macron, and France now joining this trend. President Macron's reasoning was clear: "No military nuclear power without civilian nuclear power, no civilian nuclear power without military nuclear power."[3]

Brazil is the only country with nuclear submarine capabilities that is not pursuing a commercial nuclear program. In addition, Brazil's insistence on a permanent seat on the UN Security Council speaks for particularly pronounced geopolitical ambitions that correspond to its status as a nuclear submarine power. Furthermore, only three of the eight countries with major commercial development goals are not currently pursuing military technology plans: South Korea, Saudi Arabia, and South Africa; however, these are also important regional powers that can be assumed to have military ambitions.

5.4.2 The Three Nuclear Superpowers: US, Russia, and China

Of the official nuclear powers mentioned above, three have the status of superpowers, as they currently dominate global geopolitics and will continue to do so for the foreseeable future: the US, Russia, and China. The hypothesis of economies of scope outlined above suggests that the three nuclear superpowers will continue to exploit synergies between the military and commercial sectors, albeit to different degrees.

5.4.2.1 United States

The US have traditionally had by far the greatest separation between the military and commercial sectors. Due to the development of an extensive plutonium complex in Hanford, Washington, and the availability of large quantities of highly enriched uranium (Oak Ridge, Tennessee) in the Manhattan Project, the US was not dependent on directing the development of commercial nuclear reactors toward the co-production of weapons-grade fissile material in the post-war period. Nevertheless, this "advantage" did not lead to greater economic success than in other countries. Up until today, no competitive, commercial nuclear power sector has emerged in the US (Davis 2012). Efforts to revive commercial power plant construction, which has been in decline since the 1970s, through the Energy Policy Act (EP-ACT) of 2005 have failed: Of the more than 20 applications for subsidies from the program, only two reactors emerged, that were completed in 2023 at the Vogtle site in the state of

[3] https://www.elysee.fr/front/pdf/elysee-module-16825-fr.pd

Georgia, a decade late and many times more expensive than planned. Secondly, the total collapse of the knowledge base is to be prevented by promoting the development of reactor concepts, both low-power light water reactors ("SMR" reactor concepts) and non-light water reactors. This is always justified by the need to keep the nuclear industry alive in the country or to initiate its "renaissance" (Chu 2010; Moniz 2011; Energy Futures Initiative 2017). At the same time, considerable sums are being invested into the modernization of the nuclear arsenal, including a new generation of nuclear submarines, air- and land-based nuclear warheads, and nuclear-capable aircraft (DoD 2018; U.S. Department of Defense (DoD) 2022). The planned expenditure for this over the next three decades will exceed US$1,000 billion (SIPRI 2016; 2022). In view of these sums, the subsidies for commercial nuclear power plants appear negligible.

5.4.2.2 Russia

In Russia (and previously in the Soviet Union), there were traditionally very close ties between the military sector and commercial power generation, which began in the early 1950s. Central to this was the development of a dual-use, dual-purpose reactor type, the RBMK reactor (Reaktor Bolschoi Moschtnosti Kanalny) (made famous worldwide by the Chernobyl accident), a boiling water reactor with individual pressure tubes and graphite moderation. This enables the simultaneous production of electricity and extraction of plutonium, which was also practiced at the beginning of development. Later, the plutonium required for nuclear weapons was obtained from special research facilities (Basu and Miroshnik 2019, Chap. 11). After the collapse of the Soviet Union, Russia took over most of the production and R&D capacities and also stabilized the exchange with the other former Soviet republics and satellite states, such as Hungary and Mongolia. Despite Russia's economic crisis in the transition period after 1991, the nuclear sector received considerable funding for the development of a new type of reactor, the VVER-TOI (1,200 MWe), as well as for research into non-light-water-cooled reactors (with a focus on fast reactors). At the same time, the nuclear weapons stockpile is being expanded with considerable resources (SIPRI 2022).

5.4.2.3 China

With the move from the first to the second nuclear age, China has not only succeeded in catching up with the superpowers of the Cold War but is also on its way to becoming an international technology leader thanks to its particularly ambitious approach. In 2024, 57 commercial reactors were connected to the grid (in comparison: US: 94, France: 56) and 27 were under construction; however, nuclear power only accounted for 5% of the electricity supply at 418 TWh. Over the decades, China has imported various reactor types, including the AP 1000 from the US, the VVER from Russia and the EPR from France, and has developed its own reactor type based on these, the

Hualong 1000 (light water reactor with 1,150 MWe). At the same time, China is also broadly positioned in the development of non-light-water-cooled systems and claims to be on the way to commercial plants for fast reactors (CEFR, China Experimental Fast Reactor) as well as for high-temperature reactors (HTR) and molten salt. At the same time, China is rapidly upgrading its nuclear arsenal (SIPRI 2022).

5.5 New Potential Entrant Countries ("Newbies") and Autocracy

The chapter ends with an analysis of countries that are currently preparing to go nuclear. For example, the narrative of the "renaissance" of nuclear power suggests that there are a large number of countries that are currently entering the nuclear energy market. The basic pattern of nuclear energy, "science and warfare" as well as military and commercial uses, is also on the horizon in these newcomer countries, and seemingly innocent imports of commercial nuclear technology can prepare the ground for military activities. In fact, geopolitical aspects play a significant role in the motivational situation—some countries are opting for (expensive) nuclear power plants, even though other, even low-carbon technologies are available at lower cost. These currently include Bangladesh, Türkiye, Belarus, the United Arab Emirates, and Egypt (Neumann et al. 2020; Sorge et al. 2020). They are the first newcomer countries in three decades where new nuclear power plants are actually being built or have just been completed. Other countries, such as Saudi Arabia or, since the 1970s, Poland, are considering entering the market.

A distinction must therefore be made between entry countries, where the first steps towards the construction of a nuclear power plant have actually been taken, and potentially interested countries. With regard to the latter, the World Nuclear Association lists over 30 countries that have, among other things, concluded a cooperation agreement with a supplier country or have other plans; these countries are divided into different degrees of concretization.[4] They include some countries that have been negotiating for a long time, such as Poland, where entry has been part of the political program since the 1970s. However, there are also countries where a nuclear phase-out seems less plausible, such as Libya, Sri Lanka, and Sudan.

In contrast to the narrative of a global "run" on commercial nuclear power plants with dozens of countries getting on board, the number of countries actually getting on board is small: with Türkiye, Bangladesh and Egypt, there were three (3) candidates in 2023. What has already been said also applies to these three "newbies" to a particular degree: in all three cases, Russia's Rosatom is the exporter, which also invests considerable financial resources in this form of nuclear diplomacy. Türkiye and Egypt are important regional powers, and Bangladesh, with a population of almost 200 million, also plays an important regional role in connecting South and

[4] https://world-nuclear.org/information-library/country-profiles/others/emerging-nuclear-energy-countries.aspx, updated regularly.

East Asia. It is clear that the diffusion of commercial nuclear technology and attempts by countries to enter this field are also linked to geopolitical strategies.

What both groups have in common is the low level of market-based institutions and democratic participation in most countries. The correlation between entry considerations and the lack of market-based institutions can be explained by the low competitiveness of nuclear energy, which requires considerable subsidization. With regard to the social aspects, Sorge et al. (2020) summarize the econometric results of (Neumann et al. 2020) and confirm the assumption already expressed elsewhere (Sovacool and Valentine 2010; Chung 2018): there is an of an inverse relationship between the pursuit of nuclear entry in a country and its democratic constitution.[5] The results of the econometric calculation suggest that the lower the level of democratic freedom,.[6] the more likely a country is to be classified in the group of potential nuclear newcomers. Countries with a high level of democratic freedom, on the other hand, are less likely to be classified as potential newcomer countries today.

How can it be explained that countries discussing the nuclear phase-in today tend to have autocratic social structures? The affinity between the nuclear industry and centralized political approaches is the DNA of any nuclear economy, rooted in technical complexity and high safety standards. However, leading countries in the early nuclear era were democratically constituted and ensured strong civil liberties for their populations, such as the US, the UK, France, and later Germany, Finland, etc. This has fundamentally changed for potential new entrants. It is plausible that it is easier for countries with autocratic structures to shorten or circumvent controversies in the planning and implementation process of nuclear power plants. The observed phenomenon can also be explained from a political economy perspective, as less democratic countries have fewer critical public or parliamentary debates about a controversial public investment (Fig. 5.3).

5.6 Conclusion

The development of nuclear power stems from a global geopolitical conflict, World War II, and still is linked to potential global and regional conflicts today. The latter started in Asia (Korea), Central America (Cuba) and in the middle of Europe (Berlin), and they continue to loom, as evidenced by the Russian war on Ukraine, starting in 2022. This chapter opens Part II of the book, "geopolitics," and it picks up the characterization of nuclear power as fundamentally interlinked between science and

[5] The study considers the following countries designated by the World Nuclear Association as entry-level countries: Albania, Algeria, Azerbaijan, Bolivia, Chile, Croatia, Cuba, Estonia, Ghana, Indonesia, Israel, Jordan, Kazakhstan, Kenya, Laos, Latvia, Libya, Lithuania, Mongolia, Morocco, Namibia, Nigeria, Rwanda, Paraguay, Peru, Philippines, Poland, Saudi Arabia, Serbia, Singapore, Sri Lanka, Sudan, Syria, Thailand, Tunisia, Uzbekistan, Venezuela, Vietnam, and Qatar.

[6] The level of democratic freedom is estimated by "Freedom House" in a detailed manner, see https://freedomhouse.org/countries/freedom-world/scores.

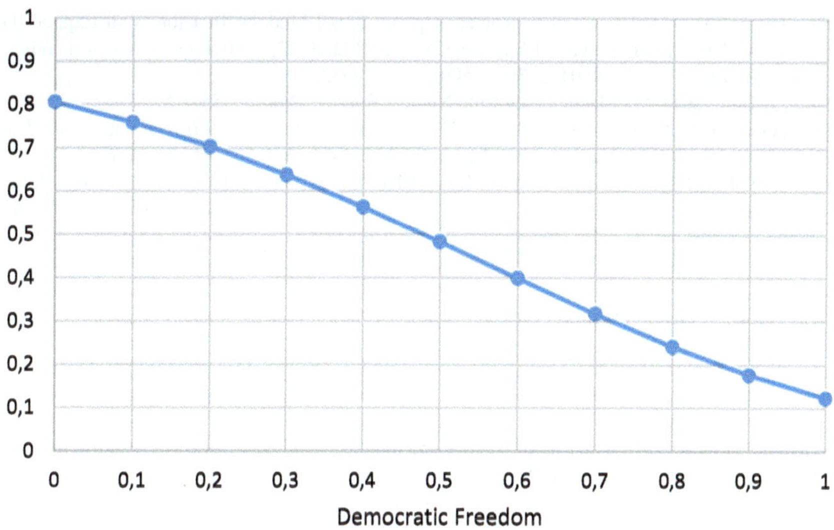

Fig. 5.3 Predicted probabilities for membership in the group of potential nuclear newcomers (in percent). *Source* (Sorge et al. 2020, 169)

warfare. Different perspectives exist on the first and the second nuclear age, but certainly regional conflicts are an integral part of the contemporary history.

The link between the three nuclear superpowers, the US, Russia, and China, and potential nuclear entry countries ("newbies") is dominated by geopolitics, too. All superpowers deploy ample "nuclear diplomacy," Russia being the most "successful" recently, with almost a monopoly on reactor exports. On the side of reactor importing countries, we find that the probability to enter the nuclear sector is inversely related to the level of democratic freedom in a country. This has been proven recently by attempts to establish more or less ambitious national science and technology systems related to nuclear power in Türkiye, Egypt, and Bangladesh.

References

Acheson-Lilienthal Report. 1946. "A Report on the International Control of Atomic Energy." Prepared for the Secretary of State's Committee on Atomic Energy. Washington, D.C. http://fissilematerials.org/library/ach46.pdf.

Baade, Fritz. 1958. *Welt-Energiewirtschaft: Atomenergie - Sofortprogramm oder Zukunftsplanung*. Hamburg: Rowohlt.

Basu, Dipak, and Victoria W. Miroshnik. 2019. *The Political Economy of Nuclear Energy*. London, UK: Palgrave Macmillan.

Bracken, Paul. 2012. *The Second Nuclear Age - Strategy, Danger, and the New Power Politics*. New York, USA: Times Books - Henry Holt and Company, LLC.

Chu, Steven. 2010. "America's New Nuclear Option: Small Modular Reactors Will Expand the Ways We Use Atomic Power." *Wall Street Journal*, March 23, 2010. https://www.wsj.com/articles/SB10001424052748704231304575092130239999278.

Chung, Ji-Bum. 2018. "Let Democracy Rule Nuclear Energy." *Nature* 555:415.

Corera, Gordon. 2009. *Shopping for Bombs: Nuclear Proliferation, Global Insecurity, and the Rise and Fall of the A. Q. Khan Network*. Paperback. New York, NY: Oxford Univ. Press.

Cox, Emily, Phil Johnstone, and Andy Stirling. 2016. "Understanding the Intensity of UK Policy Commitments to Nuclear Power." *SWPS 2016 - 16*. http://sro.sussex.ac.uk/63568/1/__smbhome.uscs.susx.ac.uk_pj55_Documents_ALL%20JOURNAL%20ARTICLES_Cox%20et%20al_2016%20Exploring%20the%20intensity%20of%20UK%20nuclear%20commitments.pdf.

Davis, Lucas W. 2012. "Prospects for Nuclear Power." *Journal of Economic Perspectives* 26 (1): 49–66. https://doi.org/10.1257/jep.26.1.49.

DoD. 2018. "Nuclear Posture Review." Washington DC, USA: U.S. Department of Defense.

Energy Futures Initiative. 2017. "The U.S. Nuclear Energy Enterprise: A Key National Security Enabler." Washington DC, USA: Energy Futures Initiative.

Fukuyama, Francis. 1992. The End of History and the Last Man. New York, NY, Hamilton.

Goudsmith, S.A. 1947. *Alsos - The Failure in German Science*. New York.

Groves, Leslie M. 1983. *Now It Can Be Told: The Story of the Manhattan Project*. New York, USA: Da Capo Press.

Hirschhausen, Christian von. 2017. "Nuclear Power in the 21st Century—An Assessment (Part I)." DIW Discussion Paper 1700. Berlin, Germany: DIW Berlin.

Hirschhausen, Christian von. 2023. *Atomenergie: Geschichte und Zukunft einer riskanten Technologie*. 1. Auflage. Reihe "Wissen." München: C.H.Beck.

Lévêque, François. 2013. *Nucléaire On/Off - Analyse économique d'un pari*. Paris: Dunod.

Lovins, Amory B., and L. Hunter Lovins. 1980. *Energy/War: Breaking the Nuclear Link*. 1st Edition edition. San Francisco: Friends of the Earth.

Mez, Lutz, ed. 1979. *Der Atomkonflikt: Atomindustrie, Atompolitik und Anti-Atom-Bewegung im internal. Vergleich*. Berlin: Olle und Wolter.

Moniz, Ernest. 2011. "Why We Still Need Nuclear Power." *Foreign Affairs* 90 (6): 83–94. http://search.ebscohost.com/login.aspx?direct=true&db=bth&AN=66803982&site=ehost-live.

Münzinger, Friedrich. 1960. *Atomkraft: Der Bau ortsfester und beweglicher Atomantriebe und seine technischen und wirtschaftlichen Probleme. Eine kritische Einführung für Ingenieure, Volkswirte und Politiker*. New York: Springer. https://link.springer.com/book/https://doi.org/10.1007/978-3-642-85689-1.

Nelson, Craig. 2014. *The Age of Radiance—The Epic Rise and Dramatic Fall of the Atomic Era*. New York, USA: Scribner.

Neumann, Anne, Lars Sorge, Christian von Hirschhausen, and Ben Wealer. 2020. "Democratic Quality and Nuclear Power: Reviewing the Global Determinants for the Introduction of Nuclear Energy in 166 Countries." *Energy Research & Social Science* 63 (May):101389. https://doi.org/10.1016/j.erss.2019.101389.

Nuttall, William J. 2022. *Nuclear Renaissance: Technologies and Policies for the Future of Nuclear Power*. Boca Raton: CRC Press. https://www.taylorfrancis.com/books/9781003038733.

Popp, Manfred. 2021. "Why Hitler Did Not Have Atomic Bombs." *Journal of Nuclear Engineering* 2 (1): 9–27. https://doi.org/10.3390/jne2010002.

Poundstone, William. 1992. *Prisoner's Dilemma*. 1st ed. New York: Doubleday.

Radkau, Joachim. 1983. *Aufstieg und Krise der deutschen Atomwirtschaft 1945-1975: Verdrängte Alternativen in der Kerntechnik und der Ursprung der nuklearen Kontroverse*. Reinbek bei Hamburg, Germany: Rowohlt.

Reed, Bruce Cameron. 2013. *The History and Science of the Manhattan Project*. 1st Edition. New York: Springer.

Rhodes, Richard, ed. 2012. *The Making of the Atomic Bomb*. 25th anniversary ed. New York, USA: Simon & Schuster Paperbacks.

SIPRI. 2016. "Global Nuclear Weapons: Downsizing but Modernizing." SIPRI for the Media. June 13, 2016. https://www.sipri.org/media/press-release/2016/global-nuclear-weapons-downsizing-modernizing.

SIPRI. 2022. *SIPRI Yearbook 2022: Armaments, Disarmament and International Security - English Summary*. Solna, Sweden: Oxford University Press. https://www.sipri.org/sites/default/files/2022-06/yb22_summary_en_v2_0.pdf.

Sorge, Lars, Claudia Kemfert, Christian von Hirschhausen, and Ben Wealer. 2020. "Nuclear Power Worldwide: Development Plans in Newcomer Countries Negligible." *DIW Weekly Report* 11/2020:163–72. https://doi.org/10.18723/diw_dwr:2020-11-1.

Sovacool, Benjamin K., and Scott Victor Valentine. 2010. "The Socio-Political Economy of Nuclear Energy in China and India." *Energy* 35 (9): 3803–13. https://doi.org/10.1016/j.energy.2010.05.033.

Stirling, Andy, and Phil Johnstone. 2018. "A Global Picture of Industrial Interdependencies Between Civil and Military Nuclear Infrastructures." SPRU Working Paper Series 2018–13 (August). Sussex, UK: University of Sussex. https://www.sussex.ac.uk/webteam/gateway/file.php?name=2018-13-swps-stirling-and-johnstone.pdf&site=25.

U.S. Department of Defense (DOD). 2022. "Nuclear Posture Review." Washington DC, USA. https://s3.amazonaws.com/uploads.fas.org/2022/10/27113658/2022-Nuclear-Posture-Review.pdf.

Part II
Geopolitics of Nuclear Power

Chapter 6
Nuclear Power Plays: Complexities of Armament, Disarmament, and (Non-) Proliferation of Nuclear Weapons

Maren Vieluf, Sophie Kretzschmar, and Elisabeth Saar

6.1 Introduction

In world politics, power is a decisive factor, and nuclear weapons guarantee their possessor a unique kind of bargaining power in their international relations. To credibly threaten the use of nuclear weapons, the core of nuclear deterrence, is to threaten to cause devastating, immediate and long-term destruction and human suffering on a scale that no other type of weapon can (cf. Schelling 1966). The devastating impacts of any use of nuclear weapons have driven both deterrence and disarmament discourses for the past decades. They will continue to be the driver of both allure and damnation of this category of weapons of mass destruction and the interlinked dynamics of (dis)armament and (non)proliferation.

Nuclear weapons were first tested by the United States of America (US) in July and then used for the first time in August 1945 in the bombings of the Japanese cities of Hiroshima and Nagasaki, causing immense loss of life and suffering. Since

The views expressed are those of the author, and do not reflect the position of the German Federal Foreign Office.

The original version of the chapter has been revised. A correction to this chapter can be found at https://doi.org/10.1007/978-3-031-99894-2_19

M. Vieluf (✉)
Institute for Political Science, University of Innsbruck, Austria / German Federal Foreign Office, Berlin, Germany
e-mail: maren.vieluf@student.uibk.ac.at

S. Kretzschmar
Nuclear Verification and Disarmament, RWTH Aachen University, Aachen, Germany

E. Saar
Department of Earth Systems Sciences, University of Hamburg, Hamburg, Germany

A. Wimmers et al. (eds.), *Nuclear Power*, Studies in Energy, Resource and Environmental Economics, https://doi.org/10.1007/978-3-031-99894-2_6

then, seven other states have tested nuclear devices[1]—amounting to a total of more than 2,000 nuclear tests.[2,3] Nuclear weapons are credited with the ability to instantly change the nature of any conflict (Schelling 1960). Yet, apart from the US, no other country has dared to use nuclear weapons in combat—leading to the assumption of the "nuclear taboo", which stipulates that nuclear weapons will not be used again due to their stigmatization (cf. Tannenwald 2007). Over the past decades, rules, norms, and legal frameworks have been developed to manage and reduce the risks associated with nuclear weapons. The global nuclear order, centered around the Nonproliferation Treaty (NPT), is shaped by a complex web of interrelated and often conflicting interests and behaviors tied to the presence and persistence of nuclear weapons. The NPT also ascribes who may possess nuclear weapons, i.e., who has the power to threaten their use. Of the nine nuclear-armed states, with a total of more than 12,100 nuclear warheads currently in their collective possession, four fall outside the NPT regime's direct influence.[4] Meanwhile, a large group of states strongly opposes nuclear weapons. A divide persists between those who support nuclear deterrence and those advocating for "Global Zero"—a world without nuclear weapons—and the surrouding nuclear politics shape international relations in many, often unseen ways.

This chapter traces the history of the nuclear age along the lines of four distinctive but interrelated and sequential concepts: armament, proliferation, nonproliferation, and disarmament, to describe how nuclear weapons have the power to shape international relations. Section 6.2 scrutinizes the developments and dynamics initiated by the discovery of nuclear fission, paving the way for the evolution of nuclear armament and proliferation. It highlights how states have navigated the complex interplay of technical capabilities and political motivations in their (non)pursuit of nuclear weapons. Section 6.3 describes the global efforts and frameworks designed to curtail the spread of nuclear weapons, with a primary focus on the NPT and its role in shaping international norms and practices, exploring both its successes and challenges. Section 6.4 describes the complexity of nuclear disarmament, outlining both its theoretical goals and practical processes. Section 6.5 concludes with a summary of current and potential future developments in the nuclear weapons landscape.

---

[1] While Israel is widely believed to be a nuclear-armed state, it has not conducted a nuclear test in accordance with its policy of nuclear opacity (see Sect. 2.3). Despite the plausible likelihood of an atmospheric low-yield nuclear test, the 1979 'Vela Incident' has never been officially confirmed as a nuclear explosion, nor has there been a consensus on its attribution to Israel, South Africa, or possibly both countries (cf. De Geer and Wright 2018).

[2] https://www.atomicarchive.com/almanac/test-sites/testing-chronology.html, last accessed on 14-02-2025. This date holds for all other online references unless otherwise indicated.

[3] Due to the secretive nature of nuclear weapons programs, it is impossible to provide a comprehensive and verified account of every initiative that was launched but never completed. South Africa is the only other country known to have built nuclear weapons but has dismantled its arsenal. With the demise of the Soviet Union, Belarus, Kazakhstan, and Ukraine temporarily had ownership over nuclear weapons stationed on their soil but transferred them to Russia (see Sect. 4.2).

[4] https://fas.org/initiative/status-world-nuclear-forces/.

6.2 Nuclear Armament and Proliferation

Nuclear armament and proliferation have played pivotal roles in shaping the global strategic environment in recent decades. Nuclear armament refers to "the process of increasing the number and strength of a country's [nuclear] weapons".[5] It is closely linked to both understandings of proliferation: On the one hand, vertical proliferation, which involves the modernization and expansion of existing nuclear weapons stockpiles, and on the other hand, horizontal proliferation, which refers to the spread of nuclear weapons, material, or technology to states and potentially non-state actors that do not yet possess nuclear weapons capabilities (Goldansky 1988).

6.2.1 Technical and Political Considerations of Armament and Proliferation

To acquire nuclear weapons, a state must meet several technical and political conditions. From a technical standpoint, the state must have mastered complex scientific and industrial processes, including the enrichment of uranium (for weapons using highly enriched uranium) or the operation of nuclear reactors and reprocessing of spent fuel (for weapons using plutonium), as well as the development of delivery systems (i.e., missiles, bombers, or submarines). Politically, nuclear acquisition often depends on a state's security perceptions, geopolitical ambitions, and domestic political consensus. The various factors that influence nuclear proliferation are debated and highly case-dependent. Most likely, however, states pursue nuclear weapons when they perceive significant security threats and see them as a valid solution (generating the willingness), have access to the necessary resources and technology (generating the opportunity), or likely both (cf. Jo and Gartzke 2007; Monteiro and Debs 2014).

While only nine states possess nuclear weapons, several other nations had or have the technical capacity to build nuclear weapons (for example, South Africa, Argentina, Brazil, Germany, Italy, Japan, South Korea, and Sweden), referred to as "nuclear latency," but have opted not to pursue nuclear ambitions (cf. Fuhrmann and Tkach 2015) often due to their commitment to international norms, treaties, their role in security alliances, but also cost–benefit and risk analyses.

[5] https://dictionary.cambridge.org/de/worterbuch/englisch/armament.

6.2.1.1 The Physics Behind Nuclear Weapons and Their Destructive Power

The explosive yield of nuclear weapons sets them apart from other weapons. Unlike conventional explosives, which derive their energy from chemical reactions, i.e., changes in the molecular (electromagnetic) structure of molecules, the energy released upon detonation from nuclear weapons stems from nuclear reactions, i.e., changes in the structure of atomic nuclei (fission or fusion reactions). Because more energy is stored more densely in atomic nuclei than in molecular bonds, more energy can be harnessed. For example, fissioning one gram of uranium U-235 releases 10 million times more energy than burning one gram of TNT. This principle is used in both civilian nuclear reactors and weaponry. Typical fission materials in nuclear weapons are highly enriched uranium (HEU, uranium with a large percentage of the isotope U-235) and plutonium (Goodwin 2021).

Explosions require a rapid release of energy. Thus, instead of controlling the energy-releasing fission reactions as in a critical nuclear reactor, a supercritical setting is specifically sought to initiate an explosive fission chain reaction, where neutrons from the fission exponentially initiate new fission reactions, leading to very high energy releases when the weapon is triggered (Goodwin 2021). To achieve such a supercritical setting, two main fission weapon designs have been developed:

- Gun-Type Design: Two subcritical configurations of fissile material are rapidly smashed together with conventional explosives. This design is easier to develop,[6] but it only works with highly enriched uranium (HEU) (Miller 2022).
- Implosion-Type Design: A subcritical configuration of fissile material is rapidly compressed with conventional explosives surrounding it. This requires more development effort due to the precise timing of the conventional explosives, but the design can also use plutonium, which allows for greater explosive yields (Miller 2022).[7]

Modern nuclear weapons, thermonuclear weapons or hydrogen bombs, utilize a two-stage design that combines fission and fusion reactions. The first stage is typically an implosion-type fission design as described above. This ignites the second stage, in which deuterium and tritium are fused into helium. The second stage not only significantly increases the energy for the explosion itself but also provides additional neutrons to achieve higher fission rates in the first stage (Schumann 2022).

The yield of nuclear weapons is usually expressed in terms of the equivalent energy released by the explosion of a thousand tons (kilotons, kT) or a million tons (megatons, MT) of TNT. The first nuclear weapon ever tested, known as Gadget in the Trinity Test of July 16, 1945, had a yield of 21 kT (Selby et al. 2021). The largest

---

[6] This design was developed early in the Manhattan Project and was never tested by the US before using it on Hiroshima.

[7] A slightly more advanced, but nowadays common nuclear weapon design increases its yield by adding tritium and deuterium to 'boost' the weapon; their fusion process releases more neutrons to increase the amount of fissioned material. The tritium must be constantly replaced due to its short half-life of about twelve years.

weapon ever tested, the Soviet Tsar Bomb, had a yield of about 50 MT,[8] which is over 100,000 times more powerful than the largest conventional explosives, with typical yields below 0.5 kT. The nuclear weapons used on Hiroshima and Nagasaki in August 1945 had yields of approximately 15 and 21 kT, respectively (Malik 1985). Today, the largest nuclear weapon in the US arsenal has a yield of 1.2 MT—over thousands of times the yield of the largest conventional explosives.[9]

This comparison underscores that nuclear weapon explosions release vastly more energy than conventional explosives, which further leads to also qualitatively different effects (cf. Glasstone and Dolan 1977). The correspondingly higher temperatures deriving from a nuclear explosion cause a significant amount of energy to be released as thermal radiation, which can ignite flammable materials and cause severe burns over considerable distances. Additionally, nuclear explosions emit both initial and residual nuclear radiation,[10] which can inflict far-reaching and long-term harm to living beings and the environment in general. This is particularly evident from the experiences of the bombings of Hiroshima and Nagasaki but also from the regions affected by nuclear testing that continue to experience high radiation levels even decades later (Abella et al. 2019).

In addition to the immediate effects of individual nuclear explosions, multiple detonations could have severe global climate consequences. Corresponding large-scale fires could release vast amounts of soot into the atmosphere, leading to decreasing global temperatures and precipitation, referred to as "nuclear winter" (Turco et al. 1983). For instance, an analysis of a hypothetical regional conflict between India and Pakistan involving the use of 100 nuclear weapons, each with a yield of 15 kT, suggests that about 255 million people could face death from famine caused by global climate disruptions, in addition to the about 27 million direct fatalities from the weapons themselves. In a worst-case scenario involving an all-out global nuclear war, over five billion people could be affected by famine due to the extensive changes in the global climate (Xia et al. 2022).

6.2.1.2 Reasons for Armament and Proliferation: Why Do States Want Nuclear Weapons?

For a long time, national security interests were seen as the sole motivation for a state's decision to pursue the possession of nuclear weapons. In situations of significant military threats—whether conventional or nuclear—the acquisition of nuclear weapons is often seen as the ultimate defense mechanism. The key rationale behind this is deterrence: by acquiring nuclear weapons, a state aims to prevent aggression by

[8] https://thebulletin.org/2021/11/the-untold-story-of-the-worlds-biggest-nuclear-bomb/.

[9] https://thebulletin.org/2024/02/why-the-biden-administrations-new-nuclear-gravity-bomb-is-tragic/.

[10] The initial radiation comes from highly energetic and penetrating light ("gamma rays") produced in the fission processes; the residual nuclear radiation ("fallout") comes from the further decaying unstable fission products.

convincing potential adversaries that the costs of conflict will outweigh any potential benefits. In this logic, simply possessing nuclear weapons acts as a deterrent, making their actual use less necessary, but always an implicit threat (cf. Waltz 1981). This could also be leveraged in case the aggressor possesses nuclear weapons, as their use lingers over any conflict involving a nuclear-armed state.

A crucial element of deterrence is credibility: adversaries must believe in both a state's capability and its willingness to follow through with its threats of actual nuclear weapons use (cf. Nye Jr. 2023). When both adversaries possess nuclear weapons, establishing and maintaining mutual vulnerability is credited with ensuring stability and peace in the international system.[11]

Nuclear deterrence also contains an inherent contradiction: states possess and threaten the use of nuclear weapons with the hope that they will never actually need to use them (cf. Aron 1980). While deterrence is often viewed through the lens of rational calculation—balancing goals, costs, and risks—it also involves significant uncertainty, misperceptions, and psychological factors. Strategic considerations, military scenarios, and security dilemmas are largely driven by national interests that are assessed in relation to other states and in the context of the broader international system. These considerations focus heavily on how potential adversaries perceive threats, with deterrence policy traditionally viewed from the perspective of a potential aggressor at the state level (cf. Mazarr 2018).

However, while the threat of mutual destruction through nuclear weapons can create stability at the strategic level by preventing large-scale wars between nuclear weapons possessor states, it can simultaneously foster instability at lower levels of conflict, often referred to as the "stability-instability-paradox" (Snyder 1965). This paradox suggests that while nuclear deterrence may help maintain peace between nuclear-armed states—contributing to a form of "nuclear peace"—it can also embolden these states to engage in limited conventional conflicts or proxy wars, particularly involving non-nuclear weapon states as their adversary. This highlights the complex role that nuclear weapons play in both preventing and encouraging certain forms of conflict.

Furthermore, while the devastating humanitarian consequences of nuclear weapons use are rarely explicitly addressed in strategic calculations, awareness of these consequences reinforces the deterrent effect. Nonetheless, discussions surrounding deterrence and security often downplay these impacts, focusing instead on state-centric concerns and rivalries. Consequently, states facing hostile nuclear-armed neighbors or rivals may feel compelled to pursue their own nuclear capabilities.

Nuclear acquisition, however, is no longer seen as driven solely by national security concerns. Scott Sagan (1996) argues that nuclear weapons are also powerful political symbols, playing a significant role in domestic politics and shaping a state's

[11] https://thethirdnuclearage.com/2022/04/22/andrew-futter-publishes-paper-deterrence-disruptive-technology-and-disarmament-in-the-third-nuclear-age-for-hiroshima-organisation-for-global-peace/.

international identity. The author identifies three interconnected reasons, or "models," for why states pursue nuclear weapons:

- Security Model: The most traditional view, where nuclear weapons are pursued to deter external threats, especially from adversaries with nuclear capabilities or superior conventional forces.
- Domestic Politics Model: In this model, internal political pressures—such as bureaucratic interests, military influence, or political competition—drive nuclear development. Nuclear weapons programs may be advanced to consolidate power or appease influential domestic groups.
- Norms Model: Nuclear weapons can also serve as symbols of modernity, prestige, and identity. For some states, possessing nuclear weapons elevates their status on the global stage and aligns them with the most powerful nations.

This view has been echoed in the literature and the models have been expanded, highlighting the complexity of motivations behind nuclear proliferation decisions. For example, Singh and Way (2004) and Jo and Gartzke (2007) emphasize the interplay between a country's domestic capacity and its likelihood of pursuing nuclear weapons. Jacques E. C. Hymans (2007) argues that national identity, particularly as perceived by a state's leadership, plays a crucial role in these decisions. In this view, the pursuit of nuclear weapons is linked to the leadership's vision of national identity and its ability to navigate complex domestic and international political landscapes.

6.2.1.3 Strategies and Barriers to the Acquisition of Nuclear Weapons

While acquiring nuclear weapons requires both technical capabilities and political willingness, empirical evidence suggests that international support is also a necessary component of the decision-making process. No nuclear-armed state developed its nuclear arsenal in isolation (Kroenig 2011). Sensitive nuclear transfers—including the exchange of materials, technology, or expertise—have played a significant role in the development and proliferation of nuclear weapons (Kroenig 2009). These dynamics of demand and supply are often driven by geopolitical alliances, covert assistance, or shared scientific expertise.

Several technical factors may hinder states' ambitions in building or acquiring nuclear weapons. First and foremost, from a practical perspective, the development of nuclear weapons relies on acquiring enough fissile material, i.e., HEU or plutonium, for which the following production pathways exist:

Weapons-grade uranium requires enrichment of the isotope U-235 (typically to above 90% U-235[12] relative to the most abundant isotope U-238, which makes up 99.3% of natural uranium).[13] The isotopes can only be separated by exploiting

[12] The exact level of enrichment needed for a nuclear weapon is not precise; potentially, lower enrichment levels would suffice (see e.g. Kemp et al. 2024).

[13] Uranium in fuel for typical nuclear power reactors is enriched to between 3 and 5%, see Chapter XXX of this book.

their slight difference in mass, as they behave chemically the same. Today, uranium isotopes are usually separated in large complexes of gas centrifuges after converting them to gaseous form (uranium hexafluoride). Gas centrifuges are technologically advanced and require significant research and development efforts, and existing designs are considered sensitive and not openly available (Kemp 2017).

Plutonium is produced by irradiating uranium U-238, the main component of natural uranium, with neutrons in a nuclear reactor. From this irradiated uranium, reprocessing facilities are needed to chemically separate the plutonium from the residual uranium or produced elements in the reactor.[14] The technologies used in the civilian nuclear fuel cycle (enrichment capabilities for reactor fuel production, reprocessing for nuclear recycling), are considered "dual-use" since they, in principle, can be used for civilian and military nuclear applications. As such, they are monitored by the International Atomic Energy Agency (IAEA) since the early 1960s for non-nuclear weapon states to ensure that they are not misused to produce materials for nuclear weapons (Schneider and Thränert 2014).

The 1970 NPT is crucial in fostering nonproliferation, as Article III requires non-nuclear weapon state party to the treaty to conclude a safeguards agreement with the IAEA, to let the agency verify that the state uses its nuclear programs only as prescribed within the treaty. These safeguards agreements allow the IAEA to apply nuclear safeguards, in which the agency collects and analyses information to verify compliance with the NPT. Some of this information is provided by the states themselves (e.g., with declarations of fissile material inventories and changes, information on the designs of nuclear facilities), some is collected by the IAEA itself (e.g., through inspections at relevant sites), and some is based on information provided by third parties (e.g., through open-source information). The measures applied during IAEA inspections include the comparison of declared information with on-site examinations, such as visual inspections (counting), measurements (destructive or non-destructive assay), or application and checking of seals (cf. IAEA 2016).

Over time, the safeguards agreements and measures have changed. A major turning point was the discovery of Iraq's clandestine nuclear weapons program in 1991, which showed that focusing on the verification of the correctness of declared materials and activities was not enough, and further capabilities for assessing completeness were required (cf. Acton and Newman 2006). For this, the "Additional Protocol" was drafted to provide the agency with extended rights, such as an expanded right to take environmental samples—eventually allowing the agency further to "detect any undeclared nuclear material or activities in the [s]tate as a whole" (IAEA 2022a). To date, 142 states have concluded an Additional Protocol with the IAEA, and the IAEA's goal is universalization of comprehensive safeguards (cf. Davenport 2024). For nuclear weapon states, Voluntary Offer Agreements (VOA) allow the agency to inspect civilian fissile materials and facilities (cf. Davenport 2024).

In addition to these obligations arising from the NPT and agreements with the IAEA, states that forego ambitions for an indigenous nuclear weapons program due

[14] https://www.ucsusa.org/resources/fissile-materials-basics.

to domestic opposition, concerns over diplomatic repercussions, or their international obligation. For many states, the potential political, economic, and social costs of developing or acquiring a nuclear arsenal outweigh the perceived security benefits. States may also assess that they do not face an existential threat necessitating nuclear weapons, or that their security can be sufficiently ensured through other means, such as strong conventional forces or international alliances. In some cases, the decision to forgo the development of nuclear weapons is also shaped by their participation in extended nuclear deterrence arrangements, whereby they rely on the so-called "nuclear umbrella" of an allied nuclear power. This reliance on allies' nuclear weapons capabilities can provide a sense of security without the need for an independent nuclear arsenal (Erästö 2023).

After 1945, the United States pioneered the stationing of nuclear weapons on foreign soil, a move soon followed by the Soviet Union and the United Kingdom. By the mid-1950s, these major nuclear powers had established extensive foreign stationing practices, driven by deterrence strategies, nonproliferation goals, and perceived geographic necessities. Geographic proximity was particularly essential early in the Cold War due to limited long-range missile technology, necessitating closer positioning of nuclear weapons to potential targets. Additionally, stationing weapons in states like Italy and Germany helped prevent these countries from developing their own nuclear arsenals, balancing security commitments with nonproliferation aims (Fuhrmann & Sechser 2014).

At the height of the Cold War, estimates suggest that over 20 states hosted foreign nuclear weapons, sometimes without their explicit knowledge. The United States had over 7,000 nuclear weapons stationed abroad, primarily in Europe and Asia, while the Soviet Union deployed around 800 in satellite states (Ibid.). With the end of the Cold War, many of these arrangements were dismantled due to high costs, strategic shifts, and disarmament efforts. Countries like Canada and Greece withdrew from these agreements, citing political and security concerns (Erästö 2023).

NATO's nuclear sharing is the only existing institutional framework for hosting foreign nuclear weapons. This arrangement allows non-nuclear NATO members to host US nuclear weapons while command and control as well as launch authority remain with the United States. Currently, Belgium, Germany, Italy, the Netherlands, and Turkey host US nuclear weapons, with ongoing modernization efforts (Kristensen et al. 2023). NATO's Art. 5 and nuclear sharing arrangements are interconnected, as the former guarantees the alliance's collective defense commitment, potentially including nuclear retaliation (NATO 2022), while the latter enhances deterrence by involving non-nuclear allies in the planning and possible delivery of US nuclear weapons. This, in turn, bolsters both the credibility and unity of the alliance's response to an armed attack.

Recently, Belarus has agreed to host Russian nuclear weapons, marking the first instance of such deployment outside Russia since the end of the Cold War (Kristensen and Korda 2023). This development has raised concerns about a potential resurgence of an arms race in Europe and reignited debates regarding the legality of foreign nuclear stationing under the NPT Art. I and II (UN Press 2023).

Moreover, states that reject nuclear weapons often support international initiatives against nuclear weapons such as the Treaty on the Prohibition of Nuclear Weapons (TPNW). This treaty, which entered into force on January 22, 2021, outlaws the development, testing, possession, and use of nuclear weapons, reinforcing global norms against nuclear armament. By endorsing or ratifying this treaty, non-nuclear weapon states affirm their commitment to disarmament and contribute to a collective effort to mitigate nuclear risks and enhance global security (Müller and Wunderlich 2020). As of March 2025, four years after its entry into force, 94 states have signed, and 73 states have ratified the TPNW.[15]

Additionally, the norm against the use of nuclear weapons, the "nuclear taboo", plays a critical role in shaping these decisions. States adhering to this taboo view nuclear weapons use as morally unacceptable and strategically destabilizing. Their commitment to this norm thereby helps maintain international peace and stability, reflecting a broad-based ethical stance against the proliferation and use of nuclear weapons (Tannenwald 2007).

Another factor influencing decision-making on nuclear weapons is risk: the risk of escalation and risks of errors, miscalculations, and unintended calculations. Since 1945, the world has been at the brink of accidental nuclear war on several occasions.

While the Cold War saw many crises where nuclear use was threatened (Sechser and Fuhrmann 2017), the most severe nuclear crisis was the Cuban Missile Crisis of 1962. Tensions escalated when the US discovered that the Soviet Union, under Nikita Krushchev, had begun constructing missile sites and deploying medium- and intermediate-range ballistic missiles in Cuba to strengthen its ties with the island nation. To prevent further shipment of missiles, US President John F. Kennedy decided to install a maritime blockade of Cuba on October 22, 1962, leading the two sides to the brink of nuclear war. Khrushchev capitulated six days later, and the Soviet missiles were removed from Cuba (cf. Burr and Kimball 2022).

On November 9, 1979, a malfunction at the North American Aerospace Defense Command (NORAD) led to a false identification of a Soviet missile attack, nearly triggering a US nuclear response (cf. Barrett 2016). Similarly, on September 26, 1983, Soviet officer Stanislav Petrov prevented disaster by recognizing a false alarm in the Soviet system, avoiding a mistaken retaliation of an imminent US nuclear attack. Both incidents underscore the crucial role of human judgment in averting nuclear catastrophe. In November 1983, the NATO exercise "Able Archer 83" was mistakenly perceived by the Soviet Union as a potential nuclear first strike, escalating tensions to perilous levels before simmering down through diplomatic engagement (cf. Fraise and Egeland 2023).

For the years between 1950 and 1988, Gregory and Edwards (1989) catalog a range of significant incidents and accidents involving bombers, submarines, and nuclear missiles that were dropped, collided, or otherwise involved in dangerous and potentially catastrophic situations. These occurrences highlight the substantial risks associated with managing and operating nuclear weapons systems, with many details of these incidents still shrouded in secrecy.

---

[15] https://www.icanw.org/signature_and_ratification_status.

In 1995, the Norwegian Rocket Incident nearly triggered a nuclear conflict when Russia misidentified a scientific rocket launch conducted by the United States and Norway as a missile attack before the falsely declared nuclear device landed in the Barents Sea, disproving an imminent nuclear attack (cf. Barrett 2016).

Such incidents have become fewer since improved communication, enhanced verification mechanisms, and more robust command and control systems help mitigate the dangers of technical failure or human error. However, risks remain, as last seen in the 2018 Hawaii false missile alert (cf. CNN 2018), the 2019 Pulwama crisis between India and Pakistan (cf. Yusuf 2019), and ongoing nuclear threats and coercion attempts in the context of the Russian war of aggression against Ukraine (cf. Horovitz and Arndt 2024; Arndt and Horovitz 2023).

6.2.1.4 Consequences of the Pursuit of Nuclear Weapons

Any decision to pursue nuclear weapons involves a complex balance of risks and rewards. States that seek to develop nuclear weapons often face significant challenges. They may encounter technology denial and international sanctions, which aim to impede their nuclear ambitions (Abraham 2010). In addition to these measures, states might experience further diplomatic coercion or isolation as the international community exerts pressure to curb their nuclear ambitions (Maller 2010). In more severe cases, the threat of military intervention can be employed to prevent or halt nuclear development (Monteiro and Debs 2014). The pursuit of nuclear weapons can thus lead to a range of negative repercussions that undermine a state's international standing and economic stability.

However, acquiring nuclear weapons or signaling such interest can offer substantial advantages. For many states, nuclear armament can enhance their international status and provide considerable leverage in global negotiations. Nuclear-armed states are often perceived to enjoy increased diplomatic influence and, importantly, greater security through deterrence (see Sect. 2.1.2). Some states engage in nuclear "hedging," i.e., maintaining the technical capability and infrastructure to develop nuclear weapons without formally pursuing them. This strategy allows these states to benefit from the political leverage associated with having latent nuclear capabilities, while avoiding the direct repercussions of full-scale nuclear development (Volpe 2023).

6.2.2 From Splitting the Atom to Bombing Japan (1938–1945)

The historical context and political motivations for developing nuclear weapons in the 1930s and early 1940s were profoundly influenced by both geopolitical tensions and scientific discoveries, with vastly differing national approaches and conditions.

In the 1930s, European scientists made significant advances in nuclear physics. The research conducted by Austrian physicist Lise Meitner, along with German chemists Otto Hahn and Fritz Strassmann, led to the discovery of nuclear fission in 1938 which paved the way for the development of nuclear weapons (Sime 1996). Considering the impending World War II (WW II) and the ambitions of the National Socialist regime in Germany, the potential of nuclear weapons was deemed of great interest by several states at the time. The German Uranium Project, also known as the Uranverein (uranium society), was initiated in 1939 to investigate the feasibility of using nuclear fission for military purposes. Under the leadership of physicists such as Werner Heisenberg and Kurt Diebner, the development of a functional nuclear reactor and research into nuclear weapons began. Despite these efforts, the Hitler regime failed to develop an operational nuclear weapon. Contributing factors included resource shortages, strategic missteps, and Allied bombing of infrastructure (Grunden et al. 2005).

The fear of the rise of fascism in Europe and the threat posed by National Socialist Germany prompted the US to initiate its own nuclear weapons program.[16] In August 1939, President Franklin D. Roosevelt and his advisors received a letter from Albert Einstein, with the support of Leo Szilard, warning of the potential dangers of a German nuclear weapons program (cf. Einstein 1939). The letter recommended a rapid increase in research funding and the beginning of uranium mining in the Belgian-colonized Congo[17] to circumvent a potential German nuclear weapon monopoly. Consequently, the Manhattan Project was established in 1942 under the direction of General Leslie Groves and scientific director Robert J. Oppenheimer, with the collaboration of scientists who had fled Europe. This top-secret initiative was aimed at developing a functional nuclear weapon first. The prospect of a weapon of such destructive power promised not only military superiority but also was envisioned to swiftly end WW II as well as secure a dominant role in the post-war order. The project utilized immense financial, material, and human resources, culminating in the first successful nuclear test in July 1945 in the desert of New Mexico and the bombings of Hiroshima and Nagasaki in August 1945 (Pyle 2013). Shaped by the experiences of both World Wars, the great powers envisioned to possess the "ultimate weapon" to prevent a recurrence of the devastating war experiences and to secure their preeminent position in the global order. With the beginning of the first nuclear age, the nuclear arms race was on (Karlsch and Zeman 2008).

[16] Refer to Chapter 11 of this book for more details on the US development efforts.

[17] Due to the intensified exploration for natural resources in occupied, colonized, and indigenous territories, Einstein surmised that, alongside regions in Canada and the former Czechoslovakia, there were substantial uranium deposits, particularly in Congo.

6.2.3 The Race is on: Proliferation and Arms Racing During the Cold War

The Soviet Union stepped up the efforts to develop its own nuclear weapons after 1945 and tested its first nuclear device on August 29, 1949, at the Semipalatinsk testing site (situated in present-day Kazakhstan). The test, codenamed "First Lightning" by the Soviets and "Joe-1" by the US, involved a 22 kT plutonium device. While the development and test were veiled in secrecy per se, President Truman publicly announced the end of the US nuclear hegemony on September 23, 1949 (Herken 1980).

The Soviet Union developed its nuclear weapons program primarily because it recognized the necessity of nuclear armament to restore and maintain a strategic balance with the nuclear-armed USA. For Soviet leader Joseph Stalin, who had initially shown little interest in developing nuclear weapons, the bombings of Hiroshima and Nagasaki underscored the urgency of countering the US nuclear monopoly. This shift in perspective was driven by the perception of an existential threat, emphasizing the need to secure national defense and ensure the survival of the state (Karlsch and Zeman 2008). In contrast to the Manhattan Project, the Soviet Union had to develop its program largely independently and under immense pressure. Soviet spies at the Manhattan Project played a crucial role in the 1940s, enabling the timely development of their nuclear weapons (Futter 2021).

Throughout the Cold War, the arms race between the United States and the Soviet Union became a defining feature. Both superpowers relentlessly competed to expand their nuclear arsenals, striving for nuclear superiority or at least parity. This intense rivalry led to the development and deployment of a vast array of nuclear weapons, including intercontinental ballistic missiles (ICBMs), submarine-launched ballistic missiles (SLBMs), and strategic bombers (Burr and Rosenberg 2010). This horizontal proliferation and diversification of arsenals peaked in the 1980s, when the combined nuclear arsenals of the two superpowers exceeded 60,000 warheads (Fig. 6.2). The doctrine of Mutually Assured Destruction (MAD) underpinned this arms race, asserting that the sheer scale of destructive power held by both sides would deter either from launching a first strike, as it would result in their own annihilation (Wohlstetter 1958).

6.2.3.1 Uranium Mining as a Prerequisite of Nuclear Weapons Development

For both the US and the Soviet Union, and any other nuclear-armed state, the production of nuclear weapons inevitably requires uranium, making uranium mining an integral part of any nuclear armament process. Although uranium is existent in various regions across the world, exploration and subsequent extraction have primarily been concentrated in territories that were (formerly) occupied, colonized, or inhabited by Indigenous communities (cf. Biswas 2021). During the entire Cold War period,

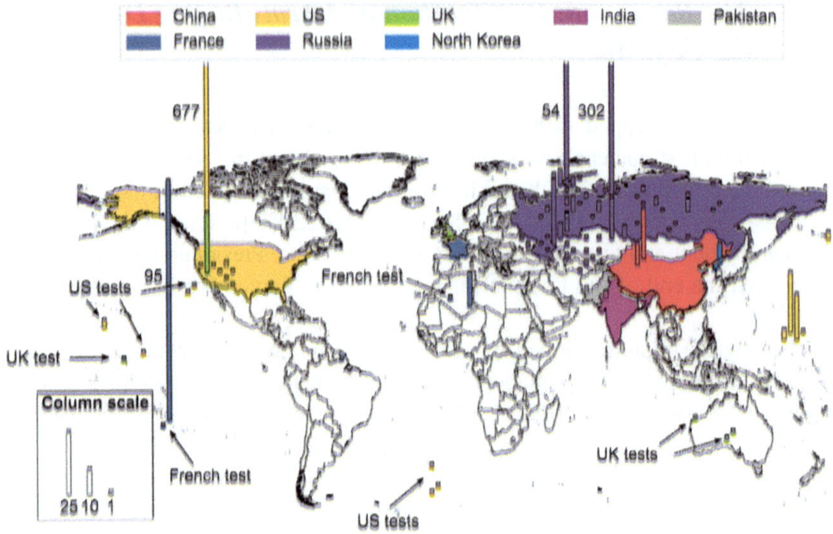

Fig. 6.1 Locations and number of global nuclear weapons tests. *Note* The color indicates the state that conducted the nuclear weapons tests, and the location is shown at the bottom of the bar (nearby locations have been clustered into one column). The height of the bar indicates the number of weapons tests. For large numbers of nuclear weapon test occurrences, the number to the left of the bar indicates the precise number. *Source* Own depiction with data taken from Yang et al. (2003), Johnston (2008), and Delbridge et al. (2023)

African states supplied 20–50% of the annual uranium supply to nuclear-armed states. One of the most important uranium suppliers for the US, besides Canada and indigenous uranium deposits, was the Shinkolobwe mine in the Haut-Katanga region of the Democratic Republic of Congo, a then Belgian colony.[18]

To gain and maintain parity with the US, the Soviet Union was urgently in need of large amounts of uranium for its nuclear weapons production, and significant effort was spent to close the uranium gap (Schütterle 2010). Domestic uranium deposits[19] and those in its Eastern European satellite states, especially in Czechoslovakia and the German Democratic Republic (GDR), were exploited. As early as the summer of 1945, military-led exploration expeditions for uranium ore were ordered and carried out. Uranium was first discovered on the Czechoslovak side of the Ore Mountains, and mining began immediately, using mainly political prisoners in the process (Ibid.). When the Soviet Union also detected uranium deposits in the GDR, extensive measures were implemented to extract it as quickly as possible. Between 1946 and 1990, uranium was mined by the company Wismut in Thuringia and Saxony,

[18] There has been no acknowledgement, let alone compensation or sustainable remediation measures, for the health consequences of radiation exposure, water pollution, and environmental damage caused by uranium mining (cf. Hecht 2012; Winde et al. 2017).

[19] The uranium deposits within the Soviet Union predominantly originated from regions in present-day Kazakhstan, Uzbekistan, Ukraine, and Tajikistan.

making the GDR and Wismut the fourth largest uranium producer in the world after the Soviet Union, the US, and Canada.[20]

6.2.3.2 The Role of Nuclear Testing in the Cold War Arms Race

Apart from mining, testing is another important component of the cold war arms race. States conduct nuclear tests for two purposes: First, to determine the performance and effects of nuclear devices under different conditions, and second, for nuclear signaling to either prove to the world that a state has managed to develop nuclear weapons or continued capability for potential use by an already nuclear-armed state. Nuclear testing is the defining criterion for nuclear weapon states under NPT Art. IX, Para. 3. The US and the Soviet Union continuously embarked on a qualitative and quantitative arms race, including the development of the hydrogen bomb, which the Soviet Union first tested in August 1953 (Wellerstein and Geist 2017). Both also began their extensive series of nuclear tests, with as many as 178 yearly tests in 1962.

Since 1945, more than 2000 nuclear weapons tests have been conducted, most of them by the US (1030), followed by the Soviet Union/Russia (715), France (210), the UK (45), China (45), North Korea (6), India (3), and Pakistan (2). The last nuclear weapon test (as of September 2025) was conducted by North Korea in 2017, but work on computer simulations of nuclear weapons tests (US Department of State 2017) and subcritical tests (NNSA 2024) continues. Noteworthy is also recent research on inertial confinement fusion, a process to initiate nuclear fusion reactions, which is not mainly conducted to gain insights into nuclear fusion for power generation but rather to study material properties relevant for nuclear weapons (cf. Mecklin 2022).

Similar to uranium mining, the history behind nuclear testing is shaped by colonial, patriarchal, and racist power mechanisms within the nuclear order and the associated systemic injustice (Teaiwa 2010; Biswas 2014). The nuclear weapons tests were conducted mainly in (formerly) colonized, peripheral, or indigenous territories (Baldus et al. 2021), including, e.g., in Kazakhstan, Australia, Algeria, the Pacific Island states and indigenous land in the US and China (Fig. 6.1). They have caused intergenerational health damage, displacements, environmental destruction on land and sea, contamination of food and water sources, and the loss of cultural heritage (Hood 2023). The consequences have largely not been addressed, redressed, or reconciled.

6.2.3.3 The Globalization of the Nuclear Arms Race

The UK began its nuclear weapons research during the 1940s, with British scientists directly involved in the Manhattan Project. As close allies, the two countries signed an agreement in 1943 to share all nuclear fission and weapons development research. However, the 1946 US Atomic Energy Act (or McMahon Act) disrupted

[20] https://www.wismut.de/de/wismut_chronik.php.

this collaboration due to fears of Soviet espionage, placing all nuclear research under civilian rather than military supervision and severing the ties between the US and the UK in this regard (Government of the United Kingdom 2006). Consequently, British scientists worked independently on their nuclear weapons program. The UK conducted its inaugural nuclear tests on October 3, 1952, aboard the HMS Plym in a lagoon of the Montebello Islands off the coast of Australia (Tynan 2016).

Shortly after, in November 1952, the US conducted its first thermonuclear test and throughout the 1950s, atmospheric testing continued, contributing to global concerns about radioactive fallout. While France conducted its first nuclear test in 1960, followed by China in 1964, the Partial Test Ban Treaty (PTBT) of 1963 banned nuclear tests in the atmosphere, outer space, and underwater. Verification measures at that time did not allow checks against underground explosions, limiting the treaty's scope. Some argue that this limitation was used to justify continued testing by the nuclear-armed states. Civil society actors and scientists worldwide were engaged in advocating for the PTBT, driven by public concerns about the global consequences of nuclear testing. For instance, traces of radioactive fallout were found in the teeth of infants, highlighting the widespread impact of atmospheric tests (Reiss 1961). By the mid-1960s, the five permanent members of the UN Security Council, the US, the Soviet Union, UK, France, and China (collectively known as the P5) had all become nuclear-armed states. This period marked the peak of nuclear tension, specifically with the Cuban Missile Crisis of 1962 standing as the closest the world came to nuclear war, while close calls and near catastrophes are a constant throughout the nuclear age (See Sect. 6.2.1.3).

During the same period, Israel's nuclear ambitions rose, and it presumably assembled its first nuclear weapon in 1967, although it has never conducted a nuclear weapon test to confirm its capabilities (see Footnote 1). Israel's nuclear program was inaugurated in the 1950s under Prime Minister David Ben-Gurion, with significant assistance from France, which provided a research reactor and plutonium production technology in 1957 (Cohen 1998). By 1958, construction of the Negev Nuclear Research Center near Dimona was underway, officially for civilian purposes (Kristensen and Korda 2022), and Israel acquired 20 tons of heavy water from Norway in 1959, enabling plutonium production by 1966.[21] Israel's policy of "nuclear opacity" or "Animut"—maximal ambiguity regarding its nuclear arsenal—allows it to maintain deterrence without openly declaring itself a nuclear—armed state (Cohen 2010), reflected in statements such as "We won't be the first to introduce nuclear weapons into the Middle East."[22] This was tested during crises such as the Six-Day War in 1967, the Yom Kippur War in 1973, and the Vela Incident in 1979 (Kristensen and Korda 2022), but remains a cornerstone of Israel's nuclear strategy today.[23]

---

[21] https://nsarchive2.gwu.edu/nukevault/ebb510/

[22] https://www.cfr.org/blog/israels-nuclear-weapons-program-and-lessons-iran.

[23] In 2023, Israel's Heritage Minister Amihai Eliyahu was suspended by Prime Minister Benjamin Netanyahu for suggesting that attacking Gaza with nuclear weapons would be an option in the ongoing conflict following the October 7 attacks by Hamas (Mecklin 2023).

6.2.3.4 Negotiating Stability: The Impact of Nuclear Arms Control and Verification

In parallel with the arms race, significant efforts were made with key treaties negotiated in the realm of arms control to enhance strategic stability and reduce the risk of nuclear conflict between the two superpowers. The Strategic Arms Limitation Talks (SALT) led to the SALT I treaty in 1972 and SALT II in 1979, which placed limits on the number of strategic ballistic missile launchers and provided verification measures to build trust for further deliberations. The Anti-Ballistic Missile (ABM) Treaty of 1972 limited the deployment of missile defense systems, reinforcing the MAD doctrine by ensuring that neither side could fully protect itself from a retaliatory strike. The Intermediate-Range Nuclear Forces (INF) Treaty of 1987 was another landmark agreement which eliminated an entire class of nuclear missiles and helped to reduce tensions in Europe. Despite facing numerous challenges and periods of tension, these arms control efforts played a crucial role in mitigating the dangers of the nuclear arms race and contributed to easing Cold War hostilities by offering avenues for building confidence and trust (Woolf 2023).

Verification efforts are key to any confidence-building measure and in the context of arms control refer to "the process of determining whether parties to an agreement are in compliance with their obligations" (cf. Hinderstein 2010). Typically, verification measures are defined in the treaty text, and the measures combine obligations to provide information (e.g., how many weapons are deployed on delivery systems) and ways for the other party to collect data (e.g., conducting inspections). The goal of verification is to build confidence in the other party's compliance, to be able to detect relevant violations early enough to respond, and thus to deter violations. Verification can also be seen as an opportunity for cooperation and collaboration, as well as being able to demonstrate one's own "good will" and assessing the other's intention for compliance (cf. UNIDIR 2010).

Agreeing on verification measures is challenged by the need to find a balance between requirements for secrecy on the one hand and the needs for transparency on the other. The history of US–Soviet/Russian nuclear arms control provides a good example of how such balance can shift (cf. Müller 1993): Verification of the SALT I treaty simply meant acknowledging that the other party would use "national technical means" (NTM), e.g., satellite imagery, to monitor the other's activities; any access to the other's territory was deemed too intrusive. Notably, the 1964 PTBT did not include a ban on underground testing, in part because the USA and Russia could not agree on verification measures that would be able to detect these tests (cf. Schmalberger 1991).

This changed with time and the accumulation of experience in bilateral arms control. The 1987 INF Treaty, banning, among others, nuclear-capable intermediate-range weapons—included more intrusive and cooperative measures, including permanent monitoring of one INF missile assembly factory for each party, and on-site inspections on Soviet and US territory. This path was continued with the 1991 START Treaty, which provided even more data sharing (including weapons and their

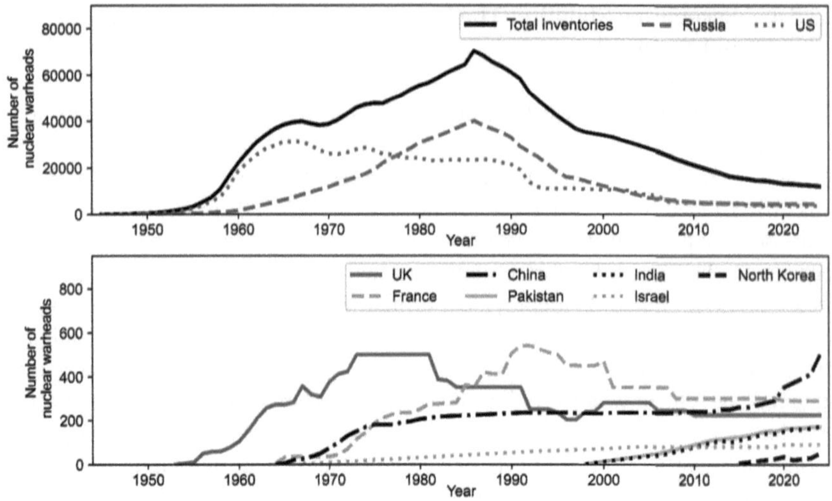

Fig. 6.2 Development of nuclear warheads. *Source* Own depiction with data taken from the Federation of American Scientists[24]

locations) and extensive on-site inspections including random short-notice challenge inspections (cf. Russell 2001).

Finally, the 2010 New START Treaty again reduced the number and complexity of the inspection system; this time, not because of mistrust but in recognition of the change in US-Russian relations (cf. Woolf 2011). The large amount of comprehensive information collected in the past, and along with it diminished concern about the other party's incentive, reduced concerns about breaches of the treaty obligation. Unfortunately, this trend has not continued to the present day: Since the Covid pandemic, Russia has suspended New START inspections, and since 2023, has even suspended its participation in New START (cf. Bugos 2023), which formally expires without the possibility of an extension on February 4, 2026. This will likely leave the US and Russia, but also the world, without an active nuclear arms control treaty for the first time in 50 years, if no successor agreement is negotiated.

6.2.4 The Nuclear Club and Its Current Members

Over time, the landscape of nuclear weapons proliferation has evolved significantly, with several new nations joining the ranks of nuclear-armed states. The so-called "Nuclear Club" now encompasses a broader range of countries (Fig. 6.2 shows the development of current estimates of nuclear weapons arsenals), each with distinct motivations, strategies, capabilities, and arsenal configuration.

---

[24] https://fas.org/initiative/status-world-nuclear-forces/

India, never having signed the NPT, embarked on its nuclear journey in the 1960s under Prime Minister Jawaharlal Nehru with a stated focus on civilian technology. In this effort, India received significant assistance from the US through the Atoms for Peace program, as well as support from Canada to develop its civilian nuclear infrastructure, enabling the production of plutonium (Mishra 2018). Despite early collaboration with Western nations, India maintained a degree of opacity regarding its nuclear capabilities, conducting its first test in 1974, known as "Smiling Buddha", framing it as a peaceful nuclear explosion (PNE), which cast doubt on the nonproliferation regime and its prospects of curbing further proliferation (O'Mahoney 2020). Officially, India claims to have conducted its first nuclear weapons test in 1998.[25]

While official numbers remain undisclosed, India has since expanded its arsenal, initially focusing on regional deterrence, particularly against Pakistan, and later also shifting its attention also toward long-range missile development with China in mind (Kristensen et al. 2024a). India has maintained a No First Use (NFU) policy since 1994 (Sundaram and Ramana 2018), though it softened this stance in 2003 and again in 2018, allowing for contingencies for nuclear response to attacks involving chemical and biological weapons. India's nuclear arsenal continues to grow and is being modernized, with increasing emphasis on missile systems, though its warheads are reportedly stored separately, with uncertain readiness levels (cf. Kristensen et al. 2024a).

Pakistan's nuclear program, which was begun in the 1950s, was first focused on civilian technology with the acquisition of a heavy-water reactor from Canada (Schalz et al. 2024). However, after India's first nuclear test in 1974, Pakistan intensified its efforts, receiving substantial support from China in the 1980s, including critical technology, expertise, and materials (Munir 2018). Technological collaboration with North Korea, along with possible financial backing from Islamic states like Saudi Arabia and Libya, further accelerated its nuclear development (Ahmed 1999). Abdul Quadeer (A. Q.) Khan, often called the "father of Pakistan's nuclear bomb," played a pivotal role through his work on uranium enrichment at URENCO, a nuclear fuel company in the Netherlands, leveraging his expertise to help Pakistan fulfil its nuclear ambitions (Squassoni 2006).

Pakistan conducted its first nuclear test in 1998, shortly after India's series of tests. Official numbers regarding its arsenal remain undisclosed. It follows a policy of "credible minimum deterrence," limiting the numbers and scenarios of use to defense and deterrence purposes while keeping weapons and delivery systems stored separately. Pakistan has continuously expanded its nuclear capabilities, particularly its delivery systems, and its Strategic Plans Division within the National Command Authority plays a key role in decision-making. Ongoing tensions with India, alongside implicit nuclear threats, remain central to Pakistan's strategic posture, as well as a steady increase in its stockpile of fissile materials.[26]

[25] https://www.fes.de/ipg/ipg4_98/artmohan.html.

[26] https://www.armscontrol.org/act/2023-10/features/pakistans-evolving-nuclear-doctrine#endnote05.

The last country to join the "Nuclear Club" was North Korea, which also began to pursue its nuclear program in the 1950s, supported initially by the Soviet Union, which provided training and technology for civilian nuclear development (Barannikova 2022). By the 1980s, North Korea had shifted focus toward developing weapons, secretly constructing facilities for plutonium production. In the 1990s, cooperation with Pakistan, particularly through the A.Q. Khan network, enabled North Korea to advance its uranium enrichment capabilities.

North Korea conducted its first nuclear test in 2006, marking its emergence as a nuclear-armed state, though the exact number of warheads remains unknown due to the secretive nature of its program. The regime's nuclear doctrine emphasizes deterrence, aiming to counter perceived threats from the US and South Korea, with a declared willingness to use nuclear weapons if attacked. North Korea has continuously developed its missile systems, now capable of reaching long-range targets, including the US mainland. Despite international sanctions and diplomatic pressure, the country steadily expands its stockpile of fissile materials, leveraging its nuclear capabilities as a central tool for both domestic legitimacy and international bargaining power (Kristensen et al. 2024b).

In summary, today's "Nuclear Club" encompasses nine members with more than 12,100 warheads in their possession,[27] as well as large stockpiles of HEU and separated plutonium that could be used to build even more nuclear weapons. For the US and the UK, relatively recent government declarations are available; for the others, expert assessments provide estimates with uncertainties between 10 and 30% (Fig. 6.3). In 2021, the global military stockpiles of separated plutonium amounted to 140 ± 10 tons, and of HEU to about 1290 ± 125 tons, most of which are held by the US and Russia (International Panel on Fissile Materials 2022).

In addition to these military stockpiles, significant amounts of fissile material potentially available for nuclear weapons production are also present in the civilian programs in form of HEU, as fuel for civilian research reactors or medical isotope production, and separated plutonium, envisaged to be used as nuclear fuel.[28] Today, global civilian stockpiles yield some 317 tons of separated plutonium and 42 tons of HEU overall, most of which are held again by the US and Russia (International Panel on Fissile Materials 2022). Significant civilian stocks of separated plutonium are also held by France and Japan, whose nuclear industries include reprocessing capabilities. For non-nuclear weapon states, such as Japan, civilian stocks are subject to mandatory monitoring under IAEA Safeguards. However, the five nuclear-armed states party to the NPT also provide these civilian stocks to IAEA Safeguards as part of a voluntary agreement with the IAEA.

[27] The total stockpiles of the US and Russia also include warheads that have been selected for dismantlement but are still intact ($\approx$ 1000 for each). Information taken from https://fas.org/initiative/status-world-nuclear-forces/.

[28] For more details on plutonium separation from spent fuel for fuel recycling, see von Hippel and Takubo (2022).

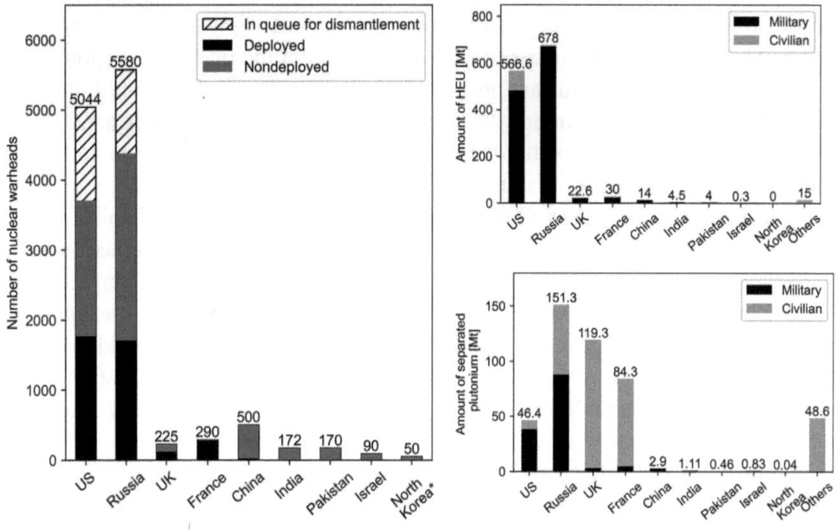

Fig. 6.3 Nuclear warhead and fissile material inventories 2025. *Sources* Bar chart showing (left) nuclear weapon inventories of the nuclear armed states (as of July 2024 taken from the Federation of American Scientists[29]) and (right) fissile material inventories in metric tons (Mt) (as of end of 2020), (International Panel on Fissile Materials 2022) for (top) HEU for civilian (grey) and military (black) purposes and (bottom) separated plutonium

6.3 Nuclear Nonproliferation

Throughout the nuclear age, both nuclear armed and non-nuclear armed states have continuously strived to constrain and hinder proliferation efforts of others. Nuclear nonproliferation, or "the effort to stop the increase in nuclear weapons,"[30] was one of the driving factors of the negotiations of the NPT in the 1960s and has stayed a core principle in the global nuclear order. Yet, motives for nonproliferation differ, ranging from a normative aspiration against the further spread of nuclear weapons to the retention of the power heralded within the possession and not wanting other states to acquire such power.

6.3.1 Nonproliferation Efforts and Treaties

The NPT is the single most important instrument and framework in this regard, prohibiting non-nuclear weapon states from acquiring or developing nuclear weapons, and obliging nuclear weapon states to not disseminate nuclear weapons to others—often referred to as the "grand bargain"—committing all signatories to

---

[29] https://fas.org/initiative/status-world-nuclear-forces/.

[30] https://dictionary.cambridge.org/de/worterbuch/englisch/nonproliferation.

pursue disarmament negotiations and guaranteeing the right to use nuclear technology for any peaceful purposes.[31] The NPT was, on the one hand, negotiated to slow down, if not hinder, further proliferation, particularly of those states that were known to have nuclear weapons development programs. On the other hand, it was negotiated to foster and institutionalize the norm against proliferation and set an international legal framework to harness efforts towards disarmament.

Another significant aspect of the nonproliferation agenda is the establishment of Nuclear-Weapon-Free Zones (NWFZs). The Treaty of Tlatelolco, signed in 1967, was a pioneering agreement that established the first NWFZ in Latin America and the Caribbean. This treaty was groundbreaking as it prohibited the testing, use, manufacture, production, and acquisition of nuclear weapons within the designated zone. The Treaty of Tlatelolco set a significant precedent for regional nonproliferation efforts and inspired subsequent NWFZ treaties in other parts of the world. By promoting the idea of nuclear weapon-free regions, the treaty contributes to global nonproliferation objectives and enhances regional security and stability. Since then, four more NWFZs have been established, covering vast regions including the South Pacific, Southeast Asia, Africa, and Central Asia. Additionally, Mongolia has declared itself a single-state NWFZ (Kutchesfahani 2019).

The Comprehensive Nuclear-Test-Ban Treaty (CTBT) and the Nuclear Suppliers Group (NSG) are additional frameworks that bolster the nonproliferation regime. The NSG, a group of nuclear supplier countries,[32] seeks to control the export of materials, equipment, and technology that could be used to produce nuclear weapons, thereby ensuring that nuclear trade does not contribute to proliferation (Kimball 2016). The CTBT, although not yet in force, introduces a ban of all nuclear test explosions and thereby surpasses the CTBT and establishes a verification system that is already in place and functioning. To this end, the Comprehensive Nuclear Test Ban Treaty Organization (CTBTO) has established and maintains the International Monitoring System (IMS), which is capable of detecting nuclear explosions with yields larger than 1 kT worldwide, and down to 0.1 kT in many environments (National Academy of Sciences 2002) with globally distributed monitoring stations that record seismic, infrasound, hydroacoustic, and radionuclide data. By combining information from these stations, nuclear explosions could be detected and located. The nuclide stations are of particular value because only they can prove that an explosion was indeed a nuclear explosion; together with atmospheric transport models and weather data, the origin of such nuclides can be deduced, and the type of nuclides allows to distinguish traces from civilian facilities, such as medical isotope production, from nuclear explosions. The IMS enabled the detection and localization of North Korea's six nuclear tests between 2006 and 2017 (CTBTO 2024).

However, the effectiveness of any such nonproliferation efforts can be undermined by discrepancies between declaratory policies and actual practices. While

[31] https://www.armscontrol.org/act/2003-12/features/nuclear-weapon-states-and-grand-bargain.

[32] The NSG was founded in 1974 by the US, the Soviet Union, the UK, France, Germany and Canada and comprises a total of 48 members today.

many states publicly support nonproliferation, their (in-)actions sometimes contradict their stated commitments. For example, all nuclear-armed states continue to modernize and expand their nuclear arsenals and certain non-nuclear-weapon states (e.g., Iran, Libya, and South Africa) have been or are suspected of pursuing nuclear capabilities under the guise of peaceful programs, others strive for nuclear latency. These contradictions highlight the ongoing challenges in achieving comprehensive and consistent nonproliferation. The recent emergence of "loose talk" about nuclear weapons (Meier and Vieluf 2021), further escalatory nuclear rhetoric, and threats are both symptoms and drivers of such damaging dynamics. Nuclear nonproliferation, therefore, remains a complex and multifaceted challenge, and its pursuit continues to require vigilance, cooperation, and the alignment of policies and practices to address the evolving landscape of nuclear risks and threats.

6.3.2 *Recent Challenges to the Nonproliferation Regime*

The global nonproliferation regime faces multiple challenges. On the one hand, today's world order is characterized by rising global tensions and increasingly nationalist-oriented leaders affecting a plethora of multilateral fora and negotiations. On the other hand, technological progress can lead to new challenges. The following section highlights three examples of recent developments that are discussed as potentially endangering the nonproliferation regime, namely Iran's nuclear ambitions, the AUKUS deal, and the "renaissance" of nuclear energy.

6.3.2.1 Iran's Nuclear Program and the JCPOA

Iran ratified the NPT in 1970, committing to remain a non-nuclear weapon state and accept IAEA safeguards under Article III. In line with this commitment, it signed a Comprehensive Safeguards Agreement (CSA) with the IAEA in 1974 and later signed the Additional Protocol in 2003.[33] Since 2002, Iran's nuclear activities have come under international scrutiny for its potential pursuit of nuclear weapons capabilities. Areas of concern include gas centrifuge uranium enrichment and uranium conversion facilities, both of which could facilitate the production of HEU, and the construction of a heavy-water reactor that could be used to produce plutonium (cf. Kerr 2024).

In 2005, the IAEA's Board of Governors found Iran to be in noncompliance with its CSA, and the case was reported to the UN Security Council in 2006. Despite diplomatic efforts, including engagement from France, Germany, and the UK, the collaboration between Iran and the IAEA remained challenging, with Iran resisting full transparency and cooperation. The UN Security Council issued six resolutions imposing sanctions on Iran to pressure compliance. In 2015, the Joint Comprehensive Plan of Action (JCPoA) was negotiated between Iran, China, France,

[33] https://www.iaea.org/newscenter/news/iran-signs-additional-protocol-nuclear-safeguards.

Germany, Russia, the UK, the US, and the EU as coordinator. This landmark agreement restricted Iran's nuclear program in exchange for sanctions relief, with Iran agreeing to limit its enriched uranium stockpile, dismantle most of its centrifuges, and halt plutonium production activities. The deal was endorsed by UN Security Council Resolution 2231, which set the framework for the implementation of the JCPOA (cf. Kerr 2024).

However, in 2018, the US, under President Donald J. Trump, unilaterally withdrew from the JCPoA, stating that the deal "failed to protect America's national security interests" and alleging that Iran "negotiated in bad faith," used its benefits to fund "terrorist proxies such as Hizballah and Hamas," and preserved its ability to develop nuclear weapons whatsoever.[34] In response, Iran began scaling back its compliance with the agreement, and tensions heightened. By 2019, the IAEA reported that Iran had exceeded JCPoA limits on the number of installed centrifuges, enriched uranium stockpile, and the level of uranium enrichment. Tehran also resumed research and development activities prohibited by the JCPoA. Furthermore, unresolved issues regarding undeclared nuclear material and related activities continue to strain relations between the IAEA and Iran, as the government in Tehran remains unwilling to fully cooperate and ensure transparency (IAEA DG 2024).

The NPT's framework lacks formal procedures to undeniably determine whether a state has violated its obligations, particularly concerning compliance with IAEA safeguards under Article III and the prohibition on acquiring nuclear weapons under Article II. This gap complicates the international community's ability to address Iran's noncompliance effectively (Kerr 2019). After a decline in cooperation and several suspicious findings regarding undeclared and highly-enriched nuclear material and sites linked to potential weaponization activities, the IAEA Board of Governors on June 12, 2025 found Iran in non-compliance with its Safeguards Agreement (IAEA 2025a). The following day, Israel attacked Iranian nuclear infrastructure, later joined by the United States in the declared effort to disrupt if not destroy the potential nuclear weapons program. Iran subsequently seized cooperation with the IAEA, severely restricting inspector access and failing to provide credible explanations for outstanding concerns (Davenport 2025). As a response to Iran's continued non-compliance with its commitments under the JCPoA, the E3 countries (France, Germany, and the UK) initiated the snapback mechanism set forth in paragraph 11 of the UN Security Council Resolution 2231 to reinstate the sanctions due to "significant non-performance" by Iran after a 30-day deadline (UK Government 2025). After threatening to leave the NPT altogether and challenges to the legality of the initiation of the snapback, Iran and the IAEA advanced on the resumption of cooperation in September 2025 (IAEA 2025b). At the time of writing (mid-September 2025), it remains unclear whether negotiations will lead to the reinstatement of sanctions after the deadline on September 27, 2025, or whether Iran will be granted an extension to return to compliance with the JCPoA.

[34] https://trumpwhitehouse.archives.gov/briefings-statements/president-donald-j-trump-ending-united-states-participation-unacceptable-iran-deal/.

6.3.2.2 Non-Nuclear Weapon States Acquiring Nuclear Submarines

As of March 2025, six nuclear-armed states (the P5 and India) are in possession of nuclear submarines. On September 15, 2021, the UK and the US announced that they would provide Australia with at least eight conventionally-armed but nuclear-powered submarines as part of the "enhanced trilateral security partnership" (Carlson 2021). The submarines provided under the agreement, referred to as AUKUS, will contain small nuclear reactors that are likely to be fueled with HEU, similar to the submarines possessed by the UK and the US. These reactors will be built into the hull of the submarine, and they are fueled by the manufacturer for their entire lifetime of about approximately 30 years, after which they will be returned to the manufacturer, i.e., the reactors will not be refueled in Australia.[35]

This type of agreement is unprecedented[36]: for the first time, a non-nuclear weapon state will acquire HEU enriched to weapon-grade levels (~93% enrichment) for military purposes; not for nuclear explosions, but for military submarines. In this case, Article 14 of the CSA signed with the IAEA allows this fissile material to be kept outside the safeguards regime; all that is required is an agreement with the IAEA to assure that the material will not be used to build nuclear weapons and to inform the Agency of the total quantity and composition of the withheld material (Hoppenbrouwers 2024).

This precedent raises proliferation concerns, since it provides weapons-usable material outside the safeguards regime to the non-nuclear weapon state, namely Australia. While it may be plausible that the HEU is virtually inaccessible without cutting the hull of the submarine (the submarines will be delivered with a fully fueled naval reactor without the possibility for refueling in Australia); further concerns go beyond the AUKUS arrangement including fears of other non-nuclear-weapons following suit and even seeking to produce their own HEU for nuclear submarines outside of nuclear safeguard regimes under Article 14. In any case, the agreements concluded with the IAEA in the AUKUS case, for which an 18-month consultation period was set to find a solution with the IAEA, will set a precedent for the future (IAEA 2022b).

6.3.2.3 A Potential Renaissance of Nuclear Energy

At the NPT Review Conference 2022, the state parties could not agree on a final document, fueling the perception of a lack of engagement of nuclear-armed states

---

[35] https://thebulletin.org/2022/11/aukus-and-nuclear-powered-submarines-lets-all-just-take-a-breath/.

[36] The only other non-nuclear weapon state with a domestic nuclear-powered submarine program is Brazil. This ambition originally dates back to the 1970s, but was only revisited after 2008 and concretized through technical cooperation with France. In comparison with AUKUS, Brazil intends to use indigenous LEU for its submarines and therefore not triggering the same nonproliferation concerns (Stewart 2022).

towards disarmament. The additional lack of support for civilian nuclear technologies for non-nuclear weapon states frustrates some non-nuclear weapon states, who increasingly feel the "grand bargain" of the NPT not being fulfilled (Notte 2024).

Those challenges to the nonproliferation regime coincide with discussions to ramp up the development and use of civilian nuclear technologies. Potential new systems, such as small modular reactors (SMRs) could present new technological and organizational challenges to the nonproliferation regime: new reactor designs would potentially require new approaches within IAEA safeguards, and new reactors or technologies could become simply more abundant and located at more complicated to reach areas—a challenge to the IAEA who already faces the need to monitor an increasing amount of fissile material while being staffed with inadequate resourcing (Rotte 2024). While there is no automatic link between civilian nuclear technologies and nuclear weapons programs, the nuclear renaissance could lead to more states being nuclear latent in the future.[37]

6.4 Nuclear Disarmament

The call for nuclear disarmament is as old as the development of the weapons themselves. With that, however, nuclear disarmament advocates sometimes mean different things: Nuclear disarmament is either referred to as the process of reducing or giving up such weapons, or it is envisioned as the goal of complete disarmament or the absolute absence of nuclear weapons, often referred to as "Global Zero".

6.4.1 *Disarmament as a Goal*

The debate about whether nuclear weapons make the world safer or less safe has persisted throughout the nuclear age. Central to this discourse are differing perspectives on how security is achieved and for whom it is guaranteed. One common thread in these debates is the consideration of risk, understood as a combination of the probability of an event and the consequences of that event. Proponents of the view that nuclear weapons enhance global security argue that their existence lowers the likelihood of global wars through deterrence, thus reducing the associated risk of nuclear extinction or large-scale conflicts. This perspective emphasizes state security, defined by military power, sovereignty, and territorial integrity (Tickner and Sjoberg 2011).

Conversely, disarmament proponents emphasize that, no matter how small, the probability of nuclear war is never zero. They argue that the catastrophic consequences of any nuclear weapons use make the mere existence of these weapons an

[37] Refer to the reports by Pistner et al. (2021, 2024) for safety and proliferation-related assessments of non-light water and so-called small modular reactors.

unacceptable risk. This view shifts the focus from state to human security, which encompasses individual well-being and considers multiple dimensions of security, including health, environmental sustainability, and economic stability. It critiques structural violence and the injustices that nuclear weapons perpetuate (Tickner and Sjoberg 2011).

While inequality and injustice are deeply intertwined with the issue of nuclear weapons, these factors have not historically held a central place in international relations or scholarly nuclear discourse (Acheson 2016). The prioritization reflects a broader historical emphasis on stability, predictability, and order, overshadowing concerns for justice, thereby sustaining an ahistorical realism that disregards the negative effects of the existence of nuclear weapons (Ray 1999). The concept of nuclear justice, primarily shaped by the testimonies of communities affected by nuclear testing, focuses on lived experiences and the humanitarian consequences of nuclear weapons. From this perspective, any compensation or reparation for the suffering caused cannot focus only on the victims and affected populations must also hold the responsible states accountable and ensure that they take responsibility for their actions (cf. Baldus et al. 2021).

Within an institutional framework, nuclear justice is discussed in regard to the asymmetry between nuclear and non-nuclear weapon states, particularly in the context of the NPT. Nuclear technology and responsibility are unequally distributed among the states, especially since only nuclear-armed states can fulfill the disarmament obligations as the sole possessors of these weapons (Biswas 2014; Egeland 2021). Even though most, if not all, states continuously state their commitment towards nuclear disarmament in national and international fora, the nuclear-armed states have a special obligation to suit their actions to these words. The goal to fully reduce the risks associated with nuclear weapons and their use can only be reached by their total elimination through concrete steps and regulations. Work towards a world without nuclear weapons hinges on the engagement, willingness, and commitments of nuclear-armed states to disarm and non-nuclear weapon states to demand for it—with the former lacking both concrete and holistic initiative, particularly beyond the P5 and including all nuclear-armed states.

6.4.2 Disarmament as a Process

The technical process of dismantling nuclear weapons itself is simple: since all nuclear weapon states have extensive knowledge and experience in maintaining their nuclear warheads, dismantling them would be straightforward. Russia and the US have demonstrated their dismantlement capabilities by reducing the large numbers of warheads from the heights of the Cold War. The open question, however, is how the role of nuclear weapons for security can be reduced, or what the conditions for nuclear disarmament would be, so that the desire for security can be met for all states without nuclear weapons, or where nuclear weapons become too much a burden, regarding it risk or costs, compared to their benefit.

A world without nuclear weapons, however, is likely to be different from today's world in which the nuclear order is so deeply embedded in the international system. Extrapolating from today's world to the world in which nuclear disarmament takes place is likely to be uncertain. Within these limits, however, it seems plausible that disarmament would take place within a bi- or multilateral agreement, and that in this context states would be required to take steps to: (1) prove that they are disarming or have already disarmed, and (2) take steps that deny them the possibility to quickly rearm. These two aspects can only be achieved if the disarmament efforts are conducted in a transparent, verifiable, and irreversible manner.

A case of nuclear disarmament can be considered verifiable if the relevant steps towards the disarmed end-state or the end-state itself can be verified properly; how much verification would be sufficient depends strongly on the specific conditions. Particularly in the case of nuclear disarmament, verification is challenged by the fact that the disarming state may still want to keep its nuclear secrets, e.g., the design of warheads. Whether for military-strategic reasons or simply to hide past issues like environmental disasters during weapons production, this interest must be balanced with the need to collect data to assess whether the state is disarming and not cheating. An additional concern is that under the NPT, non-nuclear weapons states are not allowed to see details on the nuclear weapons designs, as this would violate their nonproliferation obligations. To address these challenges, research has been conducted and is ongoing, particularly in the field of measurement techniques to authenticate nuclear warheads without revealing sensitive data (Glaser et al. 2014), or even the development and testing of complete protocols for a warhead dismantlement procedure, from the detachment of the delivery vehicle to the disposal of the separate parts (IPNDV 2020).

In addition to the dismantlement of declared warheads and materials, there is the question of how to ensure that no undeclared warheads or materials remain. Methods for reconstructing the nuclear material production and removal history in a state ("nuclear archaeology") could be useful in this regard, as they would make it possible to assess whether they are consistent with today's declared amount of material. This approach was used by inspectors to verify the disarmament in South Africa.

While there is a rich body of literature on nuclear disarmament verification, the issue of irreversibility has only recently entered the scholarly debate. Irreversible nuclear disarmament can be defined as "a feature/quality of a disarmament or an arms control process that involves limiting the capacity for re-armament, including the possible re-constitution of aspects of weapons program[s]" (Muti et al. 2023, 292). It is generally accepted that *irreversibility* is a spectrum rather than a binary. Nuclear disarmament that would only involve the dismantlement of nuclear warheads, but also include the retention of all nuclear materials and weapons facilities within the state, would possibly be associated with a lower degree of irreversibility, as the state could relatively quickly use this material to rebuild nuclear weapons. At the other end of the spectrum would be a world without any nuclear technology, similar to the world before the invention of the nuclear weapon in 1945. The level of irreversibility required and accepted would depend on the specific case and context.

Recent commentaries have also suggested that a strong degree of irreversibility could be counterproductive, as states could be less willing to join an agreement on nuclear disarmament without a certain degree of flexibility (Rodgers and Williams 2023).

Beyond these technical dimensions, the scholarly debate further emphasizes that there are also political, social, and legal factors to irreversibility, for example, nuclear disarmament embedded in a formal agreement may be considered more irreversible than one based on a unilateral decision. Nuclear disarmament cases that are strongly supported by civil society and embedded in a national identity may also be more irreversible than those where a leadership/elite has taken the decision (which could then be changed simply by another leadership).

The case of South Africa, the only country known to have dismantled a nuclear weapons program, is illustrative. Between 1979 and 1989, South Africa secretly developed a nuclear weapons program, culminating in the creation of six completed and one incomplete nuclear warhead using HEU from its domestic uranium mines and enrichment facilities. In the context of its international isolation and Cuban forces in Angola, these nuclear weapons were part of South Africa's *three-phase* doctrine, in which South Africa would first maintain strategic ambiguity, i.e., neither confirming nor denying the weapons, and possibly later, if South Africa's territory was attacked, revealing their existence to pressure international intervention (cf. Stumpf 1995). However, once the war was over and the apartheid regime dissolved, South Africa sought to participate in the international community and abandon its isolated, sanctioned position, for which joining the NPT was the first step.

In 1991, South Africa secretly dismantled its nuclear weapons and acceded to the NPT, just as IAEA inspectors were verifying South Africa's initial declarations (which included a suspiciously large amount of HEU, for which South Africa was not required to declare its origin or purpose), then-President Frederik Willem de Klerk revealed the past nuclear weapons program (cf. de Klerk 1993). The IAEA's mandate, through a UN resolution, was expanded to include: (1) inspections of the weapons complexes conducted by US nuclear weapons experts, and (2) reconciliation of many historical documents and materials to reconstruct and verify the past production of the HEU to confirm that all material had now been put under IAEA monitoring. This task was extensive, involving more than 12,000 technical documents (cf. Albright and Stricker 2016), and relied on the strong cooperation shown by the South Africans; the latter is likely to be even more important for future disarmament cases, as today's nuclear weapons programs contain much more complex and large fissile material production histories that need to be understood.

While often referred to, particularly since the Russian invasion of Ukraine since February 2022, Ukraine is not a case of disarmament but an example of the influence and importance of international norms and agreements. After the Soviet Union dissolved in December 1991, Ukraine inherited the third-largest nuclear arsenal in the world, with approximately 1,700 nuclear warheads stationed on its territory. However, lacking both the infrastructure to maintain the warheads and the political will to become a nuclear-armed state (see Sect. 2), Ukraine chose a non-nuclear path. With memories of the 1986 Chernobyl nuclear power plant disaster still fresh, Ukraine negotiated away the nuclear arsenal it inherited in exchange for security assurances in the 1994 Budapest Memorandum. These guarantees, offered by the United States, Russia, and the United Kingdom, affirmed Ukraine's sovereignty and territorial integrity in return for its commitment to nuclear disarmament.[38]

Ukraine never assumed operational control over the warheads and its decision to pursue a non-nuclear status was pivotal in shaping the global nonproliferation regime. However, the effectiveness of the security assurances has been questioned, particularly considering Russia's annexation of Crimea in 2014 and its ongoing war of aggression against Ukraine since February 2022.

Similar to Ukraine, Kazakhstan and Belarus also inherited Soviet nuclear weapons in 1991. Both nations chose to relinquish these arsenals and signed agreements that ensured the removal of nuclear warheads from their territories (cf. Potter and Matthewss 1998). Today, both Kazakhstan and Belarus are non-nuclear weapon states under the NPT. Kazakhstan, in particular, has since become a strong advocate for nuclear disarmament, by hosting negotiations on the CTBT, actively promoting the establishment of a Central Asian Nuclear-Weapon-Free Zone and TPNW[39](cf. NTI 2020; Nuclear Weapons Ban Monitor 2023).

6.5 Conclusions and Outlook

This chapter gives an introduction to the multifaceted dynamics of nuclear weapons, from their development and proliferation to efforts aimed at nonproliferation and disarmament. It traces the historical and political conditions that have shaped the global nuclear landscape, beginning with the scientific discovery of nuclear fission and the subsequent development of nuclear weapons during WW II, through the Cold War arms race, to the current complexities of the nuclear order.

Nuclear weapons have remained central to international relations, as demonstrated by the enduring dominance of nuclear deterrence strategies and failure to address and redress past harm. The possession of nuclear weapons has granted states unique forms of power, not only in terms of deterrence and defense, but also in geopolitical bargaining. At the same time, nuclear proliferation has been a major concern, as an increasing number of states have either acquired nuclear weapons or developed the technical capacity to do so. A complex interplay between motivations—including security concerns and symbolic power—drives states' decisions to acquire or forgo nuclear weapons.

The future of disarmament is uncertain, but disarmament remains a critical goal for many actors in international politics. Despite the complexity of the current geopolitical landscape, opportunities for new agreements and initiatives may arise, especially with the involvement of emerging powers and non-state actors. Moreover,

---

[38] https://www.armscontrol.org/act/2014-12/features/looking-back-ukraines-nuclear-predicament-and-nonproliferation-regime#note2.

[39] http://www.nti.org/education-center/treaties-and-regimes/central-asia-nuclear-weapon-free-zone-canwz/; https://banmonitor.org/profiles/kazakhstan.

public opinion and advocacy by non-governmental organizations, civil society, and grassroots movements will continue to play a significant role in driving disarmament efforts, as could be seen with the TPNW. The influence of these actors has historically shaped the discourse on nuclear weapons and will likely push for greater accountability and transparency in nuclear policy.

Technological advancements, on the one hand, offer new tools for verification and monitoring. On the other hand, they bring their own set of challenges: Cyber threats, the modernization of missile systems, and other emerging technologies complicate efforts to control nuclear arsenals and will require the development of sophisticated verification mechanisms to ensure compliance with nonproliferation and disarmament treaties. Simultaneously, such technologies could undermine stability by increasing the risk of miscalculations or accidental launches.

The economic implications of disarmament are another important consideration. For nuclear-armed states, the cost of maintaining an arsenal is substantial, and the financial burden of modernization projects can be overwhelming. At the same time, dismantling nuclear weapons systems involves significant costs, raising questions about the feasibility and economic sustainability of large-scale disarmament efforts.

The fear and fascination associated with nuclear weapons have shaped international politics for decades and will continue to do so for the foreseeable future. With the potential resurgence of the nuclear arms race, paired with climate change developments, overall heightened geopolitical tensions, and a potential future expansion of civilian nuclear infrastructure, more countries will consider their role in the global nuclear order and foster their nuclear latency, their participation in extended nuclear deterrence arrangements, or even consider acquiring or building their own nuclear weapons. While the vision of a world free of nuclear weapons ("Global Zero") is fraught with obstacles, it is not impossible. The future of nuclear disarmament will depend not only on the geopolitical context, but also on the ability to establish new multilateral agreements that account for the roles of emerging powers and non-state actors. Advances in technology, combined with shifting political dynamics and public advocacy, may create new opportunities for disarmament in the coming years. This is already reflected in national and international fora, for which the resolutions voted for in the 79th Session of the UN General Assembly First Committee in November 2024[40] are an example: Alongside reaffirmations, e.g., that achieving a world without nuclear weapons is a common goal shared by the international community, it was decided to support one resolution on "Nuclear War Effects and Scientific Research," effectively launching a two-year study on the effects of nuclear war.[41] Achieving a future where the salience of nuclear weapons is reduced will require sustained international cooperation, innovative verification methods, remediation of human harm, and a re-evaluation of the role that nuclear weapons play in global security. Only then, a world without nuclear weapons will become a reality rather than a distant goal.

---

[40] https://press.un.org/en/2024/gadis3754.doc.htm.

[41] https://thebulletin.org/2024/11/un-to-conduct-new-study-of-the-broad-impacts-of-nuclear-war-not-all-countries-want-to-know/.

References

Abella, Maveric K. I. L., Monica Rouco Molina, Ivana Nikolić-Hughes, Emlyn W. Hughes, and Malvin A. Ruderman. 2019. "Background Gamma Radiation and Soil Activity Measurements in the Northern Marshall Islands." *Proceedings of the National Academy of Sciences* 116 (31): 15425–34. https://doi.org/10.1073/pnas.1903421116.

Abraham, Itty. 2010. "'Who's Next?' Nuclear Ambivalence and the Contradictions of Non-Proliferation Policy." *Economic and Political Weekly* 45 (43): 48–56.

Acheson, Ray. 2016. "Foregrounding Justice in Nuclear Disarmament: A Practitioner Commentary." *Global Policy* 7 (3): 405–7. https://doi.org/10.1111/1758-5899.12338.

Acton, James, and Carter Newman. 2006. "IAEA Verification of Military Research and Development." 5. Vertic Research Reports. Vertic. https://www.vertic.org/media/assets/Publications/VM5.pdf.

Ahmed, Samina. 1999. "Pakistan's Nuclear Weapons Program: Turning Points and Nuclear Choices." *International Security* 23 (4): 178–204.

Albright, David, and Andrea Stricker. 2016. *Revisiting South Africa's Nuclear Weapons Program: Its History, Dismantlement, and Lessons for Today*. Washington, DC: North Charleston, SC: Institute for Science and International Security; CreateSpace Independent Publishing Platform.

Arndt, Anna-Clara, and Liviu Horovitz. 2023. Nuclear Rhetoric and Escalation Management in Russia's War against Ukraine: A Chronolog.

Aron, Raymond. 1980. *Clausewitz. Thinking War [Clausewitz. Den Krieg Denken]*. Propyläen Verlag. https://www.zvab.com/9783549073995/Clausewitz-Krieg-denken-Raymond-Aron-3549073992/plp.

Baldus, Jana, Caroline Fehl, and Sascha Hach. 2021. *Beyond the Ban: A Global Agenda for Nuclear Justice*. PRIF Report 2021/4. Frankfurt am Main: Leibniz-Institut Hessische Stiftung Friedens- und Konfliktforschung (HSFK).

Barannikova, Anastasia. 2022. "Korean Peninsula Nuclear Issue: Challenges and Prospects." *Journal for Peace and Nuclear Disarmament* 5 (sup1): 50–68. https://doi.org/10.1080/25751654.2022.2053409.

Barrett, Anthony. 2016. *False Alarms, True Dangers? Current and Future Risks of Inadvertent U.S.-Russian Nuclear War*. RAND Corporation. https://doi.org/10.7249/PE191.

Biswas, Shampa. 2014. *Nuclear Desire: Power and the Postcolonial Nuclear Order*. Minneapolis: University of Minnesota Press.

Biswas, Shampa. 2021. "Masculinist States, Radioactive Contamination, and Transnational Nuclear Justice: A Conversation on Building Bridges across Borders." *International Feminist Journal of Politics* 23 (1): 149–69. https://doi.org/10.1080/14616742.2020.1861962.

Bugos, Shannon. 2023. "Russia Suspends New START." March. https://www.armscontrol.org/act/2023-03/news/russia-suspends-new-start.

Burr, William, and David Rosenberg. 2010. "Nuclear Competition in an Era of Stalemate, 1963–1975." In *The Cambridge History of the Cold War: Volume 2: Crises and Détente*, edited by Melvyn P. Leffler and Odd Arne Westad, 2:88–111. The Cambridge History of the Cold War. Cambridge: Cambridge University Press. https://doi.org/10.1017/CHOL9780521837200.006.

Burr, William, and Jeffrey Kimball. 2022. "Nuclear Threats and Alerts: Looking at the Cold War Background." Arms Control Association, April. https://www.armscontrol.org/act/2022-04/features/nuclear-threats-and-alerts-lookingcold-war-background.

Carlson, John. 2021. "IAEA Safeguards, the Naval 'Loophole' and the AUKUS Proposal." *VCDNP Report*, October, 10.

CNN. 2018. "Timeline of the Hawaii False Missile Alert Shows How Drill Went Wrong." CNN, January 31. https://www.cnn.com/2018/01/30/us/hawaii-false-missile-alert-timeline/index.html.

Cohen, Avner. 1998. "Before the Beginning: The Early History of Israel's Nuclear Project (1948–1954)." *Israel Studies* 3 (1): 112–39.

Cohen, Avner. 2010. *The Worst-Kept Secret: Israel's Bargain with the Bomb.* Columbia University Press.

CTBTO. 2024. "Detecting Nuclear Tests." https://www.ctbto.org/our-work/detecting-nuclear-tests.

Davenport, Kelsey. 2024. "IAEA Safeguards Agreements at a Glance." Arms Control Association, November. https://www.armscontrol.org/factsheets/iaea-safeguards-agreements-glance.

Davenport, Kelsey. 2025. "Israel and U.S. Strike Iran's Nuclear Program." Arms Control Association. https://www.armscontrol.org/act/2025-07/news/israel-and-us-strike-irans-nuclear-program.

De Geer, Lars-Erik, and Christopher M. Wright. 2018. "The 22 September 1979 Vela Incident: Radionuclide and Hydroacoustic Evidence for a Nuclear Explosion." *Science & Global Security* 26 (1): 20–54. https://doi.org/10.1080/08929882.2018.1451050.

Delbridge, B. G., J. D. Carmichael, W. S. Phillips, K. M. Cleveland, M. L. Begnaud, and C. Gammans. 2023. "Source Characterization of the Declared North Korean Nuclear Tests From Regional Distance Coda Wave Spectral Ratios." *Journal of Geophysical Research: Solid Earth* 128 (3): e2022JB024728. https://doi.org/10.1029/2022JB024728.

Egeland, Kjølv. 2021. "The Ideology of Nuclear Order." *New Political Science* 43 (2): 208–30. https://doi.org/10.1080/07393148.2021.1886772.

Einstein, Albert. 1939. "Letter to Franklin D. Roosevelt." August 2.

Erästö, Tytti. 2023. "The Role of Umbrella States in the Global Nuclear Order." Stockholm International Peace Research Institute. https://doi.org/10.55163/VYBU7480.

Fraise, Thomas, and Kjølv Egeland. 2023. "Able Archer: How Close of a Call Was It?" *Bulletin of the Atomic Scientists* 79 (3): 155–60. https://doi.org/10.1080/00963402.2023.2199577.

Fuhrmann, Matthew, and Todd S. Sechser. 2014. "Nuclear Strategy, Nonproliferation, and the Causes of Foreign Nuclear Deployments." *The Journal of Conflict Resolution* 58 (3): 455–80.

Fuhrmann, Matthew, and Benjamin Tkach. 2015. "Almost Nuclear: Introducing the Nuclear Latency Dataset." *Conflict Management and Peace Science* 32 (4): 443–61. https://doi.org/10.1177/0738894214559672.

Futter, Andrew. 2021. *The Politics of Nuclear Weapons: New, Updated and Completely Revised*. Cham: Springer International Publishing. https://doi.org/10.1007/978-3-030-48737-9.

Glaser, Alexander, Boaz Barak, and Robert J. Goldston. 2014. "A Zero-Knowledge Protocol for Nuclear Warhead Verification." *Nature* 510 (7506): 497–502. https://doi.org/10.1038/nature13457.

Goldansky, Vitaly. 1988. "Connection between Horizontal and Vertical Proliferation of Nuclear Weapons." In *Coexistence, Cooperation and Common Security: Annals of Pugwash 1986*, edited by Joseph Rotblat and Laszlo Valki, 21–36. London: Palgrave Macmillan UK. https://doi.org/10.1007/978-1-349-19369-1_2.

Goodwin, Bruce. 2021. *Nuclear Weapons Technology 101 for Policy Wonks.* Lawrence Livermore National Laboratory.

Government of the United Kingdom. 2006. "The History of the UK's Nuclear Weapons Programme." Fact Sheet 5: The History of the UK's Nuclear Weapons Programme. hhttps://assets.publishing.service.gov.uk/media/5a7987e840f0b642860d8d88/Cm6994_Factsheet5.pdf.

Gregory, Shaun, and Alistair Edwards. 1989. "The Hidden Cost of Deterrence: Nuclear Weapons Accidents 1950-88." *Bulletin of Peace Proposals* 20 (1): 3–26.

Grunden, Walter E., Mark Walker, and Masakatsu Yamazaki. 2005. "Wartime Nuclear Weapons Research in Germany and Japan." *Osiris* 20:107–30.

Hecht, Gabrielle. 2012. "An Elemental Force: Uranium Production in Africa, and What It Means to Be Nuclear." *Bulletin of the Atomic Scientists* 68 (2): 22–33. https://doi.org/10.1177/0096340212440352.

Herken, Gregg. 1980. "'A Most Deadly Illusion': The Atomic Secret and American Nuclear Weapons Policy, 1945-1950." *Pacific Historical Review* 49 (1): 51–76. https://doi.org/10.2307/3639304.

Hinderstein, Corey, ed. 2010. *Cultivating Confidence: Verification, Monitoring, and Enforcement for a World Free of Nuclear Weapons*. Hoover Institution Press Publication, no. 596. Stanford, Calif: Hoover Institution Press.

Hood, Anna. 2023. "You Were Bombed and Now You Have to Pay for It: Questioning the Positive Obligations in the Treaty on the Prohibition of Nuclear Weapons." *Leiden Journal of International Law* 36 (2): 273–94. https://doi.org/10.1017/S0922156522000747.

Hoppenbrouwers, Alexander. 2024. "The Non-Proliferation Considerations of Nuclear-Powered Submarines." European Leadership Network. October 2024. https://europeanleadershipnetwork.org/commentary/the-non-proliferation-considerations-of-nuclear-powered-submarines/.

Horovitz, Dr Liviu, and Anna Clara Arndt. 2024. "One Year of Nuclear Rhetoric and Escalation Management in Russia's War against Ukraine: An Updated Chronology."

Hymans, Jacques E. C. 2007. *The Psychology of Nuclear Proliferation: Identity, Emotions, and Foreign Policy*. Repr. Cambridge: Cambridge University Press. https://doi.org/10.1017/CBO9780511491412.

IAEA. 2016. "Information Collection and Evaluation." Text. IAEA, June 8. https://www.iaea.org/topics/informationcollection-and-evaluation.

IAEA. 2022a. *IAEA Safeguards Glossary*. 2022 edition. Vienna: International Atomic Energy Agency.

IAEA. 2022b. "IAEA Safeguards in Relation to AUKUS. Report by the Director General." GOV/INF/2022/20. International Atomic Energy Agency.

IAEA DG. 2024. "Verification and Monitoring in the Islamic Republic of Iran in Light of United Nations Security Council Resolution 2231 (2015)." Board of Governors. https://www.iaea.org/sites/default/files/documents/gov2024-7.pdf.

IAEA. 2025a. "NPT Safeguards Agreement with the Islamic Republic of Iran. Resolution adopted on 12 June 2025 during the 1769th session. GOV/2025/38." IAEA Board if Governors, June 12. https://www.iaea.org/sites/default/files/25/06/gov2025-38.pdf.

IAEA. 2025b. "IAEA Director General's Introductory Statement to the Board of Governors". IAEA, September 8. https://www.iaea.org/newscenter/statements/iaea-director-generals-introductory-statement-to-the-board-ofgovernors-8-september-2025.

International Panel on Fissile Materials. 2022. "Global Fissile Material Report 2022 - Fifty Years of the Nuclear Non-Proliferation Treaty: Nuclear Weapons, Fissile Materials, and Nuclear Energy." International Panel on Fissile Materials. https://fissilematerials.org/library/gfmr22.pdf.

IPNDV. 2020. "Verification of Each of the 14 Steps of Nuclear Weapon Dismantlement." https://www.ipndv.org/wp-content/uploads/2020/04/WG5-Deliverable_FINAL-.pdf.

Jo, Dong-Joon, and Erik Gartzke. 2007. "Determinants of Nuclear Weapons Proliferation." *Journal of Conflict Resolution* 51 (1): 167–94. https://doi.org/10.1177/0022002706296158.

Johnston, Robert. 2008. "Database of Nuclear Tests, France." Johnston Archive. February 20, 2008. https://www.johnstonsarchive.net/nuclear/tests/FR-ntests1.html.

Karlsch, Rainer, and Zbynek Zeman. 2008. *Uranium Matters. Central European Uranium in International Politics, 1900–1960*. Central European University Press. https://ceupress.com/book/uranium-matters.

Kemp, R. Scott. 2017. "Opening a Proliferation Pandora's Box: The Spread of the Soviet-Type Gas Centrifuge." *The Nonproliferation Review* 24 (1–2): 101–27. https://doi.org/10.1080/10736700.2017.1368649.

Kemp, R. Scott, Edwin S. Lyman, Mark R. Deinert, Richard L. Garwin, and Frank N. Von Hippel. 2024. "The Weapons Potential of High-Assay Low-Enriched Uranium." Science 384 (6700): 1071–73. https://doi.org/10.1126/science.ado8693.

Kerr, Paul. 2019. "Iran's Nuclear Program: Status." Congressional Research Service.

Kerr, Paul K. 2024. "Iran's Nuclear Program: Tehran's Compliance with International Obligations." R40094. Congressional Research Service. https://apps.dtic.mil/sti/citations/ADA511010.

Kimball, Daryl G. 2016. "The Enduring Nonproliferation Value of the Comprehensive Nuclear-Test-Ban Treaty." *The Nonproliferation Review* 23 (3–4). https://www.tandfonline.com/doi/pdf/https://doi.org/10.1080/10736700.2016.1256539?download=true.

Klerk, F.W. de. 1993. "Speech by South African President F.W. De Klerk to a Joint Session of Parliment on Accession to the Non-Proliferation Treaty." In. https://digitalarchive.wilsoncenter.org/document/92687/download.

Kristensen, Hans M., and Matt Korda. 2022. "Israeli Nuclear Weapons, 2021." *Bulletin of the Atomic Scientists* 78 (1): 38–50. https://doi.org/10.1080/00963402.2021.2014239.

Kristensen, Hans M., Matt Korda, Eliana Johns, and Mackenzie Knight. 2023. "Nuclear Weapons Sharing, 2023." *Bulletin of the Atomic Scientists* 79 (6): 393–406. https://doi.org/10.1080/00963402.2023.2266944.

Kristensen, Hans M., Matt Korda, Eliana Johns, and Mackenzie Knight. 2024a. "Indian Nuclear Weapons, 2024." *Bulletin of the Atomic Scientists* 80 (5): 326–42. https://doi.org/10.1080/00963402.2024.2388470.

Kristensen, Hans M., Matt Korda, Eliana Johns, and Mackenzie Knight. 2024b. "North Korean Nuclear Weapons, 2024." *Bulletin of the Atomic Scientists* 80 (4): 251–71. https://doi.org/10.1080/00963402.2024.2365013.

Kroenig, Matthew. 2009. "Importing the Bomb: Sensitive Nuclear Assistance and Nuclear Proliferation." *Journal of Conflict Resolution* 53 (2): 161–80. https://doi.org/10.1177/0022002708330287.

Kroenig, Matthew H. 2011. *Exporting the Bomb: Technology Transfer and the Spread of Nuclear Weapons*. Cornell University Press.

Kutchesfahani, Sara Z. 2019. "Chapter 3—Regional Initiatives Toward a Nuclear Weapons Free World: The Case of Nuclear Weapons Free Zones." In *Nuclear Safeguards, Security, and Nonproliferation (Second Edition)*, edited by James E. Doyle, 59–76. Boston: Butterworth-Heinemann. https://doi.org/10.1016/B978-0-12-803271-8.00003-5.

Malik, John. 1985. "Yields of the Hiroshima and Nagasaki Nuclear Explosions." LA-8819. Los Alamos National Lab. (LANL), Los Alamos, NM (United States). https://doi.org/10.2172/1489669.

Maller, Tara. 2010. "Diplomacy Derailed: The Consequences of Diplomatic Sanctions." *The Washington Quarterly* 33 (3): 61–79. https://doi.org/10.1080/0163660X.2010.492341.

Mazarr, Michael J. 2018. "Understanding Deterrence." RAND Corporation. https://www.rand.org/pubs/perspectives/PE295.html.

Mecklin, John. 2022. "The Energy Department's Fusion Breakthrough: It's Not Really about Generating Electricity." Bulletin of the Atomic Scientists, December 16. https://thebulletin.org/2022/12/the-energy-departments-fusionbreakthrough-its-not-really-about-generating-electricity/.

Mecklin, John. 2023. "Israel's Netanyahu Suspends Minister for Suggesting a Nuclear Weapons Option in Gaza." Bulletin of the Atomic Scientists, November 5. https://thebulletin.org/2023/11/israels-netanyahu-suspends-ministerfor-suggesting-a-nuclear-weapons-option-in-gaza/.

Meier, Oliver, and Maren Vieluf. 2021. "Upsetting the Nuclear Order: How the Rise of Nationalist Populism Increases Nuclear Dangers." *The Nonproliferation Review* 28 (1–3): 13–35. https://doi.org/10.1080/10736700.2020.1864932.

Miller, Julia. 2022. "A Tale of Two Bomb Designs." Los Alamos National Laboratory. 2022. https://discover.lanl.gov/publications/the-vault/the-vault-2023/a-tale-of-two-bomb-designs/.

Mishra, Sitakanta. 2018. "India-Canada Nuclear Relations: From a Troubled Past towards a Promising Future." *Indian Foreign Affairs Journal* 13 (1): 44–50.

Monteiro, Nuno P., and Alexandre Debs. 2014. "The Strategic Logic of Nuclear Proliferation." *International Security* 39 (2): 7–51. https://doi.org/10.1162/ISEC_a_00177.

Müller, Harald. 1993. "The Evolution of Verification: Lessons from the Past for the Present." *Arms Control* 14 (3): 333–56. https://doi.org/10.1080/01440389308404044.

Müller, Harald, and Carmen Wunderlich. 2020. "Nuclear Disarmament without the Nuclear-Weapon States: The Nuclear Weapon Ban Treaty." *Daedalus* 149 (2): 171–89. https://doi.org/10.1162/daed_a_01796.

Munir, Muhammad. 2018. "Pakistan-China Strategic Interdependence: Post-9/11 Imperatives." *Strategic Studies* 38 (2): 21–42.

Muti, Alberto, Grant Christopher, and Noel Stott. 2023. "The Role of Verification in Supporting Irreversible Nuclear Disarmament." *Journal for Peace and Nuclear Disarmament*, December, 1–11. https://doi.org/10.1080/25751654.2023.2291892.

National Academy of Sciences. 2002. "Technical Issues Related to the Comprehensive Nuclear Test Ban Treaty." Washington, D.C.: Committee on Technical Issues Related to Ratification of the Comprehensive Nuclear Test Ban Treaty. https://doi.org/10.17226/10471.

NATO. 2022. "NATO 2022 Strategic Concept." NATO. https://www.nato.int/cps/en/natohq/topics_210907.htm.

NNSA. 2024. "NNSA Completes Subcritical Experiment at PULSE Facility in Nevada." Energy.Gov. https://www.energy.gov/nnsa/articles/nnsa-completes-subcritical-experiment-pulse-facility-nevada.

Notte, Hanna. 2024. "Russia, the Global South and the Mechanics of the Nuclear Order." *Survival* 66 (3): 49–57. https://doi.org/10.1080/00396338.2024.2357480.

NTI. 2020. "Central Asia Nuclear-Weapon-Free-Zone (CANWFZ)." The Nuclear Threat Initiative. https://www.nti.org/education-center/treaties-and-regimes/central-asia-nuclear-weapon-free-zone-canwz/.

Nye Jr., Joseph S. 2023. "Nuclear Ethics Revisited." *Ethics & International Affairs* 37 (1): 5–17. https://doi.org/10.1017/S0892679423000047.

O'Mahoney, Joseph. 2020. "The Smiling Buddha Effect: Canadian and US Policy after India's 1974 Nuclear Test." *The Nonproliferation Review* 27 (1–3): 161–79. https://doi.org/10.1080/10736700.2020.1803561.

Pistner, Christoph, Matthias Englert, Christian von Hirschhausen, Fanny Böse, Björn Steigerwald, Lukas Gast, and Richard Donderer. 2024. "Analysis and Evaluation of the Development Status, Safety and Regulatory Framework for So-Called Novel Reactor Concepts." BASE – Research reports on the safety of nuclear waste disposal BASE-031/24. Berlin: Federal Office for the Saftey of Nuclear Waste Management. https://www.base.bund.de/shareddocs/downloads/en/expert-info/f/final-report-novel-reactor-concepts.pdf?__blob=publicationFile&v=1.

Pistner, Christoph, Matthias Englert, Christian Küppers, Ben Wealer, Björn Steigerwald, Christian von Hirschhausen, and Richard Donderer. 2021. "Sicherheitstechnische Analyse und Risikobewertung einer Anwendung von SMR-Konzepten (Small Modular Reactors)." Vorhaben 4720F50500. Forschungsberichte zur Sicherheit der nuklearen Entsorgung. Berlin: Bundesamt für die Sicherheit der nuklearen Entsorgung. https://www.base.bund.de/SharedDocs/Downloads/BASE/DE/berichte/kt/gutachten-small-modular-reactors.pdf?__blob=publicationFile&v=6.

Potter, William C., and Jessica Matthews, eds. 1998. Nuclear Successor States of the Soviet Union: Status Report on Nuclear Weapons, Fissile Material, and Export Controls. No. 5. Carnegie Endowment for International Peace, Monterey Institute of International Studies.

Pyle, Kenneth B. 2013. *Hiroshima and the Historians: Debating America's Most Controversial Decision*. Cambridge: Cambridge University Press. https://doi.org/10.1017/9781009477482.

Ray, Aswini K. 1999. "The Concept of Justice in International Relations." *Economic and Political Weekly* 34 (22): 1368–74.

Reiss, Louise Zibold. 1961. "Strontium-90 Absorption by Deciduous Teeth." *Science* 134 (3491): 1669–73. https://doi.org/10.1126/science.134.3491.1669.

Rodgers, Joseph, and Heather Williams. 2023. "The Irreversibility Paradox: What Makes for Enduring Arms Control and Disarmament." *Journal for Peace and Nuclear Disarmament*, December, 1–19. https://doi.org/10.1080/25751654.2023.2292812.

Rotte, Ralph. 2024. "Neue Reaktortechnologien, Strategische Deglobalisierung Und 'Tech Wars': Kontraproduktive Folgen Für Das Nukleare Nonproliferations- Und Verifikationsregime?" Preprint. SocArXiv. https://doi.org/10.31235/osf.io/45wh7.

Russell, J. 2001. "On-Site Inspections under the INF Treaty A Post-Mortem." United Kingdom.

Sagan, Scott D. 1996. "Why Do States Build Nuclear Weapons? Three Models in Search of a Bomb." *International Security* 21 (3): 54. https://doi.org/10.2307/2539273.

Schalz, Max, Erik Branger, Malte Göttsche, Sophie Grape, and Cecilia Gustavsson. 2024. "Assessing Pakistan's Fissile Material Production." *Science & Global Security* 0 (0): 1–26. https://doi.org/10.1080/08929882.2024.2409524.

Schelling, Thomas C. 1960. *The Strategy of Conflict*. The Strategy of Conflict. Oxford, England: Harvard Univer. Press.

Schelling, Thomas C. 1966. *Arms and Influence*. Yale University Press. https://doi.org/10.2307/j.ctt5vm52s.

Schmalberger, Thomas. 1991. "In Pursuit of a Nuclear Test Ban Treaty: A Guide to the Debate in the Conference on Disarmament," January. https://unidir.org/publication/in-pursuit-of-a-nuclear-test-ban-treaty-a-guide-to-the-debate-in-the-conference-on-disarmament/.

Schneider, Jonas, and Oliver Thränert. 2014. "Dual Use: Dealing with Uranium Enrichment." 151. CSS Analyses in Security Policy. Center for Security Studies (CSS), ETH Zurich.

Schumann, Anna. 2022. "Fact Sheet: Thermonuclear Weapons." Center for Arms Control and Non-Proliferation. November 18, 2022. https://armscontrolcenter.org/fact-sheet-thermonuclear-weapons/.

Schütterle, Juliane. 2010. *Kumpel, Kader und Genossen: Arbeiten und Leben im Uranbergbau der DDR: die Wismut AG*. Sammlung Schöningh zur Geschichte und Gegenwart. Paderborn: Verlag Ferdinand Schöningh. https://doi.org/10.30965/9783657769223.

Selby, Hugh D., Susan K. Hanson, Daniel Meininger, Warren J. Oldham, William S. Kinman, Jeffrey L. Miller, Sean D. Reilly, et al. 2021. "A New Yield Assessment for the Trinity Nuclear Test, 75 Years Later." *Nuclear Technology* 207 (sup1): 321–25. https://doi.org/10.1080/00295450.2021.1932176.

Sime, Ruth Lewin. 1996. *Lise Meitner: A Life in Physics*. University of California Press.

Singh, Sonali, and Christopher R. Way. 2004. "The Correlates of Nuclear Proliferation: A Quantitative Test." *Journal of Conflict Resolution* 48 (6): 859–85. https://doi.org/10.1177/0022002704269655.

Snyder, Glenn Herald. 1965. *The Balance of Power and the Balance of Terror*. Chandler.

Squassoni, Sharon A. 2006. "Weapons of Mass Destruction: Trade Between North Korea and Pakistan." Congressional Research Service. https://apps.dtic.mil/sti/citations/ADA467226.

Stewart, Ian J. 2022. "Brazil Wants Special Treatment for Its Nuclear Submarine Program—Just like Australia." *Bulletin of the Atomic Scientists* (blog). June 28, 2022. https://thebulletin.org/2022/06/brazil-wants-special-treatment-for-its-nuclear-submarine-program-just-like-australia/.

Stumpf, Waldo. 1995. "South Africa's Nuclear Weapon Programme: From Deterrence to Dismantlement." https://www.armscontrol.org/system/files/ACT_South%20Africa_9601.pdf.

Sundaram, Kumar, and M. V. Ramana. 2018. "India and the Policy of No First Use of Nuclear Weapons." *Journal for Peace and Nuclear Disarmament* 1 (1): 152–68. https://doi.org/10.1080/25751654.2018.1438737.

Tannenwald, Nina. 2007. *The Nuclear Taboo: The United States and the Non-Use of Nuclear Weapons Since 1945*. Cambridge Studies in International Relations. Cambridge: Cambridge University Press. https://doi.org/10.1017/CBO9780511491726.

Teaiwa, Teresia K. 2010. "Bikinis and Other S/Pacific N/Oceans." In *Militarized Currents*, by Cynthia Enloe, edited by Setsu Shigematsu and Keith L. Camacho, 15–32. University of Minnesota Press. https://doi.org/10.5749/minnesota/9780816665051.003.0002.

Tickner, J. Ann, and Laura Sjoberg, eds. 2011. *Feminism and International Relations: Conversations about the Past, Present and Future*. London: Routledge. https://doi.org/10.4324/9780203816813.

Turco, R. P., O. B. Toon, T. P. Ackerman, J. B. Pollack, and Carl Sagan. 1983. "Nuclear Winter: Global Consequences of Multiple Nuclear Explosions." *Science* 222 (4630): 1283–92. https://doi.org/10.1126/science.222.4630.1283.

Tynan, Elizabeth. 2016. *Atomic Thunder: The Maralinga Story*. Sydney: NewSouth Publishing.

UK Government. 2025. "Joint statement by the Foreign Ministers of France, Germany and the UK on the initiation of the snapback process." August 28. https://www.gov.uk/government/news/e3-joint-statement-on-iran-initiation-ofthe-snapback-process.

UNIDIR. 2010. "Arms Control Verification." 3. Disarmament Forum. UNIDIR. https://unidir.org/publication/disarmament-forum-arms-control-verification/.

UN Press. 2023. "Deployment of Nuclear Weapons to Belarus Debated in First Committee, as Delegates Rethink Global Security." October 5. https://press.un.org/en/2023/gadis3712.doc.htm.

Volpe, Tristan A. 2023. *Leveraging Latency: How the Weak Compel the Strong with Nuclear Technology*. Oxford University Press. https://global.oup.com/academic/product/leveraging-latency-9780197669532?cc=de&lang=en&.

Von Hippel, Frank, and Masafumi Takubo. 2022. "Banning Plutonium Separation." International Panel on Fissile Materials. https://fissilematerials.org/library/rr20.pdf.

Waltz, Kenneth N. 1981. "The Spread of Nuclear Weapons: More May Be Better: Introduction." Other. Taylor & Francis. Taylor & Francis Group. World. September 1, 1981. https://doi.org/10.1080/05679328108457394.

Wellerstein, Alex, and Edward Geist. 2017. "The Secret of the Soviet Hydrogen Bomb." *Physics Today* 70 (4): 40–47. https://doi.org/10.1063/PT.3.3524.

Winde, Frank, Doug Brugge, Andreas Nidecker, and Urs Ruegg. 2017. "Uranium from Africa – An Overview on Past and Current Mining Activities: Re-Appraising Associated Risks and Chances in a Global Context." *Journal of African Earth Sciences* 129 (May):759–78. https://doi.org/10.1016/j.jafrearsci.2016.12.004.

Wohlstetter, Albert. 1958. "The Delicate Balance of Terror." https://www.rand.org/pubs/papers/P1472.html.

Woolf, Amy F. 2011. "Monitoring and Verification in Arms Control." R41201. Congressional Research Service Washington, DC. https://crsreports.congress.gov/product/pdf/R/R41201/9.

Woolf, Amy F. 2023. "The Past and Future of Bilateral Nuclear Arms Control." 9. UNIDIR Nuclear Dialogue Series. Geneva, Switzerland: United Nations Institute for Disarmament Reserach. https://unidir.org/files/2023-03/UNIDIR_past_future_bilateral_nuclear_arms_control.pdf.

Xia, Lili, Alan Robock, Kim Scherrer, Cheryl S. Harrison, Benjamin Leon Bodirsky, Isabelle Weindl, Jonas Jägermeyr, Charles G. Bardeen, Owen B. Toon, and Ryan Heneghan. 2022. "Global Food Insecurity and Famine from Reduced Crop, Marine Fishery and Livestock Production Due to Climate Disruption from Nuclear War Soot Injection." *Nature Food* 3 (8): 586–96. https://doi.org/10.1038/s43016-022-00573-0.

Yang, Xiaoping, Robert North, Carl Romney, and Paul G. Richards. 2003. "84 Worldwide Nuclear Explosions." In *International Geophysics*, 81:1595–99. Elsevier. https://doi.org/10.1016/S0074-6142(03)80279-6.

Yusuf, Moeed W. 2019. "The Pulwama Crisis: Flirting With War in a Nuclear Environment." Arms Control Association, May. https://www.armscontrol.org/act/2019-05/features/pulwama-crisis-flirting-war-nuclearenvironment.

Chapter 7
Game Theoretical Considerations of Nuclear Armament

Gero Scheck

7.1 Introduction

Nuclear weapons have the potential to kill millions of people and cause a global catastrophy in a single strike (Toon et al. 2007). Despite this enormous harmful potential, there is still an ongoing debate if the world is be better or worse off with nuclear weapons (Kydd 2019). Increasing geopolitical tension has led to increasing support for providing nuclear weapons to conflict parties as a means of preventing the invasion of their territories (Evangelista 2024). One of the key arguments in favor of nuclear armament is the theory of nuclear deterrence, which was first postulated in the 1950s, when policymakers and strategists acknowledged the technical possibility of mutually assured destruction between the United States (US) and the Soviet Union (Powell 2003). The theory states that if a country possesses a sufficiently powerful nuclear arsenal, other countries will be deterred from attacking it, whether with conventional or nuclear weapons, due to the fear of a devastating nuclear retaliatory strike. Some researchers argue that the effect of nuclear deterrence can be seen in the war in Ukraine, as both Russia and the US have omitted certain military actions that promised military advantages, but would also have escalated the situation (Bell 2024). Others argue that it is not possible to determine whether nuclear deterrence was the reason for these actions not having been performed (Sauer 2024).

Based on nuclear deterrence theory, Snyder (1961) argued that to deter Soviet aggression effectively, the US would have needed to maintain a sufficiently large stock of nuclear weapons. This would ensure the capability to perform a retaliatory strike that would render the Soviets' total expected benefits of an attack negative. Snyder argues that a theoretically optimal-sized nuclear stockpile exists, which deters Soviet aggression, while remaining cost-effective. This optimal size depends on many uncertain factors, such the military capacities of the enemy and the expected benefit

G. Scheck (✉)
Workgroup for Infrastructure Policy (WIP), Technical University (TU) Berlin, Berlin, Germany
e-mail: gfs@wip.tu-berlin.de

A. Wimmers et al. (eds.), *Nuclear Power*, Studies in Energy, Resource and Environmental Economics, https://doi.org/10.1007/978-3-031-99894-2_7

of an attack, as well as the expected valuation of losses caused by a retaliatory strike. However, such logic requires knowledge of the rationality and reasoning behind enemy weapon usage, which has been criticized because real-world actors tend to behave irrationally (Steinbruner 1983). Another factor that can influence deterrence effectiveness is threat credibility (Kraig 1999).

This chapter looks at nuclear deterrence theory from a game theoretic perspective. Its goal is to analyze how rational individuals behavior changes in geopolitical conflict when nuclear weapons are introduced and compare those results to the idea of nuclear deterrence.

7.2 Status Quo of Nuclear Weapon States

This section gives a short outline on what kinds of nuclear weapons exist, how they work and what their intended purpose is. This is followed by an overview on who nuclear weapons states are and why and how they became nuclear weapon states. This section only focusses on proliferation, for a detailed overview of current global efforts in nuclear non-proliferation and disarmament, as well as an overview of historical and current stockpiles; refer to Chap. 6 of this book.

7.2.1 Typology of Nuclear Weapons

A weapon that uses nuclear fission or fusion to cause an explosion can be classified as a nuclear weapon. Nuclear weapons can come in different functionalities, sizes, carriers, and ranges. There are three different main reaction sequences that are used in nuclear weapons that utilize different fuel types and vary in the power of their explosion, i.e., uranium and plutonium fission bombs, and thermonuclear or hydrogen bombs.

Another distinction is between so-called tactical and strategic nuclear weapons. This distinction is not clearly defined, and various definitions exist (Kristensen and Korda 2019). For this chapter, tactical nuclear weapons are defined as weapons whose main purpose is to be used on the battlefield to decide individual battles. In contrast, strategic weapons are intended to be used to attack targets outside of the battlefield to either cripple the enemy's military or economic abilities, or force them to surrender. The focus of this chapter is nuclear deterrence and mutually assured destruction, limiting this work's assessment to strategic nuclear weapons since their main purpose is deterrence. While not neglecting that tactical nuclear weapons can very well be used for deterrence, including them into this work's analysis would create strategic and diplomatic sub-scenarios that are beyond the scope of this chapter.

There are three main systems that are used to carry and, in the case of use, deliver strategic nuclear weapons: sea-based, usually submarines; air-based (bombers); and land-based systems. These systems have distinct, complementary advantages.

Submarines are survivable because they can move and hide in the ocean, bombers are flexible due to their speed and recall potential, and land-based systems allow for quick responses as they are always ready to be used. The distinct nature of these carriers makes it harder for an attacker to shut down an adversary's complete nuclear arsenal when it is divided amongst different carriers (United States Department of Defense 2020; Woolf 2021). Therefore, the use of different carriers makes a states' nuclear arsenal more resilient to attacks. For a nuclear weapon state, it is important to have a nuclear arsenal that is resilient to an attack, as this makes its threat of a retaliation strike credible. Otherwise, a state could build a big enough nuclear stockpile that is capable of eliminating an enemy's nuclear stockpile, aiming for a one-shot victory that cannot be retaliated against.

7.2.2 *Overview of the Nuclear Weapon States*

Nuclear weapons never exist in a vacuum. Their geopolitical and diplomatic effects always depend on current geopolitical situations. Therefore, arguments in favor of proliferation are usually based on how nuclear weapons can supposedly improve a state's strategic situation. Besides domestic and ideologic reasons, states build nuclear weapons to increase their own security (Sagan 1996). The primary reason for states to proliferate nuclear weapons is still under scientific debate, with some authors arguing for security, some for economic, and some for status-driven reasons (Sagan 2011).

The US was the first country to produce deployable nuclear weapons. After successfully performing the trinity test on July 16, 1945, nuclear bombs were detonated over Hiroshima and Nagasaki on August 6 and 9, 1945, respectively. Those dates mark the end of World War II and the only use of nuclear bombs in a military conflict so far. The reasons to use the bomb were most likely of a military nature, i.e., to avoid a conventional war in Japan and preserve the lives of many American soldiers. However, they were also political in demonstrating this tremendous new power to the Soviets (Walker 1990). Being able to produce nuclear bombs made the US the only nuclear weapon state until the Soviet Union successfully tested their first nuclear bomb in 1949. When there is only one nuclear weapon state, this resembles a special case, as that state has an incentive to use its dominant position and fight preventive wars to stop other states from achieving nuclear advances. American presidents at that time stated that they would not fight such a war for moral reasons, but an insufficient American nuclear stockpile to enforce such goals, as well as Soviet's dominance in conventional military forces in Europe, might have also played a role (Buhite and Hamel 1990). After successfully testing their first nuclear bomb in 1949, and subsequently quickly accumulating a nuclear stockpile, the Soviet Union became the second nuclear power.

A couple of years earlier, the Cold War started, which was heavily influenced by the presence of both states' nuclear stockpile. As both nations' stockpile grew, the term "mutually assured destruction" emerged. This term describes a state in which two nations have a large enough nuclear stockpile such that an attack by one nation would trigger a disastrous retaliatory strike from the attacked nation. The outcome of this chain of events, the complete destruction of both nations, keeps the two countries from performing an attack. During the Cold War, it was widely believed that the existence of this state of mutually assured destruction prevented the rivaling superpowers from attacking each other (Powell 2003).

The inventories of nuclear weapons of the US and the Soviet Union peaked at around 30,000 (US, 1965) and 40,000 (Soviet Union, 1985), respectively. In parallel, nuclear weapons technology also spread to other countries, sometimes with (open or secret) support or approval from the superpowers, but sometimes without. The United Kingdom and France had quickly risen to the status of nuclear powers after World War II, and China, one of the five countries with a permanent seat on the United Nations Security Council since 1945, joined in 1964. Additionally, other countries also developed a nuclear–industrial complex in the shadow of these nuclear powers, such as India, Pakistan, Israel, briefly South Africa, and most recently, North Korea (Stirling and Johnstone 2018).

7.3 Introduction to Game Theory

Game theory studies decision-making in which multiple players or agents interact. The outcomes of such interactions not only depend on one's own decisions, but also on the decision of others. Thus, game theory aims to analyze a rational individual's behavior in a situation in which they have multiple options that result in different outcomes (if the choices lead to the same outcome, they become trivial for that individual). The game in game theory is created by having more than one individual, called players, each facing one or more decisions that not only influence their own outcome, but also that of the other players. In such a scenario, players will make decisions that lead to an optimal outcome based on what they expect the other players to do.

The game defines the ruleset under which the individuals operate. Those rules define what options each Player i has in each round, when and how often they have to make a decision, and what the outcome of each course of play is. Each outcome of a game is linked to a payout for each player. Those payouts are quantifiable gains or losses for each Player i. Depending on the game, payouts are often defined as the utility U_i, amounts of money or, other quantifiable values, such as expected years in prison, see Sect. 1.3.1. Each possible sequence of decisions has a payout vector attributed to it. In general, players will behave in such a way that they maximize their individual payout. Games can vary in the number of players, the turn order, the number of rounds that are played, and the availability of information. The simplest variants are two players who choose between two possible actions each; one turn is taken simultaneously, one round is played, and all information is available to all players. One of the most well-known games in game theory is the prisoner's dilemma, see below.

7.3.1 The Prisoner's Dilemma

In the classic prisoner's dilemma, the two players are being accused of having committed a crime. They are interrogated simultaneously, but separately, and therefore, they have no way of communicating. This prevents any coordination regarding their decisions. Being very confident criminals, they also did not plan for a scenario in which they would get caught, so they have no agreements on how to act under such circumstances. In Table 7.1, player's options and their respective payouts, depending on the other player's choice, are shown. The rows list Player 1's options, the columns list those of Player 2. The table consists of vectors representing the payouts for both players given the respective combination of options. In this example, the entries represent a number of years. As players can "lose" years by having to go to prison, their payouts are negative in this case. Players aim to minimize the time they have to spend in prison and, therefore, maximize their payout.

In this example, the players can either choose to cooperate with the police and confess, or cooperate with each other and stay silent. The sentence for the crime is set at ten years. However, the police have insufficient evidence to convict them and can only have them sentenced for minor crimes, which would result in one year of prison for each player when both remain silent. However, the police offer to lower the sentence for the players as a reward when they confess. If both players confess, both will be convicted of the crime but only have to stay in prison for six years due to the reduced sentence. If only one player confesses and the other stays silent, the confessing player's sentence is reduced to zero. The player who remained silent will be sentenced to the entire ten years, as they did not cooperate with the police, but the police were able to convict them due to the other player's confession.

To analyze what each (rational) player should do in this scenario, one must identify payout maximizing strategies for each player with respect to each possible decision by the other player. The action of confessing is always better than remaining silent. Assuming that Player 2 stays silent, Player 1 faces one year in prison when they also remain silent, but faces no sentence if they confess. Therefore, if they expect Player 2 to not cooperate with the police, they should confess. However, if one assumes that Player 2 confesses, Player 1 faces ten years in prison if they remain silent, but only six years if they also confess. Therefore, it is also favorable to confess in this scenario. The same applies to Player 2 because the payout structure is equal for both players.

When a certain strategy always leads to the highest possible individual payout regardless of what the other player is doing, it is called a strictly dominant strategy.

Table 7.1 Payout Matrix for a prisoner's dilemma

		Player 2	
		Confess	Remain silent
Player 1	Confess	(–6, –6)	(0, –10)
	Remain silent	(–10, 0)	(–1, –1)

Identifying a strictly dominant strategy for all players produces a strategy combination that is optimal for all individual players. It is certain that this strategy combination is the only optimal solution of this game. It should be noted that this is not the outcome that maximizes the total welfare for all players. If both players remained silent, their total sentence would amount to two years instead of twelve years when they both confess. However, as they cannot communicate or send signals, they have no way of agreeing to this solution. Thus, as they are rational acting agents, they will always play the dominant strategy of confession, as it maximizes their individual pay-off no matter what their opponent is doing.

7.3.2 Nash Equilibrium

One way of solving a game theoretic problem is to find all stable strategy combinations. A stable strategy combination describes a situation in which none of the players can gain any advantage by choosing a different strategy. Such a combination is called Nash equilibrium, after the mathematician John Nash.

As shown above, one method to find a Nash equilibrium is to find dominant strategies for all players. As dominant strategies are always optimal for their player, each player following their dominant strategy is, by definition, a Nash equilibrium. Not all games have dominant strategies; therefore, other methods exist to find Nash equilibria in a game. However, these are beyond the scope of this work.[1]

7.3.3 Multi-Turn Games (Extensive-Form Game)

The prisoner's dilemma represents a situation resolved after a single round of decision-making. As conflicts between states are not solved after one round of strategic decisions but take place in the context of past and future decisions, a tool that allows the representation of chronological sequences of interdependent decisions is needed. Extensive-form games allow to model a multi-turn game in which the players make decisions, either one after another, called a dynamic game, or at the same time, called a simultaneous game. Each previous decision can affect both players' payoffs as well as the set of strategies available to the players in the following turn. Extensive-form games can be depicted in so-called game-trees, in which each branch represents a possible course of decisions. At the end of a branch, each player's payout of that course of decisions is indicated.

When a game consists of multiple turns, it can be split into smaller subgames. When player's strategies are Nash equilibria in each possible subgame, this is called

[1] Refer to Bartholomae and Wiens (2024) and Peters (2015) for further information on the Nash Equlibrium and a broader introduction to game theory.

a subgame perfect equilibrium. Each set of strategies that it a subgame perfect equilibrium is also a Nash equilibrium, but not vice versa.

7.3.4 Two-Stage Deterrence Game

To analyze if a state of mutual deterrence can be achieved, the escalation model developed by Zagare (1992) is applied. It resembles a two-player multi-turn game with complete information. Players are presented with the option to cooperate (C) or to defect (D). Each player's last decision determines the game's outcome and the respective payouts. Previous decisions do not affect the payouts. It is therefore sufficient to label outcomes with the last decision both players made. (XY) represents the outcome in which Player 1's last decision was to play X, and Player 2's last decision was to play Y. Note that X always represents Player 1's last choice and Y always represents Player 2's choice independent of turn order or who took the last turn. U_i (XY) represents the utility that Player i gains from the outcome XY.

7.3.4.1 Ruleset

At the start of the game, there is a state of peace. When a player chooses to cooperate (C), they will not attack the other player. In contrast, if they choose to defect (D), they will attack the other player with conventional, i.e., non-nuclear, military forces. These two options resemble a different stage of escalation. Choosing to cooperate maintains a player's peaceful status and, therefore, resembles no escalation. In contrast, a player actively participates in a conflict by choosing to defect. This move, therefore, is the first level of escalation. It is assumed that once a player has reached a certain level of escalation, they will not choose a lower level of escalation in a later stage of the game. As the game has complete information and the players are entirely rational, this is aligned with the other assumptions, as no benefits will be gained from starting a military strike and later backing down.

Players alternately make their moves, choosing between the two options, until the game ends. The game ends if it reaches a stable state, meaning that both players choose not to further escalate the game. The game can result in the absence of war, i.e., both players cooperating; war, i.e., both players defecting; or a unilateral military strike, i.e., one player cooperating and one player defecting. The game is represented by Fig. 7.1.

For simplicity, it is assumed that Player 1 makes the first move. As payouts do not depend on the order in which the moves are made, this assumption does not affect the outcome of the game. After Player 1 makes their move, the second player observes this move and chooses if they want to refrain from using their weapons (C) or if their want to perform a military strike (D).

In the case of Player 1 defecting, Player 2 also defecting (DD) resembles Player 2 answering the first attack with a retaliatory strike. This outcome resembles an active

Fig. 7.1 A two-stage deterrence game

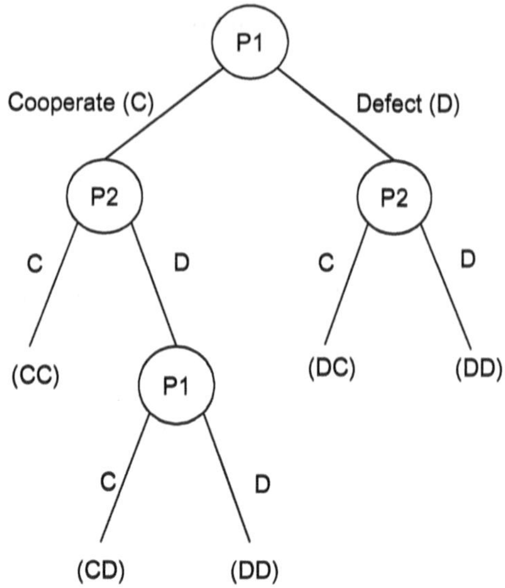

war between the two players. If Player 2 chooses to cooperate instead (DC), Player 1's attack is not countered. This outcome resembles Player 1 unilaterally attacking Player 2 who would be forfeiting without military resistance. Regardless of Player 2's choice, the state of peace is lost, and therefore, the game of deterrence ends.

If Player 1 chooses to cooperate in the first step and not attack, the same situation as at the start of the game is reached, albeit with switched roles. The game stays in the state of peace and presents Player 2 with the option to preserve that state by cooperating (CC) or seek their benefit in being the aggressor and defect from this peaceful state, i.e., attack (CD). When both players choose to cooperate, neither player has chosen to escalate the game at all; therefore, the game ends. In reality, many factors can constantly influence payoffs and might make defecting from a stable state of cooperation a viable strategy.

When Player 2 chooses to attack, Player 1 can either perform a retaliation strike (DD) or leave Player 2's attack unanswered (CD). After this, the game ends. In the case of (DD), both players have reached the maximum level of escalation and further escalation is not possible. In the case of (CD), Player 2 has reached the maximum level of escalation, and Player 1 chose not to escalate further.

All possible courses of the game are identified, the Nash equilibria, and consequentially the stable strategies, can be determined. If (CC) is the Nash equilibrium, one finds that peace is achievable. This indicates that either both players where not interested in attacking the other player whatsoever or deterrences worked. If other Nash equilibria are found, the state of peace is unstable and attacking is a viable option for at least one player.

7.3.4.2 General Assumptions

To find Nash equilibria, payoffs must be assigned to all potential outcomes. How we choose those payoffs strongly depends on the assumptions about the players that are playing and the situation they are in. The first set of assumptions is based on general preferences and is, therefore, independent of individual military capacities. First, one can assume that both players prefer a state of mutual peace rather than being unilaterally attacked by the other player. Thus, it holds that U_1 (CC) > U_1 (CD) and U_2 (CC) > U_2 (DC). Further, it is assumed that it is beneficial for both players to unilaterally attack the other player. Therefore, U_1 (DC) > U_1 (CC) and U_2 (CD) > U_2 (CC) hold true.

The second set of assumptions depends on the players' individual preferences, which influence their capability and credibility to make threats. In this model, the payout that each player receives in (DD) signals whether their threats of a retaliatory strike are capable and/or credible. For the threat to be capable, players must prefer mutual peace over mutual attacks. If this is not the case, defecting is the dominant strategy, and attacking would not be threat; it would be inevitable. Therefore, for Player i's threat to be capable, U_i (CC) > U_i (DD) must hold. For the threat to be credible, a retaliatory strike must be beneficial when compared to staying defenseless when being attacked. In this case it holds that U_1 (DD) > U_1 (CD) and U_2 (DD) > U_2 (DC). This is only the case when the attacked player has the military arsenal to successfully fight back. If retaliation only represents the loss of additional resources without any gains to be made against a superior opponent, a rational leader will not perform such an act and therefore the initial aggressor does not have to fear it. It can be shown that (CC) is only the Nash equilibrium, when both players have capable and credible threats (Zagare 1992).

7.4 Applying Game Theory to a Nuclear Conflict: The Three-Stage Nuclear Deterrence Game

When it comes to reasoning in favor of nuclear proliferation, nuclear deterrence is often used as a central argument. The underlying argument is to deter potential opponents from attacking by posing the risk of a fatal revenge strike in the case of an attack. Ideally, when all states have a sufficient stock of nuclear weapons, this causes a state of mutually assured destruction that should lead to global peace. In reality, nuclear weapons cannot deter every form of aggression. While it is debated what actions can be deterred using nuclear weapons, it is the consensus that the more existential an opposing threat is, the better nuclear deterrence works (Bell 2024).

In the following section, based on Zagare (1992) and Kraig (1999), nuclear weapons are introduced into the previously described deterrence game to analyze the effect that this additional option has on the outcome. In addition to the option for a regular military strike, nuclear powers now also have the capability to perform a nuclear strike on their opponents. In contrast to a regular military strike, a nuclear strike can cause much more damage and fatalities in a single strike.

7.4.1 Game Design and Ruleset

The extended extensive deterrence game was designed by Zarage (1992) and is depicted in Fig. 7.2. It adds the possibility to use nuclear weapons on top of the two-stage deterrence game as an additional stage of escalation. In the model, the option to use a nuclear weapon is resembled by the option to escalate (E). The options from the two-stage deterrence game, to cooperate (C) and to defect (D), do not change. Therefore, players now have up to three options. Assuming that a nuclear strike is only chosen in response to an attack, this option is only available to a player after their opponent has defected or escalated. The escalation level can thus only be increased by one point after each individual move. As in the two-stage deterrence game, the three-stage game ends when both players do not further increase their escalation level. In addition to the end states of the two-stage deterrence game, this game can also end with nuclear war, i.e., both players escalating, or one player's conventional strike (D) being answered by a nuclear strike (E).

After Player 1 defects, Player 2 has three options. If they choose to stay peaceful, no nuclear weapons are used, and the game reaches an outcome, a unilateral conventional attack performed by Player 1 (DC), already known from the two-stage game. The other two options are to engage in the conflict either with conventional or with

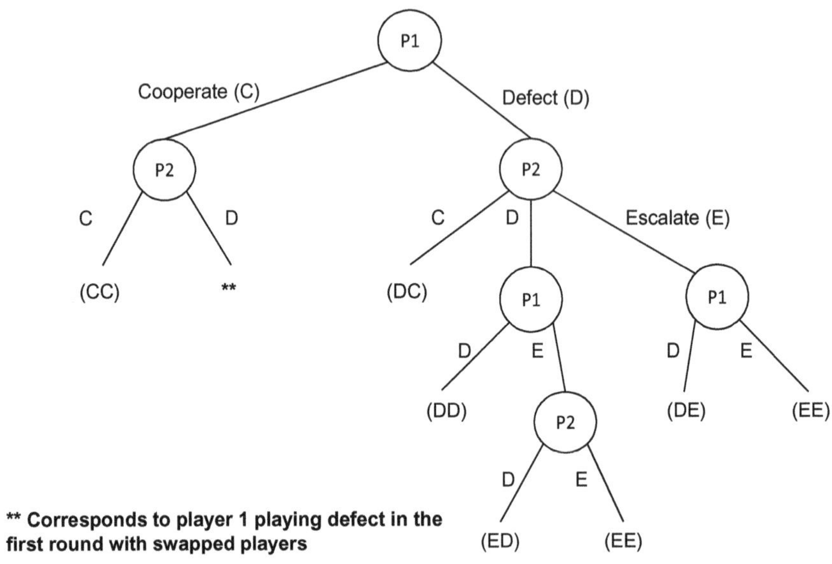

Fig. 7.2 Three-stage nuclear deterrence game

nuclear forces. In both scenarios, Player 2 increases the level of escalation, and therefore, Player 1, who is also a nuclear power, has the option to keep on fighting the war with conventional weapons (D) or use their nuclear capacity (E). If both players choose to deter from the state of peace but do not use nuclear weapons, the game enters the state of a regular war (DD) and ends. In contrast, if both players choose to use nuclear weapons (EE), a state of nuclear war or a state of mutually assured destruction is reached. When an initial nuclear strike by Player 1 (ED) or by Player 2 (DE) is not answered with a retaliatory strike, the game also ends.

As for the regular deterrence game, stable strategies must be identified through Nash equilibria. In a second step, the outcomes are compared to those of the regular deterrence game to assess whether nuclear weapons make this model world safer.

7.4.2 *Assumptions*

Building upon the assumptions from the regular deterrence game, the new options introduced in the nuclear deterrence game come with an additional set of assumptions. As in the two-stage deterrence game, some of those assumptions are of general nature, while others depend on a player's capability to perform a nuclear threat and its credibility. Starting with the general assumptions, one can assume that both players, when being attacked with conventional forces, prefer using their nuclear arsenal over regular warfare or not fighting: U_1 (ED) > U_1 (DD); U_1 (ED) > U_1 (CD); U_2 (DE) > U_2 (DD), U_2 (DE) > U_2 (DC). While this assumption is not trivial, it is necessary to ensure that the use of nuclear bombs is a considerable option. If this assumption does not hold, nuclear bombs will never be used. Therefore, they would, from a game theoretic perspective, not be a threat. It is also assumed that mutual nuclear strikes lead to an outcome that is worse than any scenario in which nuclear bombs are not used, which seems obvious. Finally, it is assumed that surrendering in the first place is generally a better option than fighting a conventional war that is answered by a nuclear strike. Thus, it holds that U_1 (CD) > U_1 (DE) and U_2 (DC) > U_2 (ED). In both options, that player loses the war, but in the latter scenario, the investment was higher and the losses were bigger.

Regarding capability and credibility, the assumption is made that if a player is not capable of performing a nuclear strike, they do not have the option (E) to escalate in the game. For a nuclear retaliatory strike to be credible from Player 1, U_1 (EE) > U_1 (DE) must hold. For a strike from Player 2 to be credible, U_2 (EE) > U_2 (ED) must hold. This assumption is not trivial, as it does not become clear immediately what potential gains from the retaliation strike would be. The credibility of a nuclear threat cannot easily be assessed. The devastating destruction and the political backlash that a nuclear attack causes can make a threat less credible. But when opposing threats increase and other options diminish, nuclear threats become more credible (Wirtz 2018).

Table 7.2 Change of outcome with respect to nuclear capability and credibility

Case	Name	Definition	Outcome
1	One-sided nuclear capability	Player 1 is not capable Player 2 is capable	Player 1 attacks Player 2 with conventional forces
2	One-sided nuclear credibility	Player 1 is capable, but not credible Player 2 is capable and credible	Player 2 attacks Player 1 with conventional forces
3	Mutual nuclear credibility	Player 1 is capable and credible Player 2 is capable and credible	The outcome depends on conventional capability and credibility, the same outcome as in the two-stage deterrence game
4	Mutual nuclear capability but without credibility	Player 1 is capable but not credible Player 2 is capable but not credible	No attacks will happen

7.4.3 Generalized Solutions

In the following, the outcome of the two-stage nuclear deterrence game, depending on the assumptions for capability and credibility, is analyzed. Therefore, the outcome for four cases is assessed, showing how those assumptions affect the outcome of the game (Table 7.2).

7.4.3.1 One-Sided Nuclear Capability

Assuming that Player 1 is not capable of performing a nuclear attack, but Player 2 is, and following a backward induction, it becomes clear that if Player 1 chooses to defect, Player 2 will choose to escalate (DE). Player 2 choosing to escalate results in a payout for Player 1 that is lower than the payouts in mutual peace (CC) and when Player 1 is attacked but concedes (DC). Therefore, it is always beneficial for Player 1 to cooperate. Knowing this, Player 2 will choose to defect and reach their optimal outcome (CD). Therefore, when only one player is capable of a nuclear threat, that player will defect from the status quo, and the other player will cooperate.

7.4.3.2 One-Sided Nuclear Credibility

In this scenario, it is assumed that both players are nuclear capable, but only Player 2 is also nuclear credible. Player 1, choosing to defect, would leave three possible answers to Player 2. They can cooperate, thereby ending the game and leading to (DC). If they choose to escalate, this leads to (DE), as Player 1 is not nuclear credible and will therefore not match Player 2's nuclear escalation. If Player 2 chooses to deter, the final outcome will be (DD). Player 1 will not choose to escalate because Player 2 is nuclear credible and, therefore, answering the escalation with escalation (EE), which has a lower payout as (DD) for Player 1. For Player 2, the payout of (DE) is higher than for (DC) and (DD); they will thus choose to escalate if Player 1 chooses to defect.

If Player 1 chooses to cooperate, Player 2 also has the option to cooperate, leading to (CC), or they can defect. If they defect, Player 1 can cooperate, defect, or escalate. Defecting and escalating will both be answered with escalation by Player 2 because U_2 (DE) > U_2 (DD) and U_2 (EE) > U_2 (ED). Therefore, Player 1 will choose to cooperate because U_1 (CD) > U_1 (DE) > U_1 (EE). Knowing this, Player 2 will choose to defect as U_2 (CD) > U_2 (CC). When only one player is nuclear credible, this player will attack the other player with conventional forces without fear of nuclear or conventional counterattacks.

7.4.3.3 Mutual Nuclear Credibility

When both players have credible nuclear threats, neither player will ever use their nuclear arsenal, as this would lead to mutual nuclear destruction. As those options never get played, the game effectively is reduced to the regular deterrence game. Therefore, its outcome depends on both players' capability and credibility of conventional threats. In the case of mutual nuclear credibility, no outcome can be achieved that is better than the outcome that the game would have without nuclear options.

7.4.3.4 Mutual Nuclear Capability Without Credibility

When both players are capable of nuclear escalation but do not have a credible threat, the state of peace is preserved. If Player 1 chooses to defect, Player 2 chooses to escalate, because U_2 (DE) > U_2 (DD), and they do not need to fear (EE), because Player 1 is not nuclear credible. Knowing this, Player 1 will not choose to defect. As the situation is now symmetric, Player 2 will also not choose to defect. Therefore, (CC) is the only outcome that will occur in this scenario, independent of the capability and credibility to deploy conventional forces.

7.5 Conclusion

The analysis shows that the introduction of nuclear weapons into the game-theoretic model preserves mutual cooperation only when a specific condition is met, with both players having non credible nuclear threats. In other scenarios, where only one player is nuclear credible, that player can wield power over the player lacking nuclear credibility. This shows that nuclear weapons do not reduce conflict in general; instead, they do so only under certain conditions. However, there are also scenarios that lead to mutual cooperation in the two-stage game, but the introduction of nuclear weapons allowed one player to defect. This shows that nuclear weapons, even if they are only used for deterrence, are not a solely defensive measure, but can enable aggressive actions that would otherwise not have been possible.

Assessing whether a nuclear threat is credible or not is a complicated and ongoing task since real-world leaders and military advisors rarely have full information when making decisions. While being complicated in a two-player scenario, it becomes increasingly impossible when the number of relevant players increases. When more players gain nuclear options, it might also become increasingly likely that the equilibrium between nuclear powers breaks and players are able to wield power in subgames where nuclear credibility is unilaterally distributed.

However, in an increasingly multipolar world, alliances can shift. When it becomes less clear who is protected by whose nuclear umbrella, more states are not protected by a capable threat, which threatens to make those states vulnerable to aggression and strengthen the few states that posses a nuclear arsenal. As can be seen in the models results, in such a scenario, the presence of nuclear weapons in their arsenal allows those states to perform conventional military maneuvers that might otherwise not have been possible. Both these scenarios show that nuclear weapons enable not only defensive deterrence, but aggression. Rather than bringing peace, they primarily bring power to those who posses them.

References

Bartholomae, Florian, and Marcus Wiens. 2024. *Game Theory and Applications: A Guide for Students and Researchers*. Classroom Companion: Economics. Wiesbaden: Springer Fachmedien. https://doi.org/10.1007/978-3-658-44608-6.

Bell, Mark S. 2024. "The Russia-Ukraine War and Nuclear Weapons: Evaluating Familiar Insights." *Journal for Peace and Nuclear Disarmament* 7 (2): 494–508. https://doi.org/10.1080/25751654.2024.2425379.

Buhite, Russell D., and WM. Christopher Hamel. 1990. "War for Peace: The Question of an American Preventive War against the Soviet Union, 1945–1955." *Diplomatic History* 14 (3): 367–84. https://doi.org/10.1111/j.1467-7709.1990.tb00096.x.

Evangelista, Matthew. 2024. "A 'Nuclear Umbrella' for Ukraine? Precedents and Possibilities for Postwar European Security." *International Security* 48 (3): 7–50. https://doi.org/10.1162/isec_a_00476.

IPFM. 2022. “Global Fissile Material Report 2022 - Fifity Years of the Nuclear Non-Proliferation Treaty: Nuclear Weapons, Fissile Materials, and Nuclear Energy.” Princeton, N.J: International Panel on Fissile Materials. https://fissilematerials.org/library/gfmr22.pdf.

Kraig, Michael R. 1999. “Nuclear Deterrence in the Developing World: A Game-Theoretic Treatment.” *Journal of Peace Research* 36 (2): 141–67. https://doi.org/10.1177/0022343399036002002.

Kristensen, Hans M., and Matt Korda. 2019. “Tactical Nuclear Weapons, 2019.” *Bulletin of the Atomic Scientists* 75 (5): 252–61. https://doi.org/10.1080/00963402.2019.1654273.

Kristensen, Hans M., and Robert S. Norris. 2017. “Worldwide Deployments of Nuclear Weapons, 2017.” *Bulletin of the Atomic Scientists* 73 (5): 289–97. https://doi.org/10.1080/00963402.2017.1363995.

Kydd, Andrew H. 2019. “The Sturdy Child vs. the Sword of Damocles: Nuclear Weapons and the Expected Cost of War.” *Security Studies* 28 (4): 645–76. https://doi.org/10.1080/09636412.2019.1631380.

Peters, Hans. 2015. *Game Theory: A Multi-Leveled Approach*. 2nd ed. 2015. Game Theory. Berlin, Heidelberg: Springer Berlin Heidelberg.

Powell, Robert. 2003. “Nuclear Deterrence Theory, Nuclear Proliferation, and National Missile Defense.” *International Security* 27 (4): 86–118.

Quackenbush, Stephen L., and Frank C. Zagare. 2016. “Modern Deterrence Theory: Research Trends, Policy Debates, and Methodological Controversies.” In *Oxford Handbook Topics in Politics*, edited by Oxford Handbooks Editorial Board, 1st ed. Oxford University Press. https://doi.org/10.1093/oxfordhb/9780199935307.013.39.

Sagan, Scott D. 1996. “Why Do States Build Nuclear Weapons?: Three Models in Search of a Bomb.” *International Security* 21 (3): 54–86. https://doi.org/10.2307/2539273.

Sagan, Scott D.. 2011. “The Causes of Nuclear Weapons Proliferation.” *Annual Review of Political Science* 14 (June):225–44. https://doi.org/10.1146/annurev-polisci-052209-131042.

Sauer, Tom. 2024. “How Useful Are Nuclear Weapons in Practice? Case-Study: The War in Ukraine.” *Journal for Peace and Nuclear Disarmament* 7 (1): 194–210. https://doi.org/10.1080/25751654.2024.2359818.

SIPRI. 2022. *SIPRI Yearbook 2022: Armaments, Disarmament and International Security - English Summary*. Solna, Sweden: Oxford University Press. https://www.sipri.org/sites/default/files/2022-06/yb22_summary_en_v2_0.pdf.

Snyder, Glenn Herald. 1961. *Deterrence and Defense: Toward a Theory of National Security*. Princeton Legacy Library. Princeton, NJ: Princeton University Press. https://doi.org/10.1515/9781400877164.

Sörenson, Karl. 2022. “Deterrence Games for the 21St Century: Representation, Theory and Evidence.” PhD Thesis, Royal Institute of Technology, Stockholm.

Steinbruner, John. 1983. “Beyond Rational Deterrence: The Struggle For New Conceptions.” In *Power, Strategy and Security: A “World Politics” Reader*, edited by Klaus Eugen Knorr, 103–25. Princeton University Press. https://doi.org/10.1515/9781400886326-007.

Stirling, Andy, and Phil Johnstone. 2018. “A Global Picture of Industrial Interdependencies Between Civil and Military Nuclear Infrastructures.” SPRU Working Paper Series 2018–13 (August). Sussex, UK: University of Sussex. https://doi.org/10.2139/ssrn.3230021

Toon, Owen Brian, Alan Robock, Richard Peter Turco, Charles Bardeen, Luke Oman, and Georgiy L. Stenchikov. 2007. “Consequences of Regional-Scale Nuclear Conflicts.” *Science* 315 (5816): 1224–25. https://doi.org/10.1126/science.1137747.

United States Department of Defense. 2020. *Nuclear Matters Handbook 2020 [Revised]*. Independently published.

Walker, J. Samuel. 1990. “The Decision to Use the Bomb: A Historiographical Update.” *Diplomatic History* 14 (1): 97–114. https://doi.org/10.1111/j.1467-7709.1990.tb00078.x.

Wirtz, James J. 2018. “How Does Nuclear Deterrence Differ from Conventional Deterrence?” *Strategic Studies Quarterly* 12 (4): 58–75.

Woolf, Amy F. 2021. "U.S. Strategic Nuclear Forces: Background, Developments, and Issues." CRS Report. https://crsreports.congress.gov/product/pdf/RL/RL33640/69.
Zagare, Frank C. 1992. "NATO, Rational Escalation and Flexible Response." *Journal of Peace Research* 29 (4): 435–54.

Chapter 8
Assessing Market Concentrations in the Nuclear Fuel Supply Chain

Benedikt Heyl and Insa Zillmann

8.1 Introduction

Nuclear power, accounting for less than 10% of global electricity production (Schneider et al. 2024), depends on an array of complex products and services that must be provided before and after the production of electricity, resulting in the definition of nuclear power as a so-called "system good," see Chap. 3 of this book. Most notably, several production steps must be conducted to provide highly asset-specific fuel to a nuclear power plant. The complexities and dependencies of the nuclear fuel supply chain pose significant challenges. From uranium mining and milling to fuel rod fabrication, each stage of the supply chain requires specialized infrastructure, advanced technologies, and robust market frameworks to function effectively.

The construction of nuclear power plants results in long-term path dependencies regarding certain technologies (Cowan 1990), and concentrated markets limit the ability to switch suppliers. Global market dynamics reveal a concentrated structure in many nuclear fuel supply chain stages, as indicated by the Herfindahl–Hirschman Index. High index values, particularly in uranium enrichment and conversion, underscore significant market concentration and reliance on a limited number of suppliers. Such dependencies create vulnerabilities, exposing countries to supply disruptions and geopolitical risks (NEA and IAEA 2023; Szulecki and Overland 2023). In the European context, these challenges are further magnified by the region's limited domestic natural uranium resource exploitation and its reliance on external suppliers for key nuclear fuel components. For example, over 99% of the European Union's (EU) uranium demand is imported, predominantly from Kazakhstan, Niger, Canada, and Russia (ESA 2024). This structural dependency underscores the urgency of addressing market concentration and diversifying supply chains.

Correspondance to editors: Alexander Wimmers (awi@wip.tu-berlin.de)

B. Heyl (✉) · I. Zillmann
TU Berlin, Berlin, Germany

A. Wimmers et al. (eds.), *Nuclear Power*, Studies in Energy, Resource and Environmental Economics, https://doi.org/10.1007/978-3-031-99894-2_8

This chapter provides a comprehensive analysis of the nuclear fuel supply chain for light-water reactors, the globally dominant reactor type. Here, we examine the four key stages—mining and milling, conversion, enrichment, and fuel rod fabrication—through the lens of global market dynamics and HHI values. We also highlight specific challenges faced by the countries of the EU, emphasizing the need for diversification, innovation, and strategic measures to enhance supply chain resilience. By identifying vulnerabilities and exploring potential solutions, this chapter contributes to the ongoing discourse on achieving energy security in times of increasing geopolitical instability.

8.2 Theoretical Background

The Herfindahl–Hirschman Index (HHI) is a statistical indicator used to measure market concentration. It illustrates how strongly a market is dominated by specific actors and whether potential dependencies exist. Competition authorities often use the HHI to evaluate the impact of mergers and acquisitions on a given market structure.

Given N firms with a market share MS, the HHI is calculated by summing the squares of the market shares of all firms in a market:

$$HHI = \sum_{i=1}^{N} MS_i^2 \tag{8.1}$$

Here, MS_i represents the market share of the i-th firm, expressed as a percentage between 0 and 100. Consequently, the HHI will always range between 0 and 10,000. The index reaches its maximum value of 10,000 in a monopoly, where one firm holds 100% of the market.[1] Conversely, in a perfectly competitive market with numerous firms, each holding a very small share, the HHI approaches zero, indicating low concentration. Market concentration is generally classified as low when the HHI is below 1,000, moderate when it is between 1,000 and 1,800, and high when it exceeds 1,800 (Rhoades 1993).

When precise HHI calculation is challenging due to limited data, such as in large markets with numerous firms, it is possible to estimate upper and lower bounds for the HHI by using the market shares of the largest $T \leq N$ firms, with N being the (unknown) total number of firms in a market. This approach relies on the known market shares, which are arranged in descending order, i.e., $MS_1 \leq MS_2 \leq \ldots \leq MS_T$. The range of the HII is estimated following the inequality depicted below (Eq. (8.2)).

[1] $HHI = (100)^2 = 10{,}000$.

$$\sum_{i=1}^{T} MS_i^2 \leq \sum_{i=1}^{T} MS_i^2 + \lim_{K\to\infty} \sum_{i=1}^{\infty} \left(\frac{1}{K}\right)^2$$
$$\leq \sum_{i=1}^{N} MS_i^2 \leq (MS_T)^2 \frac{100\% - \sum_{i=1}^{T} MS_i}{MS_T} + \sum_{i=1}^{T} MS_i^2 \tag{8.2}$$

The lower bound assumes that the unknown portion of the market is divided among an infinite number ($K \to \infty$) of firms, each contributing a negligible amount to the HHI. This scenario represents a highly competitive market structure.

The upper bound assumes that beyond T known firms, the remaining market share is distributed among the smallest possible number of additional firms, each with a share equal or smaller than MS_T. The minimum number of such firms can be estimated as

$$\frac{100\% - \sum_{i=1}^{T} MS_i}{MS_T} \tag{8.3}$$

This method of estimating HHI is particularly useful in real-world scenarios where exact market share data is incomplete, but understanding market concentration is essential for decision-making (Nes et al., 2021).

In the nuclear fuel supply chain, the HHI is an essential tool for identifying risks linked to dependence on a few dominant suppliers, such as in uranium mining or enrichment. High HHI values highlight significant dependencies that may lead to vulnerabilities in supply security. By revealing these risks, the HHI highlights the necessity of diversification strategies to reduce dependencies and improve the resilience of the supply chain.

8.3 Analysis of the Nuclear Fuel Supply Chain

The nuclear fuel supply chain is divided into four key stages: mining and milling, conversion, enrichment, and fuel rod fabrication, see Chap. 2 of this book. Each stage is critical in transforming raw uranium into nuclear fuel (WNA 2022). This chapter describes the technical background and analyzes the global market dynamics and the market concentration using the HHI within these four stages. Further downstream activities, such as power generation or nuclear waste management, are beyond the scope of this analysis. Furthermore, this analysis is limited to the provision of services for light water reactors (LWR) that, as of January 2025, account for 90% of the globally installed nuclear capacity.[2] Fuel provision for heavy water reactors, such as the CANDU reactors operational in Canada, is not discussed here. This also holds

[2] https://pris.iaea.org/PRIS/WorldStatistics/OperationalReactorsByType.aspx, accessed on 31–01-2025.

for other technological lines, such as high-temperature reactors or sodium-cooled fast reactors.

8.3.1 Mining and Milling

Mining and milling represent the first stages of the nuclear fuel supply chain, converting raw uranium resources into uranium oxide (also referred to as yellowcake due to its distinctive color), which serves as the foundational material for further processing. Before mining, the ore's quality must be analyzed to assess the economic and technical feasibility of uranium extraction, to estimate the resource quantity, and to calculate mining costs. Due to limited transparency regarding costs, the global uranium inventory fluctuates significantly each year (NEA and IAEA 2023), creating challenges for long-term planning and emphasizing the need to understand production methods and cost categories.

Uranium extraction is typically classified into three production methods, which differ in cost—the technical mining processes, the geographic distribution, and the environmental impact. The most cost-effective method for uranium mining is in situ leaching (ISL), which has production costs below 40 US-$/kgU. ISL involves injecting acid or alkaline solutions into the ground to dissolve the uranium ore. The uranium-rich solution is pumped to the surface and filtered to recover uranium oxide, allowing to skip the additional process of milling, see below.[3] This method is predominantly used in China, Kazakhstan, and Uzbekistan (NEA and IAEA 2023). While ISL minimizes surface disruption, it carries environmental risks, particularly that of groundwater contamination due to the chemicals used. In contrast, the second option groups underground and open-pit mining, which involve the physical excavation of uranium ore. These methods are more expensive, with costs up to 130 US-$/kgU, and cause significant land disturbance, requiring extensive rehabilitation efforts to mitigate environmental impacts (Larsen and Mamosso 2014). Underground mining is commonly employed in countries like Canada, Russia, and Ukraine, while open-pit mining is prevalent in Australia, Namibia, and Niger. The third method, co-product mining, differs as uranium is extracted as a by-product from mines primarily focused on other minerals or metals. This method is the most expensive, with costs reaching up to 260 US-$/kgU. It is often employed in regions with diversified mining operations, such as Canada or Brazil (NEA and IAEA 2023). Although uranium is not the primary target in these mines, the process still generates substantial environmental impacts, including land disruption, waste generation, and the added complexity of managing multiple waste streams.

Following the extraction through mining, the uranium ore undergoes a process called milling, where it is chemically processed to produce uranium oxide. Milling

[3] https://wna.origindigital.co/information-library/nuclear-fuel-cycle/mining-of-uranium/world-uranium-mining-production, accessed 16–12-2024. This date holds for all other references if not otherwise indicated.

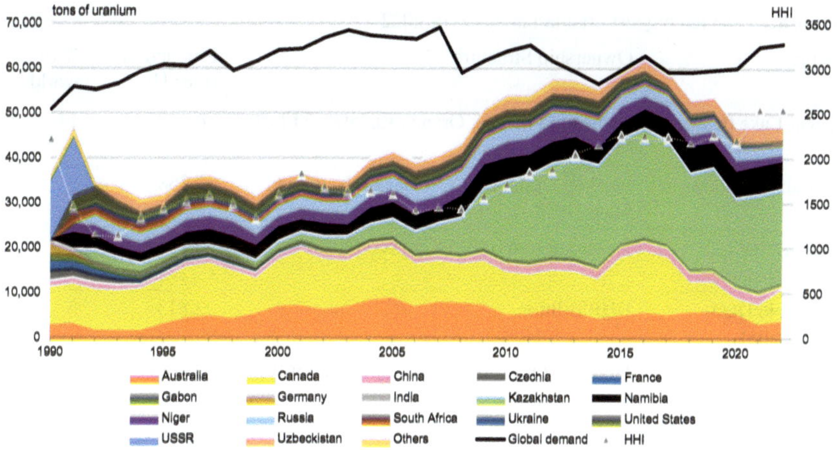

Fig. 8.1 Global uranium production and demand from 1990 to 2022. *Source* Own depiction based on data taken from (NEA and IAEA 1997–2023) and the World Nuclear Association[4]

depends on the mining method. For open-pit and underground mining, the uranium ore must be crushed, ground, leached, and separated from other solid substances. For in-situ leach mining, these steps can be skipped since the uranium is already dissolved during the leaching process. Here, the uranium is extracted as a liquid solution, precipitated, and then dried to produce yellowcake (Yim 2022).

Due to the geological distribution of uranium deposits, natural uranium cannot be mined and milled everywhere, creating inherent dependencies on specific regions with significant reserves. The largest recoverable uranium resources are concentrated in Australia, Canada, Kazakhstan, and Russia. Other notable uranium-producing countries include Namibia, Niger, the United States (US), and Uzbekistan, though their reserves are comparatively small. As shown in Fig. 8.1, global production has not been able to satisfy uranium demand for at least the past 35 years, leading to secondary supply channels being used, such as the access of stockpiles built up before 1990 or the usage of uranium from dismantled nuclear weapons (NEA and IAEA 2023).

Following the collapse of the Soviet Union and the diversification of uranium production sources to individual countries, such as Kazakhstan, Russia, and Uzbekistan, the HHI for global uranium production on a country level dropped from 2,216 in 1990 to just 1,149 in 1993, indicating a moderate market concentration. In recent years, however, with Kazakhstan significantly increasing its production value, the HHI has reached levels indicating a high market concentration, with the threshold of 1,800 being crossed for the first time in 2011 and peaking at 2,547 in 2022 (NEA and IAEA 1997–2023). This geographic concentration of mined uranium resources highlights the structural dependencies embedded in the nuclear fuel supply

[4] https://wna.origindigital.co/information-library/nuclear-fuel-cycle/mining-of-uranium/world-uranium-mining-production.

Table 8.1 The ten biggest uranium mines as of 2022

Mine	Ownership Structure	Production (tons U)	% of world
Cigar Lake (Canada)	Cameco (54.55%), Orano (40.45%), TEPCO (5%)	6,928	14
Husab (Namibia)	Swakop Uranium (CGN (90%), Namibian State (10%))	3,358	7
Inkai, sites 1–3 (Kazakhstan)	Kazatomprom (60%), Cameco (40%)	3,201	7
Olympic Dam (Australia)	BHP Billiton	2,813	6
Karatau (Budenovskoye 2) (Kazakhstan)	Uranium One (50%), Kazatomprom (50%)	2,560	5
Rössing (Namibia)	CNNC (68.62%), Iranian Foreign Investment Fund (15%), Industrial Development Fund of South Africa (10%), Namibian State (3%), Individual Shareholders (3.38%)	2,255	5
SOMAIR (Niger)	Orano (63.4%), Nigerian State (36.6%)	2,020	4
Four Mile (Australia)	Quasar	1,740	3
Central Mynkuduk (Kazakhstan)	Ortalyk (CGN (49%), Kazatomprom (51%))	1,650	3
South Inkai 4 (Kazakhstan)	Uranium One (70%), Kazatomprom (30%)	1,600	3
Top 10 total		28,125	57

Source Compiled by authors, based on World Nuclear Association and InvestingNews.com[5]

chain. However, the remaining stockpiles and the reuse of uranium from dismantled weapons and depleted fuel rods seem to limit the effect of this market concentration on uranium availability.

When analyzing the market concentration based on companies, the perception of a high concentration changes. Applying Eq. (8.2) to the market, taken from available information on the ten biggest uranium mines (as of 2022), see Table 8.1, one can approximate the HHI relative to mining companies and state actors. Accounting for the known companies, the estimated HHI lies between 402 and 424, indicating a low market concentration. However, the remaining 43% of the uranium market remains non-transparent. When accounting for the state ownership of many of the actors, such as Uranium One and TEPCO (owned by Russian state-owned Rosatom) or Orano (majority-owned by the French state), the estimated HHI is a little higher, ranging from 462 to 483. This also indicates a low market concentration.

[5] https://wna.origindigital.co/information-library/nuclear-fuel-cycle/mining-of-uranium/world-uranium-mining-production and https://investingnews.com/daily/resource-investing/energy-investing/uranium-investing/top-uranium-mines/.

The HHI calculations show that, at first glance, the market seems concentrated on a country basis (related to the locations of active mines). At closer inspection, however, the concentration drops substantially when considering the heterogeneous ownership of different mines across the producing countries. Thus, based on market actors (with limited access to reliable information), the concentration of the uranium market seems to be lower than expected.

8.3.2 *Conversion*

The conversion of yellowcake, the final product of the milling process, into uranium hexafluoride (UF_6) represents a critical stage in the nuclear fuel supply chain. This process involves the chemical transformation of uranium oxide (yellowcake, U_3O_8) into uranium hexafluoride (UF_6), which must be present in gaseous form for the subsequent enrichment stage. The yellowcake reacts with fluoride gas to produce UF_6, which is highly reactive and poses significant handling challenges. For instance, UF_6 reacts with water to form hydrofluoric acid, necessitating strict technical precautions during production and transportation. Despite these complexities, the conversion process is indispensable as it ensures that the uranium can undergo isotopic separation in centrifuges, where the concentration of Uranium-235 (U-235) is increased to reactor-grade levels, see Sect. 8.3.3. Following conversion, UF_6 is transported to enrichment facilities to complete the next stage of the nuclear fuel cycle (Yim 2022; IAEA 2023).

In 2022, the global capacity for uranium conversion was approximately 42,600 tons. Table 8.2 illustrates the distribution of active capacity among the world's only (publicly known) conversion plants. Russian state-owned Rosatom holds the largest share of production with the highest capacity utilization rate. Chinese CNNC follows, and Cameco of Canada contributes the third largest amount. French Orano operates at only 60% utilization, producing only 9,000 tons. As of 2022, ConverDyn in the United States remained inactive (Finegold 2023). The facility, known as Metropolis Works, had been idled in 2017 due to a global oversupply of UF_6 and declining demand, but successfully resumed operations in July 2023 following improved market conditions[6] and announced US government plans to build up a uranium stockpile.[7] The reopening of the Metropolis Works in 2023 supposedly added 7,000 tons of annual conversion capacity, positioning ConverDyn as the fifth-largest producer (NEA and OECD 2024). CNNC is a significant player, but the Chinese company primarily focuses on meeting domestic demand (WNA 2022).

As of 2022, the HHI value for installed conversion capacity stands at approximately 2,111, indicating a highly concentrated structure dominated by a few key

[6] https://www.world-nuclear-news.org/Articles/Idled-US-conversion-plant-preparing-for-restart and https://converdyn.com/facility/

[7] https://www.defensedaily.com/nnsa-to-make-converdyn-contract-for-uranium-reserve-official/nuclear-modernization/

Table 8.2 Conversion plants and their capacity as of 2022

Company	Location (Country)	Capacity [Tons]	Capacity Utilization [Tons]
Orano	Pierrelatte and Malvesi (France)	15,000	9,000 (60%)
CNNC	Lanzhou und Hengyang (China)	15,000	11,000 (73%)
Cameco	Port Hope (Canada)	12,500	10,600 (85%)
Rosatom	Seversk (Russia)	12,500	12,000 (96%)
ConverDyn	Metropolis (USA)	7,000	0 (0%)
Total		62,000	42,600 (69%)

Source Own depiction based on Finegold (2023)

players. In terms of production, the value is higher at 2,526, with each player supplying between 21 and 28% of global conversion services demand. Assuming the successful restart of the ConverDyn facility in July 2023, and a production of half of the theoretically available annual capacity, the HHI would decrease to 2,214—still a high market concentration and higher than if all plants utilized their full capacity.

8.3.3 Enrichment

Following conversion, UF_6 must undergo a so-called enrichment process, in which the concentration of the fissile isotope U-235 is increased to levels required for reactor-grade fuel—roughly 3 to 5% for light-water reactors which make up the vast majority of operational reactors. Mined uranium typically consists of 99.28% U-238 and only 0.72% U-235, the remainder being a very small share of U-234. This step is essential in the nuclear fuel cycle for light-water reactors and typically uses centrifugation to separate the lighter U-235 from the heavier U-238. Centrifugation involves the use of high-speed machines to separate isotopes based on their slight mass difference. Historically, enrichment was also conducted via gas diffusion. Laser-induced enrichment of U-235 steam is currently being researched but is far off from commercialization (Wimmers et al. 2025).[8] Whichever approach is chosen, the result is enriched uranium, containing 3 to 5% U-235, suitable for most nuclear reactors. Other reactor designs, such as so-called new reactor concepts, require none or even higher levels of enrichment, known as High-Assay Low-Enriched Uranium (HALEU). HALEU contains up to 20% of U-235 (see Chap. 2).

Globally, a few key players dominate the uranium enrichment sector, see Table 8.3. The Russian company Tenex, owned by Rosatom, leads the global uranium enrichment market with 46% of capacity in 2020, followed by Urenco, which operates in the UK, Germany, the Netherlands, and the US, with 30%. Orano, based in France, accounts for 12%, and CNNC, operating in China, contributes 11%. The HHI for

[8] The performance of a uranium enrichment plant is measured in separative work units (SWU), see Chap. 3 of this book.

Table 8.3 Global enrichment capacity, operational and planned

Company	Location	2013	2015	2020	2025	2030
Orano	Georges Besse I & II (France)	5,500	7,000	7,500	7,500	7,500
Urenco	Gronau (Germany); Almelo (Netherlands); Capenhurst (UK)	14,200	14,400	13,700	17,300	16,300
Urenco	Urenco (New Mexico, US)	3,500	4,700	4,900		
Tenex	Angarsk, Novo Uralsk, Zelenogorsk, Seversk; Russia	26,000	26,578	27,700	26,200	24,800
CNNC	Hanzhun & Lanzhou, China	2,200	5,760	6,300	11,000	17,000
JNFL	Japan (Rokkashu)	150	150	150	150	1,500
Others	Argentina, Brazil, India, Pakistan, Iran	75	100	66	375	525
Total separative work units / year (thousands p.a.)		51,700	58,750	55,416	62,525	66,275

Source Own depiction based on the World Nuclear Association[9]

the uranium enrichment market was 3,833, 3,342, and 3,038 in 2013, 2015, and 2020, respectively. This reflects a highly concentrated market structure dominated by a few major players. In this case, the high HHI reflects the limited number of suppliers and the significant market share held by companies like Tenex and Urenco. Planned capacity changes could reduce the HHI to 2,975 and 2,797 in 2025 and 2030, respectively. This still indicates a highly concentrated market.

This high concentration reduces market competitiveness and increases the supply chain's vulnerability, particularly during geopolitical tensions. The lack of diversification within the sector poses significant risks, particularly in the face of geopolitical tensions, making it crucial for countries and industries to explore alternative sources and invest in new (domestic) enrichment capacities.

8.3.4 *Fuel Rod Fabrication*

The gaseous enriched UF_6 must be converted into a solid form for use as nuclear fuel. It is transformed into the ceramic compound UO_2, which is known for its high heat resistance, mechanical strength, and lack of reactivity with water. From this, so-called pellets are fabricated. These are adjusted to match the geometry of the fuel cladding tubes. The tubes are filled with inert gas, such as helium. A filled tube is referred to as a fuel rod. Depending on the reactor type, the cladding tubes are made of a variation of zirconium alloys, abbreviated as Zircaloy. For instance, Zircaloy-2,

[9] World Nuclear Association. 2020. "Uranium enrichment". Accessed June 5, 2020. https://www.world-nuclear.org/information-library/nuclear-fuel-cycle/conversion-enrichment-and-fabrication/uranium-enrichment.aspx and Updated Webpage as of 11 October 2022, accessed 21–08-2024.

Table 8.4 Key players in global LWR fuel fabrication capacity (tons/year)

Country	Fabricator	Conversion [tons/yr]	Pelletizing [tons/yr]	Rod/ assembly [tons/yr]
Brazil	INB	160	120	400
China	CJN Jinzhong	800	800	800
	CJNF Jianzhong	800	800	800
	CBNF	0	0	400
	CNNFC	200	200	200
France	Framatome-FBFC	1,800	1,400	1,400
Germany	Framatome-ANF	800	650	650
India	DAE Nuclear Fuel Complex	48	48	48
Japan	Mitsubishi Nuclear Fuel	450	440	220
	Global Nuclear Fuel – Japan	0	620	630
	NFI (PWR)	0	383	284
	NFI (BWR)	0	250	250
Kazakhstan	Ulba	0	108	200
Russia	TVEL-MSZ*	1,220	1,220	1,280
	TVEL-NCCP	450	1,200	1,200
South Korea	KNFC	700	700	700
Spain	Enusa	0	500	500
Sweden	Westinghouse AB	787	600	600
UK	Westinghouse	750	400	660
US	Framatone Inc	1,200	1,200	1,200
	Global Nuclear Fuel – Americas	1,200	1,000	1,000
	Westinghouse	1,600	1,594	2,154
Total		12,965	14,233	15,576

Source Own compilation based on World Nuclear Association[10]

which contains nickel, is used in boiling water reactors, while Zircaloy-4, a nickel-free alloy, is used for pressurized water reactors (Wimmers et al. 2025). Multiple rods are assembled into bundles or assemblies designed to meet the operational and safety requirements of certain nuclear reactor designs (IAEA 2023). These assemblies are manufactured with precision and standardization to ensure compatibility with diverse reactor designs and safety systems worldwide.

To better understand the distribution of production capacities across regions and stages, Table 8.4 provides an overview of key players for LWR fuel and their contributions to conversion, pelletizing, and rod assembly.

[10] https://world-nuclear.org/information-library/nuclear-fuel-cycle/conversion-enrichment-and-fabrication/fuel-fabrication.

The global nuclear fuel rod fabrication market is dominated by a handful of key players, including Framatome (France), TVEL (Russia), Westinghouse (US), CJNF (China), and Global Nuclear Fuel (US and Japan). Grouping the production capacities to corporations,[11] the HHIs for conversion, pelletizing, and rod or assembly production are 1,931, 1,499, and 1,530, respectively. Thus, there seems to be a moderate market concentration for the last two production steps, while the fuel conversion capacity is less diverse. Unlike the earlier stages of conversion and enrichment of uranium ore, which are controlled by an even smaller number of participants, fuel rod fabrication involves a slightly broader distribution of facilities performing these steps. However, with most of the total capacity concentrated among a small number of key players, substantial dependency remains, particularly in regions heavily reliant on specific suppliers.

Furthermore, the availability of LWR fuel production capacity does not necessarily mean that every LWR can be supplied from every facility because of the high asset specificity of fuel rods. For example, Russian supplier TVEL dominates the fuel supply for Russian- (formerly Soviet) designed VVER reactors, located particularly in Eastern Europe. Alternative suppliers, such as Framatome and Westinghouse, focus primarily on Western-designed fuel assemblies. Western-designed pressurized water reactors operate with square lattice configurations, while Russian-designed VVER reactors utilize hexagonal grid configurations, requiring specially developed fuel elements to ensure compatibility with their distinct reactor cores. For example, a distinction is made between VVER-1000 and VVER-440 reactors, with the numbers denoting their differing outputs in megawatts (NEA and OECD 1999), see also Sect. 8.4.1.4. This distinction in reactor designs underscores the role of tailored manufacturing processes to meet diverse market demands. Geopolitical factors, especially in Europe, have driven efforts to diversify supply chains and reduce reliance on Russian suppliers. EU-funded initiatives, such as those in the Czech Republic and Slovakia, and the so-called "Accelerated Program for Implementation of Secure VVER Fuel Supply" (APIS), aim to enhance domestic fuel production and adapt VVER reactors for Western-manufactured fuel. These efforts are part of broader strategies to enhance energy independence in response to geopolitical tensions and mitigate supply risks (Wimmers et al. 2025).

Despite moderate concentration, the fuel rod fabrication market is undergoing significant transformation. Increasing global demand, geopolitical tensions, and the development of advanced technologies, such as accident-tolerant fuels, are reshaping the industry. Ensuring supply stability while adapting to evolving safety and performance standards remains a key challenge for manufacturers worldwide.

[11] Chinese actors are also grouped due to their common links to the Chinese state.

8.3.5 Comparison

After providing detailed assessments of each stage of the fuel supply chain for LWRs, a comparison of the market concentration indicators is merited. As of 2022, shown in Fig. 8.2, the HHI values for most steps are above or around 1,500. Conversion and enrichment services are provided in a highly concentrated market, while uranium mining and milling companies operate in a competitive environment. The arrows indicate the expected development of the HHI, i.e., the market concentration, in the coming years. For conversion and enrichment, the expansion of additional capacities by actors such as ConverDyn, or the planned reduction of capacities bringing down individual market shares in enrichment services, will likely reduce market concentration. Actual competitiveness, though, as observed in the mining and milling market, seems unlikely to be achieved in the coming decades. Fuel rod fabrication market concentration will likely increase due to planned capacity expansions of already dominant players, such as Westinghouse and Framatome. Access to information on conversion and pelletizing capacities remains limited, which is why an assessment of future perspectives is not possible in this work.

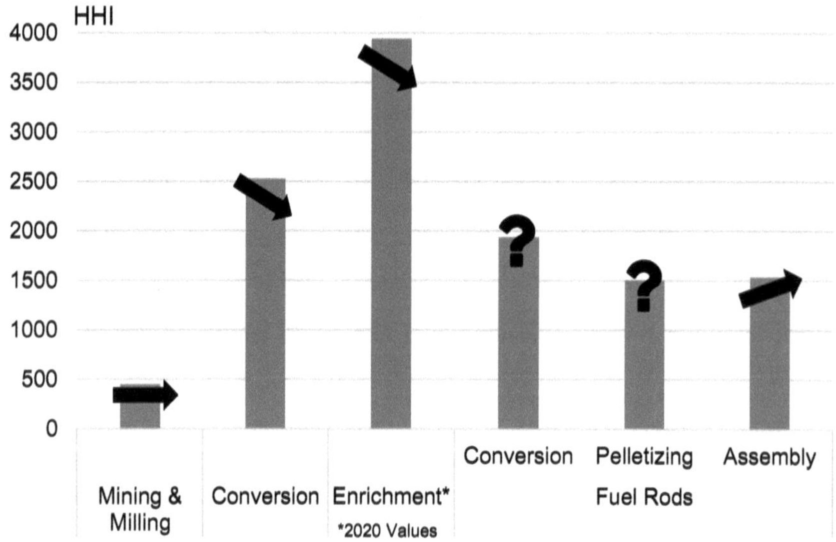

Fig. 8.2 Comparison of HHI values for the different supply chain steps as of 2022. *Source* Own depiction

8.4 The European Perspective

Following a detailed analysis of the four stages of the nuclear fuel supply chain and their respective roles in the global market, this section focuses on the European perspective. Significant dependencies often accompany high market concentration with few market participants. These dependencies entail energy and geopolitical risks, which is why it is crucial to examine them in the context of the European nuclear fuel supply chain, especially in the wake of the Russian invasion of Ukraine in 2022. Understanding these vulnerabilities provides the foundation for exploring strategies to enhance security of supply through diversification and innovation.

8.4.1 Dependencies in the European Nuclear Fuel Supply Chain

8.4.1.1 Uranium Supply

The EU's nuclear fuel supply chain faces significant dependencies, particularly in stages where domestic capabilities are either nonexistent or severely limited. Natural uranium, of which 99.5% is imported due to lacking domestic supply (Gufler and Meister 2022), is primarily sourced from a small group of countries, including Kazakhstan, Niger, Canada, and Russia. Together, these countries provided over 91% of EU uranium needs in 2023 (ESA 2024). This reliance on external sources underscores the EU's structural dependency in the early stages of the nuclear fuel cycle. Reflecting this dependency, the HHI for the origin of uranium delivered to EU utilities in 2023 is 2,305, highlighting the EU's vulnerability to supply disruptions and geopolitical risks. From 2014 to 2020, the HHI had been below 1,800, indicating a more diverse market.[12] Most notably, the most recent expansion of supply from Canada and Russia, and the reduction of deliveries from Australia, Niger, and Uzbekistan increased the concentration to an all-time high in 2023. However, information on actual supplier companies remains opaque in detail. The Euratom Supply Agency's report lists 15 individual companies (eleven when grouped by parent companies), but does not inform readers of the actual supply by individual companies (ESA 2024). Thus, similarly to the assessment in Sect. 8.3.1, the market might be more diverse than the country-based assessment indicates.

As shown in Sect. 8.3.1, the primary operators of uranium mines are Kazakhstan's Kazatomprom, which accounted for 23% of global uranium production in 2022, followed by Canada's Cameco (12%) and French Orano (11%). Russia's Rosatom, through its subsidiaries Uranium One, ARMZ, and TVEL, also plays a key role in the global uranium market. About 12% of Kazakh and approximately 90% of Russian mined uranium is exported to Europe. Additionally, Rosatom directly controls not

[12] Calculated HHI-Values: 2014: 1633; 2015: 1657; 2016: 1723; 2017: 1753; 2018: 1726; 2019: 1475; 2020: 1725; 2021: 1955; 2022: 2159.

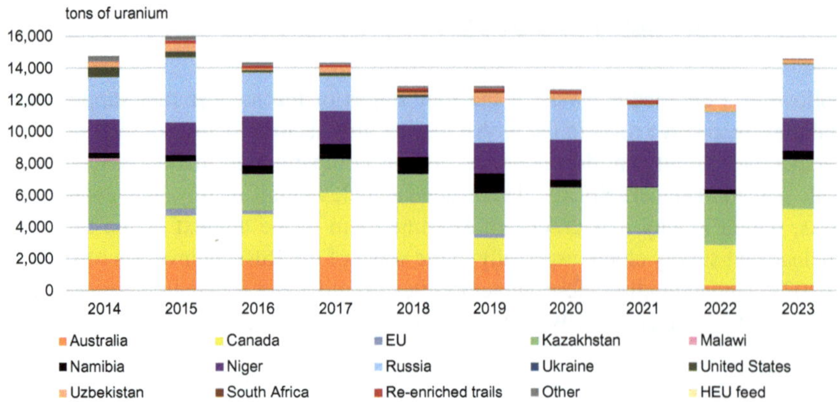

Fig. 8.3 Origins of uranium delivered to EU utilities from 2014 to 2023. *Source* Own representation based on (ESA 2024)

only mines in Russia, but also 22% of Kazakh uranium production. This means Rosatom oversees approximately 15% of global uranium production, making it the second-largest uranium producer worldwide (Gufler and Meister 2022). In 2023, Canada emerged as the EU's largest supplier of uranium, accounting for 33% of total imports, followed closely by Russia with 24%. Additionally, Kazakhstan and Niger remain key uranium suppliers to the EU, further emphasizing the geographical concentration of uranium sources and the European reliance on external partners for nuclear fuel supply (ESA 2024). Figure 8.3 illustrates the supply structure over the last decade.

8.4.1.2 Conversion Services

The EU's dependence on a few countries for natural uranium is mirrored by similar constraints in the availability of conversion facilities. The conversion of natural uranium into uranium hexafluoride (UF_6) is dominated by just four companies: Orano (France), Rosatom (Russia), Cameco (Canada), and ConverDyn (US), as shown in Sect. 8.3.2. These companies collectively supplied 97% of the EU's nuclear fuel conversion needs in 2022, and 93% in 2023. Consequentially, the market concentration is very high, indicated by an HHI of more than 2,200 for the past six years. From 2022 to 2023, the HHI decreased from 2,615 to 2,276 due to the rise of "unspecified" supply in conversion services. The volume provided by the different companies is shown in Fig. 8.4.

With Orano covering 37% of conversion demand, the EU is less dependent on external countries for conversion than for natural uranium. Furthermore, as shown in Table 8.2, Orano's capacity utilization lies at only 60%. Theoretically, European conversion demand could be fully covered by Orano. Regardless, as of today, the highly concentrated market creates notable dependencies and geopolitical risks.

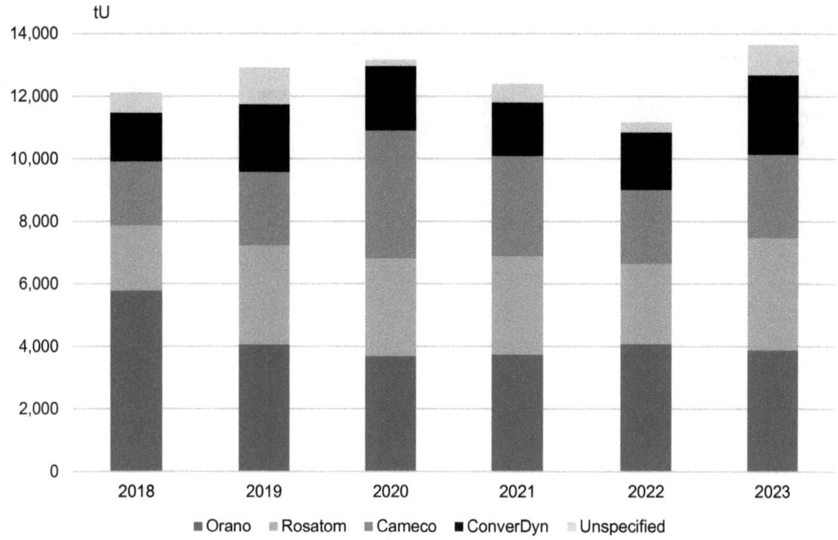

Fig. 8.4 Provision of conversion services to EU utilities from 2018 to 2023. *Source* Own depiction based on (ESA 2024)

Russia, through Rosatom, is the second-largest provider of conversion services to EU utilities, and despite the Russian invasion of Ukraine in 2022, Russian supply has increased by 45% from 2022 to 2023, while Orano's supply fell by 6% (ESA 2024). This reliance on Russian conversion capacity reflects the limited diversification in this stage of the nuclear fuel supply chain and highlights the importance of strategic measures to ensure long-term supply stability.

8.4.1.3 Enrichment Services

In uranium enrichment, which is crucial for the production of LWR fuel, there is a particularly high market concentration regarding the origin of enrichment services supplied to the EU. This is highlighted by the HHI for the origin of enrichment services delivered to the EU, which, over the past decade, has ranged from 5,263 in 2014 to 3,670 in 2021 to 4,299 in 2023, indicating a high market concentration. In 2023, 55% of enrichment services for EU utilities were supplied by EU-based providers, i.e., Urenco and Orano, see Table 8.3 and Fig. 8.5. Non-EU sources indicated in Fig. 8.5 are likely services provided by Urenco's UK facility. 38% of enrichment services originated from Russia, marking a 42% increase compared to 2022, despite Russia's aggression against Ukraine (ESA 2024). Both Orano and Urenco

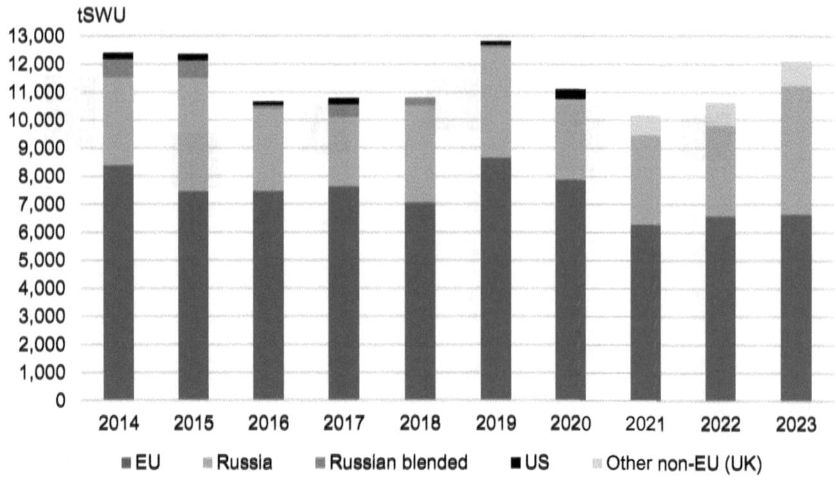

Fig. 8.5 Origin of enrichment services to EU utilities from 2014 to 2023 in thousands of separative work units. *Source* Own depiction based on (ESA 2024)

have announced plans to expand their enrichment capacities in the US and Europe (Orano 2023).[13]

8.4.1.4 Fuel Rod Fabrication Services

The EU is significantly more dependent on imports for fuel rod fabrication than for uranium conversion and enrichment services. Within the EU, Framatome and Westinghouse are the primary providers. Framatome operates facilities for LWR fuel fabrication in France, Germany, and the US, with a combined capacity of 3,250 tons per year. Westinghouse contributes approximately 3,614 tons annually from its facilities in Sweden, the US, and the UK. TVEL, a Russian state-owned manufacturer, remains a dominant global player with an annual capacity of 2,760 tons per year, supplying fuel rods not only for Russian-designed VVER reactors but also for Western reactor designs.[14] The critical role of Russian fuel rod supply became particularly evident on March 1, 2022, when a Russian transport plane was granted special permission to land in Slovakia despite an EU ban on Russian aircraft. This extraordinary measure ensured the delivery of fuel rods to Slovakia's two nuclear power plants (Schneider et al. 2024). It also underscored the EU's dependence on Russian technology. By 2023, Russia still supplied the majority of VVER fuel used

[13] https://urencousa.com/news/sustainability/2024/urenco-marks-installation-of-new-us-enrichment-capacity, and https://cdn.orano.group/usa/docs/librariesprovider9/usa-documents/news/2023/pr_orano-enrichment-increase_2023-1026.pdf?sfvrsn=88d6650b_3.

[14] https://world-nuclear.org/information-library/nuclear-fuel-cycle/conversion-enrichment-and-fabrication/fuel-fabrication.

in the EU, emphasizing the ongoing dependence on Russian expertise and resources in this essential part of the nuclear fuel supply chain.

In the EU, 19 VVER reactors are currently operational, with a total capacity of 9.9 GW. These VVER reactors can be divided into two general types, i.e., the larger VVER 1000 and the smaller VVER-440 reactors, each requiring distinct fuel rod designs. Bulgaria and the Czech Republic each operate two VVER-1000 reactors. There are four VVER 440 reactors operating in the Czech Republic, and another ten in Finland (2), Hungary (4), and Slovakia (5).[15]

Ukraine, heavily reliant on VVER 1000 fuel rods, as 13 of its 15 operational reactors are VVER-1000 designs, has actively pushed for supply diversification. US-based company Westinghouse successfully addressed this need, achieving the capacity to annually produce five or more VVER-1000 reload batches at its Swedish facility (Gunnarsson et al., 2019). In contrast, most Russian-built reactors in the EU are not VVER-1000 but VVER-440 reactors. Here, the efforts to diversify supply have been less advanced. In the 1990s, the British company BNFL developed and tested VVER-440 fuel rods, and the Spanish company ENUSA later advanced the production of fuel for these reactors. Westinghouse acquired BNFL and partnered with ENUSA to continue these efforts, enhancing the diversification of supply.[16] However, deliveries were limited to individual plants, such as the Czech Temelín plant in the early 2000s, and were discontinued after a few years (Schneider et al. 2024). Currently, Westinghouse is advancing research and verification for VVER-440 fuel under the APIS project, see Sect. 8.4.2.

Additionally, French company Framatome initiated a joint venture with Rosatom's TVEL to manufacture Russian-designed fuel assemblies at the Advanced Nuclear Fuels (ANF) plant in Lingen, Germany. While the venture was initially planned to be established in Germany, regulatory concerns, including those from the German Federal Ministry for Economic Affairs, led to the withdrawal of the application shortly before Russia's invasion of Ukraine in 2022. The joint venture was subsequently established in France, with production still planned in Lingen, focusing on hexagonal fuel assemblies for VVER reactors in Eastern Europe. The 2024 edition of the World Nuclear Industry Status Report highlights this development in the chapter on "Russian Nuclear Dependencies," noting the project's aim to reduce reliance on Russian supplies (Schneider et al. 2024, 318–31). However, the involvement of TVEL, holding a 25% stake in the venture and participating in production, raises questions about the extent of this independence.

8.4.1.5 Summary of Risks and Opportunities

Despite diversification advancements, the EU remains vulnerable to geopolitical and supply chain risks, especially tied to Russian services. These dependencies are clearly reflected in the high HHI values across different supply chain stages, consistently

---

[15] https://www.iaea.org/publications/15748/nuclear-power-reactors-in-the-world.

[16] https://info.westinghousenuclear.com/news/wec-enusa-collab-vver-440.

indicating a very high level of market concentration. The European nuclear fuel supply chain is heavily reliant on a limited number of countries for key resources, each with varying levels of geopolitical stability. Kazakhstan, the world's largest uranium producer, is considered relatively stable, but maintains close economic and political ties with Russia, which could influence its trade policies during geopolitical tensions (Siddi and Silvan 2023). Canada, another major supplier, is geopolitically stable and a reliable partner for the EU, benefiting from strong rule-of-law frameworks and secure trade relations (European Commission 2023).

Namibia and Niger, while important uranium exporters, face challenges related to political instability and governance issues, with Niger particularly susceptible to security risks due to regional conflicts and internal unrest (Anthony and Grip 2013). Furthermore, French reliance on uranium from Niger highlights potentially problematic post-colonial ties between the two countries (Yülek et al., 2023). The region in sub-Saharan Africa incorporating former French colonies is often referred to as "Françafrique." Through its state-owned company Orano, France continues to source significant amounts of uranium from Niger under favorable terms, including exemptions from export tariffs (Taylor 2019).[17] However, the 2023 military coup in Niger has raised questions about the security of uranium supplies, as the new regime imposed an export ban on uranium shortly after assuming.[18] At the end of 2024, Orano's mines were seized by the Niger government. At the time of writing, the situation remained volatile and uncertain (Wimmers et al. 2025). These developments underscore the fragility of dependencies on politically unstable regions, even when they are embedded in longstanding trade relationships.

Russia, being a major supplier across several stages of the nuclear fuel supply chain, poses significant geopolitical risks, particularly in the wake of the Russian war against Ukraine. This has heightened concerns over energy dependencies, not only just in Europe (Gufler and Meister 2022; Szulecki and Overland 2023). Uzbekistan and China, while emerging players, are heavily influenced by state-led policies, and their long-term reliability as partners remains uncertain. This complex web of geopolitical factors underscores the critical need for diversification and the development of alternative supply chains to reduce vulnerabilities.

Addressing these vulnerabilities requires coordinated efforts to expand domestic capacities, diversify supply chains, and invest in innovative technologies. The APIS program serves as a prominent example of these diversification efforts, laying the foundation for greater resilience in the European nuclear fuel supply chain, see below.

[17] https://www.wiwo.de/technologie/wirtschaft-von-oben/wirtschaft-von-oben-222-uran-aus-dem-niger-hier-braut-sich-das-naechste-energieproblem-fuer-europa-zusammen/29295338.html.

[18] https://taz.de/!5953147/

8.4.2 Diversification Strategies and Challenges

The dependencies outlined in the previous section underscore the EU's urgent need to diversify its nuclear fuel supply chain to ensure energy security and reduce geopolitical risks. To address these vulnerabilities, the EU has implemented several diversification strategies, focusing on securing alternative suppliers, adapting reactor infrastructure, and building strategic reserves (Guedes Ferreira 2024). These measures aim to strengthen the resilience of the EU's domestic nuclear fuel supply chain while reducing the reliance on a limited number of external providers. Deliveries of natural uranium and services at all stages of the nuclear fuel cycle increased in 2023, reflecting the critical importance of diversification efforts.[19] However, diversification opportunities vary significantly across the four stages of the nuclear fuel supply chain, i.e., mining/milling, conversion, enrichment, and fuel rod production.

At the mining and milling stage, the EU faces its greatest dependency, as it imports over 99% of its uranium requirements due to a lack of domestic reserves. This geological limitation necessitates stable partnerships with major uranium-exporting countries like Canada or Kazakhstan (European Commission 2023; EEAS 2023). Establishing long-term contracts with politically stable nations remains essential to ensure a reliable supply at this stage. Furthermore, European countries could reestablish closed uranium mining activities. Substantial recoverable resources have been identified in Germany, Italy, Slovakia, Spain, Sweden, and Ukraine, amongst others. However, extracting these resources is currently more expensive than purchasing uranium on the global market, making domestic European mining activities economically unattractive (NEA and IAEA 2023). A higher global market price might potentially trigger commercial activities, or geostrategic independence might be worth the additional costs in the future.

The market for uranium conversion is highly concentrated, with four companies controlling 97% of global capacity. Despite this, the EU benefits from Orano's significant market share, which reduces the reliance on non-European suppliers. Notably, in 2022, around 40% of Orano's conversion capacity remained unused, signaling an opportunity to expand production and further substitute foreign services. The EU is actively exploring ways to maximize Orano's capacity utilization and expand domestic infrastructure to reduce dependency on external providers.

The EU demonstrates stronger diversification in uranium enrichment compared to other nuclear fuel supply chain stages. European companies like Urenco, with facilities in Germany, the UK, and the Netherlands, as well as Orano, based in France, collectively meet a substantial share of the EU's enrichment needs. This localized capacity reduces some reliance on external providers. However, the global enrichment market remains highly concentrated, with Russia controlling 46% of global enrichment capacity as of 2020. In 2023, 38% of the enrichment services to EU utilities were sourced from Russia. To counter these risks, the EU is proactively

[19] https://euratom-supply.ec.europa.eu/activities/market-observatory_en.

investing in advanced enrichment technologies, particularly for HALEU.[20] Initiatives supporting HALEU development aim to strengthen EU autonomy in nuclear technology and could reduce the dependence on external suppliers, as well as prepare for increased future energy demand (Guedes Ferreira 2024). It remains questionable, though, whether the European nuclear fleet will grow substantially in the coming years (Böse et al. 2024).

Finally, the EU faces specific challenges in fuel rod fabrication, particularly for VVER reactors that historically were, and currently are, reliant on Russian suppliers. However, progress is being made: Western companies like Westinghouse and Framatome are developing compatible fuel assemblies for VVER-1000 and VVER-440 reactors, reducing the dependency on Rosatom's subsidiary TVEL. Recent agreements between Westinghouse and EU Member States, including Bulgaria, the Czech Republic, and Slovakia, have paved the way for diversification by supplying compatible nuclear fuel assemblies for the different VVER reactor types. The European Commission launched APIS to further reduce dependency.[21] Led by Westinghouse and supported by eleven partners, APIS aims to develop and license alternative VVER fuel designs, standardize regulatory frameworks, and rebuild domestic manufacturing capacities. Initial results are expected in 2025, but full implementation will require additional time due to technical and regulatory hurdles. To date, dummy fuel assemblies have been successfully tested, and most VVER-operating countries in the EU, except Hungary, have made significant progress in their diversification efforts.[22] The Framatome project in Lingen, Germany, in collaboration with TVEL, remains to be fully assessed regarding the actual inclusion of Russian actors, and Framatome's dependence on Russian know-how.

Building strategic reserves has become another essential strategy for mitigating supply risks during the transition to alternative suppliers. To date, the EU has signed 14 strategic partnerships aimed at diversifying its supply chains for critical raw materials. Over the past year, this effort has advanced further with seven newly established, mutually beneficial partnerships, including vital energy partner countries such as Australia, Norway, and Serbia. These collaborations emphasize the EU's commitment to securing supply chains while fostering cooperation with key global players (European Commission 2024). The EU has significantly increased its nuclear fuel inventories, creating a buffer for reactor operators while infrastructure adjustments are underway (ESA 2024). Reactor adaptation is also a critical aspect of diversification. For instance, the Czech Republic has made notable progress in licensing fuel manufactured by Westinghouse for its VVER-440 reactors.[23]

While these strategies represent significant progress, challenges remain. The adaptation of reactors, the qualification of alternative fuels, and the expansion of domestic fuel production capacities are complex and require time and continuous investment.

[20] https://www.enerdata.net/publications/executive-briefing/uranium-enrichment.html.

[21] https://apis-project.eu/.

[22] https://euratom-supply.ec.europa.eu/activities/market-observatory_en.

[23] https://info.westinghousenuclear.com/news/westinghouse-reinforces-its-commitment-to-energy-security-in-czech-republic.

Moreover, the global concentration of the nuclear fuel market poses inherent limitations to diversification. Despite these challenges, initiatives like APIS and investments in HALEU production mark critical steps toward a more resilient and secure nuclear energy strategy for the EU.

8.5 Conclusion and Outlook

The analysis of the nuclear fuel supply chain reveals a highly complex and concentrated market structure with significant dependencies at each stage. High HHI values in uranium enrichment and conversion reflect the global dominance of a few key players and underscore the geopolitical and supply chain risks associated with these stages. In contrast, fuel rod fabrication exhibits a more moderate HHI, suggesting a slightly broader distribution of market participants, though dependencies persist in regions reliant on specific suppliers for certain reactor types. Future research should delve deeper into the non-transparent market structures to derive a more accurate overview of these diverse markets.

In the European context, dependencies are particularly pronounced. The EU's reliance on imports for over 99% of its uranium needs, and the concentration of conversion and enrichment services to a limited number of suppliers, highlight critical vulnerabilities in the nuclear fuel supply chain. The dominance of Russian services, especially in uranium enrichment and VVER fuel rod production, amplifies these risks, necessitating urgent action to mitigate dependencies. Consequently, uranium mining and milling, as well as the production of nuclear fuel rods, represent two critical bottlenecks in the European nuclear fuel supply chain. Efforts to address these challenges are already underway. Initiatives like the APIS and investments in HALEU production represent crucial steps toward diversification and innovation. These measures, coupled with increased strategic reserves and reactor adaptations, aim to reduce reliance on dominant suppliers and might enhance the resilience of the nuclear fuel supply chain.

Looking forward, the nuclear industry faces several key challenges and opportunities. The global demand for nuclear fuel could rise, driven by the expansion of nuclear power in Asia and the development of non-light water reactor concepts that would first counter expected closures of aging reactors in North America and Europe. These trends highlight the importance of investing in domestic capacities, fostering international partnerships, and supporting research and development in advanced fuel technologies. Additionally, ensuring self-sufficiency within the nuclear fuel supply chain is vital to reduce reliance on external suppliers and to mitigate vulnerabilities associated with geopolitical tensions. At the same time, such tensions and evolving market dynamics will require continued vigilance and proactive measures to ensure supply chain stability. Previous practices of long-term contracts might have to be reassessed. The path to a secure and sustainable nuclear energy future will depend on coordinated efforts among governments, industry stakeholders, and international organizations. By prioritizing diversification, innovation, and collaboration, the nuclear fuel supply chain can evolve to meet the challenges of a rapidly changing energy landscape while contributing to the global security of energy supply.

References

Anthony, Ian, and Lina Grip. 2013. *Africa and the Global Market in Natural Uranium: From Proliferation Risk to Non-Proliferation Opportunity*. Policy Paper, No. 39. Solna: Stockholm International Peace Research Institute. https://www.sipri.org/sites/default/files/files/PP/SIPRIPP39.pdf.

Böse, Fanny, Alexander Wimmers, Björn Steigerwald, and Christian von Hirschhausen. 2024. 'Questioning Nuclear Scale-up Propositions: Availability and Economic Prospects of Light Water, Small Modular and Advanced Reactor Technologies'. *Energy Research & Social Science* 110 (April 2024): 103448. https://doi.org/10.1016/j.erss.2024.103448.

Cowan, Robin. 1990. 'Nuclear Power Reactors: A Study in Technological Lock-In'. *The Journal of Economic History* 50 (3): 541–67. https://doi.org/10.1017/S0022050700037153.

EEAS. 2023. 'EU-Kazakhstan Relations'. Fact Sheet. European Union External Action - The Diplomatic Service of the European Union. https://www.eeas.europa.eu/sites/default/files/documents/2023/EEAS-%20FACTSHEET-Kazakhstann_October2023.pdf.

ESA. 2024. 'Annual Report 2023'. Euratom Supply Agency. https://euratom-supply.ec.europa.eu/document/download/29018562-122c-4818-8774-2424fc029bf6_en?filename=ESA%20Annual%20Report%202023%20-%20Final%20draft.pdf.

European Commission. 2023. 'Joint Press Release on Canada-EU Relations', 7 March 2023. https://ec.europa.eu/commission/presscorner/api/files/document/print/en/ip_23_1486/IP_23_1486_EN.pdf.

European Commission. 2024. 'Report from the Commision to the European Parliament, the Council, the European Economic and Sicial Committee and the Committee of the Regions'. State of the Energy Union Report 2024. Brussels: European Comission. https://energy.ec.europa.eu/document/download/bd3e3460-2406-47a1-aa2e-c0a0ba52a75a_en?filename=State%20of%20the%20Energy%20Union%20Report%202024.pdf.

Finegold, Ben. 2023. 'A Nuclear Stranglehold: Russia's Unsanctionable Position in the Nuclear Industry'. Ocean Wall. https://oceanwall.com/wp-content/uploads/2023/08/OWL-Russia-Report-Final.pdf.

Guedes Ferreira, Vasco. 2024. 'Strategic Autonomy and the Future of Nuclear Energy in the EU'. Briefing. European Parliamentary Research Service. https://www.europarl.europa.eu/RegData/etudes/BRIE/2024/757796/EPRS_BRI(2024)757796_EN.pdf.

Gufler, Klaus, and Franz Meister. 2022. 'Analyse der Rosatomaktivitäten bzw. Rosatomverflechtungen mit der EU'. Wien: Bundesministerium für Klimaschutz, Umwelt, Energie, Mobilität, Innovation und Technologie. https://www.umweltbundesamt.at/fileadmin/site/publikationen/rep0814.pdf.

Gunnarsson, Lena, Carina Önneby, and Ulf Benjaminsson. 2019. 'Advancements of Westinghouse VVER-1000 Fuel Manufacturing Capacity'. In *Proceedings*, 175–78. Nesebar, Bulgaria: Bulgarian Academy of Sciences. https://inis.iaea.org/collection/NCLCollectionStore/_Public/53/086/53086889.pdf?r=1.

IAEA. 2023. 'Safety of Uranium and Plutonium Mixed Oxide Fuel Fabrication Facilities'. Safety Report. Vienna, Austria. https://www-pub.iaea.org/MTCD/Publications/PDF/PUB2040_web.pdf.

Larsen, Rasmus Kløcker, and Christiane Alzouma Mamosso. 2014. 'Aid with Blinkers: Environmental Governance of Uranium Mining in Niger'. *World Development* 56 (April):62–76. https://doi.org/10.1016/j.worlddev.2013.10.024.

NEA and OECD. 1999. 'VVER: Specific Features Regarding Core Degradation.', January, 36. https://www.oecd-nea.org/upload/docs/application/pdf/2020-01/csni-r98-20.pdf.

NEA and OECD. 2024. 'High-Assay Low-Enriched Uranium: Drivers, Implications and Security of Supply', No. 7685 (August), 72. https://www.oecd-nea.org/upload/docs/application/pdf/2024-09/nea_publication_2_2_2024-09-18_16-53-42_174.pdf.

NEA, and IAEA. 1997–2023. 'Uranium 1997–2022: Resources, Production and Demand'. The Red Book. Nuclear Energy Agency and International Atomic Energy Agency. https://www.oecd-nea.org/jcms/pl_28569/uranium-resources-production-and-demand-red-book/.

NEA, and IAEA. 2023. 'Uranium 2022: Resources, Production and Demand'. NEA No. 7634. Nuclear Energy Agency and International Atomic Energy Agency. https://www.oecd-nea.org/jcms/pl_79960/uranium-2022-resources-production-and-demand?details=true.

Nes, K., Colen, L. and Ciaian, P. 2021. *Market Power in Food Industry in Selected EU Member States*. Luxembourg: Publications Office. https://doi.org/10.2760/63613, JRC125287

Orano. 2023. 'Orano Announces 30% Increase in Uranium Enrichment Capacity by 2028'. Bethesda, Maryland. https://cdn.orano.group/usa/docs/librariesprovider9/usa-documents/news/2023/pr_orano-enrichment-increase_2023-1026.pdf?sfvrsn=88d6650b_3&_gl=1*13gn76r*_gcl_au*NTc0MDgyMzUuMTczMzY4NDY5Mw..

Rhoades, Stephen A. 1993. 'The Herfindahl-Hirschman Index'. *Federal Reserve Bulletin* Mar: 188–89. https://EconPapers.repec.org/RePEc:fip:fedgrb:y:1993:i:mar:p:188-189:n:v.79no.3

Schneider, Mycle, Antony Froggatt, Julie Hazemann, Özgür Gürbüz, Paul Jobin, Phil Johnstone, Timothy Judson, et al. 2024. 'The World Nuclear Industry Status Report 2024'. Paris, France: Mycle Schneider Consulting. https://www.worldnuclearreport.org/IMG/pdf/wnisr2024-v2.pdf.

Siddi, Marco, and Kristiina Silvan. 2023. 'Russia and Kazakhstan in the Global Nuclear Sector. From Uranium Mining to Energy Diplomacy'. 371. FIIA Briefing Paper. Helsinki: Finnish Institute of International Affairs. https://doi.org/10.13140/RG.2.2.31217.04963.

Szulecki, Kacper, and Indra Overland. 2023. 'Russian Nuclear Energy Diplomacy and Its Implications for Energy Security in the Context of the War in Ukraine'. *Nature Energy* 8 (4): 413–21. https://doi.org/10.1038/s41560-023-01228-5.

Taylor, Ian. 2019. 'France à Fric: The CFA Zone in Africa and Neocolonialism'. *Third World Quarterly* 40 (6). 1064–88. https://doi.org/10.1080/01436597.2019.1585183.

Wimmers, Alexander, Paul-Jonas Ummethun, Anke Herold, Claudia Kemfert, Björn Steigerwald, and Christian von Hirschhausen. 2025. 'Russische Exporte von Uran und Atomtechnologie in die EU – Optionen für alternative Beschaffung'. Ein Debattenbeitrag 20. Diskussionsbeiträge der Scientists for Future. https://doi.org/10.5281/zenodo.14728811.

WNA. 2022. 'The Nuclear Fuel Report: Expanded Summary'. Summary. United Kingdom. https://world-nuclear.org/images/articles/nuclear-fusel-report-2021-expanded-summary.pdf.

Yim, Man-Sung. 2022. *Nuclear Waste Management: Science, Technology, and Policy*. Lecture Notes in Energy. Springer Dordrecht. https://doi.org/10.1007/978-94-024-2106-4.

Yülek, Murat, Saim Karabulut, and Ali Osman Karcı. 2023. 'On Economic Security and the Political Economy of Neocolonialist Capitalism: The Case of France and Niger's Uranium Resources'. In *Capitalism at a Crossroads*, edited by Ali Ari, 159–80. Springer Studies in Alternative Economics. Cham: Springer International Publishing. https://doi.org/10.1007/978-3-031-23257-2_9.

Chapter 9
Nuclear Diplomacy: Analyzing Russian Foreign New Build Projects

Youssef Said Farag and Alexander Wimmers

9.1 Introduction

Nuclear diplomacy is an essential strategy of geopolitics, pursued by nuclear superpowers to export nuclear technology. In particular, Russia and China dominate global nuclear power plant construction, but with a different focus. While China almost exclusively builds domestically, Russia has emerged as the globally dominating exporter of nuclear reactors. The Russian State Atomic Energy Corporation (Rosatom) is involved in 20 ongoing reactor construction projects worldwide (Schneider et al. 2024). The Russian state has been accused of attempting to increase its geopolitical influence through its export strategy by fostering energy dependence, not only for fossil resources, but also for nuclear-related services. This can be observed in several countries where reactor projects are underway or planned, or other forms of dependency, such as fuel services, exist (Szulecki and Overland 2023).

Rosatom tailors its contractual and financial strategies to suit the specific needs and conditions of each potential host country for its reactor construction projects. Most of the projects are not formulated based on competitive tenders, but instead agreed upon between autocratic(-leaning) leaders and with substantial Russian financial assistance and assurance, against which countries falling under OECD state loan regulations cannot compete (Hickey et al. 2021). The existing literature currently lacks an analysis of the different contracts of Russian foreign nuclear projects, and there are also few economic assessments, let alone a comparison. Consequently, this analysis aims

The authors thank Carolin Grüter for her constructive comments on earlier drafts of this chapter.

Y. S. Farag · A. Wimmers
Workgroup for Infrastructure Policy (WIP), Technical University (TU) Berlin, Berlin, Germany

A. Wimmers (✉)
German Institute for Economic Research (DIW Berlin), Berlin, Germany
e-mail: awi@wip.tu-berlin.de

A. Wimmers et al. (eds.), *Nuclear Power*, Studies in Energy, Resource and Environmental Economics, https://doi.org/10.1007/978-3-031-99894-2_9

to provide a comparative multiple case study for the three ongoing Russian new build projects in Bangladesh, Egypt, and Türkiye. The projects are sufficiently similar to be comparable, e.g., in their reactor design, but differ in substantial areas, such as the project cost, the contract scheme, and funding agreements between host governments and the Russian state. This work aims to provide an overview of the three projects and garner first insights from the perspective of new institutional contract theory and first financial assessments. Notably, both assessments are limited by the lack of available data and opaque communications by host and vendor countries.

The remainder of the chapter is structured as follows: We first provide a brief introduction to the so-called nuclear diplomacy and give an overview of the organization of global nuclear power plant construction projects. This is followed up by theoretical overview of contract theory and contract designs in Sect. 9.3. Here, we also introduce the applied methodology. The structured case study analysis for all three projects is conducted in Sect. 9.4. In Sect. 9.5, we discuss the findings, while Sect. 9.6 concludes.

9.2 Nuclear Diplomacy and the Dominant Role of Russia

9.2.1 Introducing Nuclear Diplomacy

Nuclear superpowers, in particular China, Russia, and the United States (US), have developed a military–industrial complex for the production of nuclear goods and services. This began with the manufacture of nuclear weapons and later moved on to the production of infrastructure for (civilian) energy provision. Despite the shift of focus, there have always been attempts to seek synergies between military and commercial applications of nuclear power (von Hirschhausen 2017).

In addition to domestic efforts, these countries also advance their geopolitical goals through so-called "nuclear diplomacy." This term can be understood as a strategic policy for technology and foreign trade that aims at building or strengthening international relationships involving nuclear technology transfers, both civil and military. These countries attempt to use their incumbency as nuclear superpowers to pressure other countries into adapting their respective nuclear technology by selling infrastructure, such as reactors, and providing knowledge transfers. Furthermore, a geopolitical alliance policy is followed to deter potential opponents from technologically dominating foreign nuclear sectors. In other words, there is a competition for geopolitical influence in which incumbents offer favorable financial conditions and technology transfers, often to emerging economies that are willing to implement domestic nuclear industries (Bargman 1977; von Hirschhausen 2017; Szulecki and Overland 2023). Although all three countries, China, Russia, and the US, are all engaged in nuclear diplomacy, they have different focal points.

China is pursuing a particularly strong nuclear expansion strategy and has declared the ambition to become the global leader in nuclear technologies. Compared to Russia

Table 9.1 Nuclear reactors under construction as of 1 July 2024

Country	Units (Domestic Design)	Foreign vendor	Capacity (MW net)	Construction Start	Grid Connection	Units Behind Schedule
China	27 (23)	Russia: 4	29,101	2017–2024	2024 – 2029	1
India	7 (3)	Russia: 4	5398	2004–2021	2024–2027	5
Russia	6 (6)		3960	2018–2024	2025–2030	2
Türkiye	4 (0)	Russia: 4	4456	2018–2022	2025–2028	4
Egypt	4 (0)	Russia: 4	4400	2022–2024	2028–2031	
South Korea	2 (2)		2680	2017–2018	2024–2025	2
Bangladesh	2 (0)	Russia: 2	3260	2018–2019	2030–2031	2
U.K	2 (0)	France: 2	3260	2018–2019	2030–2031	2
Argentina	1 (1)		25	2014	2028	1
France	1 (1)		1630	2007	2024	1
Iran	1 (0)	Russia: 1	974	1976	2028	1
Japan	1 (1)		1325	2007	2030	1
Slovakia	1 (0)	Russia: 1[1]	440	1985	2025	1
Total	59		59,809	1976–2024	2024–2031	23

Source Compiled from (Schneider et al. 2024, 57)

and the US, Chinese nuclear power development began rather late and was built upon the adaption of existing reactor designs and substantial technology transfer. Based on this, China was able to develop a domestic concept, the Hualong One, on which the nuclear diplomacy strategy is based (von Hirschhausen 2017). However, in recent years, the focus has been on strengthening and expanding the domestic market, with 23 units under construction in 2024, see Table 9.1. The only Chinese reactor exports were completed in Pakistan and are envisioned to be expanded. In Argentina, Chinese ambitions are likely to have been halted after several years of negotiations. In the United Kingdom (UK), the state-owned China General Nuclear Power Group reduced its stake in the Hinkley Point C project while refusing to pay for additional cost increases and has fully withdrawn from the planned Sizewell C project (Schneider et al. 2024).

Traditionally, the Soviet Union had maintained a very close relationship between military and civilian nuclear activities. The organizational structures of the nuclear industry were kept under joint control, and R&D activities were coordinated in vertically integrated state organizations. After the fall of the Soviet Union, the newly emerged Russian Federation (henceforth Russia) re-established almost all of the former activities within its territory. By deploying available nuclear technology, it maintained close ties to many former Soviet countries, as well as to former allies

[1] *Note* The Slovakian project at Mochovce-4 is a Russian-designed VVER being finalized by a Czech-led consortium.

or satellites, such as Hungary (von Hirschhausen 2017). Russia is the most actively involved country regarding nuclear diplomacy. While the domestic market is also expanding moderately, with six reactors under construction, most Russian new-build projects are being pursued abroad (Schneider et al. 2024). Most of these projects are likely driven by geopolitical, rather than economic considerations, since Russia is not only providing technology, but also financial support (Lévêque 2014; Szulecki and Overland 2023).

In the form of bilateral agreements, the US has been offering countries such as Germany, Japan, and South Korea a supply of highly enriched uranium for research reactors since the inauguration of the "Atoms for Peace" program in 1953. Moreover, once countries agree to give up on enrichment and reprocessing activities, nuclear cooperation with the US can be expanded. This approach worked when the US was the dominant supplier of nuclear technology and fuel. Today, however, other major suppliers have emerged (Moniz 2011), and the US has fallen behind in the supply of nuclear reactors, primarily due to the decline of its domestic commercial nuclear technology industry (Markard et al. 2020).

On the receiving end of "nuclear diplomacy" are countries with little technical and physical knowledge that want to gradually develop commercial nuclear competencies and sometimes also militarily. These include the United Arab Emirates, Türkiye, Egypt, and, in the long term, Saudi Arabia, Jordan, and Kazakhstan. The development of a national nuclear energy industry is often also linked to regional geostrategic conflicts (Stirling and Johnstone 2018). The financial capability of these recipient countries to pay for nuclear power projects plays a less important role, resulting in various organizational and contractual schemes deployed in different countries, as discussed below.

9.2.2 *Nuclear Power Construction Projects and the Role of Russia*

9.2.2.1 Overview of Global Nuclear Power Plant Construction

As of mid-2024, 59 nuclear reactors were under construction worldwide. 46% of these (27) are in China, followed by six projects in India, six in Russia, and several other countries, see Table 9.1. Globally, 23 projects are formally delayed. Other projects, while not delayed officially, are likely to be impacted by supply chain issues caused by the Russian attack on Ukraine. Furthermore, the limited access to information, particularly on Chinese projects, hinders independent assessments. Schneider et al. (2024) provide a detailed evaluation of average reactor construction times since 1954. Of the projects completed between 2021 and 2023, only one project (Tianwan-6) was on time; every other project was delayed. European and US projects are delayed the most. The average global construction duration from 2014 to 2023 was 9.9 years—albeit with considerable variation.[2]

The above-mentioned construction durations neglect the additional time necessary for planning, licensing, site selection, procurement, contract negotiations, and other activities. Consequently, nuclear power plant construction projects can induce decade-long lead times. For example, the Hinkley Point C project in the UK was announced in 2008, but construction began only in 2018 (Johnstone and Stirling 2020). Additionally, project delays usually lead to cost increases, as observed by a number of Western projects, such as the Vogtle project in the US or the French-led projects in Finland, France, and the UK (Böse et al. 2024).

Budget overruns and construction delays are not unique to nuclear power plants but also occur at other mega infrastructure projects. This indicates that large-scale projects, in general, are challenging to plan for. However, nuclear power plants are one of the most prone types of projects to experience budget overruns, and when they do, the cost increases are substantial (Johnstone and Stirling 2020). This stems from the very high complexity of nuclear power plant construction projects, both technical and organizational. Underperforming projects are characterized by rapid developments in technologies (e.g., changing safety and security measures), many "interoperable and interdependent systems," a high number of involved organizations from different disciplines (e.g., construction engineering, nuclear engineering, and radiation specialists), a focus on minimizing cost without buffers in schedules or lacking the possibility to adapt the technology, and finally, "competitive pressures from other technologies within the same market" (Locatelli 2018, 11).

9.2.2.2 Russian Nuclear Export Activities

In the wake of the Russian attack on Ukraine in 2022, most European countries became painfully aware of their dependence on Russian energy services. With Russia amongst the leading countries for natural gas, oil, and coal exports, the country plays a significant role in global fossil energy provision. However, it is also a major player in providing nuclear fuel, constructing power plants, and providing other related services. Rosatom is the descendant of the Soviet Ministry of Atomic Energy and was inaugurated as a state corporation in 2007—meaning that the company follows direct orders from the President of the Russian Federation (Szulecki and Overland 2023).

Rosatom has since become the leading exporter of nuclear reactors and is currently involved in 19 reactor construction projects in seven foreign countries, namely Bangladesh, China, Egypt, India, Iran, Slovakia, and Türkiye (Table 9.1). Between 1950 and 2025, Russia, including the former Soviet Union, completed 65 foreign

[2] For example, Slovakia's Mochovce-4 took 38.1 years to build, the Finnish reactor Olkiluoto-3 16.6 years, Pakistan's Kanupp-2 came online after 5.6 years, and Tianwan-6 in China required only 4.7 years.

reactor projects in 13 countries (Wealer et al. 2018; Schneider et al. 2024).[3] Szulecki and Overland (2023) assess the dependence of 45 countries on goods and services provided by Rosatom or one of its subsidiaries, i.e., TVEL or AtomStroyExport. Countries with a high degree of dependence rely on Russian-related sources for more than 10% of their energy provision, such as Armenia or Bangladesh. In comparison, countries with a medium dependency rely on 4 to 10%. This includes Egypt, Türkiye, and Iran. China and India have a low degree of dependency (less than 3%), and most other countries depend on Rosatom for technical collaboration, specialized (maintenance) services, or fuel, such as Bolivia, Bulgaria, Ghana, Rwanda, or Spain. The authors critically reflect on the potential exploitation of the above-mentioned dependency on Russia—as experienced by European countries in 2022 when the natural gas supply from Russia was cut. They conclude that Russian "nuclear diplomacy" might represent both a form of soft power and an energy weapon (Szulecki and Overland 2023, 417).

In general, the dependency on one potentially opportunistic actor (Rosatom) for critical infrastructure can be regarded as risky. Countries currently involved in nuclear power plant construction projects that heavily depend on their completion and reliable operation for their energy provision are Bangladesh, Egypt, Iran, and Türkiye. Because the project in Iran has been ongoing since 1978 (and was restarted in 1996), the remainder of this work focuses on the other three countries, which are all so-called nuclear newcomer countries. This means that they operate no domestic high-capacity nuclear reactors. Notably, each of the projects is organized differently. The remainder of this paper provides an overview of the theoretical concepts used to assess all three projects and to ultimately provide an assessment of related risks to the host governments.

9.3 Theoretical Background and Methodology

9.3.1 Theoretical Background

9.3.1.1 Contract Theory

The provision of complex system goods, such as nuclear power plants, requires the availability of assets and implemented processes, and it induces a vast number of coordination needs between individual roles tasked with their provision. These coordination activities are required to ensure the timely provision of necessary assets and the completion of interlinked tasks. Coordination, which includes communication, negotiation, monitoring, and contract enforcement, induces transaction costs (Coase

[3] Some countries are former members of the Soviet Union but are listed a "foreign" regardless. Russian reactor construction projects were completed in Armenia (2), Belarus (2), Bulgaria (6), China (4), the Czech Republic (6), Finland (3), Germany (6), Hungary (4), India (2), Kazakhstan (1), Lithuania (2), Slovakia (8), and Ukraine (19).

1937). These transaction costs are influenced by factor specificity, the frequency of the transaction, and uncertainty.

- Factor specificity refers to the difference between the value of a good at its first-best use and that at the second-best use. This means that some goods have only one particular purpose, i.e., a nuclear fuel rod can only be used for one reactor type, while a regular screw can be used in many products.
- Frequency relates to the ability to repeat a given task (so-called "economies of multiples"). It partly depends on the complexity of the transaction and whether complex planning is required for the provision of said transaction. It makes a difference whether the task is to design, construct, and manufacture pencils (that are interchangeable), or if one plans to build a nuclear power plant.
- For the third factor, uncertainty, one must differentiate between environmental (or parametric) and behavioral uncertainty. Long-term projects should be hedged against potential environmental influences (e.g., tsunamis, hurricanes, drought, etc.) by reducing site-specific uncertainty. For example, seawalls should be built to protect a nuclear power plant located on the coast. Behavioral uncertainty relates to the risk of opportunistic actors exploiting information asymmetries.

Transactions involving two or more actors usually involve some form of informational advantage for one side, referred to as information asymmetries. Consequently, the principal (i.e., the "buyer") must ensure that their disadvantage is not exploited by the agent (i.e., the "seller"). This can be attempted in advance by discovery and due diligence activities, and one might attempt to cover all possibilities in a contract (the enforcement of which is guaranteed by the state). However, uncertainty and unreasonably high transaction costs make it impossible to cover all possibilities in a contract (hence the non-existence of complete contracts). Thus, the principal must monitor the activities of the agent to minimize opportunistic behavior and potential exploitation. Further, they must ensure the provision of the agreed-upon goods and services in the contractually agreed-upon quality. The main goals of a contract are thus the alignment of the principal's and the agent's motivations towards a common goal through precise incentivization, the reduction of behavioral uncertainties via fees, and the alignment of risk tolerance (Williamson 2000; Furubotn and Richter 2005).

9.3.1.2 Contract Design

Design Variables

Contracts between principals and agents can be designed in various ways. Contracts are distinguished by four main variables: the contract type, length, value, and ownership (Kim and Brown 2012).

The type of contract is primarily determined by the design of compensation. Contracts are designed as fixed price contracts, cost reimbursement contracts, or a combination. With fixed-price contracts, the final price for the product is agreed upon before delivery or the start of the project. In contrast, with cost-reimbursement contracts, the compensation is based on multiple inputs, such as time and materials. Compensation terms are crucial to consider when analyzing any contracting structure, as they determine the primary risk bearer of the project (Bajari and Tadelis 2001). Each approach has its advantages and disadvantages. For example, while a fixed-price contract tends to require a substantial amount of discovery before contract finalization (ex-ante transaction costs), the cost-reimbursement approach generally induces a high demand for monitoring (ex-post transaction costs) to ensure that the reimbursements correspond to conducted tasks. However, the principal's resources and knowledge impact the transaction costs. A combination of both designs is the target cost-based approach applied in former UK nuclear decommissioning schemes. "Here, costs [are] still reimbursed by the [principal], but the [agent] could earn fees based on achieved savings. The higher the savings, the higher the fees. However, if costs increased over a certain threshold, fees were reduced" (Wimmers and Von Hirschhausen 2024, 8). This approach combined the worst of both designs (both ex-ante and ex-post transaction costs) and ultimately led to a reorganization of the whole decommissioning organization in the UK.

The second variable is contract length, defined as the agreed-upon time to deliver the product. It is easy to specify delivery dates for simple products, but it is almost impossible to accurately forecast completion dates for highly complex and specific projects. This results in contract lengths that can span years, if not decades, for many large-scale (or "mega") infrastructure projects (Kim and Brown 2012). For these types of projects, this long duration can be problematic due to uncertainty regarding potential technological, regulatory, economic, political, and social changes, high transaction costs resulting from renegotiations, and, depending on the contract scheme, information asymmetries between principal and agent.

The value of the contract is the amount the principal pays the agent for their goods and services. Like contract length, contract value is also often adjusted throughout the project. Mechanisms can be implemented to account for necessary adaptions, such as including adjustments to meet new government regulations, changes in taxation, and scheduled value adaptions to account for inflation (Joskow 1988).

The fourth variable is the ownership structure. In recent years, integrated contracts between the principal and the agent have emerged. These integrated contracts are applied to projects designed, built, financed, and maintained by the agent, who also takes on complete responsibility for building, maintaining, and operating the asset until ownership is transferred to the principal after the contract term ends. Such contracts are usually used for large-scale, medium-to-long-term projects and require the principal to have adequate contract management skills and the agent to have a large pool of resources and expertise to bear the increased risk of financing and building a project.

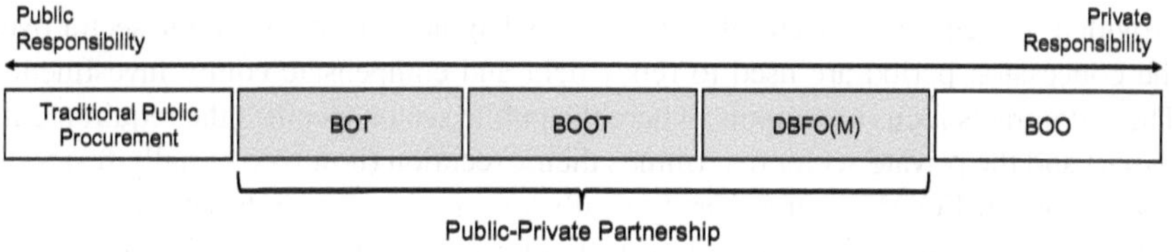

Fig. 9.1 Stylized classification of public procurement schemes. *Source* Own depiction based on (Akbiyikli and Eaton 2005, 506)

Contract Types in Public–Private Partnerships

Large infrastructure projects are usually ordered by governments and executed by private actors. In recent decades, complex contractual schemes known as public–private partnerships (PPP) have emerged. Typical arguments for implementing PPP are limited state funds and/or resources or when there is a drive for increased spending efficiency and higher quality of provided services. Ideally, a PPP will deliver better quality services (or infrastructure) for "less cost, better accountability, and [promote] public sector innovation" (Akbiyikli and Eaton 2005, 506). The degree of public responsibility depends on the exact design of the procurement scheme. Figure 9.1 gives an overview of the most common schemes introduced below.

Build-Operate-Transfer (BOT) schemes involve the creation of a project company (so-called "promoter") to finance, design, construct, and operate a facility before it is transferred into government responsibility. The promoter receives a concession to operate the facility and the official ownership remains with the government. As funding is secured through the promoter via equity contributions or loans, the financial risk to the government is reduced. However, the risk remains that the facility fails to meet demands. BOT project financing differs from conventional methods because it is based on project finance with limited or no recourse, meaning that the promoter's parent companies avoid liabilities. The project's revenue-generating ability serves as a guarantee for lenders. The promoter manages all contracts, including construction and operation. Revenue from services, like tolls or tariffs, covers debts and expenses and is intended to provide profits. At the end of the concession period, the facility is transferred to the government in good condition and free of charge. Notably, the first BOT scheme was designed for the Akkuyu nuclear power plant in Türkiye in 1984 (which did not happen), see Sect. 9.4.3 (Akbiyikli and Eaton 2005).

The Build-Own-Operate-Transfer (BOOT) scheme differs from the BOT approach in that the facility is actually owned by the specially set-up project company for the pre-defined concession period. It becomes an asset of the company for a certain period after which ownership is transferred to the government of the host country (Akbiyikli and Eaton 2005).

In Design-Build-Finance-Operate-(Maintain) (DBFO(M)) contracts, the private sector finances and provides assets, securing debt from commercial banks for most costs and covering the rest with equity. It also manages the operation and maintenance of the facility, while the public sector pays for the asset upon completion and for provided services (e.g., electricity generation). Payments from the public sector over the concession period are used to repay debt and compensate equity investments. These contracts focus on outputs, where the public sector specifies the requirements (what) and the private sector determines their execution (how), potentially fostering innovation. DBFO(M) contracts are typically long-term agreements between a public authority and a specially designed project company that handles design, financing, and operations (Akbiyikli and Eaton 2005). Some contracts also include maintenance services that could bring benefits of a "total life cycle cost" consideration, and limit shirking of the agents, albeit the inclusion induces higher ex-ante transaction costs due to extended contract negotiations, and, depending on the facility, public actors might be better equipped for such operations (Roohé 2007).

Finally, Build-Own-Operate (BOO) schemes differ from the others by the fact that the ownership of the facility need not necessarily be retransferred to the government. Under such a scheme, infrastructure is designed, constructed, operated, maintained, financed, and owned by private actors (Akbiyikli and Eaton 2005).

9.3.1.3 Organization of Nuclear Power Plant Projects

The organization of nuclear power plant construction projects (similarly to other mega infrastructure) is seldom ordered by private entities, but by state actors that take over multiple roles in the organizational scheme of the project: They are involved in procurement, contract negotiations, as well as regulatory oversight and licensing with potentially conflicting interests. Conventional nuclear construction projects involve a (vertically integrated, potentially state-owned) utility company that plans and orders a plant from a vendor that can also be state-owned. Given the high upfront costs of nuclear power plant construction projects, external funding must be ensured through third-party financiers, i.e., banks or other financial institutions. Further actors can consist of (i) an off-taker, i.e., an entity that will acquire the produced electricity; (ii) the so-called host government, i.e., the government of the country where the plant is planned to be built and that can provide revenue guarantees to the project as well as loan guarantees to financiers; and (iii) an exporting government, i.e., the government owning or overseeing the vendor. The exporting government can provide export guarantees to the vendor and might also act as a financier by providing loans to the utility (Weibezahn and Steigerwald 2024).

Weibezahn and Steigerwald (2024) provide a detailed assessment of organizational schemes to provide project funding for nuclear power plants. In the corporate financing model, the corporate lender is the primary risk bearer of the project because funds are provided to the utilities via loans and bonds (practiced in Slovakia). In a hybrid financing model (practiced in Finland), shareholders of the specially set-up limited liability company agree to provide all necessary costs for the construction and later generate revenues through fixed-price electricity purchase guarantees. In the above-described vendor-funded BOT and BOO schemes, the vendor company takes on most of the risk—in the case of state-owned vendors, this would be the exporting government (e.g., in Hungary and Türkiye, Rosatom, i.e., the Russian state, takes on full responsibility). Funds can also be provided through loan guarantees by governments (and essentially their taxpayers) (e.g., the US and the UK). If state-owned utilities provide the funds, once again, the taxpayers become the funder of last resort (e.g., in France or the Czech Republic). In "contract for difference" schemes, the plant operator is guaranteed a fixed price for generated electricity by either the government or by an off-taking utility, even if market values are below this level. Ideally, windfall profits are to be repaid to tax- or ratepayers. In a regulated asset base, the operator is guaranteed a fixed return on its investment (both planned and already applied in the UK).

9.3.2 Methodology

9.3.2.1 Multiple Case Studies

The goal of this chapter is to assess potential benefits and risks for the host countries of Russian nuclear power plant construction projects depending on the specific contract design and financing scheme. We therefore conduct a multiple case study to "predict contrasting results but for anticipatable reasons" (Yin 2014, 105). This means that before concluding the study, there are known differences between the cases, e.g., reactor designs or contractual obligations, that will lead to a variation of results. At the same time, the selected cases must have some degree of similarity to allow for comparison. First, we select the host countries Bangladesh, Egypt, and Türkiye for their sufficiently variable dependency on Russian nuclear build, the heterogenetic contract designs, similar construction start dates, and similar technologies. We then follow the step-by-step approach by Stake (2006). Here, we analyze each case based on the general energy-policy-related background and the specificities of the nuclear power plant construction projects, such as (announced) delays, cost developments, and the general (economic) performance of the project. We also compare the cases' contracts, i.e., their design, the financier of last resort, and the potential dependency on the Russian vendor. Finally, we provide a first economic assessment of the projects based on the metrics introduced below.

9.3.2.2 Economic Evaluation

One theme of this multiple case study is the economic evaluation of the projects. While the ultimate performance can only be addressed ex-post, an estimation with information available today can provide first insights into the economic efficiency and the potential benefits of the given projects. For this, we conduct a comparison of the projects using two metrics. The first is a calculation of the levelized cost of electricity (LCOE) that gives the annualized cost of electricity generated by a given technology averaged over one year (Eq. 9.1). In general, the metric is useful to compare the cost of different energy generation technologies.[4]

$$\text{LCOE} = \frac{\text{CAPEX} + \text{Fixed O\&M} + (\text{FLH} * \text{Variable O\&M})}{\text{FLH}} \tag{9.1}$$

where

- LCOE: Levelized cost of electricity in monetary unit (MU) per MWh
- CAPEX: Annualized capital expenditure calculated with constant interest rate i over total expected operation time *n* in MU per kW
- Fixed O&M: Fixed operations and maintenance cost in MU per kW
- Variable O&M: Operations and maintenance cost depending on production volume given in MU per MWh (includes fuel cost)
- FLH: Total hours of full-load operation (full load hours), often calculated by means of a capacity factor CF in percent multiplied by 8760 h.

Total CAPEX is calculated following the capital cost calculation shown in Chap. 10 and thus includes interest that accumulated during construction (IDC). Electricity generation projects are usually built with the intention of generating revenues that, in an ideal situation, cancel out the (upfront) cost and ultimately generate profit. For this, we refer to the second method, the net present value (NPV) (Eq. 9.2). The NPV is used to calculate the value of an asset by accounting for all future cash flows of an asset, both cost and (expected) revenues. For all three cases, the revenues will not be earned in a competitive market but rather through fixed tariffs. To calculate the value of the projects today, we use a discount rate calculated as the weighted average cost of capital (WACC) (Eq. 9.3). This reflects upon the different schemes of loans and lending of nuclear power project funding in the three cases.

$$\text{NPV} = -C_0 + \sum_{t=0}^{n} \frac{\text{Cash flow}_t}{(1+r)^t} \tag{9.2}$$

where

- NPV: Net present value in monetary unit (MU)
- C_0: Upfront capital expenditure
- Cash flow$_t$: Annual flow of cash (revenue minus cost) in a given year t
- n: Lifetime of the asset
- r: Discount rate

$$\text{WACC} = \frac{E}{D+E}(r_e) + \frac{D}{D+E}(r_d)(1-T_c) \tag{9.3}$$

[4] The main disadvantage of LCOE is the leveling of generation over the given timeframe, which neglects the value of flexibility or the firming capability of some technologies. LCOE further exclude external effects, such as CO_2 emissions or nuclear wastes. See Chapter 10 for further details on LCOE and other cost metrics.

Table 9.2 Universal input parameters for LCOE calculation

Parameter	Assumption
Capacity factor	90%
Fixed Operation and Maintenance Cost	81.3 US-$ per kW p.a
Variable Operation and Maintenance Cost	6.52 US-$ per MWh
Fuel Cost	11.52 US-$ per MWh
Operational Lifetime	60 years

Source Own compilation based on data taken from (Göke et al. 2025)

where

- E: Market value of the firm's equity
- D: Market value of the firm's debt
- r_e: Cost of equity
- r_d: Cost of debt
- T_c: Corporate tax rate.

For the calculation of the LCOE, input parameters not related to capital costs are assumed uniformly for all three projects. The parameters are based on average values from a literature analysis of high-capacity light water reactor projects, shown in Table 9.2 (Göke et al. 2025). Capital cost and WACC calculations vary between projects, as described in the corresponding sections below.

9.4 Case Studies

The focus of this work is to analyze the contracting structure of Rosatom's new build projects abroad. We thus conduct a three-stage analysis, starting with an economic overview for each country, including a summary of recent political or economic developments or relevant economic history, and economic metrics that evaluate how well the host country is equipped to repay foreign-currency-denominated debt. Egypt and Bangladesh must pay back substantial loans in US-$, while for Türkiye, government-owned utilities must purchase all the produced electricity in US-$ for a fixed price and term. The second is a contractual analysis, assessing the goals and benefits of each party in that agreement. This analysis is based on the theoretical framework introduced in Sect. 9.3.1, using conceptual ideas from the agent-principal framework and the New Institutional Economics. The focus is on financial and operational risks. In the third and final analysis, the focus will be on the project from the perspective of an investor, evaluating whether the project is financially profitable. LCOE and NPV will be calculated for each case study.[5]

[5] All data and corresponding calculations can be accessed via Github: https://github.com/alexwimmers/nuclearPower_TEG.

9.4.1 Bangladesh

9.4.1.1 Overview

The economic outlook for Bangladesh has been worsening over the last few years. Foreign exchange reserves have halved in only two years, from US-$40 billion in 2022 to US-$20.8 billion in early 2024. The Bangladeshi taka lost 36% of its value compared to US-$ in the same timeframe, increasing the import cost for fossil fuels. Although the country is suffering from foreign currency shortages, it has one of the lowest debt-to-GDP ratios in Asia, indicating a general capacity to take on additional debt. However, the ratio has been on an uninterrupted rise since 2016 (IMF 2023). When measuring a country's ability to repay any external debt, it is useful to consider the assessments issued by credit rating agencies. Bangladesh was rated speculative to highly speculative by the top three global credit agencies.[6] These ratings indicate a potential inability to fulfill financial obligations.

It is expected that the Bangladeshi energy demand will increase substantially over the coming decades, and correspondingly, electricity demand will also increase, requiring an estimated 61 GW of generation capacity by 2041, up from 18.7 GW in 2018 (Das et al. 2020). Currently, the electricity mix is dominated by fossil fuels, such as natural gas and oil, while renewables account for only 3%. Most fossil fuels are imported. Ambitious targets of 100% renewable power generation by 2050 stand in conflict with only one GW of installed renewable capacity (mainly solar) as of 2023 and supposed nuclear power expansion plans of up to ten GW (Das et al. 2020; Abdullah-Al-Mahbub and Islam 2023; Schneider et al. 2024).

9.4.1.2 The Rooppur Project

The Rooppur project is set to become the first commercial nuclear power plant in Bangladesh. The site for the construction of the plant was chosen as early as 1961, but plans were never followed through. After Bangladesh gained independence from Pakistan in 1971, French-Bangladeshi plans to build a 125 MW reactor were announced in 1980 but never materialized. A research reactor was commissioned for radioisotope production in 1986 in Ganakbari, some 40 km outside of Dhaka (Soner 2019). In 1999, commercial nuclear ambitions reemerged as Bangladeshi officials launched the National Nuclear Power Action Plan that led to a cooperation agreement signed with China in 2003. A plan was drafted in 2007 to build two 500 MW reactors at the site by 2015. After a revision of the plan, the capacity of the two planned reactors was increased to 1,100 MW each, and a tender was launched to which only Rosatom submitted a bid. A contract between Rosatom and the Bangladeshi Atomic Energy Commission (BAEC) was signed in 2011 for two reactors with a combined capacity of 2,200 MW to be built at the Rooppur site. Construction was to begin in 2017

[6] https://www.worldgovernmentbonds.com/credit-rating/bangladesh/. Unless otherwise noted, all references listed in footnotes were checked and last visited on 31 October 2024.

and 2018, with grid connections estimated for 2023 and 2024, respectively. Note that the regulatory agency tasked with overseeing nuclear operations and safety, the Bangladesh Atomic Energy Regulatory Authority, was set up as late as 2013 (Karim et al. 2018).

At the project, Russian actors provide turnkey services for the construction of the plant and will operate it for one year before handing operations over to the Bangladeshi staff. As of 2017, the contract was valued at US-$ 12.65 billion, 90% of which is provided by the Russian Government as a loan that shall be repaid over 28 years (including a ten-year grace period) at a maximum interest rate of 4%.[7] Despite construction works beginning as planned in 2017 and 2018, the completion dates have been postponed. The whole project completion date might be delayed to 2027. As of the time of writing in January 2025, neither reactor had been connected to the grid as originally planned.[8] Delays stem in some part from sanctions against Russia in the wake of the attack on Ukraine that limited the availability of required technology and necessitated supplier changes (Schneider et al. 2023; 2024).

9.4.1.3 Contractual Analysis

The Rooppur project is designed as a BOT contract in which BAEC acts as the principal and Rosatom is the agent. The risk of having to compensate for cost overruns lies, according to the available information, with Rosatom, given the contract was settled with a fixed price to be paid by BAEC. However, contract and cost details remain opaque, as Rosatom refuses to disclose any data.[9] Regardless, Bangladesh will have to repay the Russian loan of 90% of the project cost. While the repayment period for this loan will begin in 2027, other related payments have been conducted in Chinese Yuan to circumvent US and European sanctions against Russia. Further devaluation of the Bangladeshi taka might complicate future loan repayments, both in US-$ and Yuan. Revenues of BAEC for the loan repayment are to be generated (partly) through a power purchase agreement (PPA) with the Bangladesh Power Development Board (BPDB) at a fixed price. The exact tariff has not been disclosed but should range from 3.19 to 10 US-ct per kWh,[10] and additional delays might result in loan repayments beginning before any revenue can be generated.

The limited experience of Bangladeshi officials could pose operational and safety-related risks. While training of Bangladeshi personnel for future operations of the plant seems to be underway (Schneider et al. 2024), the limited availability of skilled personnel in general, and most notably in the overseeing agency founded in 2013,

[7] https://www.thedailystar.net/country/rooppur-nuke-plant-1265b-deal-inked-russia-192556.

[8] https://pris.iaea.org/PRIS/CountryStatistics/CountryDetails.aspx?current=BD, last accessed on 14 January 2025.

[9] https://www.tbsnews.net/bangladesh/rosatom-issues-clarification-rooppur-plant-cost-says-it-never-discloses-project-expenses.

[10] https://www.tbsnews.net/bangladesh/energy/first-unit-rooppur-nuclear-plant-be-ready-power-generation-next-year-possible.

Table 9.3 Scenario overview and resulting LCOE for the Rooppur project

Parameter/case	Best case	Medium case	Worst case
Construction Delay (Total construction time until first grid connection) in years	3 (8)	5 (10)	7 (12)
Start of electricity production	2024	2026	2028
Additional Costs incl. IDC in billion US-$	3.783	7.245	9.390
LCOE in US-$ per MWh	77.80	92.02	100.82

could facilitate opportunistic behavior by Russian actors to minimize project cost overruns and delays. Further, the plant is expected to generate 12% of the country's electricity. This could risk grid stability if the transfer of operational responsibility from Russian to Bangladeshi personnel does not go as planned or if inexperienced personnel commit operational and maintenance-related errors. Further, fuel shall be provided by Russia for the entire 60-year operational lifetime of the plant. This long-term fuel delivery contract adds further dependencies to the Bangladeshi energy supply (Szulecki and Overland 2023). All waste management obligations will likely remain with Bangladesh (Soner 2019), while others indicate towards Russian responsibilities for waste management and decommissioning (Karim et al. 2018).

9.4.1.4 Financial Analysis and Valuation

The financial analysis of the Rooppur project is conducted for three cases. In addition to the initial project cost of US-$ 12.65 billion, we assume 15, 35, and 45% cost overruns for the three cases, respectively. The best case assumes the earliest possible, and communicated, completion date of 2024 for the first reactor[11]; the second shall be connected to the grid one year later. This indicates a project delay of three years. The other two cases follow two and four years later, respectively. Table 9.3 provides an overview. IDC is to be paid upfront, and we assume the maximum interest rate of 4% as indicated in the contract, as opposed to the actual WACC for Bangladesh, see the supplementary data. For the sake of reducing the complexity of the NPV calculation, all capital costs are accounted for before the end of the construction. Decommissioning and waste management costs are neglected.

For the calculation of the NPVs, we assume three revenue scenarios per case, ranging from the proposed 32 US-$ per MWh and 100 US-$ per MWh to the Turkish Power Purchase Agreement (PPA) price of 123 US-$ per MWh, see Sect. 9.4.3. The NPV covers the period from the construction start to the final closure after 60 years of operation. The tariffs that are officially discussed are very low compared to the calculated LCOE that are themselves based on rather optimistic assumptions, such as a high capacity factor of 90%, and the uncertainty of actual project cost. Table 9.4 shows the results of the assessment.

[11] As mentioned in Sect. 9.4.1.2, this did not happen.

Table 9.4 Net present value for proposed electricity tariffs for the Rooppur project

Tariff in US-$ per MWh	32	45	100
Best Case NPV in billion US-$	−30.46	−4.29	4.56
Medium Case in billion US-$	−34.83	−8.55	0.34
Worst Case in billion US-$	−36.95	−13.28	−5.27

9.4.2 *Egypt*

9.4.2.1 Overview

The Egyptian economy is considered a leading economy in Africa. Since the 2010s, the country has been experiencing political turmoil with multiple (forced) government changes. Egypt is regularly subject to foreign currency crises and economic turmoil. The government aimed to solve the current foreign currency crisis that has been prevailing since 2022 by undertaking monetary and exchange rate changes. The Central Bank weakened the Egyptian Pound by around 35% against the US-$ in 2024 and hiked interest rates by 600 basis points to halt soaring inflation.[12] Following these moves, the country suffered from one of its worst economic crises in decades. Egypt has a negative account balance that places the country in a strong deficit that is offset by foreign exchange reserves, which are supported mainly by the Gulf states (Butter 2020). The reserves were used heavily in 2020 and 2022 to support previous currency devaluations. This puts Egypt in a position where it is constantly reliant on support from Gulf states and not able to generate the necessary amounts of foreign currency. Egypt has also been taking on large amounts of external debt to finance a variety of megaprojects, another reason for its foreign currency crisis. The country's debt-to-GDP ratio of nearly 100% is very high. This places Egypt under enormous pressure to repay its debt (IMF 2023a). Contrary to these developments, Egyptian credit ratings remain mostly positive, chiefly due to strong financial backing by the IMF and the United Arab Emirates.[13]

The current energy supply is covered mostly by fossil fuels, mainly crude oil and natural gas; 33.7 and 54% in 2020, respectively. Despite the abundant availability of fossil reserves, both gas and oil, the country is a net importer of fossil fuels because the demand for energy has risen dramatically over the past decades. Energy provision is slowly being electrified, and the share of renewables in both the energy and the electricity mix is slowly increasing. As of 2020, about 6.2 GW of renewables, consisting of 2.8 GW hydro, 1.6 GW wind, 1.7 GW solar, and 0.1 GW bioenergy, were installed on the grid, up from only 3 GW a decade ago, and expected to grow substantially in the coming years (IRENA 2023). Reported targets for energy system decarbonization are conflicting, ranging from a 42% share of renewables in electricity generation by 2035, over 58% in 2040 (Schneider et al. 2024), down to the most

[12] https://www.bloomberg.com/news/articles/2024-03-06/egypt-unexpectedly-hikes-interest-rate-to-27-25, accessed on 31 October 2024.

[13] https://www.worldgovernmentbonds.com/country/egypt/.

recently reported targets of only 40% by 2040.[14] According to some estimates, the role of nuclear power will remain limited to only 3% of power generation in 2035 (IRENA 2023).

9.4.2.2 The El-Dabaa Project

The Egyptian nuclear program was launched in 1954, and the first research reactor began operating in 1961. Ambitions were halted after the defeat against Israel in 1967, and in the 1970s, US-Egyptian bilateral agreements to supply eight nuclear power plants were first ratified and then withdrawn. A 22 MW research reactor was delivered to the outskirts of Cairo from Argentina in 1992 and remains operational today.[15] In 1984, the Egyptian government initiated the El-Dabaa project for the first time. It was halted twice, once after the Chernobyl catastrophe in 1986 and again after the Egyptian revolution of 2011 (Nasr and Ahmad 2019).

The third attempt to build a nuclear power plant at El-Dabaa was launched in 2015, when a bilateral project development agreement was signed between Rosatom and the Egyptian Nuclear Power Plant Authority NPPA. There was no competitive tender, and at the end of the year, an agreement for the construction and operation of a nuclear power plant was signed between Egypt and Russia (Hickey et al. 2021). In 2017, the plans indicated that construction of the four-unit 4400 MW pressurized water reactor would begin in 2020, with commercial operations beginning in 2026. Construction of the first unit began in 2022 and is expected to be completed by 2029, with full operation of the plant by 2030. Construction at Unit 4 began in 2024 (Schneider et al. 2024).

Rosatom, including its subsidiaries, provides vertically integrated construction services[16] and will operate the plant for the first ten years, during which Egyptian staff will be trained to operate and maintain the plant. Fuel is to be supplied by Russian TVEL for the entire 60-year lifetime of the plant. Rosatom will also provide waste management services (Schneider et al. 2024), but the waste itself will supposedly remain in Egypt. The US-$ 30 billion price tag of the project shall be funded by a 15% equity stake of the Egyptian government, and a US-$ 25 billion loan provided by the Russian state to the project at a very low interest rate of 3%. This means that ownership is retained by the Egyptian state. Repayment of the loan must begin in 2028 at the latest and shall be fully repaid within 22 years (Hickey et al. 2021).

[14] https://www.reuters.com/sustainability/climate-energy/egypt-cuts-2040-renewable-energy-target-40-keeps-focus-natural-gas-2024-10-20/.

[15] https://nuke.fas.org/guide/egypt/nuke/index.html.

[16] Note that KHNP has been contracted for some construction services, valued at US-$ 2.25 billion (Schneider et al. 2023).

9.4.2.3 Contractual Analysis

The El-Dabaa project contract is organized as a BOT scheme in which the Russian state provides 85% of the funds during construction that will be repaid by the Egyptian state from 2028 onwards with a very low interest rate of 3%. Following the agreement, ownership of the plant and all revenues will stay with the Egyptian state-owned NPPA, meaning that Rosatom acts as the agent, while NPPA is the principal. Given that all services are conducted by Russian actors, NPPA will likely experience informational disadvantages. The risk of project success lies with NPPA as the sole owner and, through the state-guaranteed loan repayment, Russian actors have a high degree of certainty that they will be repaid. However, from the limited public information, it is unclear how the agreement might change in the case of cost overruns and durations. We thus assume a fixed price contract, which would place a potential financial risk on Russian actors. However, given the Egyptian debt-to-GDP ratio and devalued domestic currency, the repayment might become increasingly challenging in the future despite the low interest rate and a grace period of eleven years. Operational risk, i.e., the dependence on Russian operators to provide electricity from the plant and maintain it accordingly, is limited to ten years. Then, Egyptian personnel should have been sufficiently trained for the plant to be operated by domestic staff. The dependence on Russian actors remains nonetheless, given agreed-upon long-term contracts for fuel delivery and waste management services.

9.4.2.4 Financial Analysis and Valuation

The financial analysis of the El-Dabaa project is conducted for three cases. We assume an initial project cost of US-$ 30 billion. The three cases include cost overruns. The best case assumes a 15% cost overrun and a one-year delay in the construction project. The medium and worst cases assume 35 and 45% cost overruns, and two and four additional years of delays, respectively. To calculate the total cost, we assume IDC, which is added to the project cost. The interest is set to 3%. The WACC would be above 8%, see the supplementary material on Github. This is favorable for Egypt. For simplicity, we assume that all capital costs are paid for at once the plant goes operational. We assume a uniform inflation rate of 4%. Table 9.5 shows the main differences for the three cases and gives LCOE estimations following Eq. 9.1. All of these calculations exclude decommissioning and waste management costs. We assume a staggered grid connection of one reactor per year.

At the time of writing, there was no information on any tariffs to be implemented to ensure revenues for the El-Dabaa project. Hasanovic (2018) assumes the average value of Egyptian household electricity prices of 400 US-$ per MWh for the constant revenue of the plant. It is uncertain whether this tariff would also be paid directly to the owner of the plant or whether it includes taxes and other components, such as grid operation fees. For the calculation of the NPV, one should thus assume that the tariff paid to the operator will be lower (but the high estimate of the Egyptian household price is included regardless). The NPV covers the period from the construction start

Table 9.5 Scenario overview and resulting LCOE for the El-Dabaa project

Parameter/case	Best case	Medium case	Worst case
Construction Delay (Total construction time until first grid connection) in years	1 (5)	3 (7)	5 (9)
Start of electricity production	2027	2029	2031
Additional Costs incl. IDC in billion US-$	7.217	15.05	19.901
LCOE in US-$ per MWh	86.73	102.82	112.76

Table 9.6 Net present value for proposed electricity tariffs for the El-Dabaa project

Tariff in US-$ per MWh	45	123.5	400
Best Case NPV in billion US-$	−76.00	−2.01	260.75
Medium Case in billion US-$	−88.85	−14.85	247.91
Worst Case in billion US-$	−102.90	−37.93	192.79

to the final closure after 60 years of operation. The values are chosen based on the tariffs in the other cases of this work, i.e., the maximum discussed tariff for the Rooppur plant and the proposed tariff for the Akkuyu plant (Table 9.6).

9.4.3 *Türkiye*

9.4.3.1 Overview

The Turkish economy has established itself as a leading emerging market economy in the twenty-first century. After experiencing substantial economic growth since the 1950s, the country has become a major trade hub between Asia and Europe (Karpat 2001). In recent years, the Turkish lira experienced very high inflation rates. Despite the devaluation of the Lira to US-$ by 700% over the past five years, foreign investments have rebounded since 2022, and foreign debt volumes remain moderate, indicating the country's ability to borrow money (IMF 2023b). In 2024, changes in monetary policies (i.e., raising interest rates and further currency devaluation) resulted in rating agencies updating their outlook on Türkiye to an overall positive assessment.[17]

Turkish energy demand is covered mainly by fossil fuels, i.e., oil, coal, and natural gas. 93 and 99% of oil and gas are imported, respectively. To reduce this dependency, Türkiye plans to explore domestic production sources (IEA 2021). Electricity supply in 2023 was covered by 58% of fossil fuel generation. Renewables, i.e., hydro, wind, and solar, covered 34%. In recent years, renewable capacities, especially wind and solar, have increased to over 27 GW. The currently installed power capacity in

[17] https://www.worldgovernmentbonds.com/credit-rating/turkey/.

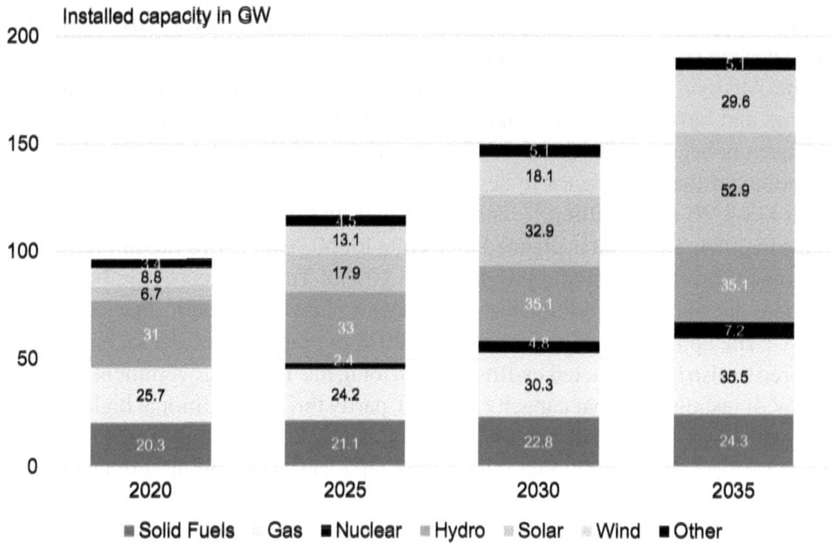

Fig. 9.2 Planned installed capacity for electricity production in Türkiye. *Source* Own depiction based on MENR (2022, 23)

Türkiye exceeds today's peak demand by 55 GW. The "Türkiye National Energy Plan" envisions a total renewable power production capacity of 122.7 GW by 2035, including biomass and geothermal plants, to cover an expected 1.5-fold increase in electricity demand, see Fig. 9.2. Nuclear power plant capacities shall reach 2.4 GW in 2025, 4.8 GW in 2030, 7.2 GW in 2035, and up to an ambitious 20 GW in 2050, according to more recent government announcements (MENR 2022; Schneider et al. 2024).

9.4.3.2 The Akkuyu Project

The Turkish Atomic Energy Commission was established in 1956. In 1962, the first research reactor, the TR-1, was commissioned and followed by a similar reactor in 1981. Today, both reactors are closed. The third research reactor, Triga Mark-2, has been operating since 1979. In addition to these research activities, from 1965 to 2000, Türkiye attempted four different commercial nuclear power plant projects, none of which reached the construction phase. The current fifth project attempt, Akkuyu, received its first license in 1976. In 2004, plans for the construction of three reactors with a combined capacity of 5 GW were announced, with all three being operational by 2012. Four years later, the tender concluded, and Russian-state-owned AtomStroyExport (a subsidiary of Rosatom) was awarded the contract. However, the tender was deemed null and void in 2009 after a lawsuit was filed against the procedure. In 2010, during a state visit by former Russian President Medvedev to Ankara, Russia and Türkiye signed an agreement on the construction and operation of the Akkuyu nuclear power plant to be commissioned between 2016 and 2019. There was no second tender (Schneider et al. 2024).

The Akkuyu nuclear power plant will consist of four VVER-1200 light water reactors and have a combined capacity of 4,800 MW. Construction of the first and last reactors began in 2018 and 2022, respectively. Most recent estimates hope for operations of the first reactor to begin in 2025, while costs have increased from an initial US-$ 20 to 25 billion.[18] The fourth and final reactor is scheduled to come online in 2028. The plant is planned to operate for 60 years with a possible extension of another 20 years and could provide up to 10% of Turkish electricity.[19]

There are plans for additional nuclear power plants to be built. For example, at Sinop, a site approval was granted in April 2024 in preparation for a second project, rumored to also be conducted by Rosatom. In total, the Turkish government envisions 20 GW of installed nuclear capacity by 2050, partly through additional high-capacity reactors, but also via small modular reactors. See Schneider et al. (2024, 192–206) and IAEA (2019) for additional information on Turkish nuclear ambitions.

9.4.3.3 Contractual Analysis

The Turkish Akkuyu power plant is the first nuclear project to follow the BOO contracting structure. Rosatom is building, funding, and will operate the plant in a vertically integrated manner (Toprak and Dal 2017). This means that Rosatom bears all financial and operational risks. The project company established to manage the project, Akkuyu Nükleer AS, was created by Rosatom, and at least 51% of shares must remain with Russian entities. Revenues from the plant operations are planned to compensate for Russian funding provisions. According to the project agreement, 70% of the power generated by the first two reactors and 30% from the other two must be purchased by the Turkish grid operator, TETAŞ, for a period of 15 years at a weighted average price of US-$ 123.5 per MWh (excl. VAT).[20] The remaining volumes are to be sold on the power market (Schneider et al. 2024).

Akkuyu Nükleer AS is responsible for nuclear decommissioning and waste management and is obliged to pay 0.0015 US-$ per kWh into a separate account for spent fuel and waste management as well as decommissioning. All low- and intermediate-level wastes will remain at the Akkuyu site or be transferred to a (currently non-existent) disposal site. Spent fuel shall return to Russia, which reduces the necessary efforts of Türkiye to provide a high-level waste storage facility (Schneider et al. 2024).

---

[18] https://interfax.com/newsroom/top-stories/103324/, and https://www.world-nuclear-news.org/Articles/Akkuyu-s-first-unit-hits-fresh-landmarks, both accessed on 30 October 2024.

[19] https://www.dailysabah.com/business/energy/turkiye-to-gain-nuclear-status-as-1st-fuel-delivery-due-for-akkuyu, accessed on 30 October 2024.

[20] Residential electricity prices were at 48 US-$ per MWh in March 2024, see https://www.globalpetrolprices.com/Turkey/electricity_prices/.

In the current BOO scheme, Rosatom bears all the financial and operational risks of the project. As Rosatom is a fully state-owned company, the final risk lies with the Russian state. In 2016, the Akkuyu project was granted the status of a "strategic investment," thereby granting the project substantial tax breaks totaling roughly US-$ 2.3 billion over the first 20 years of operation and reducing the financial risk to Russia. The contractually agreed-upon PPA ensures revenues for Rosatom, and in case of financing challenges, the tariff can be increased to 153.3 US-$ per MWh, further reducing the risk to the financier. A Turkish equity purchase of 20% ensures that after the project breaks even, revenue from electricity sales is shared between Türkiye and Russia (Hickey et al. 2021). However, the terms of the contract are only viable if the first unit is completed "within seven years of the issuance of all documents, permits, licenses, consents, and approvals" (Schneider et al. 2024, 198). The construction license was issued in April 2018, necessitating the commercial operation of Unit 1 to begin in April 2025. Further project delays could thus result in contract renegotiations. At the time of writing, Rosatom announced that the first reactor would become operational in 2025, despite ongoing supply chain disruptions.[21]

While the financial risks for Türkiye are mitigated through the BOO contract, the country might be disadvantaged in other ways. First, if the Turkish Lira keeps depreciating against the US-$, the PPA prices (defined in US-$) might become increasingly expensive. Additionally, the prices are comparatively expensive, and while the PPA will last for only 15 years, there is no guarantee that Rosatom will produce electricity at marginal prices when the PPA has run out. This relates to the second risk for Türkiye, the ownership and operation of the plant by Russian personnel. For the provision of the plant, Türkiye is the principal, ensuring oversight and regulations, while Rosatom acts as a vertically integrated agent. In the past, Turkish agencies had to interfere in dubious work practices at the site and briefly halt construction (Schneider et al. 2024). However, once the plant is completed, Türkiye becomes the sole off-taker of electricity. It will have no control over the operations (apart from safety-related access) and, consequentially, no control over the electricity output of the plant. Given historical political tensions between both countries, the potential expansion of Western sanctions, and Türkiye being a NATO member country, there remains a risk that Rosatom might use the Akkuyu plant for political bargaining, severely risking Turkish energy security (Hickey et al. 2021; Szulecki and Overland 2023). Finally, the fact that it will be Russian operators working at the plant and there will be no on-site training of Turkish personnel, Türkiye might not be able to operate the plant in the event of a potential Russian withdrawal. The establishment of a domestic nuclear industry and skilled workforce for a potential future project is complicated and requires several years to be completed.

[21] https://www.neimagazine.com/news/progress-continues-at-akkuyu-npp/?cf-view, accessed on 14 January 2025.

Table 9.7 Scenario overview and resulting LCOE for the Akkuyu project

Parameter/case	Best case	Medium case	Worst case
Construction delay (total construction time until first grid connection) in years	2 (7)	4 (9)	6 (11)
Start of electricity production	2025	2027	2029
Additional Costs incl. IDC in billion US-$	11.982	17.079	22.734
LCOE in US-$ per MWh	70.51	80.10	90.75

Table 9.8 Net present value for proposed electricity tariffs for the Akkuyu project

Tariff in US-$ per MWh	90	123.5	153.3
Best Case NPV in billion US-$	−4.95	11.77	25.61
Medium Case in billion US-$	−9.01	7.71	21.55
Worst Case in billion US-$	−11.82	4.90	18.75

9.4.3.4 Financial Analysis and Valuation

The financial analysis of the Akkuyu project is conducted for three cases. We assume an initial project cost of US-$ 20 billion. The three cases include cost overruns. The best case mirrors the currently circulated figure of US-$ 25 billion for the whole project, as mentioned in Sect. 9.4.3.2. This corresponds to a 25% cost increase. The medium and worst cases assume 35 and 45% cost overruns, respectively. To calculate the total cost, we assume IDC, which is added to the project cost. The WACC is calculated at 7%. The assumed delays of the three cases are two, four, and six years, respectively. We assume a uniform inflation rate of 4%. Table 9.7 shows the main differences for the three cases and gives LCOE estimations. All of these calculations exclude decommissioning and waste management costs. We assume a staggered grid connection of one reactor per year.

The calculated LCOE values are well below the tariffs agreed upon in the PPA. Table 9.8 shows the NPV for each of the three cases for a uniform tariff over the whole plant operations. The NPV covers the period from the construction start to the final closure after 60 years of operation. First, we assume a uniform price of the mean electricity price in Türkiye at 90 US-$ per MWh.[22] The agreed upon PPA price is 123.5 US-$ per MWh, which is chosen as the medium case, and finally, Rosatom's price increase option of 153.3 US-$ per MWh is included.

[22] We assume the average of household and business electricity prices at 48 and 127 US-$ per MWh, respectively.

9.5 Discussion

The three cases show that Russian approaches to foreign nuclear power reactor construction projects differ from country to country. However, some similarities exist, such as the (initial) provision of funds, and some form of power plant operations is always part of the contract. In the following, we assess the three projects regarding their contract design and financial aspects. We further provide a brief discussion of the sensitivities for the cost assessments.

9.5.1 Comparison of Contracts

The contractual designs reveal varying degrees of financial and operational risks, as well as energy dependency. The Bangladeshi BOT model at Rooppur places initial cost overrun risk on Rosatom through a fixed-price agreement. However, Bangladeshi reliance on a Russian loan for 90% of the project cost, to be repaid from 2027, heightens financial vulnerability for the host country. This holds especially in combination with the continued devaluation of the Bangladeshi taka. Operational challenges are also significant due to limited local expertise, as inexperienced personnel might manage a large portion of the nation's power supply, risking grid instability and potentially increasing safety-related issues. The reliance on Russia for fuel provision might induce issues related to energy security.

The BOT structure at El-Dabaa allows Egyptian operator NPPA to retain ownership, and a low-interest Russian loan could soften financial burdens. However, the repayment timeline starting in 2028 and Egypt's currency challenges increase repayment risks. Egypt is better positioned than Bangladesh for eventual operational control, as Egyptian personnel assume management after a transition period of ten years, allowing more time for training. However, this might mean a more long-term dependency on Russian personnel to provide a large part of Egyptian power. Additionally, long-term Russian fuel contracts sustain energy dependency.

Türkiye's BOO contract for Akkuyu, where Rosatom will fully own and operate the plant, mitigates financial exposure for Türkiye, but creates heightened energy-supply vulnerability, as Türkiye will rely on Russian personnel and US-$-priced electricity under a 15-year PPA. Afterwards, electricity will be provided at market values, and it remains uncertain how expensive this will be. Türkiye's position as the sole off-taker leaves it exposed to potentially high costs and political risks, with Russia maintaining operational control indefinitely—potentially a strategic leverage point given Türkiye's NATO alignment. However, the commitment to take on spent fuel storage reduces the burden of waste management for Türkiye if Russia follows through.

To conclude, all three countries rely heavily on the Russian state's ability to provide sufficient funds for the project (or to take on the financial risks associated with cost overruns). For each contract, details remain opaque, and it is uncertain how ownership and operations transfers would be handled if the host country failed to repay the loans. For Bangladesh, this risk seems rather limited, given that domestic staff shall operate the plant after only one year of Russian control, while the Egyptian ten-year training and transition period, coupled with a larger loan, might result in higher pressure for loan repayment. This coincides with the dependence of Egypt on Russian personnel, albeit it remains uncertain whether Bangladeshi staff will be trained well enough in the short timeframe to fulfill international safety guidelines and standards. While the financial risks are averted by the Turkish state, the deal itself seems disadvantageous to the host country. Reasons include the dependence on Russian operators in times of geopolitical tensions, the failure to establish a skilled domestic nuclear workforce, and the costly PPA.

9.5.2 Economic Valuation

9.5.2.1 Cost and Profitability

The LCOE of all three projects varies depending on the applied cases and, consequentially, the assumed cost overruns and construction delays, as shown in Fig. 9.3. The Akkuyu project in Türkiye has the lowest LCOE of all three projects, ranging from 70.5 to 90.7 US-$ per MWh, resulting from the comparatively low average total capital cost values of 7,763 US-$ per kW. The Egyptian project's capital costs are 10,012 US-$ per kW on average, while the Rooppur plant could cost an average of 9,321 per kW. The El-Dabaa plant is the most expensive, with worst case LCOE of 112.78 US-$ per MWh.

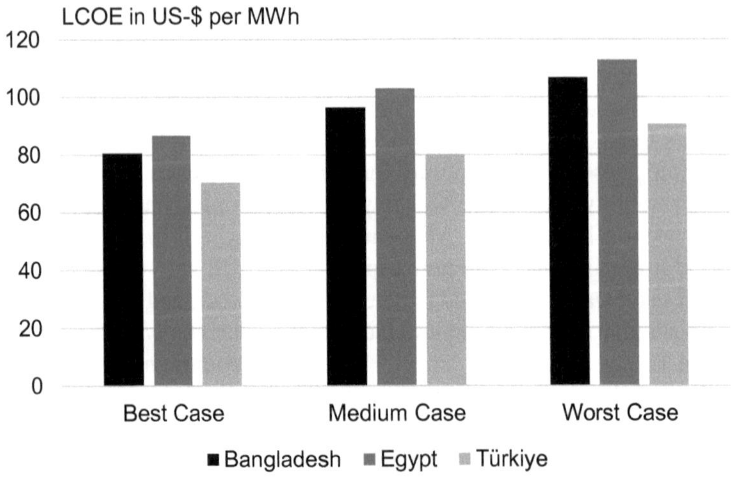

Fig. 9.3 LCOE comparison for the three projects. *Source* Own depiction

The profitability of each project depends on the assumed returns for the generated electricity. Notably, the tariffs discussed for the Rooppur plant are rather low, resulting in negative NPVs for all cases apart from the medium and best cases (little to no delays and cost overruns) paired with a high tariff of over 120 US-$ per MWh. While there was no data supplied regarding the proposed tariffs for the El-Dabaa plant, it becomes a net positive investment only at relatively high rates, i.e., such that are higher than those proposed in Türkiye, where the proposed tariff of 123.5 US-$ per MWh yields net positive results for the project. Overall, the assessments indicate that the projects will likely increase local electricity costs to ensure sufficient revenue streams either for Rosatom directly (Türkiye) or to fund loan repayments.

9.5.2.2 Sensitivity and Limitations

The financial analysis conducted in this chapter is subject to some inevitable limitations. Methodological limitations include the sensitivity of the LCOE and NPV calculations to input parameters. For example, changing the assumed (optimistic) capacity factor from 90 to 80% (11% change), increases LCOE for all the cases of the Akkuyu project by 13%, and corresponding NPVs by an average of 69%. Future research should consider further sensitivities and cases to account for potentially overly optimistic cost and performance parameter assumptions that might result in the impression of (highly) cost-efficient projects.

Additional limitations relate to the assumptions of the NPV calculation itself. Here, we assume that a uniform tariff is paid for the complete operation duration. The proposed tariffs, such as the Turkish PPA, have limited durations, and wholesale electricity prices, the other potential source of revenue, are subject to speculation. Furthermore, we assume a project-oriented view and do not differentiate the calculations of profitability for the vendor or the host country. This results in the neglect of loan payments over several decades in the NPV calculation. It is assumed that all capital costs are covered upfront during construction.

The major limitation of this study, however, lies in the limited availability of project data. Apart from individual press releases and project cost figures circulated in media reports, financial data remains opaque. Further, detailed insight into the contracts, especially the loan agreements, is lacking. Consequentially, the assessment builds on many assumptions. These are, however, skewed towards an optimistic view of the projects, such as high-capacity factors and the neglect of waste and decommissioning costs.

9.6 Conclusion

This analysis provided an overview and first contractual and financial evaluation of three Russian nuclear new-build projects in the so-called newcomer countries—Bangladesh, Egypt, and Türkiye. We show that each project is conducted in a different contractual manner. Furthermore, the economic assessment yielded variable results for each project, despite each being commissioned by the same vendor, i.e., Rosatom. All three projects are characterized by opaque contract schemes and uncertainty regarding the communicated project costs.

The notable differences in the contracts lie first with the final operator of the plant. While Bangladesh and Egypt seem to invest in establishing a skilled domestic nuclear workforce to operate their respective plants, Türkiye is completely relying on Russian operators for the whole lifetime of the plant, potentially increasing its energy-related dependency. Second, the financial obligations differ. Again, Türkiye stands out because the whole project is funded by Russia. To our knowledge, there are no loan agreements or fixed repayment schemes in place, apart from the expensive PPA for the first 15 years of operation at Akkuyu. The loan agreements in Bangladesh and Egypt seem favorable at first glance, but both countries might face challenges in repaying their debt in US-$ as both countries' currencies have been experiencing substantial devaluation. This could place future strain on both economies, resulting in funds being used for nuclear power projects that could be more efficiently spent on renewables for which both countries have high potential.

To conclude, we note a necessity to further investigate Russian foreign new-build projects. Including additional projects, e.g., those currently underway in Iran or India, and expanding the assessment to a scenario-based approach, might provide further in-depth insight into Russian nuclear diplomacy approaches and ambitions. Furthermore, an investigation into potential (geopolitical) dependencies and strategies could shed light on the motivation of the three countries in this report to conduct nuclear power plant construction projects that seem to bring more disadvantages than merits.

References

Abdullah-Al-Mahbub, Md., and Abu Reza Md Towfiqul Islam. 2023. "Current Status of Running Renewable Energy in Bangladesh and Future Prospect: A Global Comparison." *Heliyon* 9 (3): e14308. https://doi.org/10.1016/j.heliyon.2023.e14308.

Akbiyikli, Rifat, and David Eaton. 2005. "A Comparion of PFI, BOT, BOO and BOOT. Procurement Routes for Infrastructure Construction Projects." In *Proceedings of the 5th International Postgraduate Research Conference in the Built and Human Environment*, 505–24. Salford, UK. http://www.irbnet.de/daten/iconda/CIB16757.pdf.

Bajari, Patrick, and Steven Tadelis. 2001. "Incentives versus Transaction Costs: A Theory of Procurement Contracts." *The RAND Journal of Economics* 32 (3): 387. https://doi.org/10.2307/2696361.

Bargman, Abraham. 1977. "Nuclear Diplomacy." *Proceedings of the Academy of Political Science* 32 (4): 159. https://doi.org/10.2307/1173999.

Böse, Fanny, Alexander Wimmers, Björn Steigerwald, and Christian von Hirschhausen. 2024. "Questioning Nuclear Scale-up Propositions: Availability and Economic Prospects of Light Water, Small Modular and Advanced Reactor Technologies." *Energy Research & Social Science* 110 (April 2024): 103448. https://doi.org/10.1016/j.erss.2024.103448.

Butter, David. 2020. "Egypt and the Gulf. Allies and Rivals." Middle East and North Africa Programme. Chatham House. https://www.chathamhouse.org/sites/default/files/CHHJ8102-Egypt-and-Gulf-RP-WEB_0.pdf.

Coase, R. H. 1937. "The Nature of the Firm." *Economica* 4 (16): 386–405. https://doi.org/10.1111/j.1468-0335.1937.tb00002.x.

Das, N.K., J. Chakrabartty, Mrinmoy Dey, A.K. Sen Gupta, and M.A. Matin. 2020. "Present Energy Scenario and Future Energy Mix of Bangladesh." *Energy Strategy Reviews* 32 (November):100576. https://doi.org/10.1016/j.esr.2020.100576.

Furubotn, Eirik Grundtvig, and Rudolf Richter. 2005. *Institutions and Economic Theory: The Contribution of the New Institutional Economics.* 2nd ed. Economics, Cognition, and Society. Ann Arbor: University of Michigan Press.

Göke, Leonard, Alexander Wimmers, and Christian von Hirschhausen. 2025. "Flexible Nuclear Power and Fluctuating Renewables? -- An Analysis for Decarbonized Multi-Vector Energy Systems." *Energy Strategy Reviews* 60 (July): 101782. https://doi.org/10.1016/j.esr.2025.101782.

Hasanovic, Elmas. 2018. "The Politics of Egypt's Nuclear Energy Program." *The International Journal of Humanities and Social Studies* 6 (1): 91–100. https://www.researchgate.net/publication/322852556_The_Politics_of_Egypt%27s_Nuclear_Energy_Program.

Hickey, Samuel M., Salaheddin Malkawi, and Ayman Khalil. 2021. "Nuclear Power in the Middle East: Financing and Geopolitics in the State Nuclear Power Programs of Turkey, Egypt, Jordan and the United Arab Emirates." *Energy Research & Social Science* 74 (April):101961. https://doi.org/10.1016/j.erss.2021.101961.

Hirschhausen, Christian von. 2017. "Nuclear Power in the Twenty-First Century – An Assessment (Part I)." DIW Discussion Paper 1700. Berlin, Germany: DIW Berlin. https://www.diw.de/documents/publikationen/73/diw_01.c.575798.de/dp1700.pdf.

IAEA. 2019. "Turkey." Country Nuclear Power Profiles. International Atomic Energy Agency. https://www-pub.iaea.org/MTCD/Publications/PDF/cnpp2019/countryprofiles/Turkey/Turkey.htm.

IEA. 2021. "Turkey 2021." Energy Policy Review. International Energy Agency. https://iea.blob.core.windows.net/assets/cc499a7b-b72a-466c-88de-d792a9daff44/Turkey_2021_Energy_Policy_Review.pdf.

IMF. 2023a. "Arab Republic of Egypt." Country Report 23/2. International Monetary Fund. https://www.imf.org/-/media/Files/Publications/CR/2023/English/1EGYEA2023001.ashx.

———. 2023b. "Republic of Türkiye." Country Report 23/303. International Monetary Fund. https://www.imf.org/-/media/Files/Publications/CR/2023/English/1TUREA2023001.ashx.

———. 2024. "Bangladesh." Country Report 24/186. International Monetary Fund. https://www.imf.org/-/media/Files/Publications/CR/2024/English/1BGDEA2024001.ashx.

IRENA. 2023. "Socio-Economic Footprint of the Energy Transition: Egypt." Abu Dhabi: International Renewable Energy Agency. https://www.irena.org/-/media/Files/IRENA/Agency/Publication/2023/Nov/IRENA_Socio-economic_footprint_Egypt_2023.pdf.

Johnstone, Phil, and Andy Stirling. 2020. "Beyond and beneath Megaprojects: Exploring Submerged Drivers of Nuclear Infrastructures." *Journal of Mega Infrastructure & Sustainable Development* 2 (2): 220–41. https://doi.org/10.1080/24724718.2021.2012351.

Joskow, Paul L. 1988. "Price Adjustment in Long-Term Contracts: The Case of Coal." *The Journal of Law and Economics* 31 (1): 47–83. https://doi.org/10.1086/467149.

Karim, Ridoan, Mohammad Ershadul Karim, Firdaus Muhammad-Sukki, Siti Hawa Abu-Bakar, Nurul Aini Bani, Abu Bakar Munir, Ahmed Imran Kabir, Jorge Alfredo Ardila-Rey, and Abdullahi Abubakar Mas'ud. 2018. "Nuclear Energy Development in Bangladesh: A Study of Opportunities and Challenges." *Energies* 11 (7): 1672. https://doi.org/10.3390/en11071672.

Karpat, Kemal H. 2001. *The Politicization of Islam: Reconstructing Identity, State, Faith, and Community in the Late Ottoman State.* Oxford University PressNew York, NY. https://doi.org/10.1093/oso/9780195136180.001.0001.

Kim, Yong Woon, and Trevor L. Brown. 2012. "The Importance of Contract Design." *Public Administration Review* 72 (5): 687–96. https://doi.org/10.1111/j.1540-6210.2012.02537.x.

Lévêque, François. 2014. "The International Trade of Nuclear Power Plants: The Supply Side." *Revue d'économie Industrielle*, no. 148 (December), 55–68. https://doi.org/10.4000/rei.5927.

Locatelli, Giorgio. 2018. "Why Are Megaprojects, Including Nuclear Power Plants, Delivered Overbudget and Late? Reasons and Remedies." MIT-ANP-TR-172. Center for Advanced Nuclear Energy Systems, Massachusetts Institute of Technology. https://arxiv.org/pdf/1802.07312.

Markard, Jochen, Nuno Bento, Noah Kittner, and Alejandro Nuñez-Jimenez. 2020. "Destined for Decline? Examining Nuclear Energy from a Technological Innovation Systems Perspective." *Energy Research & Social Science* 67 (September):101512. https://doi.org/10.1016/j.erss.2020.101512.

MENR. 2022. "Türkiye National Energy Plan." Turkish Ministry of Energy and Natural Resources. https://enerji.gov.tr/Media/Dizin/EIGM/tr/Raporlar/TUEP/T%C3%BCrkiye_National_Energy_Plan.pdf.

Moniz, Ernest. 2011. "Why We Still Need Nuclear Power." *Foreign Affairs* 90 (6): 83–94. http://search.ebscohost.com/login.aspx?direct=true&db=bth&AN=66803982&site=ehost-live.

Nasr, Joy, and Ali Ahmad. 2019. "Middle East Nuclear Energy Monitor: Country Perspectives 2018." Annual Report. Issam Fares Institute for Pulic Policy and International Affairs. https://www.aub.edu.lb/ifi/Documents/publications/research_reports/2018-2019/20190103_middle_east_nuclear_energy_monitor_country_perspectives_2018.pdf.

Roohé, Michiel. 2007. "Effects of DBFM Contracts on Project Execution - Experiences from the Field." In *Proceedings of the Second International Conference World Of Construction Project Management*. Delft, Netherlands. https://www.irbnet.de/daten/iconda/CIB16998.pdf.

Schneider, Mycle, Antony Froggatt, Julie Hazemann, Özgür Gürbüz, Paul Jobin, Phil Johnstone, Timothy Judson, et al. 2024. "The World Nuclear Industry Status Report 2024." Paris, France: Mycle Schneider Consulting. https://www.worldnuclearreport.org/IMG/pdf/wnisr2024-v2.pdf.

Schneider, Mycle, Antony Froggatt, Julie Hazemann, Christian von Hirschhausen, M.V. Ramana, Alexander James Wimmers, Nina Schneider, et al. 2023. "World Nuclear Industry Status Report 2023." Paris: Mycle Schneider Consulting. https://www.worldnuclearreport.org/IMG/pdf/wnisr2023-v1-hr.pdf.

Soner, Abdul Malek. 2019. "Plan of Nuclear Spent Fuel and Waste Management in Bangladesh." Conference Presentation presented at the Technical Meeting on Cost Estimation Methodologies for Spent Fuel Management, Vienna, Austria, November. https://nucleus.iaea.org/sites/connect/SFMpublic/TM%20on%20Cost%20Estimation%20Methodologies%20for%20Spent%20Fuel/Malek%20Soner_AERE_Bangladesh.pdf.

Stake, Robert E. 2006. *Multiple Case Study Analysis*. New York: The Guilford Press.

Stirling, Andy, and Phil Johnstone. 2018. "A Global Picture of Industrial Interdependencies Between Civil and Military Nuclear Infrastructures." SPRU Working Paper Series 2018–13 (August). Sussex, UK: University of Sussex. https://www.sussex.ac.uk/webteam/gateway/file.php?name=2018-13-swps-stirling-and-johnstone.pdf&site=25.

Szulecki, Kacper, and Indra Overland. 2023. "Russian Nuclear Energy Diplomacy and Its Implications for Energy Security in the Context of the War in Ukraine." *Nature Energy* 8 (4): 413–21. https://doi.org/10.1038/s41560-023-01228-5.

Toprak, Sezayi, and Selman Dal. 2017. "Akkuyu Nuclear Power Plant Cost & Benefit Analysis." *Energy Policy Turkey*, no. 4, 85–91. https://dergipark.org.tr/en/pub/ept/issue/31249/340287.

Wealer, Ben, Simon Bauer, Nicolas Landry, Hannah Seiß, and Christian von Hirschhausen. 2018. "Nuclear Power Reactors Worldwide – Technology Developments, Diffusion Patterns, and Country-by-Country Analysis of Implementation (1951–2017)." Data Documentation 93. Berlin: DIW Berlin, TU Berlin. https://www.diw.de/de/diw_01.c.583380.de/publikationen/data_documentation/2018_0093/nuclear_power_reactors_worldwide_technology_developments_nd_country-by-country_analysis_of_implementation_1951-2017.html.

Weibezahn, Jens, and Björn Steigerwald. 2024. "Fission for Funds: The Financing of Nuclear Power Plants." *Energy Policy* 195 (December):114382. https://doi.org/10.1016/j.enpol.2024.114382.

Williamson, Oliver E. 2000. "The New Institutional Economics: Taking Stock, Looking Ahead." *Journal of Economic Literature* 38 (3): 595–613. https://doi.org/10.1257/jel.38.3.595.

Wimmers, Alexander, and Christian Von Hirschhausen. 2024. “Organizational Models for the Decommissioning of Nuclear Power Plants: Lessons from the United Kingdom and the United States.” *Utilities Policy* 91 (December):101843. https://doi.org/10.1016/j.jup.2024.101843.
Yin, Robert K. 2014. *Case Study Research: Design and Methods.* 5. edition. Los Angeles London New Delhi Singapore Washington, DC: SAGE.

Chapter 10
Economics of Nuclear Fission Power Plants—A Levelized Costs of Electricity Approach

Alexander Wimmers, Leonard Göke, Claudia Kemfert, and Fabian Präger

10.1 Introduction

Assessing the economic aspects of nuclear power within a renewable-based energy system requires a thorough evaluation of cost metrics and systemic challenges. A key metric is the "levelized cost of electricity" (LCOE) as a benchmark for comparing generation costs, alongside broader economic considerations such as market integration, flexibility potential and system transformation dynamics. Comparing the economic efficiency of different energy generation technologies is complex due to their heterogeneous nature. The challenge lies in balancing multiple parameters, including carbon emissions, waste management, system costs, decommissioning efforts and supply chain processes (Mai et al. 2021). Additionally, the timing of electricity generation—influenced by market values and system flexibility—plays a crucial role in assessing economic viability.

The LCOE has emerged as the dominant metric for directly comparing electricity generation technologies. LCOE determines the average cost of one unit of electricity in a given year. They include capital and operating costs as well as the average operational full-load hours of a given technology. This allows for a direct comparison of technologies. However, LCOE does not consider the intrinsic value of flexibility

A. Wimmers (✉) · F. Präger
Workgroup for Infrastructure Policy (WIP), Technical University (TU) Berlin, Berlin, Germany
e-mail: awi@wip.tu-berlin.de

A. Wimmers · C. Kemfert
German Institute for Economic Research (DIW Berlin), Berlin, Germany

L. Göke
Reliability and Risk Engineering, ETH Zürich, Zürich, Switzerland

C. Kemfert
Professorship for Energy Economics and Energy Policy, Leuphana University Lüneburg, Lüneburg, Germany

A. Wimmers et al. (eds.), *Nuclear Power*, Studies in Energy, Resource and Environmental Economics, https://doi.org/10.1007/978-3-031-99894-2_10

or the additional costs of (nuclear) waste management. Consequently, other metrics have emerged that shall be briefly discussed in Sect. 10.2.

Regardless of the chosen metric, the quality of the output is determined by the input data.[1] Consequently, to ensure that the comparison of electricity generation technologies is viable, input data must be assessed regarding the applicability to the comparison at hand (e.g., whether data is predictive or descriptive or matches the specific region). For light-water reactors (LWR) in OECD countries, Göke et al. (2025) show that predicted and actual cost data depict a substantial discrepancy. From this assessment, they derive the necessity to conduct energy system modeling analysis to incorporate further factors such as availability and flexibility and find that current new build costs for LWRs would not lead to an increase in capacity. Instead, a cost-optimizing model would rely wholly on renewables for its decarbonized energy system. This stands against other energy system models that expect an increase in nuclear power generation over the coming decades. These cost-optimizing models overwhelmingly rely on overly optimistic cost-reduction assumptions for nuclear power, as discussed by Böse et al. (2024).

From a transformational perspective, the role of nuclear power must be evaluated beyond cost metrics alone. In this chapter, we refrain from modeling entire energy systems and instead focus specifically on the costs of nuclear power generation. The goal is to provide an overview of discrepancies between reported, projected, and actual cost data for nuclear power reactors. This analysis is particularly relevant for LWRs in OECD countries and so-called new reactor concepts (SNRs, such as fast neutron reactors).

Building upon this cost assessment, this work further examines the role that nuclear power can play in a hybrid system that combines renewable and nuclear electricity generation. We also discuss the key factors that must be considered. Beyond the pure economic assessment of nuclear power via LCOEs—which must ultimately be compared to alternative electricity sources, whether fossil-based or renewable—additional energy-economic factors such as market integration and flexibility potential significantly influence and constrain the economics and integration of nuclear power plants into an energy system dominated by variable renewables, i.e., solar and wind.

The remainder of this chapter is structured as follows. Following this introduction, Sect. 10.1 presents the methodology of LCOE calculation for nuclear power, along with the economies of learning. In Sect. 10.2, we discuss reported data for high-capacity LWRs, SNRs (including reactors on fast neutron spectra (so-called "fast reactors")) and SMR concepts. Section 10.3 examines the challenges of integrating nuclear power into a renewable-based energy system, focusing on economic, technical, and systemic constraints related to flexibility, market integration, and generation characteristics. Section 10.4 concludes.[2]

---

[1] In data science, this is referred to as the GIGO principle ("garbage in – garbage out"), see Pyle (1999).

[2] Most data stem from ongoing and published research by the authors, such as Göke et al. (2025), Wimmers et al. (2024) and Präger et al. (2024). A detailed description of the different reactor

10.1.1 Calculating Levelized Cost of Electricity

The LCOE metric allows for different energy generation methods to be compared in terms of costs per generated unit of electricity over one year. LCOE includes all monetary costs, e.g., capital cost and operating cost, and accounts for a technology's availability (i.e., generating hours). However, external effects, such as nuclear waste storage, flexibility measures, and grid extensions, are not included. LCOE are calculated as shown in Eq. (10.1) (Konstantin 2017).

$$\mathrm{LCOE} = \frac{I_0 + \sum_{t=1}^{t=n} \frac{A_t}{q^t}}{\sum_{t=1}^{t=n} \frac{W_{el}}{q^t}}$$
$$q = \frac{1+i}{100} \tag{10.1}$$

where

- LCOE: Levelized cost of electricity in MU (monetary unit) per MWh
- W_{el}: Produced amount of electricity of a given year in MWh
- q: Discount factor
- i: Imputed interest
- A_t: Operating cost in MU per MWh
- t: Given year
- n: Assumed operating lifetime in years
- I_0: Investment cost in MU.

In practice, LCOE is more often calculated on an annualized basis. This gives the simplified Eq. (10.2) (Konstantin 2017).

$$\mathrm{LCOE} = \frac{\mathrm{CAPEX} + \mathrm{Fixed\ O\&M} + (\mathrm{FLH} * \mathrm{Variable\ O\&M})}{\mathrm{FLH}} \tag{10.2}$$

where

- LCOE: Levelized cost of electricity in MU per MWh
- CAPEX: Annualized capital expenditure calculated with constant interest rate i over total expected operation time n in MU per kW
- Fixed O&M: Fixed operations and maintenance cost in MU per kW
- Variable O&M: Operations and maintenance cost depending on production volume given in MU per MWh (includes fuel cost)
- FLH: Total hours of full-load operation (full load hours), often calculated by means of a capacity factor CF in % multiplied by 8,760 h.

---

technologies is provided in Chap. 2 and further technical insight is provided by Pistner et al. (2023). Refer to Steigerwald et al. (2023) for an assessment of economies of learning for SMRs. Described cost data are available as supplementary material at https://github.com/alexwimmers/nuclearPower_TEG/tree/main.

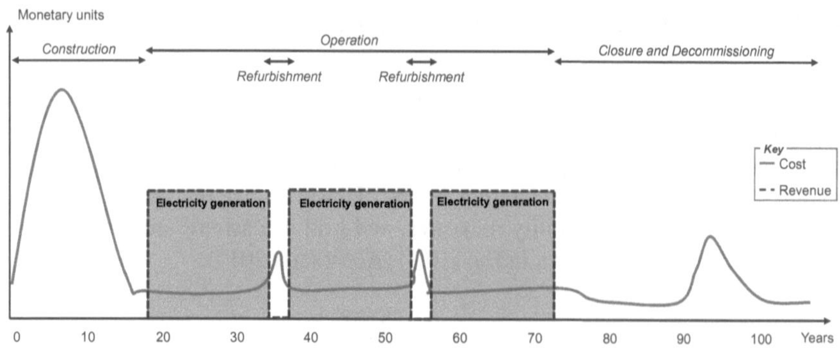

Fig. 10.1 Stylized cost and revenue streams of a nuclear power plant. *Source* Own depiction following Rubio-Varas (2020, 264)

Capital costs account for approximately 80% of total project costs for nuclear power plants. For such large infrastructure projects, capital costs are usually divided into so-called overnight construction costs (OCC), which include all costs directly related to the construction work of the project but neglect the cost of capital. For nuclear, these are the initial costs of financing and interest that accumulate during the construction time when the plant is not generating any power (or revenue) (Rubio-Varas 2020), as shown in Fig. 10.1.

Consequently, for nuclear power, capital costs are calculated following Eq. (10.3) (Rothwell 2016, 85 ff.). The average cost of capital is given by the weighted average cost of capital (WACC). For nuclear, this metric usually ranges from 5 to 10%, reflecting the risk of lending money (Polzin et al. 2021).

$$\text{TCC} = \text{OCC} * (1 + \text{IDC})$$

$$\begin{aligned} \text{TCC} &= \text{OCC} * (1 + \text{IDC}) \\ \text{IDC} &= \frac{\text{WACC}}{2} * t + \frac{\text{WACC}^2}{6} * t^2 \end{aligned} \tag{10.3}$$

where

- TCC: Total construction cost in MU per kW
- OCC: Overnight construction cost in MU per kW
- IDC: Interest during construction
- WACC: Weighted average cost of capital
- t: Construction time in years.

10.1.2 Economies of Learning

As mentioned above, LCOE is a useful metric to compare the cost of electricity production of different technologies. However, LCOE heavily depends on the input parameters. Reports by international organizations such as the International Energy Agency (IEA) often provide OCC estimates for so-called "n-th of a kind" (NOAK) costs (IEA, NEA, and OECD 2020). These types of cost data are also often provided for SMR and SNR concepts (Steigerwald et al. 2023; Wimmers et al. 2024). They assume that the repetition of processes during construction will lead to "learning effects" that will eventually reduce unit cost for a given reactor concept. In this regard, *n* refers to the number of units produced. The (often much higher) costs for the first produced unit are referred to as "first-of-a-kind" (FOAK) costs. These "economies of learning" have been proven to exist in aviation or smartphones. The n-th produced unit costs are calculated following Eq. (10.4) (Lloyd et al. 2020).

$$\begin{aligned} \mathrm{NOAK}_n &= \mathrm{FOAK} * (1 - x)^{\mathrm{d}} \\ n &= 2^d \end{aligned} \tag{10.4}$$

where

- $\mathrm{NOAK_n}$: Cost for the n-th produced unit.
- FOAK: Cost for the first produced unit.
- x: Learning rate.
- d: Doubling of production volume n.

For nuclear reactor cost projections, learning rates are assumed to range from 2 to 15% (Stewart and Shirvan 2022). However, negative learning rates have also been observed in some cases, such as in the French fleet (Grubler 2010). These economies of learning are referred to most prominently by SMR proponents who believe that modularization and mass production of reactors in factories will bring down the unit cost. However, for most reported projections, a precise value for "n" is not given. Furthermore, SMRs face the major challenge of overcoming so-called "economies of scale." This refers to the inherent disadvantage of their smaller capacity that increases their unit cost as major cost parameters, such as OCC and OPEX (material and personnel cost), do not scale linearly with the plant size. This disadvantage, however, is envisioned to be overcome by economies of multiples and learning, as shown in Fig. 10.2 (Boarin et al. 2021).

The penalty for lower capacity is given in Eq. (10.5). This penalty can then be inserted into Eq. (10.4) to calculate the necessary production volume n of a given SMR concept to reach the unit cost of a given high-capacity reactor shown in Eq. (10.6).

$$\frac{C_{\mathrm{SMR}}}{C_{\mathrm{LWR}}} = \left(\frac{S_{\mathrm{SMR}}}{S_{\mathrm{LWR}}}\right)^b \Leftrightarrow C_{\mathrm{SMR}} = C_{\mathrm{LWR}} * \left(\frac{S_{\mathrm{SMR}}}{S_{\mathrm{LWR}}}\right)^b \tag{10.5}$$

where

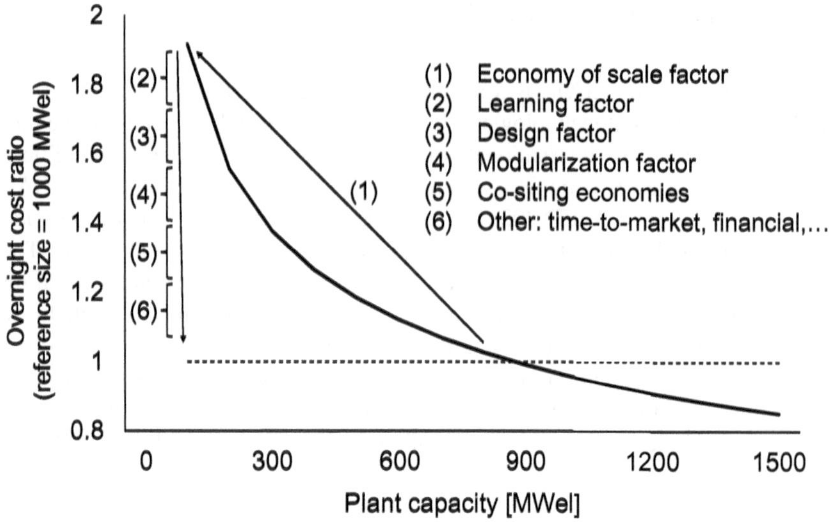

Fig. 10.2 Economy of scale effect on OCC and theoretical reduction factors for SMRs. *Source* Own depiction following Boarin et al. (2021)

- C_{SMR}: Cost of a given SMR in MU per kW
- C_{LWR}: Cost of a given high-capacity LWR in MU per kW
- S_{SMR}: Capacity (size) of a given SMR in kW
- S_{LWR}: Capacity (size) of a given high-capacity LWR in kW
- b: scaling parameter.

Inserting (10.5) into (10.4) gives the NOAK cost for an SMR by accounting for the cost disadvantages stemming from the economies of scale.

$$\begin{aligned} C_{\mathrm{SMR},n} &= C_{\mathrm{SMR},FOAK} * (1-x)^d \\ C_{\mathrm{SMR},n} &= C_{\mathrm{LWR}} * \left(\frac{S_{\mathrm{SMR}}}{S_{\mathrm{LWR}}}\right)^b * (1-x)^d \\ n &= 2^d \end{aligned} \tag{10.6}$$

where

- $C_{\mathrm{SMR,n}}$: Cost of the n-th produced unit of a given SMR in MU per kW
- $C_{\mathrm{SMR,FOAK}}$: Cost of the first produced unit of a given SMR in MU per kW
- C_{LWR}: Cost of a given high-capacity LWR in MU per kW
- S_{SMR}: Capacity (size) of a given SMR in kW
- S_{LWR}: Capacity (size) of a given high-capacity LWR in kW
- b: scaling parameter
- x: learning rate
- d: doubling of production volume n

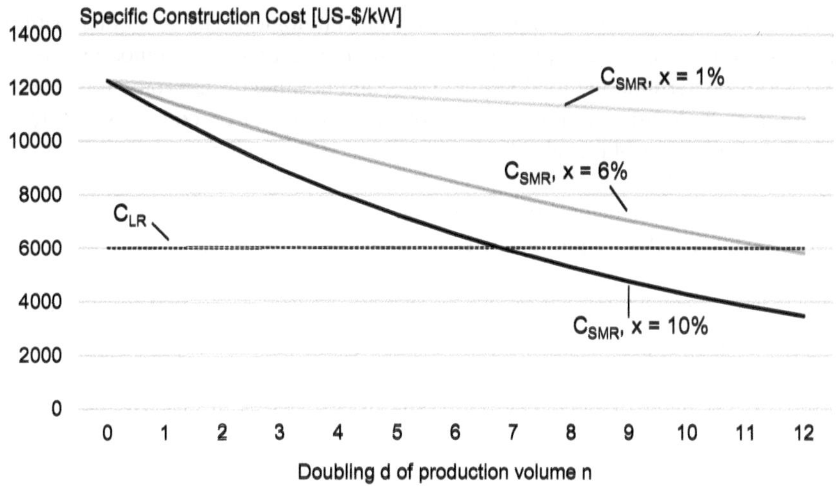

Fig. 10.3 Cost degression for OCC in SMR production economics. *Source* Taken from Böse et al. (2024, 10)

Following Böse et al. (2024), Fig. 10.3 shows the necessary production volume (i.e., the number of required doubling of the production volume) for a given SMR of 225 MWe compared to an LWR of 1,100 MWe. The OCC of the LWR is estimated at 6,000 US-$_{2018}$/kW,[3] and b is set to 0.55. Assuming a learning parameter x of 6%, about 3,000 SMRs (d≈11.55) would have to be produced to achieve specific costs (per capacity) lower than those of the LWR.

Achieving cost reductions would require serial production of reactors, which necessitates a minimum number of orders to cover upfront factory costs. However, the exact minimum is uncertain. Estimates suggest that tens to hundreds of SMR units are needed to justify serial production. Market demand is limited, as actual commitment (i.e., signed contracts) remains small (Schneider et al. 2024). SMRs may have some potential use in niche markets, such as remote, energy-intensive applications, but even in promising areas like Canadian off-grid sites, the estimated potential of 600 MWe is insufficient to justify production investments. Additionally, these remote locations are also viable for renewable energy projects, further limiting the economic prospects for SMRs (Böse et al. 2024).

10.1.3 Other Cost Metrics

LCOE capture the availability of a technology but levels this over a year. However, electricity is valued differently depending on the time of day and season. So, flexibility can also have an inherent value. Thus, there are several more sophisticated

[3] Unless otherwise indicated, all monetary values are given in 2018 values.

approaches to compare the competitiveness of technologies. Each additional parameter, however, increases complexity and uncertainty. Energy system modeling is the best way to determine the total system costs and optimal design (on a cost basis), but it is also the most complex—and seldom allows for a direct comparison of technologies. Following Mai et al. (2021, viii), Table 10.1 provides an overview of cost metrics described below.

The term comprehensive refers to a metric's ability to consider costs and system value. As discussed, LCOE refers to the total costs incurred by each unit of electricity an asset produces uniformly distributed throughout a specified evaluation period, adjusted to present a value at the base year. When consistently applied to every unit of electricity an asset generates over an evaluation period, levelized value of electricity (LVOE) equals the total system value accrued during that period, discounted to the base year. The value of a given unit of produced electricity can vary over the time of day and the season of the year.

The difference between these two metrics is the Net Value of Electricity (NVOE), which is a "comprehensive" metric. The NVOC is calculated as the difference between an asset's annual system value and annualized costs divided by the asset's installed capacity. In contrast, the system LCOE is the negative of NVOE, with an additional adjustment made using a selected benchmark price that can be arbitrarily chosen. The goal of the system LCOE is to quantify so-called "integration costs," e.g., the costs of batteries. These metrics can indicate whether a given technology is economically viable. A positive value would suggest that the system value is higher than the total costs.

Finally, system profitability metrics are derived from the ratio of total system value to cost. Examples include the system benefit–cost ratio (value/cost), system return on investment (value/cost—1), and system profit margin (1—cost/value). These metrics, which utilize the dimensionless ratio of total system value to total cost, indicate the additional value obtained for each monetary unit invested in a new project. They

Table 10.1 Overview of electricity cost metrics

Type	Metric	Conceptual expression [typical units]
Existing, Non-comprehensive	Levelized Cost of Electricity (LCOE)	Cost/Energy [MU/MWh]
	Levelized Value of Electricity (LVOE)	Value/Energy [MU/MWh]
Existing, comprehensive	Net Value of Electricity (NVOE)	(Value – Cost) / Energy [MU/MWh]
	Net Value of Capacity (NVOC)	(Value – Cost) / Capacity [MU/kW-yr]
	System LCOE	(Cost – Value) / Energy + Benchmark Price [MU/MWh]
New, comprehensive	System profitability	f(Value/Cost) [unitless]

Source Mai et al. (2021, viii)

enable meaningful economic comparisons across various technologies, regardless of the different services they offer, by standardizing all services on a common monetary basis.

The next part of this chapter quantifies the LCOE for nuclear and highlights some noteworthy developments. We will focus on available data and will, in some parts, hint toward results from energy systems modeling.

10.2 Costs of Different Technology Lines

Nuclear power is sometimes portrayed as a cheap and reliable energy generation technology. As a matter of fact, nuclear power was never able to compete in the electricity market. Historically, it was outcompeted by coal. For example, in 1957, LCOE for coal was nearly six times lower than for nuclear. While coal has become more expensive and this gap has diminished, nuclear still remains comparatively expensive (Fig. 10.4). However, today's competitors are not coal but renewables, such as wind and solar. Both technologies have experienced substantial cost reductions through the above-mentioned economies of learning, bringing LCOE down 12-fold for solar, and more than fourfold for wind since 2010, while the costs for nuclear have only been increasing.

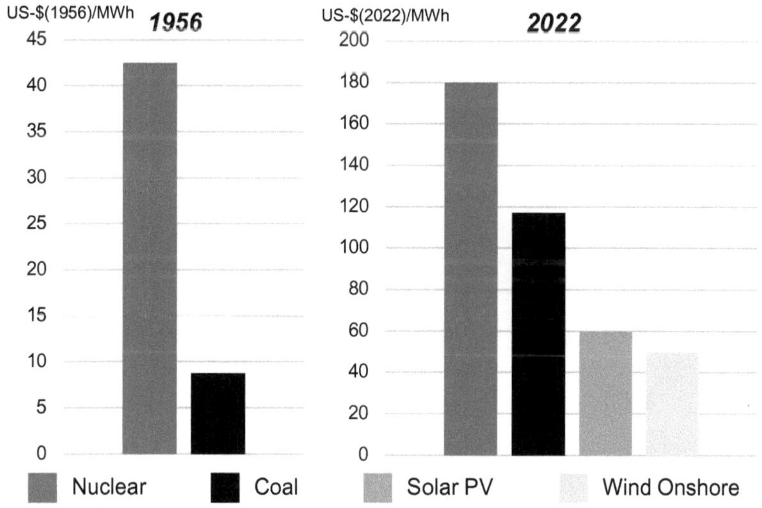

Fig. 10.4 Historical and current cost comparisons in the US. *Source* Taken from Böse et al. (2024, 7)

10.2.1 Light-Water Reactors

Göke et al. (2025) observe the discrepancy of projected or assumed OCC data for LWRs with actual project costs, especially for recent new build projects in OECD countries. From an extensive literature review, they find that this discrepancy can be observed for most of the relevant cost and input parameters used to calculate (future) LCOE for nuclear, including capacity factors, construction times and OPEX. The difference in OCC values is shown in Fig. 10.5. The mean value for actual projects lies at 9,038 US-$/kW, while projected values are much lower at 5,144 US-$/kW. The 25% and 75% quantiles limit the box plots. The x gives the mean value, and the line through the boxes identifies the median value. The whiskers show the respective maximum and minimum values.

Data for actual projects includes recently completed and ongoing projects in the United States (US) and Europe, such as Olkiluoto-3 in Finland, the Vogtle plant in Georgia, US, or Flamanville-3 in France. Each of these projects has experienced substantial delays in their construction time, and consequently, project costs have increased from earlier assumptions. Regardless, projections and assumptions in journal papers and reports by international organizations continue to assume significantly lower project costs for future LWR projects in their cost-optimizing energy system models. This inevitably leads to an overestimation of future LWR capacities

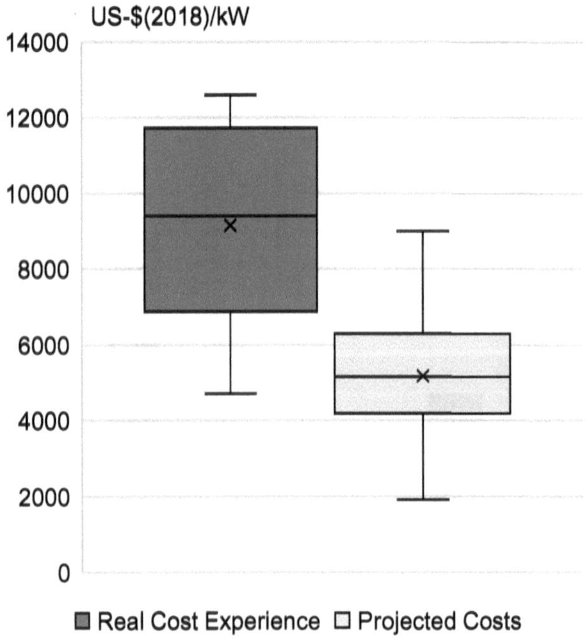

Fig. 10.5 Actual and projects OCC values for LWRs in OECD countries. *Source* Own depiction with data taken from Göke et al. (2025)

and fosters the assumption of a cost-efficient necessity of nuclear power capacity expansions for energy system decarbonization. In contrast, Göke et al. (2025) show that in order for LWRs to be viable in a future European energy system, OCC must be reduced to less than 5,000 US-$/kW—cost levels that have never been observed in Europe or the US. Their analysis assumes favorable conditions for nuclear, such as high capacity factors of 90% and full flexibility. Total system costs are lower in all fully renewable scenarios due to the high capital costs of nuclear capacity expansions.

10.2.2 So-Called New Reactor Concepts (SNRs)

SNRs incorporate reactor concepts often framed as necessary for the decarbonization of future energy systems by industry proponents. These concepts include those introduced in Chap. 2 of this book and shall only be briefly mentioned here. They are sodium- and lead-cooled reactors operating on fast neutron spectra (so-called "fast reactors," SFR and LFR), low-capacity light-water reactors ("small modular reactors," SMR), high-temperature reactors (HTR), molten salt reactors (MSR), and other even more exotic concepts. Note that in the terminology applied here, SMRs refer to LWRs with a capacity of less than 300 MWe. Other concepts can also have lower capacities but are listed as separate technologies due to their inherent technological differences.

A detailed literature review conducted by Wimmers et al. (2024) shows a discrepancy between FOAK and NOAK cost assumptions. It is important to note that none of the reactors listed in the data sets have been built and most are still many years away from being licensed yet commercialized on a large scale. Furthermore, as discussed in Sect. 1.2.1, NOAK cost reductions for nuclear reactors would optimistically require tens to hundreds of reactors of the same model to be built under similar circumstances. Cost data are normalized to 2019-USD following the cost escalation approach introduced by Abou-Jaoude et al. (2023).

In general, NOAK cost assumptions are much lower than FOAK assumptions. For SFRs, mean FOAK values lie at 9,511 USD_{2019} per kW, and HTRs are a bit cheaper at 9,439 USD_{2019} per kW. This compares to optimistic FOAK assumptions of high-capacity LWRs, that are estimated to cost an average 5,622 USD_{2019} per kW—highly questionable when current project experience is considered. Notably, NOAK costs for these reactors are not much lower at an average of 4934 USD_{2019} per kW. For all other technologies, supposed economies of learning drastically reduce NOAK cost assumptions compared to FOAK costs. The black dots in Fig. 10.6 indicate the mean values.

Despite being far from commercialization, cost projections for SNRs have been published in various reports, journal articles, and other forms of publications for decades. Figure 10.7 shows that there has been a recent increase in cost projections for these types of reactors in the past decade, potentially fueled by ongoing narratives of a supposed "renaissance" of nuclear power, as criticized by Böse et al. (2024). However, none of the reactor concepts for which cost projections were

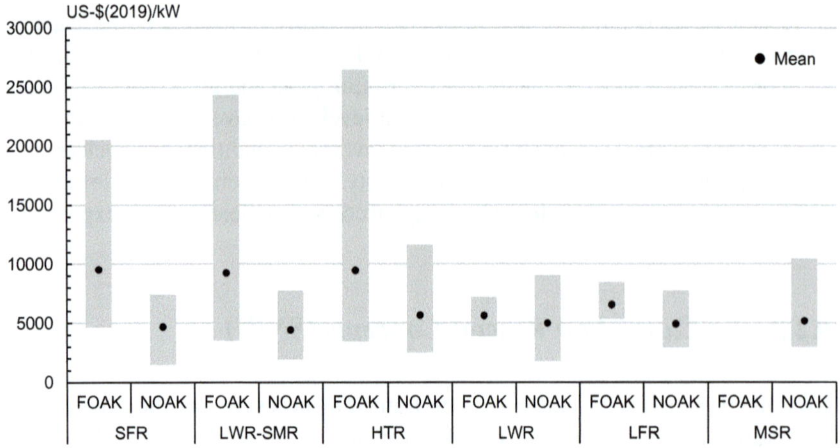

Fig. 10.6 OCC reported for different reactor types and readiness levels. *Source* Taken from Wimmers et al. (2024, 8)

published, even in the early 1990s, has been successfully commercialized. Instead, most concepts are limited to early-stage licensing or prototypes.

Wimmers et al. (2024) assess the potential economic viability of SNRs for electricity and heat provision in a future decarbonized energy system in Europe. They find that at mean FOAK cost levels, no reactor concept is competitive with renewables. Only when costs are substantially reduced could mostly HTRs and SFRs become viable mainly for the provision of high-temperature process heat for which hydrogen is the main low-carbon competitor, which is also expensive. However, this would require the limitation of construction efforts to a single design per technology, neglect of waste and decommissioning costs, a stable and reliable operation with an

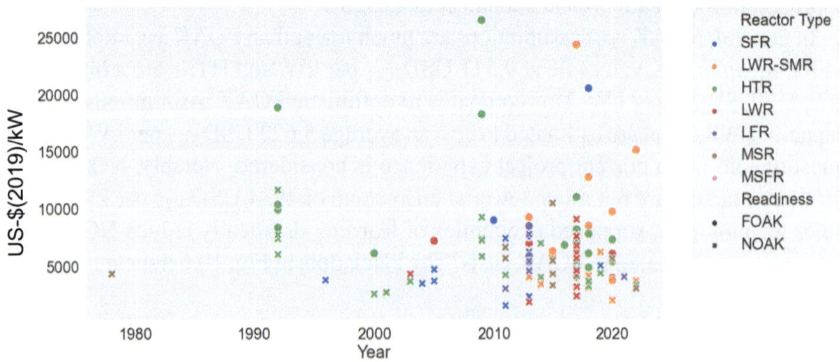

Fig. 10.7 OCC data for different reactor types and readiness levels over the year of publication. *Source* Taken from Wimmers et al. (2024, 9)

average availability of 90%, as well as limited public resistance against co-location of nuclear reactors with demand centers to minimize heat losses.

10.3 Challenges of Integrating Nuclear Power into Renewable Energy Systems

The following analysis examines the role that nuclear power can play in a hybrid system combining renewable and nuclear electricity generation. Beyond the purely economic assessment of nuclear power LCOEs, which must ultimately be compared to alternative electricity sources, whether fossil-based or renewable, additional factors such as market integration and flexibility potential influence and constrain the economic viability and the integration of nuclear power plants into an energy system dominated by variable renewables.[4]

10.3.1 From "Baseload" Electricity to Flexibility-Based Systems

To derive basic findings on this topic, it is first necessary to consider how the electricity generation-spectrum and the demand-spectrum behave within an energy system. The conventional approach to generating and supplying electricity from nuclear power is typically associated with a "baseload" generation model. However, in an energy system that is progressively relying on variable renewable energies, this baseload generation model is being replaced by a flexible, supply-based model, as shown by Hodge et al. (2020). Technical and economic factors impose limitations on the flexibility and ramping rates of nuclear power plants, as recognized by the industry (OECD/NEA 2012). These factors encompass restrictions on cycling operations, such as the need to mitigate material fatigue, as well as must-run constraints and cost considerations. The impact of these limitations becomes more pronounced when evaluating entire fleets of nuclear reactors rather than individual units (Morris 2018). Nuclear power plants usually generate electricity in the baseload spectrum with so-called "must-run constraints"[5] and must be backed up by other technologies to meet peak load demands. These technologies must be able to operate flexibly, e.g., gas turbines and hard coal-fired power plants, and, increasingly, dispatchable renewable electricity, in particular hydropower reservoirs and storage,.e.g., batteries.

[4] This section is based on Präger et al. (2024).

[5] The must-run level of the former German fleet, which was assumed to be the world's most flexible, is given with 20–50% for pressurized water reactors and 60% for boiling water reactors (Grünwald and Caviezel 2017).

Although newer reactor designs technically possess better capabilities for load-following operations (Jenkins et al. 2018), existing nuclear fleets are typically operated as continuous baseload power plants due to economic, technical, and safety considerations. Achieving flexible nuclear power, as proposed by Duan et al. (2022) for example, necessitates the integration of thermal energy storage and three to four times the steam turbine capacity of a conventional baseload nuclear power plant, assuming a reduction from 8000 to 2000 full load hours per year for flexible electricity supply. This results in worsened economics for new nuclear power plants due to increased capital cost associated with thermal energy storage and higher steam turbine capacity. Additionally, the flexibility provided by thermal energy storage is limited to a few days, constraining its effectiveness in addressing seasonal balancing requirements. Conversely, dispatchable hydropower reservoirs, bioenergy plants, and renewable electricity-based seasonal storage with hydrogen or methane offer the potential for meeting seasonal demand fluctuations more effectively, as highlighted by Bogdanov et al. (2019).

Highly renewable energy systems comprise the entire energy system, i.e., power, heat, transport, and industry, and this sector coupling enables additional flexibility and thus energy system efficiency as well as cost effectiveness (Breyer et al. 2022). The potential value of flexible nuclear power will be diminished by emerging flexibility options such as smart electric vehicle charging and vehicle-to-grid concepts which offer flexibility on a range of up to a week for significantly lower cost (Taljegard et al. 2019).

There are assertions suggesting that the continuous baseload supply provided by nuclear power holds a unique significance (Sepulveda et al. 2018). However, it is important to note that such a continuous baseload supply is not essential, as variable renewable energy sources accommodate a diverse load profile. Solar photovoltaics and wind power, in combination with short-term and seasonal storage solutions, can deliver a baseload-like generation profile at a lower cost compared to new nuclear power plants (Fasihi and Breyer 2020; Lazard 2021). Recent developments indicate that in certain regions, new renewables are cheaper than the operating marginal costs of fossil plants, let alone nuclear.[6]

10.3.2 Competing Rather Than Complementary Technology Profiles

The compatibility issues between nuclear power and variable renewable energy sources pose additional challenges, considering the projected dominance of these renewables in the electricity supply. These challenges arise from the convergence of generation profiles, making it difficult to integrate and coordinate the operation of nuclear power alongside solar and wind power (Fig. 10.8). The cumulative generation spectrum of electricity from various renewable energy units (Fig. 10.8, green

[6] https://www.bnef.com/insights/35825, accessed on 3 March 2025.

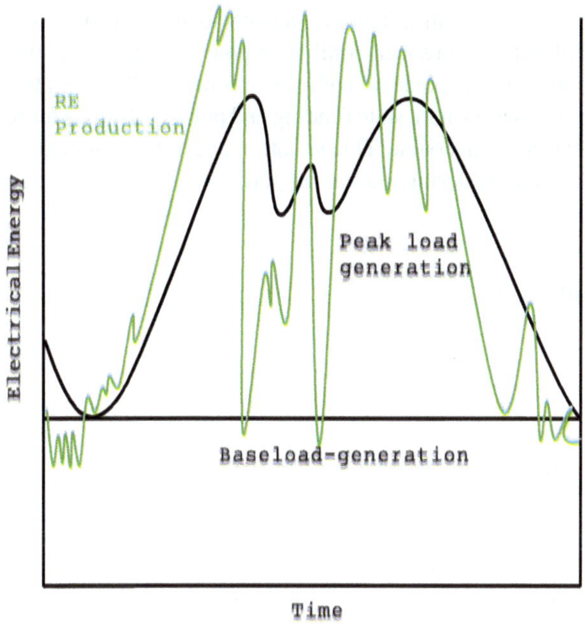

Fig. 10.8 Electricity generation profiles of renewables and nuclear power. *Source* Adapted from Präger (2025, Fig. 2.1)

line) covers all areas of the load profile, from baseload to peak load generation and excess electricity (Verbruggen and Yurchenko 2017). In an energy system predominantly reliant on nuclear power plants, renewable energy sources could theoretically provide peak load power supply when combined with storage and flexibility options. However, in such a scenario, renewables would experience constant curtailment and would be unable to realise their economic and system efficiency potential fully.

This would also result in an increased demand for storage capacity, further complicating the integration of these energy sources. Placing priority on renewable energies within the energy system would result in the virtual elimination of baseload-generation demand (Breyer, et al. 2022, Fig. 16). Integrating nuclear power plants into such a system would be challenging, as this would incur substantial economic drawbacks and technical modifications to allow their safe operation. In contrast, energy systems based primarily on variable renewable energies imply flexibility options, such as dispatchable renewables, sector coupling, power grids, and demand response (e.g., electric vehicle charging, heat pumps and electrolysers for e-fuels and e-chemicals production) (Bogdanov et al. 2021).

In addition, the future trajectory of nuclear power necessitates an examination of its historical context and its implications for socio-ecological transformation processes (Cherp et al. 2017). The expansion and perpetuation of "baseload"-generation technologies hinder the transformation towards a flexible, renewable energy system by creating a dual-system framework (fossil-fissile and renewables), which reinforces the "lock-in" effect associated with the logic of baseload-generation (REN21 2017, 158 ff.). Moreover, conflicting potentials emerge at the funding level. Both the continuation and expansion of nuclear power and the shift to a 100% renewable energy system necessitate state funding and public subsidies for research, development, and system conversion. This situation engenders competition between the two paths rather than fostering mutual support.

10.4 Conclusion

Early proponents of nuclear power envisioned the availability of electricity "too cheap to meter" (Strauss 1954, 9) generated from reactors operating on fast neutron spectra. However, over the last seventy years of commercial nuclear operations, this vision has not materialized. Instead, nuclear power has remained the most expensive electricity generation technology. What has changed are electricity markets that have been liberalized, increasing competition for nuclear. Furthermore, new competitors have emerged that are cheaper, and, most notably, available on a large-scale, i.e., renewables coupled with storage options. Extrapolating from the past shows that nuclear power cost reduction potentials have routinely been overestimated, and there is no reason to believe that projections made today, both for high-capacity LWRs and for SMR or SNR concepts, should prove otherwise.[7] Proposed economies of multiples and learning are yet to be proven. Consequentially, cost projections and assumptions used in energy system modeling should be critically assessed. It is trivial that cost-optimizing energy system models include high proportions of a given technology if costs and other parameters are assumed to be highly favorable for that technology. It is thus essential to critically reflect parameters before drawing policy implications from energy system models and scenarios.

Integrating nuclear power into a renewable-based energy system presents economic, technical, and systemic challenges. While nuclear power has traditionally served as baseload generation, the increase in electricity generation from variable renewables demands greater flexibility, which nuclear struggles to provide due to high costs, ramping limitations and must-run constraints. Economic viability depends not only on costs, but also on market integration and flexibility. Emerging storage, sector coupling, and demand-response technologies offer more cost-effective solutions. Additionally, nuclear's baseload-model conflicts with the variability of solar and wind, increasing curtailment and storage requirements. Beyond economics, nuclear expansion competes with renewables for public funding, potentially slowing the transition to a flexible, efficient, and cost-effective energy system. Given these constraints, nuclear's role in a decarbonized future remains highly uncertain as renewables and storage technologies continue to advance.

[7] For fusion technologies, cost predictions should be eyed with even more skepticism because no commercial fusion reactor has ever existed.

To conclude, for more accurate cost assessments one should include costs and revenues not included in most LCOE calculations. These include the value of flexibility (as discussed in Sect. 1.1.3) but for nuclear, most notable decommissioning, waste management, proliferation risks and insurance costs (see Chap. 12 of this book). If these were included, nuclear power would be even more expensive. Thus, to summarize, nuclear power is one of the most expensive energy generation technologies, and private investors are rightly cautious regarding investments into new build; which is why the actor base is dominated by state-owned actors, such as Russian Rosatom; see Chap. 9 of this book.

References

Abou-Jaoude, Abdalla, Linyu Lin, Chandrakanth Bolisetti, Elizabeth Worsham, Levi M Larsen, and Aaron Epiney. 2023. 'Literature Review of Advanced Reactor Cost Estimates'. INL/RPT-23–72972. Idaho National Laboratory. https://inldigitallibrary.inl.gov/sites/sti/sti/Sort_66425.pdf.

Boarin, S., M. Mancini, M. Ricotti, and G. Locatelli. 2021. 'Economics and Financing of Small Modular Reactors (SMRs)'. In *Handbook of Small Modular Reactors*, 2nd ed., 241–78. Kidlington (UK) and Cambridge (USA): Elsevier.

Bogdanov, Dmitrii, Javier Farfan, Kristina Sadovskaia, Arman Aghahosseini, Michael Child, Ashish Gulagi, Ayobami Solomon Oyewo, Larissa de Souza Noel Simas Barbosa, and Christian Breyer. 2019. 'Radical Transformation Pathway towards Sustainable Electricity via Evolutionary Steps'. *Nature Communications* 10 (1): 1077. https://doi.org/10.1038/s41467-019-08855-1.

Bogdanov, Dmitrii, Ashish Gulagi, Mahdi Fasihi, and Christian Breyer. 2021. 'Full Energy Sector Transition towards 100% Renewable Energy Supply: Integrating Power, Heat, Transport and Industry Sectors Including Desalination'. *Applied Energy* 283 (February):116273. https://doi.org/10.1016/j.apenergy.2020.116273.

Böse, Fanny, Alexander Wimmers, Björn Steigerwald, and Christian von Hirschhausen. 2024. 'Questioning Nuclear Scale-up Propositions: Availability and Economic Prospects of Light Water, Small Modular and Advanced Reactor Technologies'. *Energy Research & Social Science* 110 (April 2024): 103448. https://doi.org/10.1016/j.erss.2024.103448.

Breyer, Christian, Dmitrii Bogdanov, Manish Ram, Siavash Khalili, Eero Vartiainen, David Moser, Eduardo Román Medina, et al. 2022. 'Reflecting the Energy Transition from a European Perspective and in the Global Context—Relevance of Solar Photovoltaics Benchmarking Two Ambitious Scenarios'. *Progress in Photovoltaics: Research and Applications*, December, pip.3659. https://doi.org/10.1002/pip.3659.

Breyer, Christian, Siavash Khalili, Dmitrii Bogdanov, Manish Ram, Ayobami Solomon Oyewo, Arman Aghahosseini, Ashish Gulagi, et al. 2022. 'On the History and Future of 100% Renewable Energy Systems Research'. *IEEE Access* 10:78176–218. https://doi.org/10.1109/ACCESS.2022.3193402.

Cherp, Aleh, Vadim Vinichenko, Jessica Jewell, Masahiro Suzuki, and Miklós Antal. 2017. 'Comparing Electricity Transitions: A Historical Analysis of Nuclear, Wind and Solar Power in Germany and Japan'. *Energy Policy* 101 (February):612–28. https://doi.org/10.1016/j.enpol.2016.10.044.

Duan, Lei, Robert Petroski, Lowell Wood, and Ken Caldeira. 2022. 'Stylized Least-Cost Analysis of Flexible Nuclear Power in Deeply Decarbonized Electricity Systems Considering Wind and Solar Resources Worldwide'. *Nature Energy*, February. https://doi.org/10.1038/s41560-022-00979-x.

Fasihi, Mahdi, and Christian Breyer. 2020. 'Baseload Electricity and Hydrogen Supply Based on Hybrid PV-Wind Power Plants'. *Journal of Cleaner Production* 243 (January):118466. https://doi.org/10.1016/j.jclepro.2019.118466.

Göke, Leonard, Alexander Wimmers, and Christian von Hirschhausen. 2025. 'Flexible Nuclear Power and Fluctuating Renewables? -- An Analysis for Decarbonized Multi-Vector Energy Systems'. *Energy Strategy Reviews* 60 (July): 101782. https://doi.org/10.1016/j.esr.2025.101782.

Grubler, Arnulf. 2010. 'The Costs of the French Nuclear Scale-up: A Case of Negative Learning by Doing'. *Energy Policy* 38 (9): 5174–88. https://doi.org/10.1016/j.enpol.2010.05.003.

Grünwald, Reinhard, and Claudio Caviezel. 2017. 'Lastfolgefähigkeit Deutscher Kernkraftwerke: Monitoring'. 21. TAB-Hintergrundpapier. Berlin: Büro für Technikfolgen-Abschätzung beim Deutschen Bundestag.

Hodge, Bri-Mathias S., Himanshu Jain, Carlo Brancucci, Gab-Su Seo, Magnus Korpås, Juha Kiviluoma, Hannele Holttinen, et al. 2020. 'Addressing Technical Challenges in 100% Variable Inverter-based Renewable Energy Power Systems'. *WIREs Energy and Environment* 9 (5): e376. https://doi.org/10.1002/wene.376.

IEA. 2021. 'Net Zero by 2050: A Roadmap for the Global Energy Sector'. Paris, France: International Energy Agency. https://www.iea.org/reports/net-zero-by-2050.

IEA, NEA, and OECD. 2020. 'Projected Costs of Generating Electricity - 2020 Edition'. Paris: International Energy Agency (IEA), Nuclear Energy Agency (NEA), Organisation for Economic Co-Operation and Development (OECD). https://www.oecd-nea.org/jcms/pl_51110/projected-costs-of-generating-electricity-2020-edition.

Jenkins, L.M., Z. Zhou, R. Ponciroli, R.B. Vilim, F. Ganda, F. de Sisternes, and A. Botterud. 2018. 'The Benefits of Nuclear Flexibility in Power System Operations with Renewable Energy'. *Applied Energy* 222 (July):872–84. https://doi.org/10.1016/j.apenergy.2018.03.002.

Konstantin, Panos. 2017. *Praxisbuch Energiewirtschaft*. Berlin, Heidelberg: Springer Berlin Heidelberg. https://doi.org/10.1007/978-3-662-49823-1.

Lazard. 2021. 'Lazard's Levelized Cost of Energy Analysis'. 15.0. New York. https://www.lazard.com/media/451881/lazards-levelized-cost-of-energy-version-150-vf.pdf.

Lloyd, Clara, Robbie Lyons, and Tony Roulstone. 2020. 'Expanding Nuclear's Contribution to Climate Change with SMRs'. *Nuclear Future*, November, 39.

Mai, Trieu, Matthew Mowers, and Kelly Eurek. 2021. 'Competitiveness Metrics for Electricity System Technologies'. NREL/TP-6A20–72549. NREL - National Renewable Energy Laboratory. https://www.nrel.gov/docs/fy21osti/72549.pdf.

Morris, Craig. 2018. 'Can Reactors React? Is a Decarbonized Electricity System with a Mix of Fluctuating Renewables and Nuclear Reasonable?' *IASS Discussion Paper*, January. https://doi.org/10.2312/iass.2018.002.

OECD/NEA. 2012. *The Economics of Long-Term Operation of Nuclear Power Plants*. Nuclear Development. OECD. https://doi.org/10.1787/9789264992054-en.

Pistner, Christoph, Matthias Englert, Christian von Hirschhausen, Fanny Böse, Björn Steigerwald, and Lukas Gast. 2023. 'Analyse und Bewertung des Entwicklungsstands, der Sicherheit und des regulatorischen Rahmens für sogenannte neuartige Reaktorkonzepte'. Zwischenbericht zu AP-1 und -2 Vorhaben 4721F50501. Forschungsberichte zur Sicherheit der nuklearen Entsorgung. Berlin: Bundesamt für die Sicherheit der nuklearen Entsorgung. https://www.base.bund.de/SharedDocs/Downloads/BASE/DE/berichte/kt/zwischenbericht-gutachten-sogenannte-neuartige-reaktorkonzepte.pdf;jsessionid=1DBCEDF1CF1A46C6216FF76407B85AD1.internet982?__blob=publicationFile&v=3.

Polzin, Friedemann, Mark Sanders, Bjarne Steffen, Florian Egli, Tobias S. Schmidt, Panagiotis Karkatsoulis, Panagiotis Fragkos, and Leonidas Paroussos. 2021. 'The Effect of Differentiating Costs of Capital by Country and Technology on the European Energy Transition'. *Climatic Change* 167 (1–2): 26. https://doi.org/10.1007/s10584-021-03163-4.

Präger, Fabian. 2025. *Aus-, Ein- und Umstiegsprozesse in der sozial-ökologischen Transformation im 21. Jahrhundert*. Energiepolitik und Klimaschutz. Energy Policy and Climate Protection. Wiesbaden: Springer Fachmedien Wiesbaden. https://doi.org/10.1007/978-3-658-47014-2.

Präger, Fabian, Christian Breyer, Hans-Josef Fell, Christian Von Hirschhausen, Claudia Kemfert, Björn Steigerwald, Thure Traber, and Ben Wealer. 2024. 'Evaluating Nuclear Power's Suitability for Climate Change Mitigation: Technical Risks, Economic Implications and Incompatibility with Renewable Energy Systems'. *Frontiers in Environmental Economics* 3 (April):1242818. https://doi.org/10.3389/frevc.2024.1242818.

Pyle, Dorian. 1999. *Data Preparation for Data Mining*. San Fransisco, USA: Morgan Kaufmann Publishers.

REN21. 2017. 'Renewables 2017'. Global Status Report. Paris: Renewable Energy Policy Network for the 21st century. https://www.ren21.net/wp-content/uploads/2019/06/REN21_AnnualReport_2017_web.pdf.

Rothwell, Geoffrey. 2016. *Economics of Nuclear Power*. London, UK: Routledge.

Rubio-Varas, Mar. 2020. 'Time Is Money, but Sometimes It Costs More: An Economic History Perspective into Nuclear Projects' Pitfalls'. *Journal of Mega Infrastructure & Sustainable Development* 2 (3): 262–79. https://doi.org/10.1080/24724718.2022.2092993.

Schneider, Mycle, Antony Froggatt, Julie Hazemann, Özgür Gürbüz, Paul Jobin, Phil Johnstone, Timothy Judson, et al. 2024. 'The World Nuclear Industry Status Report 2024'. Paris, France: Mycle Schneider Consulting. https://www.worldnuclearreport.org/IMG/pdf/wnisr2024-v2.pdf.

Schneider, Mycle, Antony Froggatt, Julie Hazemann, Christian von Hirschhausen, M.V. Ramana, Alexander James Wimmers, Nina Schneider, et al. 2023. 'World Nuclear Industry Status Report 2023'. Paris: Mycle Schneider Consulting. https://www.worldnuclearreport.org/IMG/pdf/wnisr2023-v1-hr.pdf.

Sepulveda, Nestor A., Jesse D. Jenkins, Fernando J. de Sisternes, and Richard K. Lester. 2018. 'The Role of Firm Low-Carbon Electricity Resources in Deep Decarbonization of Power Generation'. *Joule* 2 (11): 2403–20. https://doi.org/10.1016/j.joule.2018.08.006.

Steigerwald, Björn, Jens Weibezahn, Martin Slowik, and Christian Von Hirschhausen. 2023. 'Uncertainties in Estimating Production Costs of Future Nuclear Technologies: A Model-Based Analysis of Small Modular Reactors'. *Energy* 281 (October):128204. https://doi.org/10.1016/j.energy.2023.128204.

Stewart, W. Robb, and Koroush Shirvan. 2022. 'Capital Cost Estimation for Advanced Nuclear Power Plants'. *Renewable and Sustainable Energy Reviews* 155 (March):111880. https://doi.org/10.1016/j.rser.2021.111880.

Strauss, Lewis L. 1954. 'Remarks Prepared by Lewis L. Straus, Chairman, United States Atomic Energy Commission, For Delivery At The Founders' Day Dinner, National Association of Science Writers, On Thusday, September 16, 1954, New York, New York.' Speech presented at the Founders' Day Dinner, National Association of Science Writers, New York, September 16. https://www.nrc.gov/docs/ML1613/ML16131A120.pdf.

Taljegard, M., L. Göransson, M. Odenberger, and F. Johnsson. 2019. 'Impacts of Electric Vehicles on the Electricity Generation Portfolio – A Scandinavian-German Case Study'. *Applied Energy* 235 (February):1637–50. https://doi.org/10.1016/j.apenergy.2018.10.133.

Toktarova, Alla, Lotta Gruber, Markus Hlusiak, Dmitrii Bogdanov, and Christian Breyer. 2019. 'Long Term Load Projection in High Resolution for All Countries Globally'. *International Journal of Electrical Power & Energy Systems* 111 (October):160–81. https://doi.org/10.1016/j.ijepes.2019.03.055.

Uddin, Kotub, Matthieu Dubarry, and Mark B. Glick. 2018. 'The Viability of Vehicle-to-Grid Operations from a Battery Technology and Policy Perspective'. *Energy Policy* 113 (February):342–47. https://doi.org/10.1016/j.enpol.2017.11.015.

Verbruggen, Aviel, and Yuliya Yurchenko. 2017. 'Positioning Nuclear Power in the Low-Carbon Electricity Transition'. *Sustainability* 9 (1): 163. https://doi.org/10.3390/su9010163.
Wimmers, Alexander, Fanny Böse, and Leonard Göke. 2024. 'Assessing the Viability of Non-Light Water Reactor Concepts for Electricity and Heat Generation in Decarbonized Energy Systems'. arXiv. https://doi.org/10.48550/ARXIV.2412.15083.
Yao, Xing, Ying Fan, Fan Zhao, and Shao-Chao Ma. 2022. 'Economic and Climate Benefits of Vehicle-to-Grid for Low-Carbon Transitions of Power Systems: A Case Study of China's 2030 Renewable Energy Target'. *Journal of Cleaner Production* 330 (January):129833. https://doi.org/10.1016/j.jclepro.2021.129833.

Chapter 11
The First Nuclear Century: A Long-Term Analysis of Nuclear Reactor Development in the US (1939–2024)

Fanny Böse

11.1 Introduction

The development of reactor technologies stems from a long and multi-faceted history, beginning with the discovery of natural radioactive decay by Ernest Rutherford and Frederick Soddy in 1902 (Rutherford and Soddy 1902). Their experimental work on the decay of thorium and radium showed the high energy potential which is connected to the process of radioactive disintegration (Soddy 1909). In 1914, the science fiction novel, "The World Sets Free," by H.G. Wells (1914), transformed this new energy potential into a utopian, but also dystopian, vision in which mankind would develop a new weapon with never-before-seen destructive potential, all while changing global power dynamics. In the early 1930s, the human-induced radioactive decay was discovered when Frédéric Joliot and Irène Curie began their theoretical and experimental work in Paris (Curie and Joliot 1934). In parallel, Leo Szilard's theoretical work on nuclear chain reactions in 1933 (Szilard 1949) and Enrico Fermi's research beginning in 1934, which focused on slowing down neutrons and irradiating heavy elements like uranium-238 (Fermi 1934; Amaldi and Fermi 1936), were essential contributions to the understanding of the fission process. Building upon these first experiments, the nuclear fission process was first experimentally shown by Otto Hahn and Fritz Strassman in Berlin in 1938, and then physically interpreted by Lise Meitner and Otto Frisch in February 1939 (Meitner and Frisch 1939). These discoveries led to tremendous technological advancements, both destructive and beneficial at the

Thanks to Zoe Felder, Annika Froese, and Sarah Glück from BASE and Christian von Hirschhausen and Alexander Wimmers from TU Berlin for their helpful comments on earlier drafts of this chapter.

F. Böse (✉)
Federal Office for the Safety of Nuclear Waste Management (BASE) in cooperation with TU Berlin, Berlin, Germany
e-mail: fab@wip.tu-berlin.de

A. Wimmers et al. (eds.), *Nuclear Power*, Studies in Energy, Resource and Environmental Economics, https://doi.org/10.1007/978-3-031-99894-2_11

same time. Today, we look back on roughly 120 years of the historical development of fission: the first nuclear century.

As the first country to develop, test, and use energy yielded from nuclear fission processes, the United States (US) plays a major role in this multifaceted history. Initial development efforts focused on fission technologies to be used as weapons during World War II (WW II) and thus led to the construction of reactors and the development of the corresponding industry complex. The connection between the military-industrial complex and civilian reactor technologies, most prominently the diffusion of light-water reactor (LWR) technologies, has been extensively covered in the literature (Cowan 1990; Cohn 1990; Reed 2013; Wealer et al. 2018). Some studies have also explored the efforts to develop other reactor technologies, particularly those with a fast neutron spectrum, which remained confined to niches or were ultimately discontinued (Pistner et al. 2024; Cochran et al. 2010). While previous work focuses either on one particular fission reactor technology[1] or a certain time period (e.g., several years to several decades), this work applies the multi-level perspective (MLP) by Geels (2002, 2004) in a long-term analysis of reactor technologies in the US. The MLP framework is useful for exploring long-term dynamics of technological development embedded into larger a socio-technical context. Moreover, the framework is useful to study the emergence and diffusion of certain technologies. It can also provide insights into why technologies remain within their niche. So far, the MLP literature has placed an emphasis on the innovation and diffusion of technologies, but according the literature on technical innovation systems, more stages are prevalent including the stagnation and decline of technologies. Thus, the analysis of this chapter couples MLP with the five stages of technological development, i.e., invention, innovation, diffusion, stagnation, and decline (Schumpeter 1934; Markard 2020). This highlights different development dynamics and the distinction of time periods. Building upon this, the analysis distinguishes five periods from 1939 to 2024. The description of each period will highlight the dynamics of the three dimensions of the MLP; the socio-technical system (ST-system), technological niches, and landscape events. Following this perspective, the chapter sheds light on the pathways of different reactor technologies, from which some stayed in niches, and others, such as the light-water cooled reactor technologies (LWR), diffused into the ST-system.

The chapter is structured as follows: Sect. 11.2 describes the key tenets of the MLP analysis, as well as the characteristics of the five proposed development phases, while also presenting the interplay of the three dimensions of the MLP for each time period. Thus, it allows one to investigate changes in the dynamics of technology development and technology use. Sections 11.3–11.7 then provide a detailed analysis of the dynamics in each of the defined time periods. The chapter concludes with the main lessons learned from this exercise and a research outlook.

[1] For the remainder of this chapter, the term "reactor" refers solely to fission reactors unless otherwise stated.

11.2 Theoretical Background

11.2.1 *Multi-Level Perspective for Technological Transformations*

The multi-level analysis is an analytical framework to study long-term transformation dynamics that can integrate various disciplines, including evolutionary theory, technological studies, structural theory, and neo-institutional theory (Geels 2002; Berkhout et al., 2004; Geels and Schot 2007). This framework has already been applied to the music sector, when the breakthrough of rock "n" roll was analyzed (Geels 2007), and to the energy sector to analyze the niche of renewables in the Dutch electricity system (Verbong and Geels 2007). The following highlights key tenets of the theory and defines essential terms.

To analyze long-term dynamics of socio-technical configurations, the MLP distinguished between three levels, which are described below as follows:

(1) ST-system: An ST-system is where technologies can be described as artifacts that are fully embedded into society and fulfill societal functions, such as the provision of energy, transportation, communication, and housing, among others. The embeddedness of technology into society becomes observable as technology uses lead to elements of a society being adapted to its use, e.g., standards, regulations, user practices and cultural values. Consequentially, changes in the use of existing technologies or the development of new ones can transform elements of society and vice versa (Geels 2002). Thus, a ST-system comprises the "linkages between elements that are necessary to fulfill societal functions" (Geels 2004, 900). The elements of an ST-system can be grouped into three dimensions: first, a network of human actors and social groups that, for example provide, distribute, and finance resources required for technological development and use. Secondly, these actors share a set of rules that encompass formal, normative, and cognitive rules. Third, the ST-system also inhibits the material and technical elements. Within literature, it is often referred to the so-called socio-technical regime defined as "deep structure" of ST-systems (Geels 2002, 1260). These deep structures guide the actions of actors and, therefore, can create a stability of the ST-system and a resistance to change. A regime can consist of practices, routines, expectations, and rules (carried out by actors) and can be grouped into different sub-regimes, such as technical regime, science regime or policy regime (Geels 2004). However, the chapter will make use of the broader term of ST-system and its dimensions, namely actors, sets of rules, and technical elements.

(2) Technological niches represent new technologies, which are under development to fulfill a societal function such as the supply of electricity. Several niche technologies compete against each other before a dominant design occurs. The technological niches co-exist with the ST-system under public subsidies and in niche markets, protecting them from market pressure.

(3) Landscape events are external factors, e.g., war, oil crises, or climate change that occur outside the system's control and pressure the ST-systems, but also the technological niches. The pressure can have destabilizing or stabilizing effects on the technological development at the niche level, but also at the level of the ST-system (changes in laws, rules, institutional structures, funding, actor's expectations, prices, etc.).

The ST-system is dynamic over the long-term, therefore, its configurations can change from which new technologies may be adopted. While others may phase out, new laws will be passed, new institutions will be created, organizational structures will change, or new networks of actors will evolve. However, according to the MLP framework, the elements of the ST-system have a stable configuration, which creates resistance to change. For example, existing infrastructure is a sunk investment that can foster resistance towards new investments. Further, incumbent actors have vested interests and existing regulations and standards can guide actions in certain directions. Consequently, the ST-system moves along technological trajectories that lead to incremental change and improvements building upon existing knowledge and competencies. As long as challenges are expected to be overcome within the existing regime, incremental change is dominant, and innovation happens along technological trajectories and path dependencies. The ST-regime is therefore "dynamically stable" (Geels 2004, 910).

In dynamically stable configurations of ST-regimes, the system stays aligned with existing technologies, while radical innovations remain in technological niches. However, there are pressures on ST-systems, in particular via landscape events and internal forces. Internal factors are challenges that occur within the ST-system, e.g., technical challenges that trigger new technological developments. Both events affect the system's properties. They can destabilize the deep structure of ST-systems and may create a so-called "window of opportunity" in which (radical) new technologies can emerge and ultimately diffuse into the system (Geels 2004, 914). However, landscape events can also stabilize existing ST-system configurations (Geels 2011). Figure 11.1 shows the stylized emergence of novel technologies from a niche into the ST-system induced by a landscape event.

At the niche level, several co-existing niches can compete against each other before one might become dominant. Only when a technology is "fully developed" it can take advantage of an open window of opportunity (Geels and Schot 2007, 405). Whether or not niche innovations are sufficiently developed is not objective, and perceptions differ between niche actors and regime actors. According to Geels and Schot (2007), several indicators can be used to assess the readiness of a technological niche:

- learning processes have stabilized → dominant design occurs,
- powerful actors have joined the support network,
- price/performance have improved and strong expectations of further improvement (e.g. learning curves),
- already used in market niches.

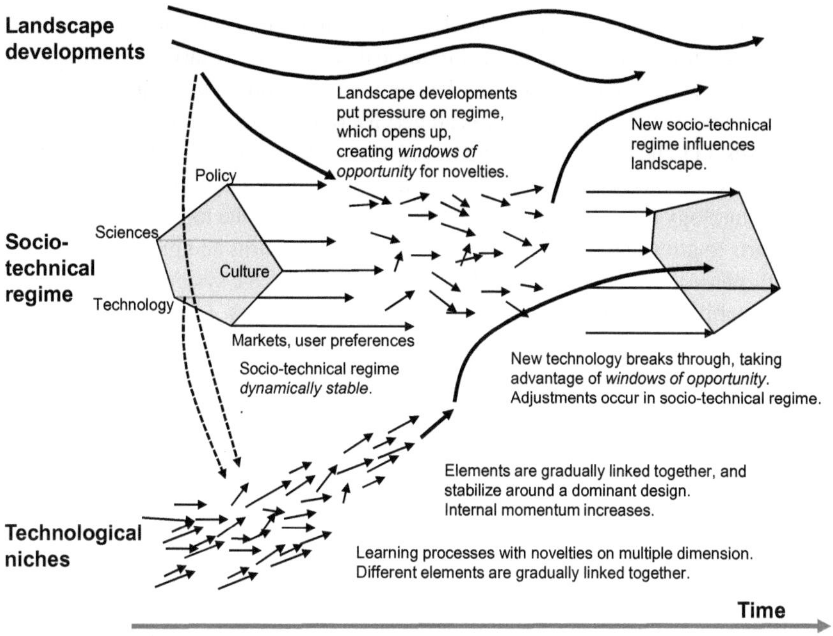

Fig. 11.1 Multi-level analysis framework (*Source* Own depiction following Geels [2002, 1263])

If a technology is not ready to break through and requires further development, the window of opportunity might close before it can move into the ST-system. Thus, technological niches can also remain niches for a long time, especially when there is a mismatch with the existing ST-system (Geels 2007).

11.2.2 Stages of Technological Development

Within literature, the MLP approach is often used with the focus to analyze the stages of invention, innovation, and diffusion, which also follows the main stylized phases proposed by Schumpeter (Schumpeter 1934). Hereby, the invention can be understood as new idea for a product, or in this case a reactor design that requires further experiments and testing relating to materials, components, and safety issues, among others. For energy technologies, the subsequent innovation process can be described by innovation stages of a so called "innovation chain," starting with basic science and moving on to applied science, which involves engineering and the adaptation of a potential market fit. Before the "commercial" stage (diffusion) can be entered, the technology enters a demonstration phase in which the product design properties and costs are tested and potentially adapted (Stern 2007; Grubb et al. 2014).

However, there are additional stages of development that are important to capture the long-term dynamics of technologies and not only include its deployment and use, but also its potential discontinuation. Therefore, literature of technological innovation systems refers to the term "technological life cycle" (Markard 2020, 1) and proposes five stages of technological development: invention, innovation, diffusion, stagnation and decline. To cover these stages, Markard (2020) distinguishes between four phases: (1) The first phase (formative phase) is characterized by the initial invention. It also encompasses the innovation process, thus testing, refining, and adapting a given technology to fit market conditions. (2) Afterwards, the technological deployment starts to grow and is characterized by increasing returns and increasing demand (Growth phase). This is similar to the diffusion into the ST-systems and the subsequent restabilization and alignment toward this technology. (3) In the third phase (saturation phase), growth ends as countervailing forces (e.g., socio-political resistance, ecological factors, costs, etc.) dampen the growth, and the technology will enter a saturation phase. (4) Finally, an additional phase can occur that marks the decline of a technology deployment, potentially leading to complete phase out.

As described in the following, we propose five phases of reactor technology development in the US building upon the technological life cycle and the MLP framework (see Fig. 11.2).

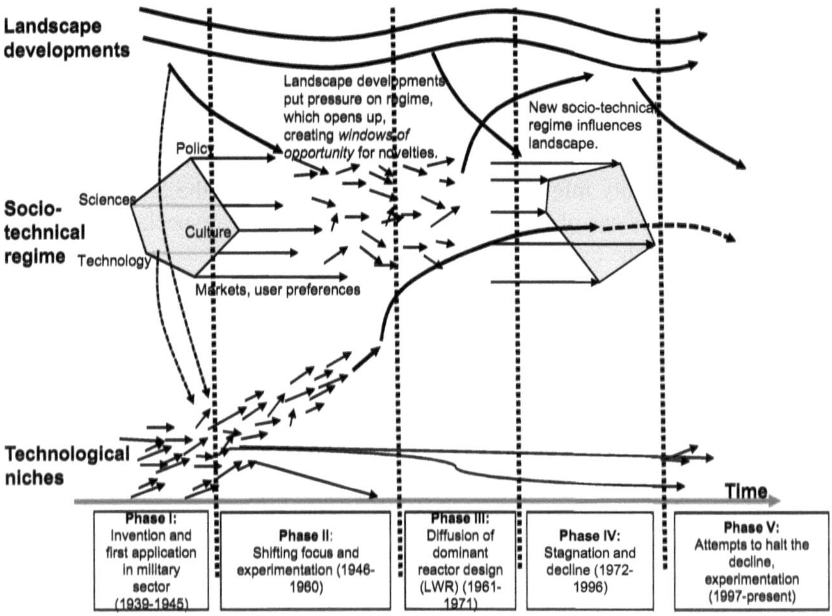

Fig. 11.2 MLP and technical life cycle stages of reactor technologies (*Sources* Own depiction following Geels [2002] and Markard [2020])

11.2.3 Application of MLP Levels to the Nuclear Case

The following will explain briefly how the distinction between time periods was chosen and how the description of each phase is conducted.

As stated above, the chapter analyzes the history of nuclear fission development from its invention to innovation, to first applications in the US, and the subsequent diffusion of one reactor technology in the energy sectors by covering a time span of 85 years. To capture the full technological life cycles, a special focus will be placed upon each of the five stages of technological development. However, the stages of stagnation and decline were combined and an additional phase was added to highlight recent developments. A summary of the proposed phases is shown in Table 11.1.

As the chapter applies the MLP framework to the nuclear case in the US, it is useful to specify some of the MLP terms toward the case study. As described above, in ST-systems, technology satisfies different societal functions. Nuclear power reactors are used in a range of applications in the energy and military sector, but also for medical applications such as isotope production. However, the focus of this chapter is placed on reactor technologies which fulfill the demand for energy and national security. Consequentially, reactors are relevant for electricity production and military applications (e.g., the production of fissile materials for nuclear weapons and reactors for submarine propulsion or remote military bases). If the ST-system adopts a certain reactor technology for energy production, it will lead to changes in policy and laws, regulations, science, and relevant knowledge. Moreover, nuclear power will compete against other electricity generating technologies in energy markets. To provide the required resources for technology development, multiple actors are involved. This includes national laboratories, public authorities for regulation and infrastructure development, but also private actors and environmental groups. On a deeper level, ST-systems also reflect narratives and expectations. Finally, the use of reactor technology can create internal dynamics within the ST-system, such as economic or technical challenges, that can lead to adaptions of the system.

On the niche level, the innovation process of different reactor technologies will be identified. Generally, the innovation process of reactor technologies is very lengthy and complex. They require highly specific knowledge from different disciplines, but also large and complex test and experimentation infrastructure with specific quality requirements related to nuclear safety, nonproliferation of nuclear weapons-usable material, and radiation protection (OECD/NEA 2007). Usually, the innovation process involves the construction of a prototype version for testing materials and components. However, this may lead to design changes. Afterwards, a demonstration plant is constructed, which has higher capacity, often in similar size as envisaged for commercial use. Before a reactor project can begin construction, the design has to be licensed and a construction permit has to be issued.

All of these stages involve a set of actors including public authorities, industry, and private entities. All of this represents a high sunk cost of a ST-system during reactor technology development. Following the MLP rationale, different technologies compete against each other before one dominant technology designs occurs that can

Table 11.1 Proposed phases of reactor technology development in the US

	Period	Main developments
Phase I	1939–1945	Describes the very beginning of the discovery of the fission process, which led to the first use of a nuclear bomb during World War II shortly after. This includes the build-up of large infrastructure for R&D and production facilities for fissile materials. Consequently, this allowed for the build-up of specific know-how and production techniques
Phase II	1946–1960	Describes the transfer of knowledge from military control to civilian agencies, all while emphasizing the military's role as an important amplifier towards reactor development for their own uses, but also for electricity production. Moreover, this phase includes the emergence of (public) organizations and private actors that aim to expand nuclear power for energy production. Finally, the technological uncertainty and experimentation with different reactor technologies are illustrated
Phase III	1961–1971	Describes the brief time period of LWR diffusion in the electricity sector, when many orders were placed and an increasing number of reactors were constructed. This development is known as the "bandwagon market." Other reactor technologies remained in their respective technological niches
Phase IV	1972–1996	Describes how the growth of nuclear power deployment diminished and multifaceted challenges for the industry occur as a result of ST-dynamics and landscape events. This also affects the niche developments facing budget reductions and project cancellations
Phase V	1997–2024	As LWR technology continues to stagnate and decline, attempts are made to reverse this trend by pushing new builds of inferior, but dominant, reactor technologies, extending the lifetimes of the existing fleets, and renewing attempts to develop other reactor technologies, all while facing some unsuccessful attempts

Source Own depiction

diffuse into the ST-system. There are "niche markets" in which innovative designs can be deployed without markets pressure, for example the military sector or under substantial governmental funding (Kanger 2021).

At the landscape level, the analysis focuses on external events that occurred outside of the US or in a broader global context, and which impact the ST-system's priorities towards electricity and military sectors. This can lead to changes in the techno-economic or sociopolitical environment for reactor development, which translates into formulated laws, and changes in funding, as well as the establishment or cancellation of reactor projects.

By taking this long-term and multi-dimensional perspective, it enables us to show the mechanisms that drove or hindered the innovation and diffusion process of reactor technologies in the US.

11.3 Phase I (1939–1945): Invention and First Application in Military Sector

The discovery of the fission process swept over to the US in 1939 when Niels Bohr attended a nuclear physics conference in Washington, DC (Gosling 1999). This had profound socio-technical consequences, especially at a time when the landscape event of WW II pressured the US (as well as other countries) to make national security an urgent and top political priority. Thus, the discovery of the nuclear fission process and its potential use as a weapon were focused upon the US and elsewhere. For example, in Germany, the "Uranverein" (uranium association) was set up in 1939 to explore the necessary resources for military uses of nuclear power (Walker 2024). Until 1942, initial US research was organized into several civilian institutions, both decentralized and in secrecy (Reed 2013). The US Navy was also interested in nuclear power as an alternative to diesel propulsion of its submarines. After a meeting with Enrico Fermi in March 1939, Ross Gunn, a civilian physicist at the Naval Research Laboratory (NRL), initiated research on uranium separation (Ahern 2003). However, the efforts of the US Navy remained small-scale and competed with the grand scale of the US Army's Manhattan Project (Ahern 2003; Reed 2020a).

The uncertainties about national security grew after scientists feared that the Germans would develop an atomic bomb. Thus, pressures to develop a US nuclear bomb increased. After presidential approval, the Manhattan Project was established in 1942 (Reed 2013). Subsequently, a military niche led by the US Army was created. It resulted in large investments in nuclear research; the production of fissile materials, including large-scale facilities and national laboratories; and even the construction of whole new cities. The goal was to pool the best national and international researchers under the leadership of J. Robert Oppenheimer. This concentration allowed monitoring by the US Army at a limited number of locations (Groves 1983; Rhodes 2012; Di Nucci 2019). The apparatus included many actors, including scientists, military staff, private suppliers, and public authorities. The whole project was shrouded in secrecy and was given a high national priority due to the perceived urgency (Groves 1983; Bird and Sherwin 2006; Rhodes 2012; Reed 2013).

Thus, the project required research and industry infrastructure. For example, the Metallurgical Laboratory at the University of Chicago, which later became the Argonne National Laboratory, was established to demonstrate a nuclear chain reaction. This was achieved at the so-called Chicago Pile 1 (CP-1) (200 W) under the supervision of Enrico Fermi on December 2, 1942 (Cochran et al. 2010). Furthermore, the bomb required fissile material. Two pathways to a bomb were developed, ultimately leading to the construction of two major fissile-material production complexes: The "Clinton Engineering Works" for uranium enrichment to isolate uranium-235 near Oak Ridge, Tennessee, and the "Hanford Engineering Works" that constructed reactors for plutonium separation in Hanford, Washington. Smaller amounts of plutonium were produced by the air-cooled and graphite-moderated reactor X-10 (1000 kW), which went critical on November 4, 1943, at the Oak Ridge National Laboratory. The main purpose was to separate plutonium and to test and train personnel (Reed 2021; Rosenthal 2010). However, larger amounts of plutonium were required for the production of a bomb. Thus, the company, DuPont, constructed several higher capacity reactors at the Hanford site, e.g., reactor B (250 MWth), a water-cooled, graphite-moderated reactor fueled by natural uranium (DOE 2017; Taylor and Freer 2002). Finally, to design and test the bomb, a new city was created in Los Alamos, New Mexico to incorporate a secret laboratory. In total, US$2.3 billion[2] were spent on the Manhattan Project, and thousands of people were recruited (Groves 1983). At the Hanford site, an average workforce of 22,500 people was maintained and peaked at 45,000 in 1944. At the Clinton Site, the number of construction workers peaked at 45,000 in 1944, and by 1945, the Clinton Engineering Works employed 80,000 people (Reed 2020b).

During that time, the national innovation system had a close interaction and cooperation between science and industry that was embedded into military secrecy and control. Fueled by a mission to serve the highest geopolitical priority, many resources were pooled. This ultimately led to the rapid advancement in knowledge, skills, and production techniques (Di Nucci 2019). The Manhattan Project fulfilled its purpose: A design for a bomb was developed, sufficient fissile material was produced, and a prototype was tested. This resulted in the first use of fission as a weapon in the bombings of Hiroshima on August 6 and of Nagasaki, Japan on August 9, 1945. In response, WW II ended five days later (Reed 2013).

The Manhattan Project caused the ST-system of the US to make tremendous investments into nuclear power, thus, facing high sunk costs. This was used to continue the production of nuclear weapons at established sites. However, during the Manhattan Project, research was already directed towards other applications of nuclear power. For example, in addition to an informal group of scientists, including Leo Szilard, Author Compton, Enrico Fermi, Eugene Wigner, and others, both General Groves and Colonel Nichols pressured for the "peaceful" use of nuclear power after the war had ended. For this, the provision of expertise and know-how created during the Manhattan Project had to be preserved. So, in April 1945, work began at the Chemical Division of the Metallurgical Laboratory (today: Argonne National Laboratory) to develop a device to generate "peacetime electricity" under the supervision of Farrington Daniels who had attended several meetings of the above-mentioned group of scientists. The efforts involved civilian industry actors from Monsanto,

[2] Assuming 1945 values and adjusting for inflation to 2024, this corresponds to approximately US$40.1 billion.

General Electric, Allis-Chalmers, and Westinghouse (O. B. Daniels 1978, 222–226). The so-called "Daniel's Pile" was designed as a high-temperature reactor cooled by helium and moderated with beryllium-oxide, allowing for outlet temperatures between 1,500 and 2,000 °C (F. Daniels 1944).[3] Furthermore, a scientific committee under the lead of Richard Tolman suggested governmental funding for the experimentation of using fission to propel submarines and ships and provide electricity to military sites and remote cities and settlements, e.g., in Alaska (Tolman's Committee 1944).

The ST-system of the US became aligned towards producing fissile materials via fission reactors and the design and construction of nuclear weapons. Through the MLP lens, these military activities can be characterized as niche developments that have resulted from the pressure of the landscape event of WW II under full governmental funding.

11.4 Phase II (1946–1960): Shifting Focus Towards Civilian Purpose and Technological Uncertainty

After the war ended in 1945, the deadly power of nuclear power had been shown, and remaining nuclear advocates aimed to demonstrate that nuclear power could also be used for "peaceful" purposes. The military niche underwent a major restructuring of nuclear oversight to enable nuclear power to potentially diffuse into the ST-system as a commercial technology for energy production. The Atomic Energy Law of 1946 shifted military control towards a newly established governmental agency, the Atomic Energy Commission (AEC), which would later split into the Department of Energy and the Nuclear Regulatory Commission. The AEC was given the oversight and control of plants, equipment, and research facilities of the Manhattan Project (Buck 1983). This includes national laboratories like Los Alamos National Laboratory (LANL), Argonne National Laboratory (ANL, near Chicago, previously named Metallurgical Laboratory), and the Oak Ridge National Laboratory (ORNL, previously called the Laboratories at Clinton Engineering Works) (DOE 2017).

Subsequently, the national laboratories broadened their research to incorporate reactor development for "peaceful uses," although military use cases remained a major focus. Thus, the AEC managed and funded almost all niche developments in which various reactor technologies evolved. For example, initial designs for reactors with a fast neutron spectrum were drafted. These reactors would use less uranium resources due to their higher utilization rates and would have the ability to breed plutonium (Cochran et al. 2010). Furthermore, it was expected that this type of reactor would lead to unlimited energy that would be "too cheap to meter" (Strauss 1954, 9). This technology route was explored in parallel to the cancellation of the

[3] After technical challenges amassed, and the Atomic Energy Commission (AEC) had taken over from the Manhattan Committee, in 1947, the project was scrapped without having been built, despite protests from the team and industry representatives (O. B. Daniels 1978).

Daniel's Pile. The Clementine Experimental Reactor, a mercury-cooled fast neutron spectrum reactor (25 kWth) at Los Alamos, became critical at the end of 1946. The development and construction of the first sodium–potassium-cooled experimental reactor, the EBR-I (200 kWe), was authorized at the National Testing Station in 1947 with the goal to breed plutonium and to generate electricity (Cochran et al. 2010). In 1951, the EBR-I reached first criticality and lit four 200-W light bulbs, showing its capability to generate electricity. Moreover, the feasibility of plutonium breeding was announced by the AEC in 1953. However, the reactor was closed in 1963 after a partial core meltdown during an experiment (Cochran et al. 2010).

The military sector continued to play an important role as amplifier for reactor development. The US Navy aimed to develop a reactor for submarine propulsion to extend the duration of dives compared to combustion propulsion. However, as described above, initial priority was given to the development of a nuclear bomb. In 1946, following the end of WW II, the naval propulsion program of the US Navy, led by Hyman Rickover[4] was initiated, and the establishment of a dedicated research facility followed. Previous research had been conducted at the Naval Propulsion Laboratory. As part of the new program, the Bettis Atomic Power Laboratory was founded in 1949 and operated by Westinghouse, which designed the first LWR for submarine propulsion (Hewlett and Duncan 1974). Due to uncertainty about which reactor design would be best suited for nuclear propulsion, two designs were developed in parallel, with the goal of becoming operational as quickly as possible. The target date was set to January 1955 (Hewlett and Duncan 1974). Two designs emerged: Westinghouse's LWR technology, and a sodium-cooled and beryllium-moderated reactor developed by General Electric (S2G). As of 1950, it had become clear that the sodium-cooled reactor design (also referred to as MARK A) was experiencing delays compared to Westinghouse's designs (referred to as MARK I and II). So, in 1954, the first nuclear-powered US submarine, the "Nautilus," went into operation with an LWR (Hewlett and Duncan 1974).

The time of experimentation for different reactor technologies and the shifting focus toward energy production was embedded into the context of the emerging landscape event of the Cold War. Thus, national security remained a high political priority. In 1949, the detonation of the first Soviet nuclear bomb displayed another landscape event that showed the Soviet Union would also progress with nuclear power and that a technological monopoly for the US was not guaranteed. Moreover, a third overlapping landscape event occurred with the Korean War in the following year. In response, the US-Soviet tensions increased and created a state of "national emergency" that led national policy to demand more nuclear weapons. Subsequently, the production of more fissile material with additional Hanford-type reactors grew (Hewlett and Duncan 1972). In turn, the budget requests of the AEC increased threefold from 1950–1954 onwards to cover the expansion of facilities (Hewlett and Holl 1987).

[4] Rickover was an engineer-turned-naval officer who had previously worked on the Daniel's Pile (O. B. Daniels 1978).

With the increased focus on military aspects, the AEC's mission to demonstrate the feasibility of reactors for energy production stood in competition with plutonium production and naval propulsion. Even more so, the new Eisenhower Administration in 1952 had not only strong priorities towards nuclear defense, which required more fissile material for additional weapons, it also wanted to reduce federal expenditures. As a result, both military and non-military projects were assessed, and the budget for the AEC envisaged for 1954 was cut short by US$800 million.[5] This included a cut of US$176 million[6] of the budget for reactor development facilities (Hewlett and Holl 1987). In response, the AEC cut its reactor development program to include only one reactor project: a sodium-cooled, graphite-moderated reactor (SGR),[7] which was considered to be the nearest to actual operation (Hewlett and Duncan 1974). However, the Eisenhower administration cancelled this envisaged project. Military projects were also subject to budget cuts. A potential candidate for cancellation was the US Navy's nuclear propulsion program for aircraft carriers. A propulsion reactor for aircraft carriers had to be more powerful than those designed for the submarine, and thus, it could also be interesting for the application of energy production. Westinghouse had begun designing such a reactor in 1950 (Hewlett and Duncan 1974). Due to the budget reduction, this prototype was at stake. This led to intense debates between the AEC, the National Security Council, and the Eisenhower Administration. After the SGR proposal of the AEC had already been cancelled, Westinghouse's land-based LWR prototype was the only reactor project that they could hope for to be used for energy production in the near term.

As a compromise, it was suggested to cancel the aircraft carrier project and choose a project that was less costly and would demonstrate the "peaceful" use of the atom (Hewlett and Duncan 1974, 199). As a result, in 1953, the project for reactor propulsion in aircraft carriers was canceled, and instead, a "smaller" project was prioritized. Subsequently, the members of the AEC proposed a cheaper reactor project based on the Westinghouse LWR design for the aircraft carrier (Hewlett and Holl 1987). With President Eisenhower's support, the project's construction was initiated in the summer of 1953. Consequentially, the technological decision to follow through with the Shippingport project was made even before the first nuclear submarine, the Nautilus, went into operation in 1954. Only limited experience had been gathered with LWR technology, but compared to other projects, Westinghouse's MARK I and MARK II designs had progressed faster. The LWR technology also had the advantage of showing a non-dual-use reactor to secure American leadership to control nuclear deployment nationally and internationally for "peaceful" uses (Eisenhower 1953; Cowan 1990). Finally, under the lead of Rickover, in cooperation with Duquesne

[5] Roughly US$9.3 billion as of 2024.

[6] Roughly US$2 billion as of 2024.

[7] By 1953, several experimental facilities were constructed at national laboratories. For energy production, the AEC considered various designs but faced considerable technological uncertainty. However, they proposed to finance a sodium-cooled and graphite-moderated reactor (75 MWe) (Hewlett and Holl 1987).

Light Company of Pittsburgh and Westinghouse, Shippingport was contracted in 1953 (Hewlett and Holl 1987).

After the Shippingport project was established, Eisenhower's speech, "Atoms for Peace," pressed for the fast commercialization of nuclear power, and the Atomic Energy Act of 1954 followed. This law further adapted the institutional framework to encourage private-sector investment. It facilitated private involvement by declassifying reactor technology information; providing access to research and development infrastructure; and offering financial support through a newly established development program by the AEC, the Power Reactor Demonstration Program (PRDP). Under this program, the AEC invited the private industry to place bids for demonstration reactor projects (Hewlett and Holl 1987).[8]

Between 1955 and 1957, three financing rounds of the PRDP were realized in which different reactor types were funded. The proposals mainly aimed to construct LWR reactors, although some other reactor types were funded, in particular the SFR (Fermi-I)[9] and an HTR (Peach Bottom) (Allen 1977). As Table 11.2 shows, most of the initial orders between 1955 and 1959 were already announced even before Shippingport began operating. Therefore, order decisions could not be based on the project cost data or operational experience. In total, thirteen nuclear power plants had been ordered by 1959 by several utilities and different contracted reactor vendors, most with LWR technology and very low capacities.

11.5 Phase III (1961–1971): Diffusion via the Bandwagon Market and Acceleration of LWR

By the early 1960s, it had become clear that lower capacity LWRs had such high capital costs that orders of reactors rapidly decreased. As a response, utilities and manufacturers pressured the AEC to expand the PRDP program and to include subsidies for higher-capacity reactors (>400 MWe) in order to take advantage of economies of scale (reducing costs per power capacity) (Burness et al. 1980). Subsequently, the AEC used parts of the already appropriated budget in 1962 to fund two reactors under this program: Connecticut Yankee (aka Haddam Neck Plant), 575 MWe, ordered in 1962; and San Onofre 1, 430 MWe, ordered in 1963 (ERDA 1975; Burness et al. 1980). Both reactors were delivered by Westinghouse under a turnkey contract selling the reactor at fixed prices (153 US-$/kW and 180 US-$/kW)[10] to the utilities.

The nuclear reactor vendors expected cost reductions through learning by doing, and offered NPPs at intentionally low prices to compete with fossil fuel alternatives

[8] Already in 1951, the AEC began sharing some classified information with four industry groups to study the development of different reactor types on paper, which had shown that governmental support would be required for private participation (Hewlett and Duncan 1972).

[9] Fermi-I aimed to breed more fuel than consumed and was a project of a consortium of many utilities.

[10] Approximately 1,586 US-$/kW and 1,843 US-$/kW as of 2024.

Table 11.2 US orders of NPPs between 1955–1959

Project	Owner	Reference unit power (MWe)	Type	Contractor	Public announcement	Contract award	Initial cirticality	First grid connection
Shippingport Atomic Power Station	Duquesne Light Company and AEC	90	PWR	Westinghouse, Stone & Webster (S&W)	October-1953	July-1953	December-1957	December-1957
Indian Point Station, Unit 1	Consolidated Edison Co	265	PWR	Babcock & Wilcox (W&B)	February-1955	February-1955	August-1962	September-1962
Dresden Nuclear Power Station, Unit 1	Commonwealth Edison Co	200	BWR	GE, Bechtel	April-1955	February-1955	October-1959	April-1960
Yankee Nuclear Power Station	Yankee Atomic Electric Co	175	PWR	Westinghouse, Stone & Webster (S&W)	April-1955	June-1956	August-1960	November-1960
Enrico Fermi Atomic Power Plant, Unit-1	Power Reactor Development Company	60.9	SFR	Power Reactor Development Company, Commonwealth Associates	April-1955	March-1957	August-1963	August-1966
Hallam Nuclear Power Facility	Consumers Public Power District and AEC	75	SGR	Atomics International, Bechtel	April-1955	September-1957	August-1962	May-1963
Elk River Nuclear Plant	Rural Cooperative Power District and AEC	22	BWR	Allis-Chalmers, Sergent and Lundy (S&L)	February-1956	June-1958	November-1962	August-1963

(continued)

Table 11.2 (continued)

Project	Owner	Reference unit power (MWe)	Type	Contractor	Public announcement	Contract award	Initial circiticality	First grid connection
Piqua Nuclear Power Facility	City of Piqua, Ohio and AEC	11.4	OMR[11]	Atomics International, Holmes and Narver	February-1956	June-1959	June-1963	November-1963
Pathfinder Atomic Power Plant	Nothern States Power Company	58.5	BWR	Allis-Chalmers, Pioneer Service and Engineering	February-1957	May-1957	March-1964	July-1966
Carolinas-Virginia Tube Reactor	Carolina-Virginia Nuclear Power Associates, Inc	17	HWR	Westinghouse, Stone & Webster (S&W)	August-1957	January-1959	March-1963	December-1963
Peach Bottom Atomic Power Station, Unit 1	Philadelphia Electric Company	40	HTR	General Atomic Corp., Bechtel	January-1958	November-1958	March-1966	January-1967
Humboldt Bay Power Plant, Unit 3	Pacific Gas & Electric Company	65	BWR	GE, Bechtel	February-1958	February-1958	February-1963	April-1963
Big Rock Point Nuclear Plant	Consumers Power Company Michigan	75	BWR	GE, Bechtel	December-1959	December-1959	September-1962	December-1962

Source Compiled from ERDA (1975)

[11] OMR: organically cooled and moderated.

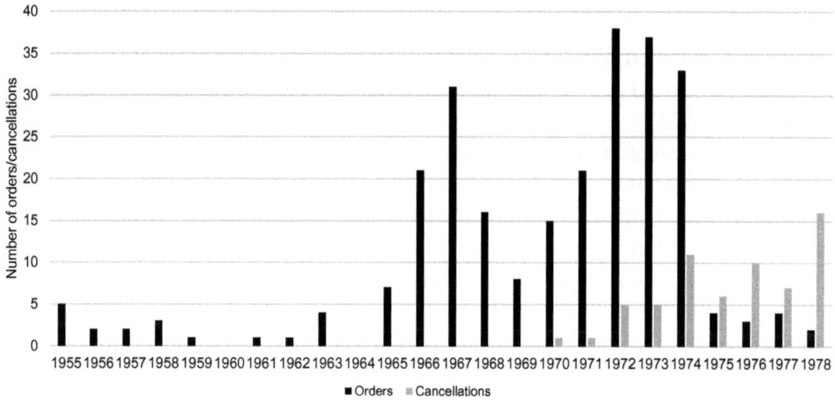

Fig. 11.3 Orders and cancellations of reactor projects in the US (1955–1978) (*Source* Own depiction based on Burness et al. [1980])

(Cohn 1990). In response, this triggered several utilities to order reactors, in particular, after the announcement that General Electric (GE) would set a fixed price at 132 US-$ /kW[12] at Oyster Creek, which was cheaper than comparable coal plants in this area (Burness et al. 1980). During this turnkey era between 1962–1966, the two main suppliers for reactors in the US, GE and Westinghouse, sold 13 plants and were able to serve an increasing demand for nuclear power plants. This newly emerged market is also referred to as the "Great Bandwagon Market" (Burness et al. 1980; Cowan 1990). By 1967, 45 GWe of nuclear capacity had been ordered, while less than 2 GWe were in operation—the orders were thus based on very limited empirical cost data availability. As reactor costs surged, GE and Westinghouse suffered from major economic losses estimated at US-$800 million to 1 billion[13] (Cohn 1990). Therefore, GE withdrew from turn-key contracts in 1966 and Westinghouse followed soon after (Burness, Montgomery, and Quirk 1980).

After several years of decreasing order numbers, they grew again between 1970 and 1972 (see Fig. 11.3). Burness et al. (1980) argue that the expectations of the utilities were that nuclear would become competitive with the (fossil) alternatives very soon. However, the number of orders created bottlenecks in supplying components and regulatory procedures. The upscaling of reactor capacity was designed to bring costs per capacity unit down, but it also created new engineering problems. As a result, costs further increased and delays occurred. The Great Bandwagon Market peaked in the mid-1970s. In total, 259 reactors were ordered, but faced increasing cancellations beginning in the 1970s (Burness et al. 1980; Jurewitz 2002).

Despite the diffusion of LWRs in the US, other reactor technologies tried to compete. In the late 1960s, GE attempted to enter the market with a high-temperature reactor (HTR). By 1975, GE had orders for seven HTR power plants, of which one

[12] Approximately 1,352 US-$/kW as of 2024.

[13] Approximately US-$ 7.7 to 9.6 billion as of 2024.

330 MWe reactor came into operation at Fort St. Vrain in 1976. However, all other orders were canceled. Fort St. Vrain remained the only larger-scale HTR in the US. The plant was closed in 1989 and has been refurbished into a gas power plant (Cowan 1990; IAEA 2022).

In parallel, research and development of other reactor concepts went on at the National Laboratories, in particular the sodium-cooled reactor with a fast neutron spectrum (SFR). One expected that the SFR would substitute the LWR technology and the US would move towards a so-called "Plutonium Economy" (Seaborg 1970; AEC 1974). This would be enabled through "breeding," in which an SFR would breed plutonium that, after costly reprocessing and fuel production, could be reused (Bunn et al. 2005). Building upon the EBR-I, the EBR-II (20 MWe, critical in 1963) was also developed by ANL at the National Reactor Testing Station. This research facility included reprocessing in a Fuel Cycle Facility (FCF) that fabricated fresh fuel from 1964 to 1969. However, in 1967, the EBR-II was redesigned to become an experimental irradiation facility (Cochran et al. 2010).

Moreover, the private industry, with the support of the AEC, aimed to construct and operate an SFR called Enrico Fermi Atomic Power Plant or Fermi 1 (66 MWe). 34 companies participated in the Power Reactor Development Company (PRDC) under the funding of the AEC's PRDP of 1955 to show that more fuel can be bred than it consumed. The leader of the project was Cecil A. Cisler, the CEO of Detroit Edison. As a former AEC member, he had been involved in the development of SFRs. The construction of Fermi 1 started in 1959, and the first criticality was achieved in 1963. However, a partial meltdown occurred in 1966. After an initial repair, technical problems with sodium remained. Furthermore, the reactor showed a very low average capacity factor of 3.4%, and thus, the extension of the operation license was denied, and the plant was closed in 1972 (Cochran et al. 2010). Parallel research was conducted on other concepts, such as molten salt reactors (MSRs), e.g., the Molten Salt Reactor Experiment at Oak Ridge National Laboratory with 8 MWth, also operating from 1965 to 1969 (Cochran et al. 2010).

As shown above, the LWR technology dominated the deployment of reactors, and others stayed in their niches at different levels of "readiness." LWR vendors, as well as the utilities, had hoped to reduce costs and thus created an environment of expectations, self-reinforcing, and cheaper prices by taking economic losses. The diffusion of LWR can only be partially explained by the MLP rationale, where a landscape creates a window of opportunity for a niche to break through. Instead, its diffusion was primarily driven by a ST-system that had nested a strong (military) niche. Therefore, the diffusion is characterized by strong political funding and growing industry commitment. Changes in capacity and orders reflected expectations that the technology would eventually integrate into the system and that costs would decrease due to economies of scale and learning effects. However, instead of decreasing, costs increased, along with project cancellations.

11.6 Phase IV (1972–1996): Stagnation and Decline

In 1970, a flurry of project cancellations occurred. The expectation of cheap electricity from nuclear power had increasingly faded away. Instead, the planning of new NPPs was delayed in many cases and often canceled. New orders decreased after peaking in 1972 and completely stopped by 1980. From 1970 to 1999, 124 projects were cancelled (Jurewitz 2002). This marked a significant shift from the previous growth phase within the Great Bandwagon Market, which had faded quickly by the mid-1970s. Figure 11.4 shows the dynamic of reactor construction projects in the US until the last project that was fully connected to the grid in 2024, the Vogtle project in Georgia. This only shows projects for which construction operations had begun; projects that were terminated in earlier project stages are not included.

The reasons for the reversed trend are multifaceted and include internal dynamics of the ST-system and the destabilization through landscape events. The fast adoption of the ST-system to LWR technology was realized by expanding orders of reactors with increasing electrical capacity. This posed new challenges for materials and incurred increased safety measures. These changes pressured the AEC, which had limited personnel able to conduct licensing, causing delays in the licensing processes. At the time, the licensing procedure was non-standardized and conducted on a case-by-case approach (NRC 2024).

Furthermore, public opposition increased as the public, the Congress, and the State and Federal agencies were increasingly concerned by the thermal pollution of rivers from nuclear power plant (NPP) operations. Consequently, the AEC was pressured to limit thermal pollution by changing requirements for NPP licensing to add cooling towers and ponds, further increasing construction costs. Moreover,

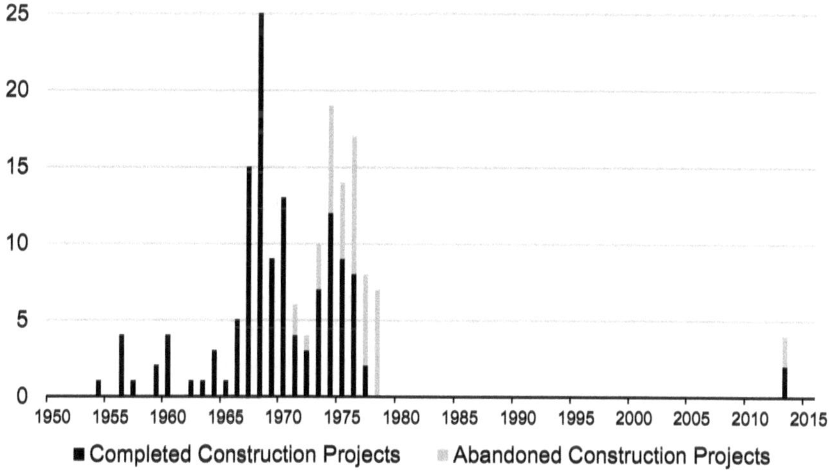

Fig. 11.4 Completed and abandoned reactor construction projects in the US (*Source* Own depiction with data taken from Schneider et al. [2024])

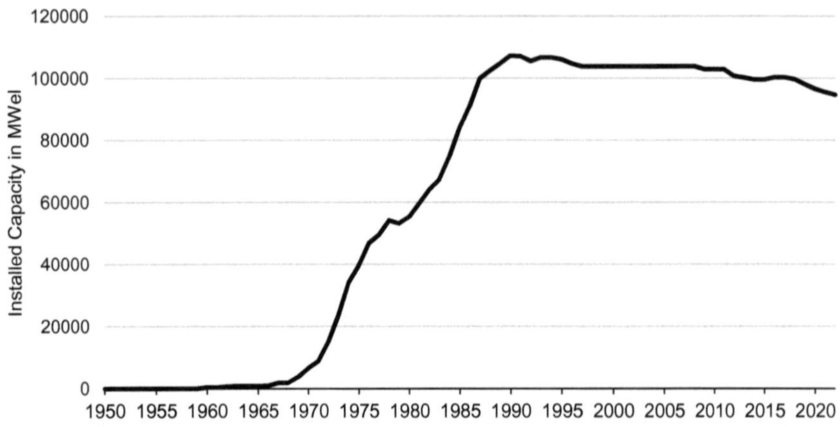

Fig. 11.5 Evolution of cumulative installed nuclear power capacity from 1955 to 2020 in the US (*Source* Own depiction with data taken from the IAEA's PRIS)[14]

the AEC faced increasing criticism regarding radioactive discharges from routine operations. Different social and environmental groups, as well as scientists, criticized AEC standards for operational discharges to be insufficient. This growing opposition regarding nuclear power was embedded into larger environmental movements in the ST-system. As a result, the National Environmental Policy Act of 1970 mandated that Federal agencies assess the environmental impact of their activities. The AEC was reluctant to realize these assessments for NPPs. Environmental movements took the case to court for the Calver Cliff Plant. Although they lost the case, the public trust in the AEC decreased and the licensing process was stretched even further (NRC 2024).

These internal tensions of the ST-system led to the reorganization of the AEC, which was split into the Nuclear Regulatory Commission (NRC) and the Department of Energy (DOE) in 1973. In addition to sociopolitical pressures and institutional restructuring, the economic perspective was likewise challenging. As costs rose due to licensing delays and public opposition, the utilities also faced economic challenges resulting from the Oil Crisis of 1973 that increased electricity prices and reduced energy demand. Furthermore, as the utilities had built up overcapacities due to the rapid ordering of NPPs, they now faced economic challenges. Consequently, there were no additional NPP orders, and many others were canceled as project costs increased (Burness et al. 1980; Harris and Wynne 1989). From the mid-1980s, US nuclear power capacity had stagnated at around 90 NPPs and 100 GWe (see Fig. 11.5).

Parallel to the LWR diffusion, niche developments continued, but also faced cancellations. After EBR-1, Fermi-1 and the operational experiences of EBR-II, a demonstration project for fast reactors was planned. The Clinch River Demonstration Breeder Project (350 MWe) was authorized as a joint project of the AEC and several

[14] https://pris.iaea.org/PRIS/CountryStatistics/CountryDetails.aspx?current=US (last accessed on 13-11-2024).

utilities in 1969 and was majorly funded by the Federal Government. However, the project struggled with public and political opposition from 1972 onwards due to the high cost, proliferation risks, and safety concerns (Cochran et al. 2010). These challenges within the niche were deepened by an external landscape event in 1974, during which India tested a nuclear bomb using plutonium that had been separated with US assistance. In response, the proliferation concerns increased and led to the postponement of commercial reprocessing in 1977. Moreover, the cost continued to rise from an initial cost estimate of the AEC from US-$699 million in 1972 to US-$4 billion in 1983.[15] As a consequence, Congress cut off funding for the Clinch River project and halted the licensing process in 1983.

The EBR-II reactor and the reprocessing plant were converted to research facilities for the Integral Fast Reactor Program. For several years, ANL was able to acquire some government funding for the program. In 1994, the Clinton Administration discontinued the project and Congress cut all funding. The EBR-II was decommissioned that same year. The reprocessing plant was redesigned to condition radioactive waste (Cochran et al. 2010).

Following the incidents at Three Mile Island and Chernobyl in 1979 and 1986, respectively, public opposition against nuclear power deepened, and environmental protests increased. However, the number of reactor orders had already declined before these events occurred due to unfavorable economics (Jurewitz 2002). Finally, the funding for nuclear research had been reduced to zero in 1998 (Abdulla et al. 2017). The budget cuts for fission were part of general federal budget cuts during the decade that led to a significant reduction in DOE overall spending on R&D (Joskow 2001). As part of this shift, the DOE began reallocating its R&D resources away from coal and nuclear programs to efforts that worked toward energy efficiency and renewable energy technologies. The decline of nuclear power research and construction projects was complemented by broader energy policy of the 1990s, which saw the Energy Policy Act of 1992 and subsequent deregulation efforts aimed at restructuring the energy market to favor competition and price-driven energy supply (Joskow 2001).

11.7 Phase V (1997–2024): Efforts for a "Renaissance" in the Presence of High Technological Uncertainty

From the late 1990s, one could observe increased attempts to reemphasize the future role of nuclear power, labeled the "nuclear renaissance" by proponents (Jurewitz 2002, 208). Different factors are suggested in the literature, such as increasing fossil fuel prices and threatened energy shortages, shifting public attitudes, and increasing awareness of climate change (Jurewitz 2002; Nuttall 2022).

The latter displays the landscape event of climate change that requires increasing attention towards the reduction of carbon emissions, and can thus be characterized as a destabilization factor for the fossil-dominated energy sector of the US.

[15] Approximately US-$5.2 and 12.6 billion as of 2024.

In 1997, President Clinton requested a report from the Committee of Advisory on Science and Technology (PCAST) to revisit the national energy R&D portfolio. The report stressed global challenges, which included the increasing need to decarbonize energy systems, but also the prevention of nuclear proliferation, and the importance of maintaining global leadership in science and technology. Therefore, PCAST recommended an increase in budgets for fission and fusion alongside renewables and energy efficiency measures. Further, the DOE should establish a new initiative called "Nuclear Energy Research Initiative" (NERI) to reinvigorate R&D efforts (PCAST 1997). Moreover, geopolitical priorities regarding the role of nuclear power were also rearticulated: The US aimed to recover its leadership in nuclear energy, confronting not only energy security, but also competition from nations like Russia and China, which had become global leaders in nuclear construction projects (Schneider et al. 2024).

In response, there was political will to halt the decline of US nuclear power plant technology by incentivizing new construction projects for the dominant high-capacity LWR technology and to foster the development of alternative reactor designs on the niche level. Subsequently, laws were established to increase funding for public authorities and to establish new initiatives. In particular, the Energy Policy Act (EPAct) of 2005 provided significant financial support for the construction of new LWRs in the form of "cost-overrun support," production tax credits, and loan guarantees. It also included substantial funding for a gas-cooled HTR titled "Next Generation Nuclear Plant Project" and extended the Price-Anderson Act to 2025, the Federal back-stop for nuclear accident liabilities (U.S. Congress 2005, 794). Moreover, the Global Nuclear Energy Partnership Program was initiated in 2006 to expand research on nuclear reactors with a fast neutron spectrum, all while focusing not on breeding, but "burning" radioactive material (Cochran et al. 2010, 105).

As a result of the EPAct, 31 applications were made by several utilities for the construction of new nuclear power units, see Table 11.3.[16] However, only the Vogtle project in Georgia and the VC Summer project in South Carolina began construction. Moreover, the construction at the VC Summer units was discontinued after substantial cost increases, delays, and a corruption scandal.[17] The two units at Vogtle faced high cost increases, leading core vendor Westinghouse to file for Bankruptcy in 2017 (Schneider et al. 2018). Additionally, some existing reactors struggle to operate economically on competitive markets and have been taken off the grid before their license termination, or are backed up with specially designed subsidies (Bah 2023). As a result of the lack of newbuilds, the nuclear fleet is aging. There are also attempts to push inevitable reactor closures into the future by creating economic incentives for reactor lifetimes extensions, via, e.g., the 2022 Inflation Reduction Act (U.S. Congress 2022). This shows how the dominant and diffused technology struggles to stay in the ST-system.

[16] The Nuclear Energy Institute (n.d.) cited by Bob Bellemare (2007) New Nuclear Plants Coming to the United States? UtilitiPoint Daily IssueAlert 17-01-2007. Archived at https://web.archive.org/web/20070926225149/http://www.utilipoint.com/issuealert/article.asp?ID=2789.

[17] Refer to Welton and Harrison (2024) for a detailed account of the „Nukegate" scandal.

Table 11.3 Plans and applications for reactor new builds as a result of EPAct, as of January 2007

Company	Site	Design (No. of reactors)	Expected Begin of Construction and Operating License Procedure
Alternate Energy Holdings	Bruneau, Idaho	TBD	2008
Amarillo Power	Vicinity of Amarillo, Texas	ABWR (2)	After 2007
Constellation (Unistar)	Calvert Cliffs, Maryland, or Nine Mile Point, New York	EPR (5)	Q4 2007
Dominion	North Anna, Virginia	ESBWR (1)	November 2007
Duke	William States Less, Cherokee County, South Carolina	AP1000 (2)	October 2007
	Davie County, North Carolina	TBD	TBD
	Oconee County, South Carolina	TBD	TBD
Entergy	River Bend, Louisiana	ESBWR (1)	May 2008
	Grand Gulf, Mississippi	ESBWR (1)	November 2007
Exelon	Clinton, Illinois	TBD	TBD
	TBD, Texas	TBD	2008
Florida Power & Light	TBD	TBD	2009
NRG Energy/ STPNOC	South Texas Project	ABWR (2)	2007
Progress Energy	Harris & Levy County, Florida	AP1000 (2) & TBD (2)	October 2007 & July 2008
South Carolina Electric & Gas	VC Summer, South Carolina	AP1000 (2)	October 2007
Southern Company	Vogtle, Georgia	AP1000 (2)	March 2008
Texas Utilities	TBD, Texas	TBD (2–5)	2008
Tennessee Valley Authority (TVA)	Bellefonte, Alabama	AP1000 (2)	October 2007

Source Taken from UtiliPoint.[18] Abbreviations used: ABWR = Advanced Boiling Water Reactor (GE Hitachi Design); AP1000 = Advanced Pressurized Water Reactor (Westinghouse Design); EPR = European Pressurized Water Reactor (Areva/Framatome Design); ESBWR = Economic Simplified Boiling Water Reactor (GE Hitachi Design); TBD = To be determined

At the niche level, the development and planned construction of other reactor technologies are reemphasized in the DOE's national labs in cooperation with private industry on a cost-sharing partnership (U.S. Congress 2005, 910). So far, none of these projects have succeeded. The NGNP development project that was initiated in

[18] Bellemare, op.cit.

2005 was discontinued in 2011 after no significant advances (INL 2011, 38). The DOE also promotes the deployment of so-called “Small Modular Reactors” (SMR), i.e., reactors with a capacity of less than 300 MWe (Böse et al. 2024). In particular, the mPower reactor, proposed by Babcock & Wilcox, and the NuScale VOYGR each received grants of up to US-$226 million in 2012 and 2013, respectively. However, both projects have failed. Babcock & Wilcox struggled to find investors, and the project was terminated in 2017 (Thomas and Ramana 2022). Nuscale proceeded with the design and received a NRC license in 2023 for a plant consisting of a maximum of twelve individual units of 50 MWe each. It is pursuing a new design license for a module capacity upgrade to 77 MWe. NuScale had partnered with Utah’s Associated Municipal Power Systems in the so-called “Carbon Free Project” to construct a reactor in Idaho. However, the project was canceled in November 2023 for economic reasons (Böse et al. 2024).

Nevertheless, experimentation with reactor designs has reemerged. Similar to the 1950s, the DOE (former AEC) seeks to develop reactors for energy production in conjunction with private industry. Thus, in 2020, the DOE initiated the “Advanced Reactor Demonstration Program” (ARDP) with the goal of funding two demonstration projects that could begin operation in 2035 (DOE 2020, 1). Proposals from newly founded companies dedicated to developing alternative (non-light water) reactor designs were made and have received funds. In particular, these are X-energy for an HTR and TerraPower for an SFR. In addition to the DOE’s commitment, the federal energy policy decided to provide additional funds via the “Bipartisan Infrastructure Law” that was passed in late 2021. This new legislation allows the ARDP to spend US-$3.2 billion on the two demonstration projects (Pathway 1, DE-FOA-0002271) from 2022 to 2027 (U.S. Congress 2021). In addition, the DOE is allowed to request additional funds from Congress.

Finally, the military niche continues to play a role in the development in reactor technologies, although their involvement is mostly limited to energy provision at military sites. In 2009, the National Defense Authorization Act called for a feasibility study on deploying SMRs at military bases, further highlighting the US military’s interest in nuclear energy for strategic needs.[19] Subsequently, military departments are looking for SMR reactor deployment for remote areas, including the US Department of Defense (DoD), which initiated Project Pele to support the construction of a prototype LWR reactor design called BWRx (1–5 MWe) at INL.[20]

---

[19] https://ndupress.ndu.edu/Portals/68/Documents/stratforum/SF-262.pdf (last accessed on 13-11-2024).

[20] https://www.defense.gov/News/Releases/Release/Article/3915633/dod-breaks-ground-on-project-pele-A-mobile-nuclear-reactor-for-energy-resiliency/ and https://www.world-nuclear-news.org/articles/work-starts-at-us-prototype-microreactor-site.

11.8 Conclusion

Nuclear reactor development has been full of non-linearities, unexpected outcomes and unfulfilled expectations since the 1940s. This is particularly true for the earliest and largest country of reactor developments—the US. In this chapter, (almost) a century of reactor development in the US is analyzed via the multi-level perspective and by focusing on technological niches, socio-technical system, and overarching landscape developments. The analysis helps to understand the discrepancy between the high expectations of the 1940s and '50s, the relatively short growth phase that followed, the stagnation and decline in the 1970s, and the current attempts to revitalize the sector (while relying on reactor concepts of the 1940s).

The analysis shows that the development of US nuclear power reactor technology was shaped by a combination of unique external landscape events that shifted political priorities and created or increased internal challenges, such as escalating costs, delays, and public opposition. The conceptual approach identifies key periods that can be distinguished by special events relating to innovation (or absence thereof) in the socio-technical system. Table 11.4 gives a brief overview. It was find that in Phase I (1939–1945), the establishment of the Manhattan Project during World War II created a powerful niche for nuclear power, driven by significant investments in large industrial facilities to produce fissile materials, research, and a pool of highly skilled labor.

In Phase II (1946–1960), the nuclear niche evolved from the military to a civilian-controlled niche overseen by the AEC. Regardless, in the early 1950s, the development of reactors for commercial application was dominated by military demands for nuclear weapons, submarines, and aircraft carriers. The US Navy's strong interest in nuclear propulsion drove the near-term deployment of reactors, where the first nuclear-powered submarine was put into operation in 1954. The Cold War and national reorientation towards cutting the AEC's budgets and focusing on weapons manufacturing further limited the progress of commercial development of reactors. However, the geopolitical "Atoms for Peace" strategy created enough political urgency and political will to establish the Shippingport reactor based on LWR technology originally intended for marine propulsion. During that time, technological uncertainty prevailed, and research and development were conducted for various reactor types. However, the LWR emerged as the dominant player in the move of nuclear power into the ST-system of the US energy market.

Phases III (1961–1971) and VI (1972–1996) mark a change of technological strategy, with the light-water bandwagon taking off (Phase III) and breaking down in Phase IV, as early as the mid-1970s. Phase III includes the short period of time for the diffusion of the LWR technology during the Bandwagon Market throughout the 1960s. Reactor vendors Westinghouse and GE took economic losses by selling reactors at fixed prices. In turn, the private industry became increasingly excited about nuclear power, hoping that it would become economically competitive in the

Table 11.4 Overview of Phases I–V

Phase	Landscape	ST-System characteristics	Niche
Phase I (1939–1945)	WW II and Manhattan Project	– Buildup of large-scale research infrastructure and production facilities for fissile material production including whole new cities – Led by military-state involvement, including many scientists and workers and families – Shared vision of national security priority and secrecy and urgency	– Research at labs of different universities and of national institutions; – Establishment of the Manhattan Project
Phase II (1946–1960)	Cold war and technical competition between the US and the Soviet Union	– Gradual conversion of military research labs to commercial services – Focus shifts mainly on electricity production, some other applications, e.g. medical, material sciences – Privately managed R&D systems require substantial public funding to survive	– Competing reactor technologies – Research in large "national labs," e.g., non-light water reactors – Some research on military applications remaining in niche
Phase III (1961–1971)	No particular landscape events	– Private reactor firms establish (loss-making) business models – Light-water (unexpectedly) becomes dominant reactor design – Bandwagon market is established between mainly private reactor vendors and utilities	– Non-light water reactors remain in the niche (mainly R&D in national labs)

(continued)

Table 11.4 (continued)

Phase	Landscape	ST-System characteristics	Niche
Phase IV (1972–1996)	Breakthrough of natural gas, restructuring of electricity markets	– Stagnation of installed capacity – Decline of orders and increase of project cancellations – Non-light water R&D also in crisis after abandonment of "fuel cycle"	– Niche developments decline after project cancellations and budget cuts – Some fundamental research on "advanced" reactors
Phase V (1997–2024)	Climate change considerations	– Attempts to revive the S&T system through systematic subsidy schemes, e.g. EPAct (2005) – DOE employs additional efforts to develop "SMR concepts," i.e., light water reactors with low power ratings – National research labs continue some fundamental research	– Reemphasize on diverse reactors demonstration programs to enhance public–private partnerships – Military R&D still plays a certain role as a "niche"

Source Own depiction

near future. In parallel to the diffusion process of the LWR, there were also expectations that reactors with a fast neutron spectrum would supersede the LWR once they became available, and that the "plutonium economy" would establish itself. However, Phase IV shows that the diffusion phase entered a saturation phase in which orders declined and installed capacity stagnated. Ongoing projects experienced cost increases and delays as well as increasing public opposition due to concerns regarding operational safety and thermal pollution. The short growth phase had revealed the technology's misfit with the existing ST-system evident in cost challenges, regulatory limitations, and public opposition. These factors contributed to a stagnation in nuclear development, leading to a decline of orders of LWR and non-LWR projects and R&D funding and nuclear industry since the mid-1970s.

The final Phase V (since 1997) shows the diverse attempts to halt the decline of nuclear power and reemphasize technological experimentation of different reactor technologies at the niche level. Despite the significant political push of the 2005 Energy Policy Act, which aimed to reinvigorate nuclear expansion, the decline of US nuclear technology continued, underscoring the deep-rooted challenges in revitalizing a declining industry. The recent efforts to (re)promote various reactor concepts, e.g., SMRs and non-light water reactor concepts, reflect the policy push towards maintaining or even expanding the nuclear fleet to reassert leadership in the global nuclear technology market. It also shows the technological uncertainty resulting from the decline of high-capacity LWRs and their potential replacement. Nuclear power remains a highly complex and capital-intensive system, with high sunk costs and stringent safety requirements that create a strong path dependency. As such, nuclear energy can be characterized as a "lock-in" technology, where past investments and decisions continue to shape its future, presenting both opportunities and constraints for innovation and expansion.

The MLP is a useful framework to show that in the US, reactor technologies had experienced dynamics which can be classified as niche-competition that led to a subsequent diffusion of a technology. However, the classic reasoning of MLP, that diffusion becomes possible because landscape events disrupt existing ST systems and thus open up a "window of opportunity," cannot be directly applied. Instead, the landscape events of WW II clearly disrupted the ST system and nested a unique and powerful niche, which resulted in mass destructive weapons. However, the development of reactor technologies for energy production, and the diffusion into the ST-systems competing with other energy producing technologies, was clearly pushed politically, as well as by some private industry actors, all while taking high economic risks. Thus, the diffusion was initiated by a hopeful industry and reactor vendors who had gained experience in the military niche and took economic losses in expectation of decreasing costs.

Overall, the reactor technology development in the US was shaped by geopolitical context, strong government funding, and the sunk cost of the Manhattan Project. The Cold War further fueled the desire for nuclear leadership in the geopolitical context, creating a strong political will for nuclear power. However, the diffusion process, led by private industry accepting losses, ultimately revealed the misfit between the technology and the ST-system. This resulted in stagnation and a decline in the nuclear industry, including the collapse of leading reactor vendors. Given that in the mid-2020s, not a single large-scale reactor is being built, it is likely that nuclear power will face further marginalization in a future Phase VI, say between 2025 and 2040. Even though some reactor designs might be developed, the share of nuclear power in the energy mix is likely to diminish. Identifying parallels and discrepancies with other major nuclear reactor countries (such as Russia, China, and France) seems a promising way forward to compare the results, and to identify country-specific developments and perspectives.

References

Abdulla, A, M J Ford, M G Morgan, and D G Victor. 2017. 'A Retrospective Analysis of Funding and Focus in US Advanced Fission Innovation'. *Environmental Research Letters* 12 (8): 084016. https://doi.org/10.1088/1748-9326/aa7f10.

AEC. 1974. 'Liquid Metal Fast Breeder Reactor Program. Volume IV. Environmental Statement'. WASH--1535(Vol.4)(Draft 2), 4200424. https://doi.org/10.2172/4200424.

Ahern, Joseph James. 2003. '"We Had the Hose Turned on Us!": Ross Gunn and the Naval Research Laboratory's Early Research into Nuclear Propulsion, 1939-1946'. *Historical Studies in the Physical and Biological Sciences* 33 (2): 217–36. https://doi.org/10.1525/hsps.2003.33.2.217.

Allen, Wendy. 1977. 'Nuclear Reactors for Generating Electricity: U.S. Development from 1946 to 1963'. R-2116-NSF. The Rand Corporation. https://www.rand.org/content/dam/rand/pubs/reports/2007/R2116.pdf.

Amaldi, Edoardo, and Enrico Fermi. 1936. 'On the Absorption and the Diffusion of Slow Neutrons'. *Physical Review* 50 (10): 899–928. https://doi.org/10.1103/PhysRev.50.899.

Bah, Muhammad Maladoh. 2023. 'State and Federal Nuclear Support Schemes in Dynamic Electricity Market Conditions: Insights from NYISO and PJM'. *Energy Policy* 182 (November):113764. https://doi.org/10.1016/j.enpol.2023.113764.

Berkhout, Frans, Adrian Smith, and Andy Stirling. 2004. 'Socio-Technological Regimes and Transition Contexts'. In *System Innovation and the Transition to Sustainability*, by Boelie Elzen, Frank Geels, and Ken Green, 3335. Edward Elgar Publishing. https://doi.org/10.4337/9781845423421.00013.

Bird, Kai, and Martin J. Sherwin. 2006. *American Prometheus: The Triumph and Tragedy of J. Robert Oppenheimer*. First Vintage books edition. New York: Vintage Books.

Böse, Fanny, Alexander Wimmers, Björn Steigerwald, and Christian von Hirschhausen. 2024. 'Questioning Nuclear Scale-up Propositions: Availability and Economic Prospects of Light Water, Small Modular and Advanced Reactor Technologies'. *Energy Research & Social Science* 110 (April 2024): 103448. https://doi.org/10.1016/j.erss.2024.103448.

Buck, Alice L. 1983. 'A History of the Atomic Energy Commission'. DOE/ES-0003/1. https://www.osti.gov/opennet/servlets/purl/16130958-BUEwKo/16130958.pdf.

Bunn, Matthew, John P. Holdren, Steve Fetter, and Bob van der Zwaan. 2005. 'The Economics of Reprocessing Versus Direct Disposal of Spent Nuclear Fuel'. *Nuclear Technology* Vol. 150 (June). https://fetter.it-prod-webhosting.aws.umd.edu/sites/default/files/fetter/files/2005-NT-repro.pdf.

Burness, H. Stuart, W. David Montgomery, and James P. Quirk. 1980. 'The Turnkey Era in Nuclear Power'. *Land Economics* 56 (2): 188. https://doi.org/10.2307/3145862.

Cochran, Thomas B., Harold A. Feiveson, Walter Patterson, Gennadi Pshakin, M.V. Ramana, Mycle Schneider, Tatsujiro Suzuki, and Frank von Hippel. 2010. 'Fast Breeder Reactor Programs: History and Status'. Research Report 8. Princeton, N.J: International Panel on Fissile Materials. http://fissilematerials.org/library/2010/02/fast_breeder_reactor_programs_.html.

Cohn, Steve. 1990. 'The Political Economy of Nuclear Power (1945–1990): The Rise and Fall of an Official Technology'. *Journal of Economic Issues* 24 (3): 781–811. https://doi.org/10.1080/00213624.1990.11505071.

Cowan, Robin. 1990. 'Nuclear Power Reactors: A Study in Technological Lock-In'. *The Journal of Economic History* 50 (3): 541–67. https://doi.org/10.1017/S0022050700037153.

Curie, Irène, and Frédéric Joliot. 1934. 'I. Production Artificielle d'éléments Radioactifs - II. Preuve Chimique de La Transmutation Des Éléments'. *Journal de Physique et Le Radium* 5 (4): 153–56. https://doi.org/10.1051/jphysrad:0193400504015300.

Daniels, Farrington. 1944. 'Suggestions for a High-Temperature Pebble Pile'. Technical Report N-1668b. University of Chicago. https://www.osti.gov/servlets/purl/4359817.

Daniels, Olive Bell. 1978. *Farrington Daniels. Chemist and Prophet of the Solar Age*. Madison, Wisconsin.

Di Nucci, Maria Rosaria. 2019. 'From Military to Early Civilian Applications: An Appraisal of the Initial Success of the Light Water Reactor Technology'. In *The Technological and Economic Future of Nuclear Power*, edited by Reinhard Haas, Lutz Mez, and Amela Ajanovic, 7–34. Energiepolitik Und Klimaschutz. Energy Policy and Climate Protection. Wiesbaden: Springer Fachmedien Wiesbaden. https://doi.org/10.1007/978-3-658-25987-7_2.

DOE. 2017. 'Annual Report on the State of the DOE National Laboratories'. U.S. Department of Energy. https://www.energy.gov/downloads/annual-report-state-doe-national-laboratories.

DOE. 2020. 'Advanced Reactor Demonstration'. Funding Opportunity Announcement DE-FOA-0002271. U.S. Department of Energy. https://www.grants.gov/web/grants/view-opportunity.html?oppId=326997.

Eisenhower, Dwight D. 1953. 'Atoms for Peace'. Speech at the 470th Plenary Meeting of the United Nations General Assembly, December 8. https://www.iaea.org/about/history/atoms-for-peace-speech.

ERDA. 1975. 'Status of Central Station Nuclear Power Reactors: Significant Milestones'. Technical Report ERDA-30(4/75). Washington, D.C.: Energy Research and Development Administration. https://doi.org/10.2172/4221760.

Fermi, Enrico. 1934. 'Radioactivity Induced by Neutron Bombardment'. *Nature* 133 (3368): 757–757. https://doi.org/10.1038/133757a0.

Geels, Frank W. 2002. 'Technological Transitions as Evolutionary Reconfiguration Processes: A Multi-Level Perspective and a Case-Study'. *Research Policy* 31 (8–9): 1257–74. https://doi.org/10.1016/S0048-7333(02)00062-8.

Geels, Frank W. 2004. 'From Sectoral Systems of Innovation to Socio-Technical Systems'. *Research Policy* 33 (6–7): 897–920. https://doi.org/10.1016/j.respol.2004.01.015.

Geels, Frank W. 2007. 'Analysing the Breakthrough of Rock "n" Roll (1930–1970) Multi-Regime Interaction and Reconfiguration in the Multi-Level Perspective'. *Technological Forecasting and Social Change* 74 (8): 1411–31. https://doi.org/10.1016/j.techfore.2006.07.008.

Geels, Frank W. 2011. 'The Multi-Level Perspective on Sustainability Transitions: Responses to Seven Criticisms'. *Environmental Innovation and Societal Transitions* 1 (1): 24–40. https://doi.org/10.1016/j.eist.2011.02.002.

Geels, Frank W., and Johan Schot. 2007. 'Typology of Sociotechnical Transition Pathways'. *Research Policy* 36:399–417. https://doi.org/10.1016/j.respol.2007.01.003.

Gosling, F.G. 1999. 'The Manhattan Project: Making the Atomic Bomb'. DOE/MA-0001–01/99. Department of Energy. https://www.osti.gov/opennet/manhattan-project-history/publications/DE99001330.pdf.

Groves, Leslie M. 1983. *Now It Can Be Told: The Story of the Manhattan Project*. New York, USA: Da Capo Press.

Grubb, Michael, Jean-Charles Hourcade, and Karsten Neuhoff. 2014. *Planetary Economics*. Routledge - Taylor & Francis Group.

Harris, Malcom C., and Kevin J. Wynne. 1989. 'Explaining the Decline of Nuclear Power in the United States: Evidence from Plant Canellation Decisions'. *The Journal of Energy and Development* 14 (2): 253–67. https://www.jstor.org/stable/24807937.

Hewlett, Richard G., and Francis Duncan. 1972. 'A History of the United States Atomic Energy Commission - Volume II 1947/1952'. Atomic Shield. U.S. Atomic Energy Comission. https://www.energy.gov/sites/default/files/2013/04/f0/Hewlett%20and%20Duncan%20-%20Atomic%20Shield%20%28complete%29.pdf.

Hewlett, Richard G., and Francis Duncan. 1974. *Nuclear Navy (1946–1962)*. Chicago and London: The University of Chicago Press. https://www.energy.gov/sites/prod/files/2013/08/f2/HewlettandDuncanNuclearNavyComplete.pdf.

Hewlett, Richard G., and Jack M. Holl. 1987. 'A History of the United States Atomic Energy Commission, 1952-1960: Volume 3'. DOE/NBM-7010972, 6150636. https://doi.org/10.2172/6150636.

IAEA. 2022. 'Power Reactor Information System PRIS'. The Database on Nuclear Power Reactors. International Atomic Energy Agency. 2022. https://pris.iaea.org/pris/home.aspx.

INL. 2011. 'NGNP Project - 2011 Status and Path Forward'. Idaho National Laboratory. https://inldigitallibrary.inl.gov/sites/sti/sti/5250275.pdf.

Joskow, Paul L. 2001. 'U.S. Energy Policy During the 1990s'. *NBER Working Papers*, NBER Working Papers, September. https://ideas.repec.org//p/nbr/nberwo/8454.html.

Jurewitz, John L. 2002. 'The U.S. Nuclear Power Industry: Past, Present, and Possible Futures'. *Energy & Environment* 13 (2): 207–37. https://www.jstor.org/stable/43734471.

Kanger, Laur. 2021. 'Rethinking the Multi-Level Perspective for Energy Transitions: From Regime Life-Cycle to Explanatory Typology of Transition Pathways'. *Energy Research & Social Science* 71 (January):101829. https://doi.org/10.1016/j.erss.2020.101829.

Markard, Jochen. 2020. 'The Life Cycle of Technological Innovation Systems'. *Technological Forecasting and Social Change* 153 (April):119407. https://doi.org/10.1016/j.techfore.2018.07.045.

Meitner, Lise, and O. R. Frisch. 1939. 'Disintegration of Uranium by Neutrons: A New Type of Nuclear Reaction'. *Nature* 143 (3615): 239–40. https://doi.org/10.1038/143239a0.

NRC. 2024. 'A Short History of Nuclear Regulation (1946–2024)'. Brochure NUREG/BR-0175. NUREC-Series Publications. Washington, D.C: U.S. Nuclear Regulatory Commission. https://www.nrc.gov/reading-rm/doc-collections/nuregs/brochures/br0175/index.html.

Nuttall, William J. 2022. *Nuclear Renaissance: Technologies and Policies for the Future of Nuclear Power*. Boca Raton: CRC Press. https://www.taylorfrancis.com/books/9781003038733.

OECD/NEA. 2007. *Innovation in Nuclear Energy Technology*. Nuclear Development. Paris: Nuclear Energy Agency, Organisation for Economic Co-operation and Development. https://www.oecd-nea.org/upload/docs/application/pdf/2019-12/6103-innovation-technology.pdf.

PCAST. 1997. 'Report to the President on Federal Energy Research and Development for the Challenges of the Twenty-First Century'. President's Committee of Advisors on Science and Technology (Panel on Energy Research and Development). https://fire.pppl.gov/PCAST_1997.pdf.

Pistner, Christoph, Matthias Englert, Christian von Hirschhausen, Fanny Böse, Björn Steigerwald, Lukas Gast, and Richard Donderer. 2024. 'Analysis and Evaluation of the Development Status, Safety and Regulatory Framework for So-Called Novel Reactor Concepts'. BASE – Research reports on the safety of nuclear waste disposal BASE-018/24. Berlin: Federal Office for the Saftey of Nuclear Waste Management. https://www.base.bund.de/SharedDocs/Downloads/BASE/DE/fachinfo/fa/Abschlussbericht_neuartige_Reaktorkonzepte_2024.pdf?__blob=publicationFile&v=5.

Reed, Bruce Cameron. 2013. *The History and Science of the Manhattan Project*. 1st Edition. New York: Springer.

Reed, Bruce Cameron. 2020a. 'Organizing: Coordinating Government and Army Support 1939–1943'. In *Manhattan Project*, by Bruce Cameron Reed, 91–147. Cham: Springer International Publishing. https://doi.org/10.1007/978-3-030-45734-1_4.

Reed, Bruce Cameron. 2020b. 'Piles and Secret Cities'. In *Manhattan Project*, by Bruce Cameron Reed, 149–69. Cham: Springer International Publishing. https://doi.org/10.1007/978-3-030-45734-1_5.

Reed, Bruce Cameron. 2021. 'Manhattan Project'. In *Encyclopedia of Nuclear Energy*, 78–91. Elsevier. https://doi.org/10.1016/B978-0-12-819725-7.00042-8.

Rhodes, Richard, ed. 2012. *The Making of the Atomic Bomb*. 25th anniversary ed. New York, USA: Simon & Schuster Paperbacks.

Rosenthal, Murray W. 2010. 'An Account of Oak Ridge National Laboratory's Thirteen Nuclear Reactors'. ORNL/TM-2009/181. Oak Rdige Tennessee: Oak Rdige National Laboratory - Managed by UT- Battelle for tthe DOE. https://citeseerx.ist.psu.edu/document?repid=rep1&type=pdf&doi=b6e5436fc57a3e6ab7c46934415081 2c0f8063e2.

Rutherford, E., and Frederick Soddy. 1902. 'The Radioactivity of Thorium Compounds. II. The Cause and Nature of Radioactivity'. *J. Chem. Soc., Trans.* 81 (0): 837–60. https://doi.org/10.1039/CT9028100837.

Schneider, Mycle, Antony Froggatt, Julie Hazemann, Özgür Gürbüz, Paul Jobin, Phil Johnstone, Timothy Judson, et al. 2024. 'The World Nuclear Industry Status Report 2024'. Paris, France: Mycle Schneider Consulting. https://www.worldnuclearreport.org/IMG/pdf/wnisr2024-v2.pdf.

Schneider, Mycle, Julie Hazemann, Tadahiro Katsuta, Andy Stirling, Ben Wealer, Frogatt Antony, Phil Johnstone, et al. 2018. 'The World Nuclear Industry - Status Report 2018'. Paris, London. https://www.worldnuclearreport.org/IMG/pdf/wnisr2018-v2-lr.pdf.

Schumpeter, Joseph A. 1934. *Theory of Economic Development*. Cambridge MA (USA): Harvard University Press.

Seaborg, Glenn T. 1970. 'The Plutonium Economy of the Future'. Presented at the Fourth International Conference on Plutonium and Other Actinides (5 October 1970), Santa Fe, New Mexico. http://fissilematerials.org/library/aec70.pdf.

Soddy, Fredric. 1909. *Interpretation of Radium: Being the Substance of Six Free Popular Experimental Lectures*. Lectures at the University of Glasgow. S.l.: Forgotten Books. https://archive.org/details/interpretationof00soddrich/page/n7/mode/2up.

Stern, Nicholas. 2007. *The Economics of Climate Change: The Stern Review*. Cambridge: Cambridge University Press. https://doi.org/10.1017/CBO9780511817434.

Strauss, Lewis L. 1954. 'Remarks Prepared by Lewis L. Straus, Chairman, United States Atomic Energy Commission, For Delivery At The Founders' Day Dinner, National Association of Science Writers, On Thusday, September 16, 1954, New York, New York.' Speech presented at the Founders' Day Dinner, National Association of Science Writers, New York, September 16. https://www.nrc.gov/docs/ML1613/ML16131A120.pdf.

Szilard, Leo. 1949. Improvements in or relating to the Transmutation of Chemical Elements. GB630726A, filed 28 June 1934, and issued 28 July 1949. https://worldwide.espacenet.com/patent/search/family/010124701/publication/GB630726A?q=pn%3DGB630726.

Taylor, Bryan C., and Brian Freer. 2002. 'Containing the Nuclear Past: The Politics of History and Heritage at the Hanford Plutonium Works'. *Journal of Organizational Change Management* 15 (6): 563–88. https://doi.org/10.1108/09534810210449523.

Thomas, Stephen, and M. V. Ramana. 2022. 'A Hopeless Pursuit? National Efforts to Promote Small Modular Nuclear Reactors and Revive Nuclear Power'. *WIREs Energy and Environment*, January. https://doi.org/10.1002/wene.429.

Tolman's Committee. 1944. 'Piles of the Future Review'. Review Report N-1447n. Oak Rdige National Laboratory. https://atomicinsights.com/wp-content/uploads/Tolman-Piles-of-Future.pdf.

U.S. Congress. 2005. *Energy Policy Act of 2005*. https://www.govinfo.gov/content/pkg/PLAW-109publ58/pdf/PLAW-109publ58.pdf.

U.S. Congress. 2021. *Infrastructure Investment and Jobs Act (IIJA) - "Bipartisan Infrastructure Law"*. https://afdc.energy.gov/laws/infrastructure-investment-jobs-act.

U.S. Congress. 2022. *Inflation Reduction Act. Public Law 117–169*. https://www.govinfo.gov/content/pkg/PLAW-117publ169/pdf/PLAW-117publ169.pdf.

Verbong, Geert, and Frank Geels. 2007. 'The Ongoing Energy Transition: Lessons from a Socio-Technical, Multi-Level Analysis of the Dutch Electricity System (1960–2004)'. *Energy Policy* 35 (2): 1025–37. https://doi.org/10.1016/j.enpol.2006.02.010.

Walker, Mark. 2024. *Hitler's Atomic Bomb: History, Legend, and the Twin Legacies of Auschwitz and Hiroshima*. 1st ed. Cambridge: Cambridge University Press.

Wealer, Ben, Simon Bauer, Nicolas Landry, Hannah Seiß, and Christian R. von Hirschhausen. 2018. 'Nuclear Power Reactors Worldwide: Technology Developments, Diffusion Patterns, and Country-by-Country Analysis of Implementation (1951–2017)'. DIW Data Documentation 93. Berlin: Deutsches Institut für Wirtschaftsforschung (DIW). http://hdl.handle.net/10419/179000.

Wells, Herbert George. 1914. *The World Sets Free: A Story of Mankind*. London: MacMillan and Co, Limited. https://archive.org/details/interpretationof00sodd/page/n9/mode/2up.

Welton, Shelley, and Conor Harrison. 2024. 'Lessons in Climate Derisking: The United States' Failed Nuclear Renaissance'. https://doi.org/10.2139/ssrn.4973695.

Chapter 12
Uncovering Hidden Nuclear Subsidies and Other Forms of State Support

Doug Koplow

12.1 Introduction

Nuclear power is a complex system good. Several chapters of this book have demonstrated the complex nature of the production chain and the concentration of the individual market segments (see Chaps. 2, 8 and 9 of this book). While nuclear power is often portrayed as an inexpensive and reliable energy source, it has never been able to economically compete with other technologies. Historically, the challenge has come from coal; today, it is coming increasingly from renewables (Böse et al. 2024). The cost of nuclear power has been increasing in OECD countries, mainly due to the escalating costs of new build projects (Göke et al. 2025). The expansion of competitive electricity markets has also contributed to economic challenges for nuclear power plants (Bah 2023). While state involvement has always been a central element of the nuclear production chain, it has increased alongside rising competitive challenges. State supports are directed to multiple parts of the nuclear system good, including the fuel supply chain; new reactor funding; continued operation of existing nuclear power plants, particularly in deregulated electricity markets; and the "back end" activities such as decommissioning and waste disposal. States also backstop or fully absorb key operating risks such as accidents. While the program details differ by country, the general subsidies are common across the globe.

This chapter is based on the author's chapter "Nuclear Economics and Finance" in the 2023 edition of the World Nuclear Industry Status Report (WNISR) (Schneider et al. 2023, 334–410). For this reason, online references are provided with additional detail in this chapter compared to the others of this book. Refer to the original chapter for supplemental information and additional references on the topics covered here. The WNSIR series offers insights into the global status of the nuclear industry, including overall trends and challenges, as well as detailed country reports on current developments.

D. Koplow (✉)
Earth Track, Cambridge, MA, USA
e-mail: dkoplow@earthtrack.net

A. Wimmers et al. (eds.), *Nuclear Power*, Studies in Energy, Resource and Environmental Economics, https://doi.org/10.1007/978-3-031-99894-2_12

Recent trends in these areas are explored in Sect. 12.2 of this chapter. Section 12.3 then analyzes the recent economic performance of key players in the sector, many of which are public firms. Section 12.4 is dedicated to the "hidden costs" of nuclear power such as accident liabilities, post-closure costs, and impacts on weapons proliferation risks. These factors are seldom accounted for in levelized cost of energy (LCOE) estimates for nuclear, skewing comparisons with alternative electricity generation technologies. Limited published data, uncertainties on the scale and timing of costs, and a lack of empirical project data make these factors challenging to quantify accurately. However, in most cases, these costs are at least partially supported by the state, and in some cases, the costs and risks are entirely shifted from the operator onto the government. Even where commercial operators are financially responsible, contributions are often insufficient.

12.2 State Involvement

12.2.1 State Ownership in the Nuclear Fuel Supply Chain

State involvement has been a significant part of the nuclear power industry since its inception. David Newbery at the University of Cambridge wrote that "[n]o nuclear power station has ever been constructed without some (and usually extensive) risk mitigation, either by public ownership or under regulatory guarantees" (Newbery 2021, 231). This involvement seems to be growing, as patterns in new construction and ownership of multiple parts of the nuclear supply chain point to an increasing role of state policy and state funding over market economics. These policies have been implemented instead of, or sometimes on top of, strategies to price carbon. Although carbon pricing would also benefit nuclear relative to incumbent fossil competitors, it would bring climate and other benefits compared to nuclear-specific support schemes by equally advantaging efficient end-use of energy and renewable energy generation.

The world's nuclear newbuild mainly involves two centrally planned economies—China and Russia. Reactor construction starts have been increasingly concentrated into two buckets: new projects within China and projects in other countries implemented by Russia. An even more pronounced concentration is evident with reactor manufacturing, as both countries have promoted construction in third countries using their equipment. Except for Pakistan, where it has built all six operating units, China has not yet had widespread success internationally. However, it continues to make incremental progress such as having its Hualong One (HPR-1000) design approved by British regulators.[1] Based on data from 2023, most reactor start-ups have been occurring in China (53% of unit count and 55% of capacity from 2018–2022). Russia follows with 14% of unit counts and 7% of capacity, and Türkiye comes in third with 11% of construction starts, or 12% by nameplate capacity (second position). Other

[1] World Nuclear News, "UK regulators approve China's UK HPR1000 design," 7 February 2022, accessed on 05-02-2025.

large reactor construction starts since 2013 have been limited: four each in India and the United States (US), two of which were canceled midway; three each in South Korea and the United Arab Emirates; and two each in the United Kingdom (UK) and a handful of other countries, including projects implemented by Russia in nuclear newcomer countries Bangladesh, Belarus, and Egypt.

The role of Chinese and Russian equipment can be seen in the reactor designs implemented in new construction projects. Chinese reactors (sometimes derived from originally Western designs) made up more than 40% of the total over the past five years. The values for Russian reactors were 37% for the 2013–2022 period (35% of nameplate capacity), rising substantially to 50% (number of reactors) and 46% (capacity) for the past five years.

During that same period, France was the source of only two reactors and South Korea supplied only one. Indeed, two-thirds of reactor construction starts since the beginning of 2000 took place either in China and Russia or were implemented by the Russian industry, "where the costs of financing depend on their government's ability to raise money rather than on market interest rates. In the case of these countries—besides China and Russia, in particular, Bangladesh, Egypt, India, and [Türkiye]—the government undertakes the projects, electricity demand continues to soar, and public opinion has little chance to interfere with government plans" (Rubio-Varas 2020, 268).

Global uranium enrichment capacity is also heavily controlled by the state-owned enterprises of Russia (46% of 2020 capacity) and China (10.5%). Even Western capacity is mostly government-owned, with involvement by the UK, France, the Netherlands, and Japan. Governments control nearly 90% of global enrichment capacity.[2]

12.2.2 *State Credit Support in the Production Chain*

These trends highlight government-led nuclear investments at such a scale that industry expert Geoffrey Rothwell concluded in 2022 that "[p]rivately-owned equity companies in the nuclear sector are no longer competitive in international markets" and that "China and Russia are in the process of putting the West's nuclear industry out of business" (Rothwell 2022, 3–5). Within China, civilian nuclear cooperation is viewed as an integral part of its Belt and Road Initiative, a major foreign policy initiative investing in scores of countries around the world and in alignment with the government's view of China as a rising global power. Under this initiative, at least 25 countries are participating in some nuclear reactor cooperation. The US National Academy of Sciences notes that there may be a security risk if these engagements lead to weaker Nuclear Cooperation Agreements, particularly with nations hosting their first nuclear infrastructure (Committee on Laying the Foundation for New and Advanced Nuclear Reactors in the United States et al. 2023). The US Department of

[2] See Chap. 8 on a detailed assessment of the nuclear fuel supply chain.

Energy is equally blunt: "Russia—a nation that has 'weaponized' its energy supply as an instrument of coercion—dominates nuclear markets" (U.S. DOE 2020, 6).

China's investments beyond Hinkley Point C in the U.K. (in which China halted funding in December 2023)[3] are slated to ramp up quickly, with 30 reactor projects planned abroad by 2030 and an associated investment of more than CN-¥1 trillion (US-$$_{2019}$145 billion). Helping to propel these deals are China's "advanced nuclear technology, competitive prices, and lavish financing," with subsidized sovereign credit covering 80% of the cost of projects in Pakistan and 85% in Argentina.[4] Much of the capital provided to Pakistan carries interest rates of only 1 or 2%; the interest on the Argentina project is 4.5% (Bowen and Apostoaei 2022). In comparison, the annualized inflation rate in Pakistan hit a record 38% in April 2023.[5] China also has a history of "debt-for-equity swaps," where defaults on large infrastructure projects are forgiven in return for control over recipient countries' strategic assets, including ports and mines. This approach could be extended to reactor loans, or China could use ongoing construction and maintenance of reactors as a lever in political disagreements.[6] However, Chinese latitude for growth may be affected by the inclusion of crucial providers such as China National Nuclear Group Corporation (CNNC) and China General Nuclear Power Corporation (CGN) on US government blacklists.[7]

Russia's success in marketing nuclear reactor projects has also relied on low financing rates, favorable terms on fuel supply, and even agreements for the spent fuel to be returned to Russia, precluding the need for customers to develop local disposal capabilities. Project data suggests Russian lending rates of 3 to 5%, depending on the partner country and date of origination. Russia has also taken equity interests in some projects (Bowen and Apostoaei 2022). These subsidies advance Russian foreign policy objectives by enabling them to play "an important role in the customer country's critical energy infrastructure for the life of the plant" (Committee on Laying the Foundation for New and Advanced Nuclear Reactors in the United States et al. 2023, 179) lasting many decades—a very powerful lock-in effect. However, Russia's role as a leading nuclear vendor has been harmed by its invasion of Ukraine, with the cancellation of planned Russian-built nuclear plants in Finland, Jordan, and Slovakia.

---

[3] Alex Lawson, "Hinkley Point C could be delayed to 2031 and cost up to £35bn, says EDF", The Guardian, 23 January 2024, see https://www.theguardian.com/business/2024/jan/23/hinkley-point-c-could-be-delayed-to-2031-and-cost-up-to-35bn-says-edf, accessed 05-02-25.

[4] U.S. DOE, "Restoring America's Competitive Nuclear Energy Advantage—A Strategy to Assure U.S. National Security", 2020, see https://www.energy.gov/articles/restoring-americas-competitive-nuclear-energy-advantage, accessed 07-21-23.

[5] Al Jazeera, "Pakistan inflation hits record for second consecutive month", 2 June 2023, see https://www.aljazeera.com/news/2023/6/2/pakistan-inflation-hits-record-for-second-consecutive-month, accessed 07-21-23.

[6] Lami Kim, "Nuclear Belt and Road and U.S.-South Korea Nuclear Cooperation", Center for Strategic & International Studies, 24 April 2023, see https://www.csis.org/analysis/nuclear-belt-and-road-and-us-south-korea-nuclear-cooperation, accessed 05-03-25.

[7] Bureau of Industry and Security, "Supplement No. 4 to Part 744 – Entity List", U.S. Department of Commerce, 25 February 2025, see https://www.ecfr.gov/current/title-15/part-744/appendix-Supplement%20No.%204%20to%20Part%20744, accessed 28-02-25.

Rosatom was earlier excluded from bids in the Czech Republic following the disclosure of Russian involvement in a 2014 explosion at a Czech ammunition depot (Szulecki and Overland 2023).

In contrast, countries like Bangladesh, China, Egypt, and Türkiye, all with existing nuclear deals with Russia, have largely stayed the course, as has, so far, as the only country within the EU, Hungary (Bowen and Apostoaei 2022; Szulecki and Overland 2023).[8] While the US sanctions list may complicate some of these projects, it does not seem to have shut them down (Prokopenko 2023). Efforts to move away from Russia have also affected other parts of the fuel chain, with nearly every country looking to diversify supply within days of the invasion. Implementation can be challenging, though at least five of the six countries operating VVERs in Europe (Bulgaria, Czech Republic, Finland, Hungary, Slovakia, and Ukraine) have signed new supply contracts for fuel fabrication since February 2022 and many are working on an EU-effort to support the buildup of alternative sources.[9]

Nuclear-related credit support from other countries does exist, though at a much smaller scale than what is at play with Russia and China. Table 12.1 summarizes credit support originating from twenty development and export credit providers around the world during 2008–2022 as compiled by US NGO Oil Change International (OCI) and published in its Public Finance for Energy Database.[10] Given the challenges of obtaining credit data (and particularly the terms of that support), the database likely provides only a partial picture of what is happening. State bank guarantees to domestic projects or vendor financing by state-owned vendors may not appear. US government-subsidized credit to the Vogtle reactors in the US state of Georgia is an example, with US-$$_{2019}$ 12 billion in guarantees not captured on the export-focused OCI tabulation.[11] In addition, reporting may not be consistent across all countries, including different time frames or the treatment of commitments that have not been finalized.

Ideally, this information should be readily available so taxpayers can see patterns of government-provided credit support across industries and evaluate outcomes. At present, however, the gaps can sometimes be large. This is most evident regarding lending data for Russian projects. The OCI database tabulates Russian credit support to civilian nuclear projects totaling US-$263 million,[12] while Columbia University's

[8] Refer to Chap. 9 of this book for an assessment of Russian construction projects in Bangladesh, Egypt, and Türkiye.

[9] Westinghouse Electric Company, "Westinghouse-led Project will Secure VVER Fuel Supply in Europe and Ukraine", Press release, 6 July 2023, https://info.westinghousenuclear.com/news/westinghouse-led-project-will-secure-vver-fuel-supply-in-europe-and-ukraine, accessed 02-03-25. See also Chap. 8 of this book for European diversification strategies.

[10] Oil Change International, "Public Finance for Energy Database", Oil Change International, see https://energyfinance.org/#/, accessed 21-07-2023.

[11] U.S. Department of Energy, "Vogtle", Undated, see https://www.energy.gov/lpo/vogtle, accessed 11-09-2023.

[12] If not otherwise indicated, all monetary values are provided for 2022.

Table 12.1 Patterns in sovereign credit support to nuclear power, 2008–2022

Origin Country	Institution	Recipient Country	Project	Funding Mechanism (in million US-$)			
					Equity	Guarantee	Loan
Australia	EFIC	UK	Consulting on Nuclear Waste Management			0.25	
Brazil	BNDES	Brazil	Eletronuclear				21.5
Canada	EDC	Canada	Cameco Corporation				39.3
		South Korea	Tyne Engineering Corporation				0.75
	SDTC	Canada	Deomonstration of Fusion Energy Technology	9.5			
China	CDB	UK	Hinkley Point C Nuclear Project				7,800
	Chexim	Pakistan	Karachi Nuclear Power Complex (K-2/K-3)				6,500
	Sinosure	Argentina	Argentina Nuclear Plant			4,700	
EU	EBRD	Ukraine	Nuclear Power Plant Safety Upgrade Program				414.30
	EIB	Finland	TVO Safety Improvements				112.22
		France	AREVA Uranium Enrichment Facility—A		286		278
		Italy	ENEA—Divertor Tokamak Test Facility				280.77
		Netherlands	Urenco Uranium Enrichment Facility II				242
		Slovakia	SE Safety Improvement				68.26
		UK	Urenco Uranium Enrichment Facility II and II C				227
France	COFACE	China	Taishan Nuclear Power Plant			2,448.52	
India	EXIM India	Bangladesh	Bangladesh India Exim Bank LoC 2017 (Rooppur nuclear power plant)				666.67
Italy	SACE	Slovakia	Mochovce 3&4 Project			540	

(continued)

Table 12.1 (continued)

Origin Country	Institution	Recipient Country	Project	Funding Mechanism (in million US-$)			
					Equity	Guarantee	Loan
Japan	JBIC	Canada	joint investment in Uranium One Inc. (U1) (UUU.TO)		274.56		
		US	Equity Participation in NuScale Power, LLC				110
South Korea	Kexim	UAE	Barakah Nuclear Power Plant (5,600 MW)				3,100
		Jordan	Research-and-Training Reactor Construction Project				12.80
Russia*	EXIAR	India	Kudankulam Nuclear Power Plant			102.44	
	VEB	Russia	Multipurpose Research Reactor				160.50
UK	UKEF	China	Sale of pumps and spares to Chinese Nuclear Energy Industry Corp			9.1	
		Switzerland	Pumps (Arcade UK Ltd)			0.1	
US	EXIM	Mexico	Nuclear Power			64.48	
		UAE	Abu Dhabi Nuclear Plant				1,988.22
		US	Consult Services for Nuclear Power Station			1.15	
	OPIC	Ukraine	Energoatom Central Spent Nuclear Fuel Storage Facility			250.00	
Total				10	561	8,116	22,053
				30.7 billion			

Source Oil Change International's Public Finance for Energy Database.[13] *Note* * Lending data for the Russian Federation is missing some significant project lenders, with the result that the US-$263 million should not be considered representative for Russian nuclear lending activities. Abbreviations used: EXIM = Export–Import Bank of the United States; EIB = European Investment Bank; Sinosure = China Export and Credit Insurance Corporation; EXIM India = Export–Import Bank of India; Kexim = Export–Import Bank of Korea; OPIC = Overseas Private Investment Corporation; JBIC = Japan Bank for International Cooperation; CDB = China Development Bank; Chexim = Export–Import Bank of China; EXIAR = Export Insurance Agency of Russia; SACE = Servizi Assicurativi del Commercio Estero; VEB = Russian Development Bank; EBRD = European Bank for Reconstruction and Development; COFACE = Compagnie Française d'Assurance pour le Commerce Extérieur; UK = United Kingdom; US = United States; UAE = United Arab Emirates; EU = European Union

[13] Oil Change International, "Public Finance for Energy Database", see https://energyfinance.org/, accessed 21-07-2023.

Center on Global Energy Policy highlights individual lending and equity commitments orders of magnitude larger, at US-$75 billion (Bowen and Apostoaei 2022). The US DOE has an even higher value: US-$$_{2020}$133 billion[14] in foreign orders (not all of which are likely to reach fruition). Most of these orders probably have some associated sovereign credit support (U.S. DOE 2020). The disparity across data sources seems to arise primarily from project finance provided by Rosatom or its subsidiary Atomstroyexport, which is not captured within the OCI dataset. Credit from Chinese institutions to nuclear power equaled US-$19 billion according to the Oil Change database, by far the largest country provider in their tabulation. Estimates by Bowen and Apostoaei (2022) were US-$22 billion, much more closely aligned with the OCI dataset than the estimates for Russia.

State investment in reactor projects through subsidized interest rates, non-market lending terms, or equity infusions enables the granting countries to buy market share and create competitive barriers to non-nuclear substitutes and more market-oriented nuclear power competitors. OECD countries, for example, participate in the 1978 "Arrangement on Officially Supported Export Credits," which limits subsidies on credit support. Specifically, the arrangement requires interest rates tied to national Treasury rates using the Commercial Interest Reference Rates (CIRR); the use of risk premia for default risk; caps directed trade at no more than 85% imported content, and local content at no more than 30% (before 2021); and repayment periods of no more than 18 years for nuclear projects (albeit a longer duration is allowed for other projects). Some flexibility in the schedule of repayments is permitted, but only if the weighted-average loan duration does not exceed nine years and the maximum repayment term is 15 years (Bowen and Apostoaei 2022). This last provision prevents backloading loan payments or interest to the very end of the lending period, a practice that increases the net present value of the subsidy.

The Export–Import Bank of the United States (EXIM Bank) financed "most of the world's nuclear projects sold internationally from 1965 to 1985" (Rubio-Varas 2020, 265). Repayment periods for nuclear were twice as long as for other EXIM Bank projects, with grace periods of up to nine years. In recent decades, this funding for reactor projects has stopped. However, since 2020, the OCI data highlight US-$2.3 billion in credit support for US equipment and services to foreign nuclear projects. Their dataset also captures material credit support to nuclear exports by the European Investment Bank (EBRD) towards France, India, Italy, Japan, and South Korea. Industry and government officials have been working to increase the role of these institutions as a source of subsidized credit to the nuclear energy industry. Large nuclear funding packages have started to inch forward in the past few years. The US has issued letters of interest for US-$4 billion in financing for SMRs in Poland (US-$3 billion from EXIM Bank and US-$1 billion from the Development Finance Corporation), the same for SMRs in Romania, and supposed EXIM Bank support for 85% of the cost of the first two reactors (estimated cost US-$10 billion) of a multi-reactor project in Ukraine.

[14] Approximately US$$_{2022}$163.5 billion.

12.2.3 State Support for Continued Operation of Nuclear Reactors

For decades, proponents have characterized nuclear power as expensive to build, but relatively inexpensive to run. The characteristics driving this claim are low operating costs in comparison to other power sources, a long service life for reactors, and high load factors that enable the high initial investment to be spread over many kWh, thereby reducing the fixed costs per unit of energy produced. Indeed, data on the US fleet collected by the Nuclear Energy Institute, a trade association, suggests real cost declines in fuel, operating, and capital costs since 2012. US load factors have also remained high; those outside of the US are not as high, but still robust in comparison to most competing forms of electricity generation (Schneider et al. 2023).

Despite this long-held view, growing market pressure from less expensive competitors poses a significant and continuing threat to the nuclear power sector. With competition in the US initially arising from low-cost natural gas, falling costs for wind and solar represent a growing competitive risk for nuclear globally, especially during certain periods of the year or times of day (Bah 2023). This resulted in reduced wholesale prices and, coincidingly, declining revenues for nuclear power plants that have historically competed by running at full capacity all the time. In Finland, for example, the Olkiluoto-3 nuclear reactor commenced commercial operation after years of delays in mid-April 2023. Yet, a month later, surging renewables production and negative wholesale power prices forced curtailment of nuclear generation.[15] Similar cuts have been made at Spanish reactors.[16]

In the short term, these pressures vary. For example, the Russian invasion of Ukraine drove European power prices up sharply, and increased exports of LNG from the US to replace disrupted European supplies did the same in the US. However, over longer market cycles, competitive pressures have resulted in even operating reactors—where all or most of the original invested capital has already been paid off—closing or threatening to close. In the US, for example, 13 reactors closed between 2013 and 2022 (including three reactors that had ceased electricity production in 2009 and 2012). The cost pressures are evident primarily in competitive power markets; in regulated markets, higher costs are passed through to customers (Bah 2023).

In other sectors of the economy, innumerable facilities have shut down temporarily or permanently when changing market conditions have rendered their products too expensive or no longer desired by consumers. Permanent shutdowns happen routinely, and rarely is this done because the facility is no longer physically able to manufacture its product. These closures are viewed as the normal functioning of market forces, retiring obsolete assets to make way for competitive new ones.

---

[15] YLE, "Finnish nuclear plant throttles production as electricity price plunges", see https://yle.fi/a/74-20032375, accessed 24-11-2023.

[16] Pablo Bronte, "España modula sus reactores para ayudar a las renovables", Montel, 8 September 2023, see www.montelnews.com/es/news/1521095/espaa-modula-sus-reactores-para-ayudar-a-las-renovables, accessed 11-09-2023.

Nuclear plant closures have been framed entirely differently. Arguing that plant closures would drive up carbon emissions, and that their product, labeled as "low-carbon, reliable power," was not being properly valued by the market, the industry has tagged the closures as premature. It has lobbied for, and increasingly often successfully obtained, subsidies to remain in operation. Below are some examples of operating conditions in key countries and associated state subsidies to keep the sector going. Additional information can be found in Schneider et al. (2023, 2024).

12.2.3.1 United States

Threats of reactor closures led to the implementation of state-level per MWh subsidies for 19 US reactors in New York, Illinois, New Jersey, and Connecticut. Payments last five to twelve years, depending on the state and the specific program. The Connecticut program also covers one reactor in New Hampshire. Total subsidies through these programs are estimated at more than US-$15 billion by 2030 and will further increase should the programs be extended. A similar program in Ohio was ultimately rescinded upon discovery that US-$60 million in bribes had been directed to politicians in return for passage of the US-$1 billion subsidy to nuclear reactors.

Production Tax Credits (PTC) have historically been implemented to subsidize the deployment of new capital-intensive infrastructure in the power sector, with the idea that the capital investment was the main hurdle in developing new capacity that could then compete on its own. More recently, policies have extended PTCs beyond the eligibility time-period set in the original statutes, resulting in a structure similar to operating subsidies. The Zero-Emission Nuclear Production Credit supplemented an earlier PTC for new plants to include existing nuclear plants. It offers a maximum of US-$15 per MWh for plants operating from 2024 to 2032. In some circumstances, these may be combined with large subsidies also being offered for hydrogen produced from so-called "clean" (low carbon) energy, which would include nuclear.[17] The flexibility to sell the tax credits to parties with higher tax liability, or in some cases to have their value paid directly to the operator by the Treasury, has increased the subsidy value to investors and plant owners. A different program, the Civil Nuclear Credit (CNC), funded a national pool of US-$6 billion in subsidies to keep economically distressed reactors from closing. The structure is more efficient than the flat per-kWh payments many states have implemented, as reactors must bid the minimum subsidy they claim to need to continue operating, and where there are multiple bids, they are tendered against each other. There is an initial five-year program window with an extension option. The Diablo Canyon plant in California was awarded US-$1.1 billion to remain in operation in the initial round (2022). The CNC application requires disclosure of other supports, though receiving subsidies from other programs will not lead to disqualification in the tender. Further, restrictions on using the CNC

[17] Marie Saprie, "A Closer Look at the Final Hydrogen Regs", Tax Notes, 20 January 2025, see https://www.taxnotes.com/featured-analysis/closer-look-final-hydrogen-regs/2025/01/17/7qbw7 accessed 01-30-2025.

in regulated markets have been relaxed. The provision is now also open to reactors that closed after 15 November 2021 (Bah 2023). This second change facilitated eligibility for restarting the Palisades reactor in Michigan despite safety issues. The operator ultimately received a US-$15 billion loan through a different program, the US Department of Energy's "Energy Infrastructure Reinvestment Program." In addition, the US Department of Agriculture provided US-$1.3 billion in grants to two rural cooperatives to purchase the nuclear energy the restarted Palisade reactor will eventually produce.[18]

12.2.3.2 France

In the 2020s, the French reactor fleet needed to stabilize operations following widespread and unplanned corrosion problems that, together with other aging issues, climate-induced operating restrictions, strikes, and more, resulted in roughly half of the country's 56 reactors being taken offline for extended periods in 2022. Maintenance costs also surged. Revenue losses from reduced production were compounded by two other factors. First, price caps were implemented to buffer damage to consumers following energy market dislocations from the Russian invasion of Ukraine. Second, a 2010 agreement called ARENH (Accès Régulé à l'Électricité Nucléaire Historique), implemented with the EU to address concerns about EDF becoming a monopoly supplier, mandated EDF to sell 100 TWh, then about a quarter of its production, to competitors at a fixed price. In 2023, this portion of production earned EDF revenues only about 1/10th of its market value.[19] Total losses for 2022 were €17.9 billion (US-$8.8 billion), contributing significantly to rising total net debt of €64.5 billion (US-$67.9 billion) (EDF 2023). Facing worsening fundamentals, the French government renationalized EDF and announced a restructuring of its nuclear activities in June 2023. Government ownership will reduce borrowing costs because rates will follow the sovereign, not the business. Large cash infusions from the government are expected, supporting both the continued operation of existing reactors and new construction projects.[20]

[18] Sonal Patel, "DOE Finalizes $1.52B Palisades Loan for First-Ever U.S. Nuclear Plant Recommissioning", 30 September 2024, see https://www.powermag.com/doe-finalizes-1-52b-palisades-loan-for-first-ever-u-s-nuclear-plant-recommissioning/, accessed 23-10-2024.

[19] Silvia Aloisi and Forrest Crellin, "Explainer: Why a French plan to take full control of EDF is no cure-all," Reuters, 13 July 2022, see https://www.reuters.com/business/energy/why-french-plan-take-full-control-edf-is-no-cure-all-2022-07-07/, accessed 21-07-2023.

[20] In June 2024, S&P Global remarked in a financial note on EDF that "We continue to view the likelihood of extraordinary French government support to EDF as very high, making it the highest among our rated EU utilities." S&P Global, "EDF Outlook Revised To Positive On Recovering Domestic Nuclear Activity; Ratings Affirmed At 'BBB/A-2'", 5 June 2024, see https://disclosure.spglobal.com/ratings/en/regulatory/article/-/view/type/HTML/id/3192062, accessed 08-11-2024.

12.2.3.3 Belgium

A framework agreement between the government and utility Engie to restart and extend the operations of two reactors was formally signed in December 2023. The deal includes a price floor using a Contract-for-Difference approach. It caps the waste liability for all Engie reactors in Belgium (not just the ones being restarted/lifetime extended). "As a result of the transfer of all nuclear waste liabilities to the Belgian government, Engie will no longer be exposed to the evolution of future costs related to the management of waste,"[21] where "evolution" presumably means "large increases in unpredictable costs."

12.2.3.4 Japan

Since the Fukushima disaster began in 2011, Japan has spent approximately US-$\$_{2024}$75.37 billion (JP-¥$_{2024}$11,403 billion) in compensation and remediation costs as of November 2024, according to TEPCO. Most of this has been funded by the national government, thus the taxpayer.[22] Total clean-up cost is estimated to be much higher, up to US-$\$_{2021}$322–758 billion, see Sect. 12.4.3. Operators have applied for the restart of 25 reactors. As of July 2023, seventeen had received regulatory approval, but only ten had resumed operation, increasing to fourteen by the end of 2024.[23] To expedite the restart of reactors shuttered since 2011 by required safety and security upgrades, the Japanese government is including nuclear for the first time in the second round of its "long-term decarbonized power supply auction," which provides a 20-year fixed revenue stream to winning bidders. Subsidized safety upgrades supporting up to two GW in capacity are included in the second round of this auction, scheduled for 2025.[24] Earmarking funds for nuclear operations would reduce funding available for non-nuclear bidders.

---

[21] World Nuclear News, "Agreement reached for extended operation of Belgian reactors", 29 June 2023, see https://www.world-nuclear-news.org/Articles/Agreement-reached-for-extended-operation-of-Belgia; and Engie, "ENGIE signs a final agreement with the Belgian government on the extension of Tihange 3 and Doel 4 nuclear reactors", see https://en.newsroom.engie.com/news/engie-signs-a-final-agreement-with-the-belgian-government-on-the-extension-of-tihange-3-and-doel-4-nuclear-reactors-6aef-314df.html, both last accessed 06-03-2025.

[22] https://www.tepco.co.jp/fukushima_hq/compensation/images/jisseki01-j.pdf, last accessed 06-03-2025.

[23] US Energy Information Administration, "Since the 2011 Fukushima accident, Japan has restarted 14 nuclear reactors", 10 January 2025, see https://www.eia.gov/todayinenergy/detail.php?id=64204#, last accessed 06-03-2025.

[24] "METI eyes upping target capacity and creating dedicated nuclear category in the 2nd long-term decarbonization auction", Japan Energy Hub, 3 July 2024. https://japanenergyhub.com/news/2nd-long-term-decarbonization-auction-outline/, last accessed 31-10-2024.

12.2.3.5 European Union

Price caps to buffer customers from sharply rising power and fuel prices following the Russian invasion of Ukraine were adopted across the EU. They have been quite costly. As of June 2023, the subsidy had hit €$_{2023}$758 billion (US-\$$_{2023}$820 billion) across 29 European countries.[25] Value-Added Tax (VAT) cuts and price caps played a significant role. The subsidies allowed consumers to continue to purchase energy products despite the scarcity-induced price surges. They also reduced or waived taxes and operating fees on energy providers, including power stations. Prices continued to rise through most of 2022 but began to stabilize in 2023, partly because of the EU market interventions.[26] It is noteworthy that wholesale prices during the high-price period between the second quarter of 2022 and the first quarter of 2023 were consistently higher in France than in Germany.

12.3 Economic Performance of Key Players

Implicit in increasing government interventions to bolster the nuclear sector is that key firms were not doing particularly well without this government support. This section provides an overview of six major international players and potential nuclear builders: Rosatom (Russia), CGN/CNNC (China), EDF (France), Westinghouse (US/Canada), KEPCO (South Korea), and TEPCO (Japan). Also included is US utility Constellation, a main beneficiary of state-level subsidies.

12.3.1 Électricité De France (EDF)

Électricité de France (EDF) is the largest reactor operator in the world, but due to financial challenges—predominantly from problems with prolonged reactor outages and needed repairs to its fleet of operating reactors, as well as cost overruns at its construction projects in France and the UK—the company was fully renationalized in October 2022. The immediate taxpayer cost from reversing partial shareholder ownership was nearly €10 billion (US-$10.5 billion). EDF debt levels were very high, and state ownership was viewed as a credit positive by rating agencies. After large losses in 2022, the company swung to profit during 2023 due to higher power prices and reduced nuclear power plant outages. Also benefitting EDF were two government decisions: a reduction in required discounted power sales under the

---

[25] Bruegel, "National fiscal policy responses to the energy crisis", 26 June 2023, see https://www.bruegel.org/dataset/national-policies-shield-consumers-rising-energy-prices, last accessed 20-10-2024.

[26] Eurostat, "Electricity & gas hit record prices in 2022", Statistical Office of the European Union, 26 April 2023, See https://ec.europa.eu/eurostat/web/products-eurostat-news/w/DDN-20230426-2, last accessed 20-10-2024.

ARENH agreement from 2022 levels, thereby increasing market-priced sales by 20 TWh; and an agreement for the government to subsidize power consumers if prices rose above €325/MWh (US-$342/MWh).[27]

However, long-standing tensions over pricing between the French government and EDF managers remained. This was complicated by the expiration of the ARENH agreement in 2025, and determining what should replace it, while still satisfying EU market competition requirements for some semblance of avoiding overt state support. A replacement plan was reached at the end of 2023. Instead of price caps, the new agreement uses profit sharing between EDF and consumers if prices rise above the targeted average. The agreement will apply to all of EDF's French nuclear production.[28]

EDF net debt reached nearly €65 billion (US-$68.7 billion) at the end of 2022, dropping marginally (to $€_{2024}$54 billion) by mid-2024.[29] Funding additional capital spending with more debt was not feasible, and thus, management focused on boosting revenues. Reactors play a central role in these tensions, as the government wants EDF to build at least six new reactors in France at an estimated cost of more than $€_{2023}$50 billion (US-$\$_{2023}$54.7 billion).[30] The government has proposed zero-interest loans for these units.[31]

EDF is also behind and over budget on nuclear projects outside of France, including the UK and Finland. A 2025 review by the French Court of Accounts noted a doubling of estimated costs on EDF's Hinkley C project in the UK, to $£_{2015}$31–34 billion (Cour des comptes 2025).[32] The company's Flamanville-3 EPR reactor was finally connected to the grid in December 2024, more than twelve years late and at a price tag of at least $€_{2015}$13.2 billion.[33] This is not a complete estimate, however. In 2020, the Court of Accounts calculated financing and other additional

[27] "EDF swings to profit thanks to price increases," Reuters, 27 July 2023, see https://www.reuters.com/business/energy/edf-swings-back-profit-h1-thanks-price-increases-2023-07-27/; and "France cuts volume of nuclear power EDF must sell cheaply in 2023", 27 October 2022, see https://www.reuters.com/business/energy/france-reduces-volume-nuclear-power-utility-edf-must-sell-cheap-2023-2022-10-27/, both accessed on 05-02-2025.

[28] World Nuclear News, "Agreement on post-ARENH nuclear electricity pricing", 15 November 2023, see https://www.world-nuclear-news.org/Articles/Agreement-on-post-ARENH-nuclear-electricity-pricin, accessed 12-01-2025.

[29] EDF Group, "2023 Management Report", March 2024, see https://www.edf.fr/sites/groupe/files/2024-03/edf-annual-results-2023-management-report-2024-03-01.pdf; and EDF, "Financial and Extra-Financial Performance, Key Figs. 2024", see https://www.edf.fr/en/the-edf-group/dedicated-sections/investors/financial-and-extra-financial-performance/financial-results, both last accessed 03-05-2025.

[30] Sarah White and Leila Abboud, "EDF chief and French government clash over strategy", Financial Times, 24 September 2023, see https://www.ft.com/content/eaee5a3d-ba76-40ba-8300-082f586574e8, accessed 27-09-2023.

[31] Benjamin Mallet, "Exclusive: France is weighing zero-interest loan for 6 nuclear reactors, sources say", Reuters, 27 November 2024, see https://www.reuters.com/business/energy/france-is-weighing-zero-interest-loan-6-nuclear-reactors-sources-say-2024-11-27/, last accessed 03-05-2025.

[32] Approximately $£_{2023}$40–43 billion (US-$\$_{2023}$50–53 billion).

[33] Approximately $€_{2023}$15 billion (US-$\$_{2023}$16 billion).

costs would reach €$_{2015}$6.7 billion[34] for an operations startup then scheduled for 1 July 2023 (Cour des comptes 2020). The total would thus reach €$_{2023}$23 billion (US-$$_{2023}$24.6 billion). However, financing and other cost estimates have not been updated to the revised startup schedule.

12.3.2 Rosatom

State-owned Rosatom controls most aspects of civilian nuclear power and military applications in Russia. The Rosatom Overseas subsidiary, "involved in Rosatom's corporate development, with stated goals of building up Rosatom's overseas nuclear power plant order portfolio and maintaining Russia's leading position in the global nuclear market," is already under US sanctions. However, despite the invasion of Ukraine, Rosatom's claimed exports rose 15% in 2022, and foreign orders were reportedly stable at US-$200 billion. The sanctions may be of limited efficacy so far, leading to the Biden administration imposing sanctions on the leadership of Rosatom in January 2025.[35] Whether the sanctions remain in place in the Trump administration remains to be seen. Russian nuclear fuels, though technically banned in the US, continue to be delivered as normal under special waivers.[36] Although the development of substitutes, such as enriched uranium through Urenco, will increase supply security over time, the US and Europe purchased US-$1.7 billion worth of Russian fuels and related products (of US-$2.2 billion total) during 2022. French imports of enriched uranium from Russia nearly tripled between 2021 and 2022, see Chap. 8 of this book. The firm reports its revenues have continued to climb during 2024, noting the importance of domestic defense work to its operations.[37] However, imports of uranium and nuclear fuel from Russia by the US and EU were down overall in 2024.[38]

---

[34] Approximately €$_{2023}$8 billion (US$$_{2023}$8.6 billion).

[35] U.S. Department of State, "Further Curbing Russia's Efforts to Evade Sanctions and Perpetuate its War against Ukraine: Fact Sheet," 12 April 2023, see https://2021-2025.state.gov/further-curbing-russias-efforts-to-evade-sanctions-and-perpetuate-its-war-against-ukraine/, and U.S. Department of State, "Sanctions to Degrade Russia's Energy Sector, Fact Sheet", 10 January 2025, see https://2021-2025.state.gov/office-of-the-spokesperson/releases/2025/01/sanctions-to-degrade-russias-energy-sector/, both last accessed 06-03-2025.

[36] Reuters, "Russia continues uranium deliveries as usual, and can supply to U.S. under special waivers – Rosatom," 18 November 2024, see https://www.reuters.com/markets/commodities/russia-continues-uranium-deliveries-usual-can-supply-us-under-special-waivers-2024-11-18/, last accessed 06-03-2025.

[37] Nuclear Engineering International, "Rosatom reports growth despite challenges", 18 October 2024, see https://www.neimagazine.com/news/rosatom-reports-growth-despite-challenges/?cf-view, last accessed 06-03-2025.

[38] Dmitry Gorchakov, "EU and US reduce Russian uranium and nuclear fuel purchases in 2024", 13 January 2025, Bellona, see https://bellona.org/news/nuclear-issues/2025-01-eu-and-us-reduce-russian-uranium-and-nuclear-fuel-purchases-in-2024, last accessed 06-03-25.

12.3.3 *China General Nuclear Power Group (CGN) and China National Nuclear Corporation (CNNC)*

CGN is rated well by rating agency Fitch due to its linkage to the Chinese government, which "appoints its key management and controls its major business and financial decisions." The company receives frequent funding from the government, including both direct support of CN-¥$_{2023}$2.3 billion (US-\$$_{2023}$325 million) in 2023 and CN-¥2.4 billion (US-\$356 million) in 2022) and equity infusions.[39] Both firms also receive significant government support from VAT collections, which are mostly refunded to the reactors. This provided more than US-\$350 million to CGN and US-\$360 million to CNNP in 2022.[40] Subsidized state credit is common for state-owned enterprises in China. CGN and the other major Chinese nuclear player, CNNC, have both been blacklisted by the US government, complicating their international expansion in the nuclear sector.

12.3.4 *Westinghouse*

Westinghouse Electric Company, a subsidiary of Toshiba at the time, declared bankruptcy in March 2017 due to losses on reactor projects in the US. The Vogtle reactors in Georgia and the VC Summer plant in South Carolina experienced large cost overruns and delays. Westinghouse agreed to pay US-\$$_{2017}$3.7 billion to Vogtle and US-\$$_{2017}$2.17 billion[41] to Summer to walk away from the projects. Westinghouse was purchased in 2018 from Toshiba by Brookfield Business Partners (BBP) and Brookfield Asset Management (BAM), both part of the Canadian holding company Brookfield Corporation. The estimated value at that time was US-\$$_{2018}$4.6 billion.[42] In October 2022, Brookfield Renewable Partners (51%) and nuclear fuel supplier Cameco (49%) purchased Westinghouse from BBP and BAM for US-\$7.9 billion.[43]

[39] Fitch Ratings, "China General Nuclear Power Corporation", 5 January 2023, see https://www.fitchratings.com/research/corporate-finance/china-general-nuclear-power-corporation-05-01-2023; and Fitch Ratings, "Fitch Upgrades China General Nuclear Power to 'A + '; Negative Outlook", 25 November 2024, see https://www.fitchratings.com/research/corporate-finance/fitch-upgrades-china-general-nuclear-power-to-a-negative-outlook-25-11-2024, both last accessed 6 March 2025.

[40] Stephen Ezell, "How Innovative Is China in Nuclear Power?", Information Technology & Innovation Foundation, June 2024, see https://www2.itif.org/2024-chinese-nuclear-innovation.pdf, accessed 10-11-2024.

[41] Approximately US-\$$_{2022}$4.4 and 2.6 billion.

[42] Approximately US-\$$_{2022}$5.4 billion.

[43] Tom Hals, "Toshiba reaches \$2.2 billion deal over SCANA's South Carolina nuclear project", 27 July 2017, see https://www.reuters.com/article/us-toshiba-accounting-westinghouse-scana/toshiba-reaches-2-2-billion-deal-over-scanas-south-carolina-nuclear-project-idUSKBN1AC3DN/ and Antoine Gara and Myles McCormick, "Westinghouse to be sold for \$7.9bn in sign of nuclear power revival", Financial Times, 11 October 2022, see https://www.ft.com/content/46df2aa9-0963-47a6-881c-f715a18a8527, accessed 20-08-2023.

The service portion of the business looked attractive since Westinghouse reactor technology is used in a substantial share of the world's nuclear reactors.

Westinghouse also has major investments in reactor development, through the AP100 PWR and the AP300, a smaller version of the AP-1000 currently under development. Both have continuing challenges in terms of production and deployment, and the reactors are not a standard line of business for Brookfield Renewable Partners. Westinghouse has been actively promoting its various reactor designs internationally, especially in Europe, including in Bulgaria, the Czech Republic, Poland, and Ukraine. In November 2024, the company signed an engineering services contract (jointly with South Korea's Hyundai Engineering & Construction) to develop two AP-1000 reactors at the Kozloduy nuclear power plant in Bulgaria.[44] A memorandum of understanding for four AP300 SMRs in the UK was also signed with Community Nuclear Power in early 2024, allowing site development and licensing activities to continue.[45]

12.3.5 *Korea Electric Power Corporation (KEPCO)*

KEPCO is majority-publicly owned, with 32.9% ownership by the Korea Development Bank and 18.2% ownership by the Government of Korea as of the end of 2022. The KEPCO Act requires state ownership of at least 51%, and shareholder rights by both state owners are "exercised by the Ministry of Trade, Industry and Energy in consultation with the Ministry of Economy and Finance."[46] Nuclear activities are primarily conducted within the Korea Hydro & Nuclear Power (KHNP) subsidiary.

Coal continues to be the largest share of generating assets owned by KEPCO, though nuclear is second and has grown from 34% of assets in 2012 to 38% in 2021. During this period, the firm's operating margin was hurt by high and volatile coal and LNG prices, with operating losses in four of the five years through 2022. Operating losses in 2022 were nearly US-$25 billion, the largest ever. The utility had been restricted from increasing rates by the government and claims its electricity prices are half those in Japan—which, by international standards, are themselves very high. One result was surging debt, which hit ₩200 trillion (US-$149 billion) for the first time

[44] Fitch Ratings, "Fitch Upgrades China General Nuclear Power to 'A + '; Negative Outlook", 25 November 2024, see https://www.fitchratings.com/research/corporate-finance/fitch-upgrades-china-general-nuclear-power-to-a-negative-outlook-25-11-2024, last accessed 06-03-2025.

[45] Sonal Patel, "UK's First SMR Nuclear Project to Showcase Four Westinghouse AP300 Reactors", Power Magazine, 8 February 2024, see https://www.powermag.com/uks-first-smr-nuclear-project-to-showcase-four-westinghouse-ap300-reactors/, last accessed 06-30-2025.

[46] KEPCO, "Form 20-F," filing to the US Securities and Exchange Commission, Korea Electric Power Company, 29 April 2023, see https://www.sec.gov/Archives/edgar/data/887225/000119312523124899/d427740d20f.htm, accessed 27-09-2023.

in its history and approaching the limits of its bond issuing capacity.[47] Losses within KHNP were over ₩$_{2023}$1 trillion (US-$\$_{2023}$763 million) on sales of ₩$_{2023}$4 trillion (US-$\$_{2023}$3 billion) during the first half of 2023.[48] Five price increases were allowed by the regulators in 2023 and have helped to stabilize revenues. KHNP had a profit for both the full year 2023 and the first nine months of 2024 (KEPCO 2024a, 2024b). Debt levels are down from 2022, though still high at roughly ₩$_{2023}$150 trillion in 2023 and 2024. They are expected to remain at that level.[49]

KEPCO's debt levels would likely be unsupportable were it a private company. With majority state ownership, "KEPCO has benefited from the sovereign ratings umbrella and received a rating that is six to eight notches higher than its baseline credit assessment, despite having difficulties with cash flows and debt service. This has enabled KEPCO to have good financing options despite weak underlying financial fundamentals" (KEPCO 2023, 47). Credit rating firm Fitch notes that the ratings on KEPCO are equal to those of the state (AA-/Stable) "under Fitch's Government-Related Entities (GRE) Rating Criteria, reflecting a very strong likelihood of government support."[50]

12.3.6 *Constellation Energy Corporation (CEG)*

Constellation Energy is the largest operator of nuclear power plants in the US. It was created through a tax-free spinoff of the power generation business out of Exelon, which took effect in February 2022. While some regulated generation transmission and distribution assets remained within Exelon (EXC), all of Exelon's nuclear assets were spun into the new company. CEG includes many non-nuclear generation assets as well. Both are private firms. Constellation controlled more than 20 percent of US nuclear capacity in 2019. Calculations made by Schneider et al. (2023) indicate nearly 40 percent of its reactors (comprising nearly 1/3 of its production capacity) received

[47] Jung Suk-yee, "KEPCO Debt Surpasses 200 Trillion Won for First Time in History", Business Korea, 23 August 2023, see https://www.businesskorea.co.kr/news/articleView.html?idxno=200147, accessed 27-09-2023.

[48] KEPCO, "Korea Electric Power Corporation and Its Subsidiaries, Interim Consolidated Financial Statements - For the six-month periods ended June 30, 2023 and 2022," Korea Electric Power Company, 11 September 2023, p. 47.

[49] S&P Global, "Korea Electric Power Corp.'s Proposed U.S. Dollar Senior Unsecured Notes Rated 'AA'", 3 February 2025, see https://disclosure.spglobal.com/ratings/pt/regulatory/article/-/view/type/HTML/id/3317948, last accessed 06-03-2025.

[50] Fitch Ratings, "Korea Electric Power Corporation," 19 July 2024, see https://www.fitchratings.com/research/corporate-finance/korea-electric-power-corporation-19-07-2024, last accessed 06-03-2025.

state support as of 2021.[51] Further, of the twelve reactors receiving targeted state (subnational) support to nuclear power in 2021, nine were owned by Constellation. State subsidies are a significant contributor to returns on Constellation's nuclear business.

12.3.7 Tokyo Electric Power Company (TEPCO)

The Government of Japan has held a majority stake in TEPCO since the firm's impending collapse soon after the Fukushima accident in March 2011. The government has targeted a profit margin of JP-¥450 billion ($\text{US-}\$_{2023}$3 billion) as part of the reconstruction plan, but the utility has fallen short in recent years. In 2022, TEPCO experienced its first net loss in a decade and was unable to fund its annual contribution to a Fukushima victim's fund. The firm needed US-$3 billion in emergency loans from a consortium of banks in early 2023 to help stabilize the utility in the face of surging fuel prices and a weakening Yen.[52] TEPCO returned to profitability in 2023 (TEPCO 2024).

Cleanup challenges at Fukushima remain daunting. In 2023, the first batch of radioactive wastewater from Fukushima was released into the sea despite public and international opposition and commercial concerns about markets for Japanese seafood. Decommissioning the Fukushima Daiichi plant has "barely progressed" and removal of the melted nuclear fuel did not even begin until September 2024, with decommissioning expected to take at least 30 years.[53] Estimates on the costs to deal with the accident cleanup costs and victim compensation continue to rise. Schneider et al. (2017) noted a doubling of estimates in 2016 to JP-¥_{2016}22 trillion ($\text{US-}\$_{2022}$223 billion). In 2019, a private think tank, the Japan Center for Economic Research, said that compensation, decommissioning, and decontamination costs were expected to reach JP-¥_{2019}41 trillion ($\text{US-}\$_{2022}$430 billion) for a scenario in which Fukushima water was diluted and discharged into the sea.[54] As of June 2023, the amount paid out by TEPCO in compensation alone already totaled JP-¥_{2023}10.8 trillion ($\text{US-}\$_{2023}$77.5 billion) (TEPCO 2023).

---

[51] Nuclear Energy Institute, "U.S. Nuclear Plant Owners and Operators," January, 2024, see https://www.nei.org/resources/statistics/us-nuclear-plant-owners-and-operators, last accessed 06-03-2024.

[52] Jun Watanabe, "Japan's TEPCO to receive $3bn in emergency loans from top lenders", Nikkei, 15 February 2023, see https://asia.nikkei.com/Business/Energy/Japan-s-TEPCO-to-receive-3bn-in-emergency-loans-from-top-lenders, accessed 20-09-2023.

[53] Mari Yamaguchi, "A robot begins removal of melted fuel from the Fukushima nuclear plant. It could take a century", Associated Press, 10 September 2024, see https://apnews.com/article/japan-fukushima-reactor-melted-fuel-robot-9ffc309fb072580bee0161e8a24c8490, last accessed 06-03-2025.

[54] Kiyoshi Takenaka, "More nuclear challenges await Japan after Fukushima water release," Reuters, 24 August 2023, see https://www.reuters.com/world/asia-pacific/more-nuclear-challenges-await-japan-after-fukushima-water-release-2023-08-24/, accessed 27-09-2023.

Litigation over the accident at Fukushima continues. A case brought to court by shareholders led to a 2022 civil court ruling of US-$95 billion in liability for four former TEPCO executives for damages associated with the Fukushima accident—the first time a court had found top executives liable. Courts in 2019 had found the executives not criminally liable in the accident. This original verdict of not guilty was upheld by the Tokyo High Court in 2023.[55]

12.4 Hidden Costs

12.4.1 Nuclear Decommissioning

Facility decommissioning is an Asset Retirement Obligation (ARO). AROs are not unique to nuclear power and commonly include substantial end-of-lifetime costs that require outlays after revenues from business operations have ceased. This creates a high risk of "liability dumping," where assets are stripped from the firm, and end-of-life costs are shifted to taxpayers through bankruptcy or abandonment. Corporate actions such as structural separation, sales, or spinoffs may be used to isolate lines of business with high liabilities from more profitable ones. Large closure costs, radioactivity, and an extremely long period for which waste must be managed create more significant challenges for the nuclear sector than for most other sectors with AROs. Because the decommissioning costs are incurred at the end of a facility's lifetime, and LCOE calculations discount those costs (often over many decades), decommissioning appears immaterial to the sector's initial investment decisions. At a 10% cost of capital, for example, the International Energy Agency (IEA) estimates the cost of decommissioning for a newbuild plant at about US-$0.01/MWh, or a maximum (at a 3% cost of capital) still below US-$0.40/MWh for all countries but the Slovak Republic (IEA, NEA, and OECD 2020). The IEA's cost models suggest that an increase in decommissioning costs from 15 to 25% of total investment costs will raise generating costs by only 1%. When accruals are collected throughout the entire operation of the plant, the share of electricity prices is small (IEA 2022).

Unfortunately, adequate funds are often not collected during the full operation of the facility. In other cases, collected funds have been misappropriated due to structural weaknesses in controls. The near-disappearance of long-term liabilities in levelized cost models (due to discounting practices) can create problems because the liabilities cost hundreds of millions of dollars per reactor. Further, past experiences suggest that the prospective estimates are bad predictors of actual costs that will be incurred, and

---

[55] -Reuters, "Tokyo court orders ex-Tepco execs to pay $95 bln damages over Fukushima disaster", as published on CNN Business, 13 July 2022, see https://www.cnn.com/2022/07/13/business/tokyo-court-fukushima/index.html; and Tom Bateman and Sakura Murakami, "Tokyo court upholds not guilty verdict for ex-Tepco execs over Fukushima disaster," Reuters, 18 January 2023, see https://www.reuters.com/world/asia-pacific/tokyo-court-upholds-not-guilty-verdict-ex-tepco-execs-over-fukushima-disaster-2023-01-18/, both accessed 27-09-2023.

funding levels often lag far behind even the estimated costs of what will be needed. Decommissioning cost estimates for advanced gas-cooled reactors in the UK doubled from 2004 to 2021 to £$_{2021}$23.5 billion (US-\$$_{2021}$32 billion), and are at significant risk of rising further (House of Commons 2022). A detailed reactor-level analysis estimates decommissioning costs for Germany, Italy, and Lithuania at €$_{2020}$5.98 (US-\$$_{2020}$6.82), €$_{2020}$14.05 (US-\$$_{2020}$16.02), and €$_{2020}$13.78 (US-\$$_{2020}$15.71) per MWh, respectively, for high-capacity commercial reactors[56]—significantly larger than the IEA estimates and at a level that could affect the competitiveness of nuclear, especially on wholesale markets. Though countries have implemented a wide array of mechanisms to fund decommissioning costs (discussed below), in all cases, taxpayers end up covering shortfalls, and these shortfalls can be very large. Decommissioning programs need to address a handful of core elements effectively, each discussed in turn.

12.4.1.1 Who Must Cover the Cost?

In most countries, utilities are legally obligated to fund decommissioning. This structure theoretically aligns with the "polluter-pays" principle—built into the IAEA Joint Convention on the Safety of Spent Fuel Management and on the Safety of Radioactive Waste Management—which links the costs of addressing pollution to the products, services, and companies that generated it (Bärenbold et al. 2024).

Despite this objective, however, decommissioning costs in many jurisdictions have been shifted in part, or fully, onto the state, creating a de facto subsidy to nuclear. A recent IAEA review of decommissioning funding in 58 countries (covering fuel chain facilities and research reactors as well as power plants) indicated that although more than 60% of the countries officially rely only on user funding, many do not (see Table 12.2) (IAEA 2023). Twelve countries rely solely on government funding for decommissioning, and five are reliant on some government funding either due to statutory structure or a shortfall in collected funds. For all groups, long-term shortfalls in funding are backstopped by governments, making funding adequacy of great importance and largely defeating the purpose and distorting the reality of nominal user funding. For an analysis of the organization of decommissioning funds, see Bärenbold et al. (2024).

12.4.1.2 How Much Funding Must Be Set Aside?

Decommissioning funding targets are set by statute or regulation, or by the utilities. Unfortunately, there remains limited empirical data on decommissioning: as of July 2024, only 23 out of more than 200 closed reactors had been decommissioned. These were concentrated in only four countries: the US (17 reactors), Germany (4),

[56] See supplementary material at https://github.com/alexwimmers/nuclearPower_TEG for data and calculation.

Table 12.2 Funding mechanisms for nuclear waste management and decommissioning

Funding Mechanisms	Funding management of spent fuel and high-level radioactive waste		Funding of decommissioning	
	Number of countries	Share of countries	Number of countries	Share of countries
Government Funding	10	21.3%	12	24.5%
Government Funding b/c State Owned Enterprise	4	8.5%	2	4.1%
Some Government Funding by Design or Shortfall*	6	12.8%	5	10.2%
User Funding Only*	27	57.4%	30	61.2%

*Note that all countries rely on governments to cover any shortfalls
Source Author compilation using data from (IAEA 2023)

Japan (1), and Spain (1). Of the closed reactors, a scant 4.4% of the sites (11 reactors) have been returned to greenfield sites for unrestricted use (Schneider et al. 2024). As a result, decommissioning estimates most often rely on engineering models, some based on studies done many decades ago. Underfunding of decommissioning reserves is a concern in many countries.[57] These risks are exacerbated by core assumptions in the modeling, given that "[s]mall changes in the assumptions of the rates have tangible effects on the present value of the financial resources and hence the amount of funds that need to be set aside; in particular, when the rate of return (discount rate) is prone to overestimation and the cost escalation rate to underestimation" (Besnard et al. 2019, 78).[58]

Funding targets in the US rely either on a regulatory minimum or a site-specific engineering estimate, both of which use cost models. These do a reasonable job estimating mean funding levels, but even with a 25% contingency margin, under-funding is common. Decommissioning cost estimates for US reactors span a range of US-\$478–1,435/kWe for publicly owned reactors and US-\$615–2,148/kWe for investor-owned reactors. The reasons behind the much higher cost projections for investor-owned utilities are unknown (Moriarty 2022; Lordan-Perret et al. 2023). Detailed European case studies highlighted substantial aggregate shortfalls between provisioned funds for decommissioning and the expected costs. This gap amounted to estimated €$_{2020}$ 10 billion in France (US-\$$_{2023}$ 10.9 billion), €$_{2020}$6 billion in Germany

[57] Doug Koplow, "B5: Other Federal Interventions Into Energy Markets", pp.10–11, in "Federal Energy Subsidies: Energy, Environmental and Fiscal Impacts—Report and Appendices," The Alliance to Save Energy, April 1993, B5-10–11, see https://www.earthtrack.net/sites/default/files/library/FedSubAppB5.pdf, accessed 22-07-2023.

[58] A critical analysis of the funding structure and accounting procedures of the UK's Nuclear Decommissioning Authority is provided by Thomas (2023).

(US-$$_{2023}$6.6 billion), and US-$3.3 billion in Sweden (Wimmers et al. 2023). Provisioning estimates in EU countries generally refer to the projected cost of the service, not how much cash has already been set aside; thus, even if provisioning is not showing a shortfall against other estimates of expected need, there could still be funding gaps.

In France, several reports over the past decade have seriously questioned the adequacy of decommissioning provisions. In 2015, independent financial analyst AlphaValue published a commercial report entitled "Électricité de France – What a mess!" It concluded that "EDF's dismantling costs are underprovided by a factor of three on a 'private funding' basis per MW installed." AlphaValue further noted that: "[b]ased on today's levels, the German peers end up with more than €4 [billion] per reactor by 2025 (€2.5 [billion] current actuarial), while EDF may only have set aside €1 [billion] for that rainy day. At current levels and discount rates, EDF would achieve the €4 [billion] per reactor mark by [...] 2057. [...] This is [eleven] years behind their best case scenario: a 60-year reactor life." (Rodriguez 2015, 27).

A 2017 report by the French National Assembly on nuclear decommissioning confirmed the conclusions of AlphaValue. First, decommissioning is assumed to take more time than anticipated. The authors note that the temptation will, therefore, be great for the utility (EDF) to spread out the dismantling over time to compensate for the low level of provisions. Secondly, provisions are among the lowest in the OECD, with no safety net in case of cost variances. Thirdly, decommissioning cost is underestimated because several items are not included in the estimates and an optimistic discount rate of 4.4% was applied. Finally, foreign dismantling operations are considered to be more expensive, questioning EDF's cost assumptions in general (Assemblée Nationale 2017).

Data on US shortfalls reflect the gap between Nuclear Decommissioning Trust Fund (DTF) balances (i.e., actually funded) and anticipated decommissioning costs (Lordan-Perret et al. 2023). These exceed US-$17 billion nationally, which the Callan Institute translates to about US-$1 billion per year in extra collections to fill the gap during the remaining operating license of the reactors (Moriarty 2022). Outside of the US, much greater transparency on changes in provisioning estimates over time and the portion of those estimates that have actually been funded would be useful.

12.4.2 Nuclear Waste Management

Management of nuclear waste shares many of the challenging economic attributes of decommissioning liabilities. One is the scale of the task. While other industries also have waste, the technical requirements for nuclear waste are bigger, more expensive, and extend over a much longer period. Ideally, the costs of managing these wastes also follow the polluter-pays principle, in that funding is supposed to come from the operator during the facility's lifetime. As with decommissioning (though even more so), the costs are highly uncertain, and estimating the required funding is challenging to do with any accuracy. Another similarity: with repository opening dates so far in the future, LCOE models tend to discount away the necessary accruals, thereby minimizing the impact of this complex problem on current economic decisions and selected energy pathways.[59] Nuclear waste funds also require careful segregation and management to ensure they are available to achieve the purpose for which they were collected. Funds are generally supposed to be ring-fenced to prevent diversion, though this has not always happened in practice (OECD/NEA 2013).

The fission process yields spent fuel that is often managed on-site while its radioactivity decays (close to half of spent fuel worldwide is currently stored on-site); low- and intermediate-level radioactive waste, for which there are some existing repositories; and other radioactive waste streams (see Chap. 16). Government policies that pay most or all of the cost of managing nuclear waste create a subsidy for nuclear power. This comes partly through the repository contributions that were not charged to nuclear power plant owners and customers like they should have been. More importantly, however, the long-tail risk from the uncertainty and high potential costs of managing nuclear waste would, by itself, chill investor excitement about the sector and confidence in the predictability of returns. This would be reflected in a higher cost or reduced availability of capital. At least as far back as 1980, it was recognized that "[g]overnment responsibility for ultimate waste disposal removes significant uncertainties from those investing in nuclear power production" (Bowring 1980). The simple willingness by governments to take legal responsibility for the waste constitutes a very large subsidy to the sector.

Strategies for waste management via reprocessing and reuse have been expensive and of concern, especially in terms of proliferation risks involving large quantities of weapons-usable materials (plutonium). Most countries are focused instead on deep geological repositories (see Chap. 16). While more than half of the countries plan to pay these costs solely through fees on generators, a great deal of residual risk and cost remains with taxpayers.

Globally, the construction of deep repositories for radioactive waste has suffered considerable delays. There is no sufficient worldwide reference for the planned concept, although Finland plans to begin operating their repository sometime in the mid-2020s (Scheer et al. 2024).[60] As a result, the true cost will only be known decades after the last nuclear power plant has been decommissioned—at a point when the owners will most likely have ceased to exist in their present form. Therefore, the owners will not be able to make any additional payments.

The enormous scale of anticipated costs compounds this risk. Estimates for the US reach as high as US-$168 billion; for France, US-$35 billion, including high-, intermediate-, and low-level radioactive waste. For Canada, the estimate is CA-$26 billion (US-$19 billion); US-$140 billion for Germany (all radioactive waste

[59] Refer to Schulze et al. (1981) for a theoretical assessment of just discount rates in an intergenerational context.

[60] World Nuclear News, "Successful start to trial run at Finnish repository", 20 September 2024, see https://www.world-nuclear-news.org/articles/successful-start-to-trial-run-at-finnish-repository, last accessed 03-06-2025.

streams); roughly US-$17 billion for Sweden (SKB); and US-$21 billion for Switzerland (Schneider et al. 2023). While these figures do not all estimate the exact same services over the same time frame, it is clear the repositories are expensive and, based on cost escalation patterns for projects of this scale, costs are likely to rise.

Even where some of the countries assign responsibility to utilities (such as Sweden and Finland), those responsible for building and managing the repository through a multi-utility partnership are themselves largely state-owned. While the challenge of construction of deep geological storage has been known for decades, delivery of a viable option continues to be kicked further into the future nearly everywhere. Many countries still have not set timelines for repository completion, have pushed forward previously set targets, or have set deadlines that remain decades into the future, see Chap. 16 in this book.

Interestingly, the push for so-called small modular reactors (SMR) may exacerbate the nuclear waste management challenge. Thorium-232/uranium-233 fuel chains will not produce less radioactive waste than uranium fuels, as is commonly perceived. Some designs appear to produce more waste than current light-water reactors. Analysis of three distinct SMR designs, for example, found that "relative to a gigawatt-scale [pressurized water reactor], these reactors will increase the energy-equivalent volumes of [spent nuclear fuel], [and] long—[…] and short-lived [low- and intermediate-level waste] by factors of up to 5.5, 30, and 35, respectively. These findings stand in contrast to the waste reduction benefits that advocates have claimed for advanced nuclear technologies" (Krall et al. 2022, 2). Waste streams may also be more chemically unstable with a higher risk of recriticality.

12.4.3 Accident Liability

Insurance helps to protect individuals and businesses against catastrophic losses. The cost of insurance premiums signals which activities are riskier and where investments in training, safety equipment, or product enhancements could reduce insurance costs, while also providing improved safety for workers and customers. Because not every energy pathway entails the same risks to people, property, and the environment, accurate insurance coverage pricing can incentivize shifts to lower-risk options, including enhanced nuclear plant safety. Where government rules allow some firms and activities to operate with insufficient insurance limits to realistically address damages they can cause others, this constitutes an operating subsidy.

Inadequate or subsidized insurance to the nuclear sector is common worldwide. This includes offsite damages from accidents at nuclear plants or fuel chain facilities, as well as risks during transportation. Required coverage levels for smaller reactors (even if grouped on a single site), fuel cycle facilities, and transport tend to be lower, and sometimes significantly lower, than for conventional nuclear power plants.

Focusing on reactor accidents as an example, liability requirements for offsite damage are set by domestic statute. Additional tiers may be provided by national governments once the operator liability limit is reached, and then by a third tier of coverage provided by a series of international treaty agreements (which include the Paris Convention, Vienna Convention, various Joint Protocols and Supplementary Conventions). The treaties help to set a coverage floor and to standardize coverage somewhat across very diverse countries. However, even the total coverage in the US, which is the largest liability pool for nuclear accidents in the world, is well below the expected damages from even a moderate accident. For example, the Japanese Government estimated the cost of the 2011 Fukushima disaster at US-$\$_{2021}$223 billion, about 16 times the total domestic US insurance pool of US-$\$_{2024}$16.1 billion[61] (CRS 2024). In 2019, an independent assessment by the Japan Center for Economic Research (JCER) came up with a cost range of US-$\$_{2021}$322–758 billion, with the higher end including US-$\$_{2021}$370 billion for tritium removal (Schneider et al. 2021). This range represents 20–47 times the total US nuclear insurance pool. Even where accident insurance is required of operators, the cost of the coverage is sometimes paid for by the state. While this structure is helpful to protect parties who may be harmed in an accident, it results in larger subsidies to the operator and weak price signals to invest in safety-enhancing upgrades and training or shift to other energy sources.

OECD-NEA data indicate that Germany, Japan, and Russia have instituted unlimited liability on operators. Finland has unlimited operator liability for damages inside the country only, and Sweden provides it internally and to selected neighboring countries.[62] Financial guarantees on operators for accidents are often much lower, however, and provide a useful proxy for the financial exposure for reactor owners to cover costs following an incident. Unlimited coverage in countries where state ownership dominates is a little different from a state guarantee. In other nations, it will still be limited by firm bankruptcy, as occurred with TEPCO in Japan following the Fukushima accident in March 2011 (OECD/NEA 2024).

Table 12.3 summarizes liability coverage in US dollars that applies to the largest countries, sorted by nuclear-generating capacity. Many countries have less than US-$1 billion in coverage from all sources, and some important nuclear power countries appear to have very low liability capacity. This includes the equivalent of less than US-$715 million in India, less than US-$400 million in South Korea, and only US-$150 million in China.

12.4.4 Security and Proliferation

Radioactivity, and the connection between the nuclear fuel chain (and associated knowledge, skills, equipment, and technology) and potential weapons proliferation, both create unique risks associated with nuclear energy (see more detailed assessments of proliferation risks in Part III (Geopolitics) of this book). When such costs

[61] Approximately US-$\$_{2021}$13.9 billion. Note that as of October 2020 (before adjustments to the Price-Anderson-Act), the maximum covered cost was about US-$\$_{2021}$13.5 billion (U.S. NRC 2021).

[62] "Unlimited liability for damage suffered in Sweden, Denmark, Finland and Norway and other States with no nuclear installations" (OECD/NEA 2024, 24).

Table 12.3 Maximum liability coverage for nuclear accidents as of 2024

Country	Share of 2023 Global Nuclear Capacity (Operating, LTO, Under Construction)	Operator Amount In Million US-$ (Note 1)	Domestic Supplement In Million US-$, after operator amount exhausted	International Treaty Supplement In Million US-$	Max, all tiers In Million US-$ (Note 2)
United States	21.55%	16,097	Guarantee of operator amount; additional with Action by Congress	136	16,233
China	17.25%	42	111 m; additional with State Council approval	NL*	153
France	14%	751	1,288	322	2,361
Japan	7.34%	UNL/777	Action by Diet	136	UNL/913
Russia	6.78%	UNL	Available but amount not listed	NL	UNL
South Korea	6.34%	397	NL	NL	397
Canada	3.03%	718	Action by Parliament	136	854
India	2.85%	178	397	136	711
United Kingdom	2.03%	1,288	1,288	322	2,897
Spain	1.58%	1,288	1,288	322	2,897
Sweden	1.54%	UNL to selected neighbors, else reciprocity/ 1,288	1,288	322	UNL/2,897
Finland	0.98%	UNL inside; 1,288 max outside	1,288		UNL/2,897
Czech Republic	0.87%	339	NL	NL	339
Belgium	0.87%	1,288	1,288	322	2,897
Switzerland	0.66%	UNL/1,288	1,288	322	2,897
Slovakia	0.61%	322	NL	NL	322

(continued)

Table 12.3 (continued)

Country	Share of 2023 Global Nuclear Capacity (Operating, LTO, Under Construction)	Operator Amount In Million US-$ (Note 1)	Domestic Supplement In Million US-$, after operator amount exhausted	International Treaty Supplement In Million US-$	Max, all tiers In Million US-$ (Note 2)
Hungary	0.43%	132	226	NL	358
Germany	/	UNL/2,862	2,682	322	5,686

LTO = Long-term outage; NL = Value not listed in source documents; UNL = unlimited liability per country statutes

Sources Author compilation using data from (OECD/NEA 2024; CRS 2024). *Notes* 1. For countries with unlimited operator liability by statute, the maximum required financial security operators need to demonstrate is also shown. 2. The maximum pool shown reflects the sum of operator's liability plus domestic and international supplements. For countries with unlimited operator liability, the financial security amount is incorporated into the quantified totals; and where there is preferential treatment for some countries, the amount available generally is used. Values not listed are assumed to be zero.

are shifted to the public, rather than recovered from the nuclear sector and reflected in prices, competitors to nuclear are disadvantaged.

In addition to the risk of diversion of weapons-related materials from the nuclear fuel chain, civilian nuclear activities can also contribute to latent proliferation. More facilities, associated training of more people, supply chains with potential dual use, and an ability to conceal military activities under the cloak of civilian power production, can all shorten the timeline for a country to become a nuclear weapons state. The IEA acknowledges that if many SMRs were built around the world, this would increase proliferation risks (IEA 2022).

These risks appear more significant for some reactor designs than for others. The most problematic are designs that entail or encourage the production, shipment, and use of direct weapons-usable materials, like separated plutonium or high-enriched uranium. The use of high-assay low-enriched uranium (HALEU)[63] is projected to grow to support a number of the current reactor concept designs and is an area of particular concern "because of the potentially greater attractiveness of this material for nuclear weapons compared with the low-enriched uranium used in light water reactors" (National Academies of Sciences, Engineering, and Medicine 2023, 182).

The Union of Concerned Scientists assessed the main emerging reactor technologies for their impact on plant safety and proliferation risks. It is notable that most of the designs were significantly worse in terms of safety, and generally worse in proliferation and terrorism risks. Reliance on HALEU is a common thread for many of

[63] HALEU is enriched to just below 20% in fissile Uranium-235 considered the limit of low enriched uranium. However, technically, the 20% enrichment level constitutes a significant step towards direct weapons usable high enrichment levels.

these (Lyman 2021). This is supported by other literature, e.g., Lehtveer and Hedenus (2015) or Pistner et al. (2024).

In terms of plant security, the design basis threat defines the core security risks for which reactors are expected to prepare. It is clear that these parameters do not adequately reflect current threats. Nuclear plants have been targets in the Russian invasion of Ukraine,[64] and damage to essential safety-relevant equipment affecting electricity supply or cooling chains, including intentional damage, has been, and remains, a continuing concern. Even in 2019, drone attacks significantly damaged oil facilities in Saudi Arabia; capabilities have greatly expanded since then. That same year, the US NRC declined to demand key nuclear facilities, including nuclear power plants, develop a capability to defend against unmanned aerial vehicles. In September 2019, a number of drones flew around restricted portions of the Palo Verde Nuclear Power Plant in the US State of Arizona—an intrusion that plant owners were unable to stop (Schneider et al. 2022). Reported drone overflights of nuclear power plants seem to be growing. One analysis of US reactors found drone sightings between December 11 and 17, 2024 equal to all of the reported sightings in 2024 up to that point.[65] War game scenarios indicate drone risks at nuclear power plants to be a significant security risk.[66]

Efforts to quantify the cost of addressing specialized security risks of the nuclear fuel cycle are sparse, perhaps because spending is "spread across a hard-to-fathom number of budget lines." One recent estimate puts it at US-$4 billion per year to the US industry, with an additional US-$1.1 billion annually spent on international operations by the US government.[67] Similar costs incurred by other countries would be additional. Stirling and Johnstone (2024) disentangle the intertwined spending and budgetary shifts in the UK civil and military sectors. They find that at least £2 billion of taxpayer money dedicated to civilian purposes is covertly used for military applications, upping annual nuclear military spending by 16–20%.

---

[64] For example, in February 2025, the sarcophagus of the Chornobyl Unit 4 was hit by a Russian drone and damaged. Daria Tarasova-Markina, Tim Lister and Christian Edwards, "Ukraine says Russia drone attack hits Chernobyl nuclear plant, radiation levels normal", CNN, 14 February 2025, see https://edition.cnn.com/2025/02/14/europe/russia-ukraine-drones-chernobyl-intl-hnk/index.html, last accessed on 18-02-2025.

[65] Howard Altman, "Nuclear Power Plants Report Massive Uptick In Drone Sightings", The War Zone, 21 December 2024, see https://www.twz.com/news-features/massive-uptick-in-official-drone-sightings-by-nuclear-power-plants, last accessed on 03-06-2025.

[66] Henry Sokolski, "Memo to Trump: Address the new threat of drone-vulnerable nuclear reactors", Bulletin of the Atomic Scientists, see https://thebulletin.org/2025/01/memo-to-trump-address-the-new-threat-of-drone-vulnerable-nuclear-reactors/, last accessed 03-06-2025.

[67] Michal Barnard, "Nuclear Security Represents $4 Billion Annual Subsidy In US, Trillion All Costs For Fleet For Full Lifecycle," Clean Technica, 29 March 2021, see https://cleantechnica.com/2021/03/29/nuclear-security-represents-4-billion-annual-subsidy-in-us-trillion-for-fleet-for-full-lifecycle/, accessed 23-07-2023.

12.5 Conclusion

The nuclear industry's growing reliance on substantial state involvement to bolster finances, subsidize operations, and shift risks obscures the true costs of nuclear power. Subsidized finance, as well as hidden costs, such as decommissioning, nuclear waste management, accident liability, and security concerns, are often excluded from levelized cost comparisons, leading to an underestimation of the economic burden on the public and distorting the competitive landscape for alternative energy sources.

The increasing role of state actors across the nuclear value chain, including subsidies to technology development, new plant construction, and continued operation to existing facilities, highlights the inability of nuclear power to compete in competitive energy markets. Key players in the sector are either state-owned or kept afloat through substantial governmental aid.

Better information on the scale of these subsidies across countries would enable a more accurate comparison across energy options. Gathering the data is challenging, though the pattern is consistent when it has been done. For example, decommissioning presents a significant financial burden, with historical data showing that actual expenses frequently surpass initial estimates and unfunded costs are often shifted to taxpayers. Nuclear waste management adds another layer of complexity. The long-term nature of waste disposal, particularly for high-level radioactive waste, means that the costs and technical challenges will persist for many decades, if not centuries. The reliance on government-backed solutions for waste management reduces the financial incentives for nuclear operators to fully internalize these costs, leading to potential underinvestment in safer, more efficient waste management technologies or alternative energy resources that do not generate such challenging waste products. This situation is further complicated by the security risks associated with the nuclear fuel cycle, including the potential for weapons proliferation. The growing interest in SMRs and the potential use of HALEU fuel present new proliferation risks. Current regulatory frameworks likely need improvement in order to address this adequately. The high share of public ownership of the nuclear industry, and the economic challenges facing both public and private nuclear companies, exacerbate risks to taxpayers. Absent ever larger public subsidies, the economic challenges for nuclear expansion are likely for the foreseeable future.

Given these challenges, two key policy recommendations emerge. First, more stringent regulatory frameworks that require nuclear operators to fully internalize the costs of decommissioning, waste management, and security are needed. This could involve stricter financial provisions to estimate the costs of decommissioning, ensuring that funds are adequately safeguarded throughout a plant's operational life outside the control of the plant owner or operator, and mandating insurance products for a group of reactors to create a fund for cost overruns that is adequately large to cover most funding shortfall scenarios.

Second, international cooperation and strengthened non-proliferation agreements are essential to address the latent proliferation risks associated with emerging nuclear technologies. Expanding the scope and enforcement of treaties, like the Paris and

Vienna Conventions, to cover new reactor designs and fuel cycles would be one important step to ensure that the international community is better equipped to manage the security risks posed by a global expansion of nuclear power. The current costs of monitoring proliferation need to be more carefully tracked as well.

Nuclear power can theoretically provide benefits through low-carbon energy. However, these advantages must be weighed against the significant, and often underappreciated, financial, environmental, and security costs. A more transparent accounting of these costs, coupled with robust policy frameworks, is essential to ensure that policymakers, investors, and utility customers are well-informed about the actual cost of nuclear pathways when weighing investment decisions both in nuclear and other low-carbon alternatives.

References

Assemblée Nationale. 2017. 'Rapport d'Information déposé en application de l'article 145 du règlement par la mission d'Information relative à la faisabilité technique et financière du démantèlement des installations nucléaires de base'. Paris: Commission du développement durable et de l'aménagement du territoire. http://www.assemblee-nationale.fr/14/pdf/rap-info/i4428.pdf.

Bah, Muhammad Maladoh. 2023. 'State and Federal Nuclear Support Schemes in Dynamic Electricity Market Conditions: Insights from NYISO and PJM'. *Energy Policy* 182 (November):113764. https://doi.org/10.1016/j.enpol.2023.113764.

Bärenbold, Rebekka, Muhammad Maladoh Bah, Rebecca Lordan-Perret, Björn Steigerwald, Christian Von Hirschhausen, Ben Wealer, Hannes Weigt, and Alexander Wimmers. 2024. 'Decommissioning of Commercial Nuclear Power Plants: Insights from a Multiple-Case Study'. *Renewable and Sustainable Energy Reviews* 201:114621. https://doi.org/10.1016/j.rser.2024.114621.

Besnard, Manon, Marcos Buser, Ian Fairlie, Gordon MacKerron, Allison Macfarlane, Eszter Matyas, Yves Marignac, Edvard Sequens, Johan Swahn, and Ben Wealer. 2019. 'The World Nuclear Waste Report. Focus Europe'. Berlin, Brussels. https://worldnuclearwastereport.org/wp-content/themes/wnwr_theme/content/World_Nuclear_Waste_Report_2019_Focus_Europe.pdf.

Böse, Fanny, Alexander Wimmers, Björn Steigerwald, and Christian von Hirschhausen. 2024. 'Questioning Nuclear Scale-up Propositions: Availability and Economic Prospects of Light Water, Small Modular and Advanced Reactor Technologies'. *Energy Research & Social Science* 110 (April 2024): 103448. https://doi.org/10.1016/j.erss.2024.103448.

Bowen, Matt, and Alec Apostoaei. 2022. 'Comparing Government Financing of Reactor Exports. Considerations for U.S. Policy Makers'. Columbia Center of Global Energy Policy. https://www.energypolicy.columbia.edu/wp-content/uploads/2022/08/NuclearFinance-CGEP_Report_111022-1.pdf.

Bowring, Joseph. 1980. 'Federal Subsidies to Nuclear Power: Reactor Design and the Fuel Cycle'. Analysis Report. United States Department of Energy. https://www.earthtrack.net/sites/default/files/EIA%20Nuclear%20Subsidies%201980.pdf.

Committee on Laying the Foundation for New and Advanced Nuclear Reactors in the United States, Nuclear and Radiation Studies Board, Board on Energy and Environmental Systems, Division on Engineering and Physical Sciences, Division on Earth and Life Studies, National Academy of Engineering, and National Academies of Sciences, Engineering, and Medicine. 2023. *Laying the Foundation for New and Advanced Nuclear Reactors in the United States*. Washington, D.C.: National Academies Press. https://doi.org/10.17226/26630.

Cour des comptes. 2020. 'La filière EPR'. 9 July 2020. https://www.ccomptes.fr/fr/publications/la-filiere-epr.

Cour des comptes. 2025. 'La Filière EPR: Une Dynamique Nouvelle, des Risques Persistants'. Rapport de suites. https://www.ccomptes.fr/sites/default/files/2025-01/20250114-La-filiere-EPR%20-une-dynamique-nouvelle-des-risques-persistants_0.pdf.

CRS. 2024. 'Price-Anderson Act: Nuclear Power Industry Liability Limits and Compensation to the Public After Radioactive Releases'. In Fous. Congressional Research Service. https://crsreports.congress.gov/product/pdf/IF/IF10821.

EDF. 2023. 'Consolidated Financial Statements at 31 December 2022'. Financial Report. https://www.edf.fr/sites/groupe/files/2023-02/annual-results-2021-consolidated-financial-statements-2023-02-17.pdf.

Göke Leonard, Alexander Wimmers, and Christian von Hirschhausen. 2025. Flexible nuclear power and fluctuating renewables? — An analysis for decarbonized multi-vector energy systems, *Energy Strategy Reviews 60*, https://doi.org/10.1016/j.esr.2025.101782

House of Commons. 2022. 'The Future of the Advanced Gas-Cooled Reactors - Third Report of Session 2022–23'. HC 118. London: House of Commons, Committee of Public Accounts. https://committees.parliament.uk/publications/22301/documents/164944/default/.

IAEA. 2023. 'Global Status of Decommissioning of Nuclear Installations'. No. NW-T-2.16. IAEA Nuclear Energy Series. https://www-pub.iaea.org/MTCD/Publications/PDF/PUB2023_web.pdf.

IEA. 2022. 'Nuclear Power and Secure Energy Transitions - From Today's Challenges to Tomorrow's Clean Energy Systems—Executive Summary'. International Energy Agency. https://www.iea.org/reports/nuclear-power-and-secure-energy-transitions/executive-summary.

IEA, NEA, and OECD. 2020. 'Projected Costs of Generating Electricity - 2020 Edition'. Paris: International Energy Agency (IEA), Nuclear Energy Agency (NEA), Organisation for Economic Co-Operation and Development (OECD). https://www.oecd-nea.org/jcms/pl_51110/projected-costs-of-generating-electricity-2020-edition.

KEPCO. 2023. 'Korea Electric Power Corporation and Its Subsidiaries, Interim Consolidated Financial Statements - For the Six-Month Periods Ended June 30, 2023 and 2022'. Financial Report. Korea Electric Power Corporation. https://home.kepco.co.kr/kepco/EN/ntcob/list.do?boardCd=BRD_000259&menuCd=EN030201.

KEPCO. 2024a. 'Korea Electric Power Corporation and Its Subsidiaries, Consolidated Financial Statements. For Each of the Two Years in the Period Ended December 31, 2023'. Interim Consolidated Financial Statements. Korea Electric Power Corporation. https://home.kepco.co.kr/kepco/cmmn/fms/FileDownSecure.do?atchFileId=d587b3bda67cbf8f5807e90ea688ad6c10dc8f761ab9c125ff7091d2f9d6fbb8a2&fileSn=cc47fb4d2fb9e1a9b3a50c2f51543817d8.

KEPCO. 2024b. 'Korea Electric Power Corporation and Its Subsidiaries, Interim Consolidated Financial Statements. For Each of the Nine-Month Periods Ended September 30, 2024 and 2023'. Interim Consolidated Financial Statements. Korea Electric Power Corporation. https://home.kepco.co.kr/kepco/cmmn/fms/FileDownSecure.do?atchFileId=8d7691d26d5edda9e2b8e2e8759bb09b0cb7a16f9df7522a367a406727b1223b58&fileSn=50a64e78a55f2b3329433887a893bbe799.

Krall, Lindsay M., Allison M. Macfarlane, and Rodney C. Ewing. 2022. 'Nuclear Waste from Small Modular Reactors'. *Proceedings of the National Academy of Sciences* 119 (23): e2111833119. https://doi.org/10.1073/pnas.2111833119.

Lehtveer, Mariliis, and Fredrik Hedenus. 2015. 'Nuclear Power as a Climate Mitigation Strategy – Technology and Proliferation Risk'. *Journal of Risk Research* 18 (3): 273–90. https://doi.org/10.1080/13669877.2014.889194.

Lordan-Perret, Rebecca, Rebekka Bärenbold, Hannes Weigt, and Robert Rosner. 2023. 'An Ex-Ante Method to Verify Commercial U.S. Nuclear Power Plant Decommissioning Cost Estimates'. *Energy Economics* 127 (November):107023. https://doi.org/10.1016/j.eneco.2023.107023.

Lyman, Edwin. 2021. '"Advanced" Isn't Always Better: Assessing the Safety, Security, and Environmental Impacts of Non-Light-Water Nuclear Reactors'. Union of Concerned Scientists. https://doi.org/10.47923/2021.14000.

Moriarty, Julia. 2022. '2022 Nuclear Decommissioning Funding Study'. Callan Institute. https://www.callan.com/blog-archive/2022-ndt-study/.

National Academies of Sciences, Engineering, and Medicine. 2023. *Merits and Viability of Different Nuclear Fuel Cycles and Technology Options and the Waste Aspects of Advanced Nuclear Reactors*. Washington, D.C.: National Academies Press. https://doi.org/10.17226/26500.

Newbery, David. 2021. 'The Cost of Finance and the Cost of Carbon: A Case Study of Britain's Only PWR'. *Economics of Energy & Environmental Policy* 10 (2). https://doi.org/10.5547/2160-5890.10.2.dnew.

OECD/NEA. 2013. 'The Economics of the Back End of the Nuclear Fuel Cycle'. 7061. Nuclear Energy Agency. https://www.oecd-nea.org/upload/docs/application/pdf/2019-12/7061-ebenfc.pdf.

OECD/NEA. 2024. 'Nuclear Operators' Third Party Liability Amounts and Financial Security Limits'. Nuclear Energy Agency/Organisation for Economic Co-operation and Development. https://www.oecd-nea.org/upload/docs/application/pdf/2024-02/2020.10_operators_liability_amount_table_general_final_clean_v2_2022-06-30_13-34-34_716.pdf.

Orano. 2023. 'Annual Activity Report 2022'. https://cdn.orano.group/orano/docs/default-source/orano-doc/finance/publications-financieres-et-reglementees/2022/orano_annual-activity-report_2022_online.pdf.

Pistner, Christoph, Matthias Englert, Christian von Hirschhausen, Fanny Böse, Björn Steigerwald, Lukas Gast, and Richard Donderer. 2024. 'Analysis and Evaluation of the Development Status, Safety and Regulatory Framework for So-Called Novel Reactor Concepts'. BASE – Research reports on the safety of nuclear waste disposal BASE-031/24. Berlin: Federal Office for the Saftey of Nuclear Waste Management. https://www.base.bund.de/shareddocs/downloads/en/expert-info/f/final-report-novel-reactor-concepts.pdf?__blob=publicationFile&v=1.

Prokopenko, Alexandra. 2023. 'Rosatom: A Difficult Target - Russia's Global Energy Role'. Working Paper No. 1. Energy Innovation Reform Project. https://policycommons.net/artifacts/4319479/rosatom/5129090/.

Rodriguez, Juan Camillo. 2015. 'Éléctricité Des France - What a Mess!' AlphaValue.

Rothwell, Geoffrey. 2022. 'Projected Electricity Costs in International Nuclear Power Markets'. *Energy Policy* 164 (May):112905. https://doi.org/10.1016/j.enpol.2022.112905.

Rubio-Varas, Mar. 2020. 'Time Is Money, but Sometimes It Costs More: An Economic History Perspective into Nuclear Projects' Pitfalls'. *Journal of Mega Infrastructure & Sustainable Development* 2 (3): 262–79. https://doi.org/10.1080/24724718.2022.2092993.

Scheer, Dirk, Frank Becker, Thomas Hassel, Peter Hocke, Thorsten Leusmann, and Volker Metz. 2024. 'Trittsicherheit auf Zukunftspfaden? Ungewissheitsbewältigung bei der Entsorgung hochradioaktiver Abfälle'. In *Entscheidungen in die weite Zukunft*, edited by Anne Eckhardt, Frank Becker, Volker Mintzlaff, Dirk Scheer, and Roman Seidl, 113–40. Energiepolitik und Klimaschutz. Energy Policy and Climate Protection. Wiesbaden: Springer Fachmedien Wiesbaden. https://doi.org/10.1007/978-3-658-42698-9_7.

Schneider, Antony Froggatt, Julie Hazemann, Ali Ahmad, Mariana Budjeryn, Yuichi Kaido, Naoto Kan, et al. 2021. 'World Nuclear Industry Status Report 2021'. Paris: Mycle Schneider Consulting. https://www.worldnuclearreport.org/IMG/pdf/wnisr2021-hr.pdf.

Schneider, Mycle, Antony Froggatt, Julie Hazemann, Özgür Gürbüz, Paul Jobin, Phil Johnstone, Timothy Judson, et al. 2024. 'The World Nuclear Industry Status Report 2024'. Paris, France: Mycle Schneider Consulting. https://www.worldnuclearreport.org/IMG/pdf/wnisr2024-v2.pdf.

Schneider, Mycle, Antony Froggatt, Julie Hazemann, Christian von Hirschhausen, M.V. Ramana, Alexander James Wimmers, Michael Sailer, et al. 2022. 'World Nuclear Industry Status Report 2022'. Paris: Mycle Schneider Consulting. https://www.worldnuclearreport.org/IMG/pdf/wnisr2022-hr.pdf.

Schneider, Mycle, Antony Froggatt, Julie Hazemann, Christian von Hirschhausen, M.V. Ramana, Alexander James Wimmers, Nina Schneider, et al. 2023. 'World Nuclear Industry Status Report 2023'. Paris: Mycle Schneider Consulting. https://www.worldnuclearreport.org/IMG/pdf/wnisr2023-v1-hr.pdf.

Schneider, Mycle, Antony Froggatt, Julie Hazemann, Tadahiro Katsuta, M. V. Ramana, Juan Camillo Rodriguez, Andreas Rüdinger, and Agnès Stienne. 2017. 'The World Nuclear Industry Status Report 2017'. Paris, France: Mycle Schneider Consulting. https://www.worldnuclearreport.org/IMG/pdf/20170912wnisr2017-en-lr.pdf.
Schulze, William D., David S. Brookshire, and Todd Sandler. 1981. 'The Social Rate of Discount for Nuclear Waste Storage: Economics or Ethics?' *Natural Resources Journal* 21 (4): 811–32. https://digitalrepository.unm.edu/cgi/viewcontent.cgi?article=2065&context=nrj.
Stirling, Andy, and Phil Johnstone. 2024. 'Illuminating the "UK Nuclear Complex". Implications of Hidden Links between Military and Civil Nuclear Activities for Replacing Negative with Positive Irreversibilities around Nuclear Technologies'. 2. York IND Research Report. University of York. https://www.york.ac.uk/media/politics/documents/yorkindproject/2024%20Report2%20Stirling%20&%20Johnstone%20Civil%20and%20military%20link%20in%20the%20UK%20nuclear%20weapons%20complesx.pdf.
Szulecki, Kacper, and Indra Overland. 2023. 'Russian Nuclear Energy Diplomacy and Its Implications for Energy Security in the Context of the War in Ukraine'. *Nature Energy* 8 (4): 413–21. https://doi.org/10.1038/s41560-023-01228-5.
TEPCO. 2023. 'Baisho Kin no Oshiharai Jokyo" ["Current status of Compensation paid so far"]'. Tokyo Electric Power Company Holdings. https://www.tepco.co.jp/fukushima_hq/compensation/images/jisseki01-j.pdf.
TEPCO. 2024. 'FY2023 Financial Results (April 1, 2023 - March 31, 2024)'. Financial Report. Tokyo Electric Power Company Holdings. https://www.tepco.co.jp/en/hd/about/ir/library/presentations/pdf/240430_01-e.pdf.
Thomas, Steve. 2023. 'The Nuclear Decommissioning Authority's Annual Report 2022/3'. *nuClear News* 143 (December). https://www.no2nuclearpower.org.uk/wp/wp-content/uploads/2023/11/nuClearNewsNo143.pdf.
U.S. DOE. 2020. 'Restoring America's Competitive Nuclear Energy Advantage. A Strategy to Assure U.S. National Security'. United States Department of Energy. https://www.energy.gov/articles/restoring-americas-competitive-nuclear-energy-advantage.
U.S. NRC. 2021. 'The Price-Anderson Act: 2021 Report to Congress'. NUREG/CR-7293. United States Nuclear Regulatory Commission. https://www.nrc.gov/docs/ML2133/ML21335A064.pdf.
Wimmers, Alexander, Rebekka Bärenbold, Muhammad Maladoh Bah, Rebecca Lordan-Perret, Björn Steigerwald, Christian von Hirschhausen, Hannes Weigt, and Ben Wealer. 2023. 'Decommissioning of Nuclear Power Plants: Regulation, Financing, and Production'. DIW Data Documentation 104. Berlin: DIW Berlin, German Institute for Economic Research. https://www.diw.de/documents/publikationen/73/diw_01.c.864222.de/diw_datadoc_2023-104.pdf.

Chapter 13
The Economic Value of Plutonium

Christian von Hirschhausen

13.1 Introduction

Since its discovery in 1940 at the University of California (Berkeley), plutonium has been the subject of investigation across multiple disciplines, including technology and innovation studies and political science. However, contributions within the field of the resource and environmental economic literature is still scarce. In addition, the energy literature on nuclear power does not consider plutonium specifically, e.g. Davis (2012) or Lévêque (2015). However, interest in the topic is increasing, driven by a variety of factors. In the context of the low-carbon energy transformation and climate change mitigation, consideration is rising for non-traditional nuclear technology, including so-called new reactor concepts (SNRs) such as fast neutron reactor concepts and SMR ("small modular reactors") non-light-water concepts. Some expect it to become competitive at some point in time (IAEA 2017, 2018). In particular, Russia and China are currently advancing their fuel cycles quite ambitiously, with a reactor (CFR 600) and two new nuclear fuel reprocessing facilities in China, and one new fast neutron reactor (BN 800 at Beloyarsk-4, online since 2015) and one in planning (BN 1200) in Russia. Other countries are either pursuing

This chapter updates the DIW Discussion Paper No. 2011. It is an output of a transdisciplinary research project on nuclear waste disposal in Germany ("TRANSENS"), funded by the Federal Ministry for Economic Affairs and Energy (BMUV) and the Volkswagen Foundation on behalf of the Ministry for Science and Culture of Lower Saxony (NMWK) (2019–2024). The paper was first presented to the Committee for Environmental and Resource Economics of the German Verein für Socialpolitik, (AURÖ) in September 2021. Thanks to Björn Steigerwald, Gero Scheck and Ben Wealer for research assistance, and comments and discussions, and Amy M. Kagan for proofreading. The usual disclaimer applies.

C. von Hirschhausen (✉)
Workgroup for Infrastructure Policy (WIP), Technical University (TU) Berlin, Berlin, Germany
e-mail: cvh@wip.tu-berlin.de

German Institute for Economic Research (DIW Berlin), Berlin, Germany

A. Wimmers et al. (eds.), *Nuclear Power*, Studies in Energy, Resource and Environmental Economics, https://doi.org/10.1007/978-3-031-99894-2_13

reprocessing of plutonium (Japan) or are considering it (Korea, United States (US)). Plutonium is also an important resource for space missions (through radioisotope thermoelectric generators, RTGs), boosting interest in the start-up scene in the US and elsewhere (Gates 2021). Recent research has emerged on reducing existing plutonium stockpiles through a process of partitioning and transmutation, sometimes called "plutonium burning" (MIT 2018, 82).

The renewed interest in plutonium and the fast neutron reactors that both generate and use large quantities of plutonium can also be attributed to structural changes in the nuclear industry. To some extent, it recalls the 1940s and 1950s, when the plutonium route was considered to become the pathway to roll out commercial nuclear power as a low-cost by-product of nuclear weapons under very favorable conditions (Strauss 1954). Consequently, current debates very much resemble those in the early years of nuclear power, when fast reactors were considered the natural pathway forward, and to solve "civilization's energy problems for millennia" (von Hippel et al. 2019, 2). Today, the plutonium dream is coming back, because the dominant form of using nuclear energy, the once-through fuel light water reactor based on uranium, has never become economically competitive in any energy system (Davis 2012; MIT 2003; 2018; Wealer et al. 2021). Alternatives are now being explored, under the heading of the "Generation IV International Forum" (GIF), to develop replacements for the light water reactors currently in the market. The GIF has a special focus on fast reactors, in particular the sodium-cooled reactor.

After decades of unsuccessful attempts to use plutonium for electricity generation, have resource and energy economic conditions changed sufficiently to reverse this trend?

In this chapter, I reconsider economic issues related to plutonium, and present insights from ongoing research by the author and the research group, among others, on the economics of nuclear power (Wealer et al. 2021; Böse et al. 2024), and on the technical developments in the context of the SNR developments. I take the environmental dangers and security issues (proliferation) as given and only explore the resource and energy economic aspects, comparing them with traditional nuclear power from light-water cooled reactors. After a brief recap of the physical and technical basics (Sect. 13.2), Sect. 13.3 provides a survey of the development of plutonium, and then presents an analytical framework taken from Bunn et al. (2003), i.e., a comparison between the economics of the traditional nuclear energy, the light-water reactor (LWR) using a once-through fuel process, with the plutonium route using a fast reactor with plutonium reprocessing. Until now, the plutonium route has not been able to establish itself as an economic form of electricity due to technical problems, security issues (proliferation) and subsequently high costs. Sections 13.4, 13.5, and 13.6 address one economic parameter, and whether these could change the relative economics going forward, i.e., in favor of plutonium breeding and burning:

- In Sect. 13.4, I analyze whether resource economic foundations have changed that would make the use of plutonium economic vis-à-vis its backstop resource, natural uranium.

- In Sect. 13.5, I discuss recent trends in SMR concepts ("small modular reactors") with low capacities (< 300 MW_{el}), including reactors based on a fast neutron spectrum, checking under which conditions these innovations might become competitive with existing nuclear facilities.
- In Sect. 13.6, I discuss the option to abate plutonium, through processes of partitioning and transmutation, and whether these provide an option to reduce the negative environmental impact, or, eventually, even combine it with additional energy production.

The last section concludes and provides a research outlook.

13.2 Background

Within the nuclear industry, plutonium plays a very specific role: It is the only element that can contribute to large-scale energy production by generating new fuel from spent fuel (uranium 238) through a process called "breeding." This comes with additional environmental and economic costs, though.

13.2.1 Pu-239

Plutonium (Pu) is a heavy metal with 94 protons in the core and consists of notoriously unstable phases with temperature, pressure, chemical additions, and time (Morss et al. 2006, 814). It sits near the middle of the family of actinides, i.e., heavy metals with an atomic number of 89 (actinide) and above, which are spontaneously inflammable in contact with air. Plutonium is a so-called "trans-uranium" element, i.e., heavier than the heaviest naturally occurring chemical product, uranium (with an atomic number of 92 protons).

Plutonium emerges in the process of nuclear fission of enriched uranium (e.g. 3–5% of uranium-235) with thermal ("slow") neutrons. About 1% of the original uranium is converted to plutonium, mainly plutonium-239, plus some other isotopes such as Pu-238, Pu-240, and Pu-244. Pu-239, by far, is the most commonly produced isotope, has a half-life of 24,110 years, and thus, is very long lived.

Plutonium can be generated in much larger quantities when "fast" neutrons (not moderated) fission uranium-238, which makes up 99.7% of uranium and is not fissionable with thermal neutrons (Neles and Pistner 2012, 38). This process includes several steps by which fast neutrons fission uranium-238, which then becomes plutonium-239, via several steps, including the formation of neptunium and two beta decays; here, too, a smaller share of even higher isotopes (Pu-240, Pu-241, and Pu-242) occurs (Neles and Pistner 2012, 37):

$$^{238}_{92}U + ^{1}_{0}n \rightarrow ^{239}_{92}U \rightarrow ^{239}_{93}Np + \beta^- \rightarrow ^{239}_{94}Pu + 2\beta^-$$

Provided there is a sufficient flow of neutrons, more plutonium can be generated than is used as fuel, a process called "breeding." "If the ratio of final to fissile content is less than one, the reactors are burners, consuming more fissile material (U-235, Pu, and minor actinides) than they produce (fissile Pu)" (WNA 2021). Both processes require complex reprocessing facilities to separate the transuranium metals from each other and the rest of the material.

The research team of Glen Seaborg and Edwin McMillan, at the University of California at Berkeley, isolated plutonium for the first time on December 14, 1940, by deuteron bombardment of uranium in the cyclotron. This research was continued in the context of the Manhattan Project, including work by Enrico Fermi and Leo Szilard at the University of Chicago to develop nuclear weaponry to be used in World War II. The first nuclear test, July 1945, used plutonium as fuel, as did the nuclear bomb on Nagasaki, August 9, 1945.[1]

13.2.2 Plutonium in the Nuclear Power System

13.2.2.1 The System Good Nuclear Power

As explained in detail in Chap. 3, the nuclear power sector consists of a front-end (mining and processing, enrichment, fuel fabrication), the nuclear power plant itself, and a back-end (reprocessing, interim storage, final disposal). In addition, (high-level) uranium and plutonium are used for the production of nuclear weapons. Within that system, plutonium plays a critical role in the development of specific nuclear power plants, e.g. fast neutron reactors, the reprocessing, and the subsequent reuse for fuel fabrication in the form of mixed uranium–plutonium-oxide fuel (MOX-fuels). Plutonium also poses specific challenges to interim storage and final disposal, due to the long half-life and the high activity of its fission products (technetium, iodine, cesium).

13.2.2.2 Two Common Designs for Nuclear Energy Production

Light Water Reactor (LWR) and the Once-Through Fuel Process

In the light-water reactor, uranium-235 is split, and the uranium-238 catches some neutrons that generate the plutonium-239. In what has become the dominant way of generating electricity, the spent fuel is put in water pools ("spent fuel pools") on site, to be cooled for about five years, after which it is placed in interim storage (~ 20–40 years), and should then be placed in a geological repository. This is called the "once-through" process (Fig. 13.1).

[1] The other weapon-grade fuel developed was highly enriched uranium (HEU), which was used in the Hiroshima bombing.

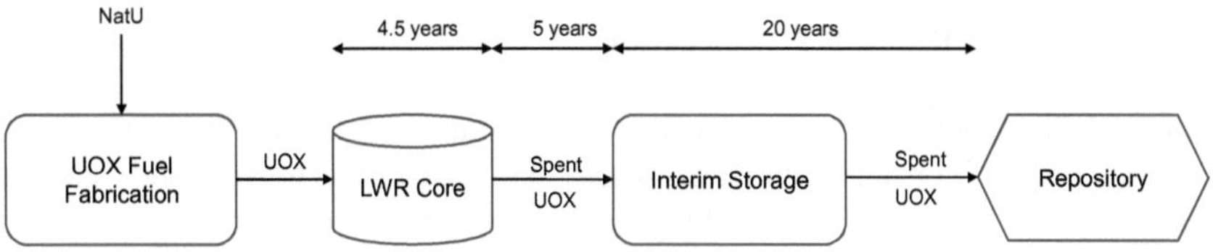

Fig. 13.1 The once-through process (*Source* DeRoo et al. [2011, 827])

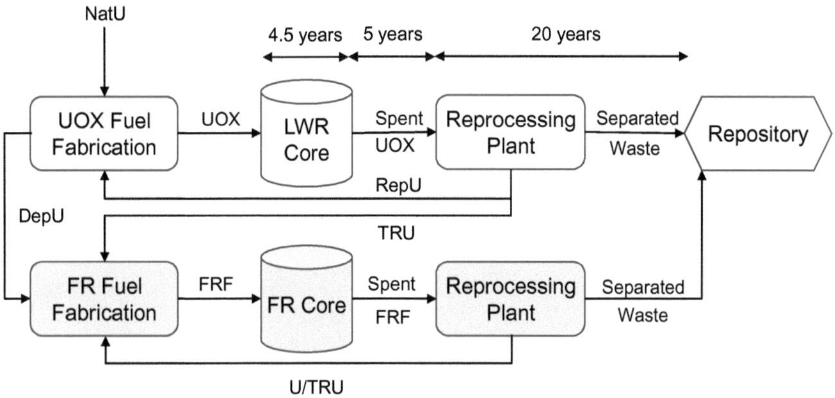

Fig. 13.2 The fast reactor with reprocessing (*Source* DeRoo et al. [2011, 828])

Fast Reactor (FR) and Plutonium Processing

The specifics of plutonium consist of the fact that it can be used to generate additional fuel, theoretically generating up to 60–100 times more heat. This requires the reprocessing of the plutonium-containing spent fuel, the fabrication of a specific mixed-oxide (MOX) uranium–plutonium fuel for the fast reactor, the neutron chain reaction in a fast reactor, and subsequent reprocessing (Fig. 13.2). This is called the “fast reactor route/cycle,” which also referred to as the “plutonium route.” Thus, in addition to the fast reactor, a specific piece of equipment is required: the reprocessing plant.[2]

[2] “Reprocessing was developed to produce a fuel for the fast reactors. Later, when this did not develop into industrial scale, reprocessing and production of MOX-fuels was continued with the justification to substitute for uranium fuel for Light Water Reactors (LWRs)” (Besnard et al. 2019, 55).

13.2.2.3 Current Stocks of Plutonium

Actors dealing with plutonium involve both commercial nuclear companies, e.g. electricity generation, and public institutions in the nuclear sector (research reactors, military, etc.). The first 100 tons of plutonium were produced for military purposes. In the late 1960s, with commercial nuclear electricity generation at scale, the stocks of commercial (civilian) plutonium started to pile up. Today, there are about 300 t of plutonium for commercial use stored in other countries (e.g. France, UK, and China, (von Hippel et al. 2019, 8)). Since the end of the cold war, about 180 tons of previously military plutonium was disassembled. Thus, today the supply of plutonium is composed of (Fig. 13.3):

- Military plutonium, either in operational warheads (~ 20 tons) or in previously disassembled warheads (~ 180 t).
- Civilian (or: commercial) plutonium, based on reprocessing of spent fuel from breeder commercialization and conventional light-water power reactors (~ 300 tons).

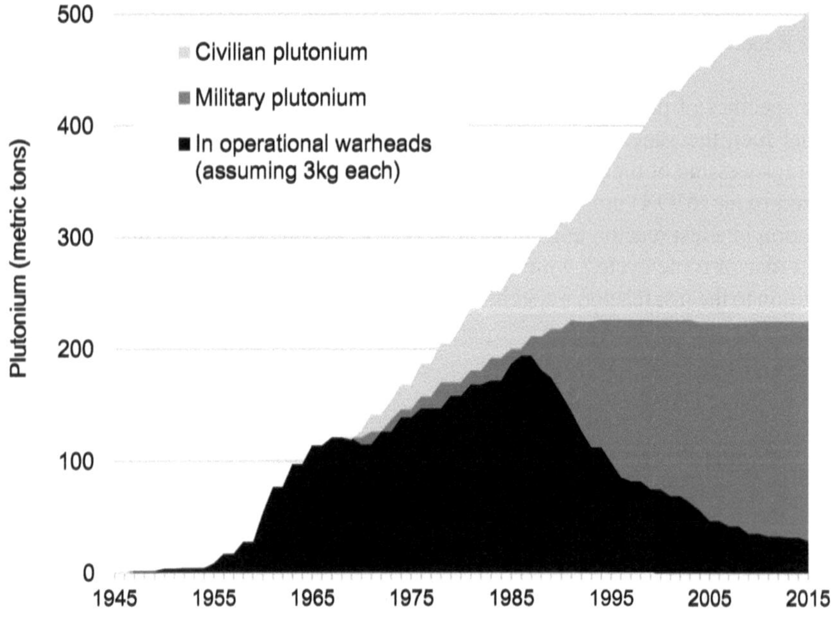

Fig. 13.3 Global stock of separated plutonium (*Source* Own depiction based on [von Hippel et al. 2019, 8])

13.2.3 Environmental Aspects and Proliferation

13.2.3.1 Toxicity and Radioactivity

Plutonium is very toxic, radioactive, and can cause harmful radiation effects, mainly through alpha-radiation (helium particles) and gamma (electromagnetic) radiation. Plutonium can be inhaled and then accumulates in the bones, the liver, and lymph nodes, with long retention times (~ 50–100 years in bones), leading to cancer risk.[3]

Plutonium has a very long half-life period of 24,110 years and therefore needs (like many other radioactive products) very long-term safe depositories. In the decay process, fission products include mainly technetium (Tc-99), iodine (Io-129), and caesium (Cs-135), likewise very nuisive when brought in contact with human beings or the environment.[4]

Reprocessing is a particularly complex process that involves risks for workers and the environment. In addition to plutonium, the other minor actinides separated in the process, too, are highly radioactive. At present, about 15% of the world's spent nuclear fuel is reprocessed (Besnard et al. 2019, 55).[5]

13.2.3.2 Proliferation

Small amounts of plutonium (few kilograms) are sufficient to produce a nuclear bomb. Not only can countries working on (clandestine) military plutonium reactors get a hold of weapon grade plutonium relatively easily, but individuals or terrorist groups can as well. The danger of controlling proliferation was raised after World War II (Acheson-Lilienthal Report 1946), and attempts began to control proliferation. However, until today, the dangers of proliferation persist, which is why arguments can be made, independent of economic considerations, to ban plutonium processing and recycling.[6]

---

[3] See https://www.bfs.de/EN/topics/ion/effect/radioactive-materials/plutonium/plutonium_node.html (last accessed 2025-09-11), and https://world-nuclear.org/information-library/nuclear-fuel-cycle/fuel-recycling/plutonium.aspx (last accessed 2025-09-11).

[4] If not mentioned otherwise, most of the physical and chemical information in this chapter is taken from Morss (2006).

[5] "Reprocessing involves the dissolution of spent fuel in boiling concentrate of nitric acid, followed by the phycio-chemical separation of plutonium and uranium from the dissolved fuel. This difficult, complex, expensive and dangerous process results in numerous nuclear waste streams, very large releases of nuclide waste to air and sea, and large radiation exposures to workers and to the public." (Besnard et al. 2019, 54).

[6] See for details (von Hippel et al. 2019) and the work by the International Panel on Fissile Materials (IPFM).

13.2.4 Environmental Economic Assessment

The environmental economic literature does not deal with the specific externalities of plutonium. However, it can be assumed that the plutonium route generates more negative externalities than the once-through route. In addition, proliferation risks, though difficult to assess quantitatively, are particularly high. Thus, in 2007, the UK's Royal Society warned that the potential consequences of a major security breach or accident involving the UK's stockpile of separated plutonium "are so severe that the Government should urgently develop and implement a strategy for its long term use or disposal" (Besnard et al. 2019, 55).

The only environmental economic reference to plutonium, the classical textbook by Tietenberg and Lewis, seems to share this assessment, qualifying plutonium as a substance where "the optimal level of the pollution may be zero, or close to it" (Tietenberg and Lewis 2016, 374).[7]

13.3 Economic Aspects

In this section, I focus on the economic benefits of the plutonium route, i.e., its potential role as a producer of heat and electricity. I first analyze the economic development of plutonium reactors in the past, and then present an analytical framework to compare them with the dominant route, the light-water once-through reactors.

13.3.1 Three Phases of Development

13.3.1.1 Phase I: Planning for the "Plutonium Economy" (1946–1970s)

The early enthusiasm about the economic perspective of nuclear power was based on the assumption that breeders were the way forward to use the scarce resource uranium adequately (Szilard 1945, 1947; von Hippel et al. 2019, 2). The initial focus on the fast reactor was based on the much higher energetic use of uranium, by activating uranium isotope U-238, and the assumed shortage of uranium ore. Szilard (1947) observed that if more than one atom of plutonium could be produced from U-238 for every atom of plutonium fissioned, "the resource base for nuclear power would become U-238 and it would be possible to produce about 100 times more energy from the same amount of uranium. Therefore, it was considered that if what Szilard called the plutonium 'breeder' reactor could be designed, civilization's energy problems

[7] "This situation occurs when the damage caused by even the first unit of pollution is so severe that it is higher than the marginal cost of controlling it. … This circumstance seems to characterize the treatment of highly dangerous radioactive pollutants such as plutonium." (Tietenberg and Lewis 2016, 374).

would be solved for millennia" (von Hippel et al. 2019, 2). The Swiss economist, Edgar Salin (1955), too, placed significant hopes on the development of nuclear fission and breeder technology, that would lead to the end of the coal era, and open up a new step of industrial revolution. Nothing short of a "plutonium economy" was expected.[8]

Until the 1970s, longer-term planning of nuclear power relied on the breakthrough of the plutonium route. In the US, Glenn Seaborg, the "father" of plutonium back in 1940, had become Chairmen of the Atomic Energy Commission that foresaw fast neutron reactors to become the dominant source of nuclear energy once implemented (Fig. 13.4). Similar expectation prevailed in Europe, too (Midttun and Baumgartner 1986). In Germany, significant resources went into the fast breeder research, mainly in Karlsruhe, as the basis for demonstrator and commercial reactors to be built later on, e.g. the development of a commercial breeder reactor in Kalkar (North Rhine Westphalia, NRW) (Radkau 1983). Table 13.1 presents a list of experimental fast breeder reactors developed mainly in that period.

13.3.1.2 Phase II: Technical and Economic Obstacles (1970s–…)

Until today, the expected economic breakthrough of the plutonium reactor has not come about. In addition to the general economic risks of nuclear power (such as the absence of insurance against accidents), specific obstacles have prevented the diffusion of fast reactors thus far. Pistner et al. (2017) provide a comprehensive survey of prototype and demonstrator fast reactors (Table 13.2). None of these reac tors succeeded in providing reliable electricity over significant periods, such that demonstrators could be scaled and become commercial reactors. In sum, economic cost escalations have led to the abandonment of most of the pilot plants. The World Nuclear Association (2021) concluded that fast reactors are "expensive to build and operate, including the reprocessing."[9]

The literature identifies a number of technical obstacles that explain the low load factors, and high costs (Ramana and Suchitra 2007; Suchitra and Ramana 2011; Pistner and Englert 2017; MIT 2018; von Hippel et al. 2019). Most fast reactors use

8

> Thirty years from now this same man-made elment can be expcted to be a predominant energy source in our lives. … Plutonium, as the key to electrical energy production in the future, is thus a vital element of the overall economic well-being of this country. … I wish to speak of plutonium as the energy cornerstone of our future economy and to speak of that economy as the plutonium economy of the future. (Seaborg 1970, 1/2).

9 "Today there has been progress on the technical front, but the economics of FNRs (fast neutron reactors) still depends on the value of the plutonium fuel which is bred and used, relative to the cost of fresh uranium. … They are however expensive to build and operate, including the reprocessing, and are only justified economically if uranium prices are reasonably high, or on the basis of burning actinides in nuclear wastes." (WNA 2021).

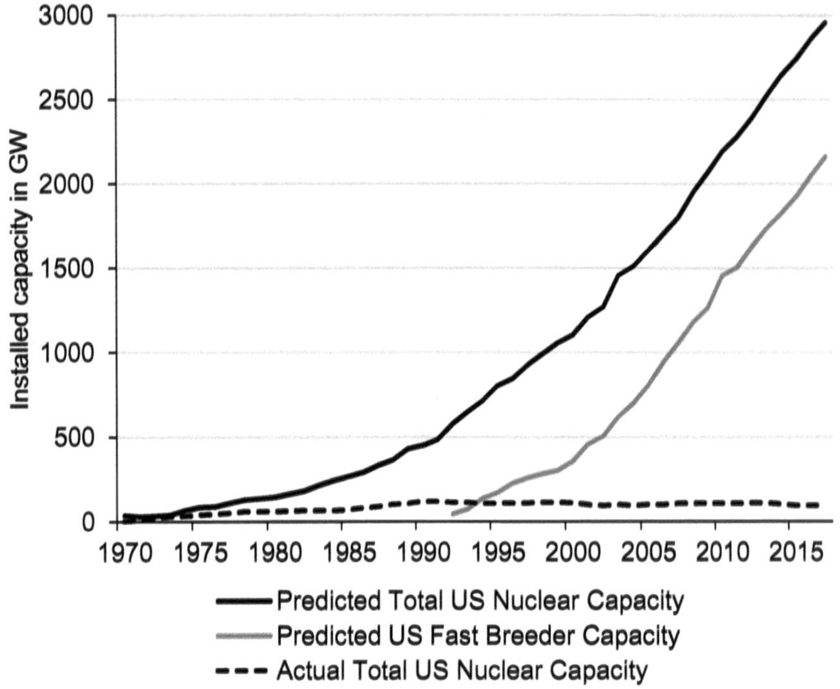

Fig. 13.4 US Atomic Energy Commission (AEC) growth projections for US nuclear generating capacity, and the role of fast reactors therein, in 1974 (*Source* Own depiction based on von Hippel et al. [2019, 33])

sodium as coolant, which in case of leakages leads to immediate fire; other coolants such as lead or helium also have their specific problems. Transuranium physics and the translation into industrial machinery turned out to be more complicated than expected. Corrosion was and remains a major problem. In addition, plutonium does not behave as properly as expected as a fuel, so that enriched uranium often had to be used. As a result, construction costs for the plutonium reactor tend to be significantly higher than for the light water reactors, at least 25% higher, often more; operational costs, too, are higher because of the fuel treatment.

The high costs of reprocessing added to the unfavorable economics seem to prohibit an economic use of plutonium. Initially believed to decrease with increasing use, reprocessing costs have stayed high and presented, in combination with other factors, a breakthrough of the fast breeder (Pistner and Englert 2017; von Hippel et al. 2019). In France, costs for plutonium and uranium recycling tend to be five times more than the savings in enriched fuel costs (Commissariat Général du Plan 2000); by today, with lower capacity factors due to loss of foreign clients, this ratio is likely to have gone up. In Japan, the Japan Atomic Energy Commission (JAEC) estimated costs of recycling about ten times higher than the savings in low enriched uranium (LEU) fuels (JAEC 2011). Case studies in India confirm these results (Ramana and Suchitra 2007; Suchitra and Ramana 2011).

Table 13.1 List of experimental fast reactors

Country	Reactor	Construction start	Power (MWth)	Operation
China	CEFR	2000	65	Since 2010
France	Rapsodie	1962	40	1967–1983
Germany	KNK-II	1975	52	1977–1991
UK	DFR	1954	60	1957–1977
India	FBTR	1972	40	Since 1985
Italy	PEC	1974	120	(1)
Japan	JOYO	1970	140	1977–2007
Russia	BR-10 BOR-60	1956 1964	55 9	1959–2002(2) Since 1958(3)
US	EBR-I EBR-II Fermi FFTF	1947 1958 1956 1970	1,2 62,5 200 400	1951–1963(4) 1963–1994 1965–1972 1980–1992

Notes (1) Construction interrupted in 1987; (2) Core meltdown; (3) 1955 partial core meltdown; (4) 1955 partial core meltdown

Source Compiled from (Pistner and Englert 2017, 42), with further references

Note, however, that the early phase of what was expected to become the plutonium economy was based on a purely technical vision, and ignored economic facts. Thus, already in the early days of research on sodium-cooled fast reactors, however, some skepticism prevailed among economists. The President of the Kiel Institute for the World Economy (IfW Kiel), Fritz Baade, published a critical assessment of the plutonium route in 1958 already, suggesting that it was expensive and not competitive. While he shared the high technical expectations, he stressed that, back in the 1950s, plutonium had a negative value, due to high processing costs, that even the extraordinary energetic characteristics would not compensate.[10] Baade relied on extensive economic analyses carried out already, both on a light water reactor in the US (Shippingport), and a plutonium-reactor in the UK (Calder Hall)[11]

10

> Therefore, it can be stated with great certainty about reactors of the breeder type: 1. that in the future they could make a decisive contribution to atomic energy, which would decisively relieve the energy balance of the world and could achieve a multiple of the performance of the classical energy source coal; … 3. that today these reactors are still far from the possibility of producing energy at a price comparable to coal-fired power plants. … Many readers may be surprised to learn that in the case of a nuclear power plant, which is often presented in popular articles as the solution to the atomic energy problem, one must actually reckon with the possibility that the cost of separating out the incubated plutonium may become so high as to create a debit unless reprocessing processes can be found which will work much more cheaply than those now in use." Own translation from Baade. (1958, 138/139)

[11] Allocating the plutonium costs to the military (and making several very optimistic assumptions), Jukes (1956, 186, cited in Baade (1958)) calculated costs of 0.76 UKpence/kWh (6.5 Pfg/kWh) for a (hypothetical) Calder Hall plant with 150 MWel (the original was 60 MWel). This would already be twice the price of electricity from coal. On the contrary, Loebl (1956) estimated fuel costs at 11.45

Table 13.2 Prototype and demonstrator fast reactors

Country	Reactor	Construction start	Power (MWth)	Operation	Average load factor
Germany	SNR-300	1973	762	No operation license	
France	Phoenix	1968	563	1973–2009	~ 0.50
UK	PFR	1966	650	1974–1983	0.07
India	PFBR	2003	1250	Since 2012	
Japan	Monjou	1985	714	1994–2016	1996–2010; out of operation after accident
Kazakhstan	BN-350	1964	750	1972–1999	0.85
Russia	BN-600 BN-800	1967 2006	1470 2100	Since 1980 Since 2016	0.74 (1982–2009) 0.71
US	CRBRP	1982		Abandoned in 1983	

Source Taken from (Pistner and Englert 2017, 42)

13.3.1.3 Phase III: Renaissance of the Plutonium Economy?

Despite its technical and economic failure so far, one observes a renaissance of the idea of fast reactors, both in countries that were previously experienced with the technology (e.g. Russia, India, Japan) and others (such as China and the USA). Fundamental research, experimental reactors and demonstrator reactors are again being explored, under the heading of the "Generation VI International Forum" (GIF), to develop successful replacements for the light water reactors currently in the market that are more economic, secure, etc. The GIF has a special focus on fast reactors, and among them, "the sodium-cooled reactor is considered the most technically advanced of the Generation IV reactors" (Frieß et al. 2021, 31).

Table 13.3 provides the presentation of the World Nuclear Association of fast reactor design for near- to mid-term development. Even though some of the projects are planned in the future, with uncertain starting dates, concrete work is ongoing for some of them. Russia is particularly active, with a high-power demonstrator already running (BN-800, breeder, but potentially also a burner), a concept under development for commercial use (BN-1200, breeder), and two reactors of smaller power rating (BREST, MBIR). China has entered the sector with an experimental reactor (CEFR), followed by two breeder reactors with higher power (CFR-600), and the purchase of two reprocessing plants, one from Russia and one from France (Zhang 2020). Even in the US, breeder and burner technologies with fast neutron

Pfg/kWh for a total cost of 17.84 Pfg/kWh. Admiral Hyman Rickover, the "father" of the Nuclear Navy and subsequent manager of the fist commercial nuclear power plant (Shippingport, PA, USA) also reports costs corresponding to 21,8 Pfg./kWh (all citations from Baade (1958, 123sq.)).

reactors are being explored in the context of SMR-concepts (PRISM, ARC-100), and a traveling wave reactor (TWR). Other countries and regions (e.g. the European Union) are also active (Murakami 2021).

Note that ten out of the 14 concepts use fast sodium-cooled technology, i.e., the one initially proposed by Szilard and others at the end of the Project Manhattan.

Table 13.3 Current and planned fast reactor activities according to the World Nuclear Association

Reactor	Type, coolant	Power (Thermal or Electric) in MW	Fuel (Future)	Country	Expected operation
PRISM & Natrium	Demonstration, pool, sodium	840/311	Metal	US	From 2020s
ARC-100	Prototype, pool, sodium	260/100	Metal	US	Working with GEH
Allegro	Experimental, loop, gas	50–100 MWth	Oxide (carbide)	Visegrád countries with France	From 2030
MYRRHA	Experimental, Pb-Bi	57 MWth	Oxide?	Belgium with China	Early 2020s
ALFRED	Demonstration, lead	300/120	Oxide	Romania with Italy & EU	From 2025
BN-1200	Commercial, pool, sodium	2800/1220	Oxide, Nitride	Russia	From 2030
BREST-300	Demonstration, loop, lead	700/300	Nitride	Russia	From 2021
MBIR	Experimental, loop, sodium (Pb-Bi, gas)	100–150 MWth	Oxide	Russia	From 2020 (under construction)
CDFR-600	Demonstration, pool, sodium	600 MWe	Oxide	China	From 2023 (under construction)
CDFR-1000	Commercial, pool, sodium	1000–1200 MWe	Metal	China	From 2034
PGSFR	Prototype, pool, sodium	392/150	Metal	South Korea	From 2028
JSFR	Demonstration, loop (now pool), sodium	500 MWe	Oxide	Japan	From 2025?
TWR	Prototype, sodium	600 MWe	Metal	China with US	From 2023?
FBR 1&2	Commercial, sodium	600 MWe	Oxide, Metal	India	?

Source (WNA 2021)

Other types of fast reactors in the program include gas- and lead-cooled reactors (Locatelli et al. 2013; Pistner and Englert 2017; GIF 2018; WNA 2020a).

13.3.2 Analytical Framework by Bunn et al. (2003)

The value of plutonium mainly depends on the costs of the next best alternative: If this alternative is expensive, plutonium, in conjunction with reprocessing, can be used economically. However, if the so-called "once-through fuel" process is cheaper, e.g. because uranium is relatively cheaply available, the value of plutonium is negative. The literature contains several approaches to modeling the two idealtype concepts for nuclear electricity generation (Bunn et al. 2003; De Roo and Parsons 2011; Suchitra and Ramana 2011). In this subsection, I present the model developed by Bunn et al. (2003) that results in breakeven conditions for the choice between the two routes. The plutonium route is benchmarked to the traditional light-water, once through route, and identifies conditions under which the former could become economic (which it is not at present).[12]

13.3.2.1 Comparing Once-Through Process and Plutonium Recycling

The economic analysis take into account the fuel costs of reprocessing and recycling the spent fuel from a light reactor, including eventual disposal costs for waste (right-hand side of Eq. (13.1)), with the alternative, once-through, interim storage and direct disposal of spent fuel (left-hand side of Eq. (13.1)). At equilibrium, both costs should match:

$$\begin{bmatrix} \textit{cost of interim storage} \\ \textit{and disposal of spent fuel} \end{bmatrix} = \begin{bmatrix} \textit{cost of reprocessing} \\ \textit{and disosal of wastes} \end{bmatrix} - \begin{bmatrix} \textit{value of recovered} \\ \textit{plutonium and uranium} \end{bmatrix} \tag{13.1}$$

The cost of interim storage and disposal of spent fuel, the cost of reprocessing, and the cost of disposal of waste are constants, estimated from a complex techno-economic analysis (Bunn et al. 2003). The value of the recovered fuels (plutonium and uranium) is a function of the uranium price, because the recovered fuels displace a certain amount of low-enriched uranium (LEU); it also takes into account the cost of producing the fuels.

$$\begin{bmatrix} \textit{value of recovered} \\ \textit{plutonium and uranium} \end{bmatrix} = \begin{bmatrix} \textit{cost of LEU fuel made} \\ \textit{with natural uranium} \end{bmatrix}$$

[12] This analysis does not include energy economic evaluation vis-à-vis other fuels, neither is an energy-environmental cost–benefit analysis intended.

$$- \begin{bmatrix} \textit{cost of equivalent fuel made} \\ \textit{with recovered Pu and U} \end{bmatrix} \tag{13.2}$$

Bunn et al. (2003, 14 sq.) apply the following values (in 2000 values):
~ Cost of interim storage: US-$200/kg.
~ Present value of the cost of direct disposal of spent fuel (including transportation): US-$400/kg.
~ Cost of reprocessing: US-$1,000 kg.
~ Present value of disposing radioactive waste from reprocessing: US-$200/kg.

In that case, the costs of interim storage and disposal of fuel (US-$200 + US-$400) are lower than the costs of reprocessing and disposal of waste (US-$1,000 + US-$200); therefore, the value of recovered plutonium and uranium would have to be US-$600 to break even.

13.3.2.2 Value of Recovered Plutonium and Uranium

How can the value of the recovered uranium and plutonium be assessed? 1 kg of spent fuel yields about 95% uranium and 1% plutonium, in addition to 4% radioactive fission products and minor actinides (Bunn et al. 2003, 14). The value of the recovered uranium can be estimated with the price of the fresh uranium produced from ore. The value of plutonium consists of the economies of displacing LEU as fuel, minus the cost of producing the fuel (MOX, including final disposal):

$$\begin{aligned} \textit{Value of recovered plutonium} &= \textit{Cost of displaced LEU} \\ &\quad - \textit{cost of MOX} \end{aligned} \tag{13.3}$$

6 kg of fuel has to be processed to produce 1 kg of fresh fuel. Fabricating the fresh MOX costs US-$1,500, and the costs of the (displaced) LEU are the following:

~ Uranium: 7 kg for enrichment to (7 × 0,7% = 4,9%)@ $ 50 /kg	= $350
~ Conversion: 7 kg@ $ 5/kg	= $35
~ Enrichment: 6 separating work units (SWU)@ $ 100/SWU	= $600
~ Fabrication: 1 kg@ $ 250/kg	= $250
	$1,235

The cost of producing MOX is $1,500/kg, which is higher than the $1,235 of savings on LEU:

$$\textit{Value of recovered plutonium} = \$1235 - \$1500 = -\$265. \tag{13.4}$$

According to Bunn et al. (2003), given these estimates for the variables, at a price of $50/t of uranium, the value of plutonium is negative.

13.3.2.3 Breakeven Price of Uranium

The overall value of the recovered plutonium and uranium becomes:

$$\begin{aligned}&\left[recovered\ uranium\&plutonium\right] = \left[ecovered\ uranium\right] + \left[recovered\ plutonium\right]\\&0,95C_U + \left[\frac{(cost\ of\ LEU) - (cost\ of\ MOX)}{6}\right] = 0,95C_U + \left[\frac{(7C_U + 885) - 1500)}{6}\right]\\&= 2,12C_U - 102,5\end{aligned} \tag{13.5}$$

Finally, the initial Eq. (13.1) becomes:

$$\begin{aligned}\left[\begin{matrix} cost\ of\ interim\ storage \\ and\ disposal\ of\ spent\ fuel \end{matrix}\right] = \left[\begin{matrix} cost\ of\ reprocessing \\ and\ disosal\ of\ wastes \end{matrix}\right] - \left[\begin{matrix} value\ of\ recovered \\ plutonium\ and\ uranium \end{matrix}\right]\\ [600] = [1200] - [2.12C_u - 102.5]\\ 2.12C_u = 702.5\\ C_u = 702.5\end{aligned} \tag{13.6}$$

The uranium price required to break even is US-$332/t, about seven times higher than the prevailing one at the time.

13.3.3 Three Critical Parameters: Uranium Price, Construction Cost, Disposal Cost

The analytical framework is useful to identify three critical parameters that will be studied in the next three sections: the breakeven price of uranium, the capital cost of the respective reactor technologies, and the cost of interim and final storage.

13.3.3.1 Breakeven Price of Uranium

One can display the above analysis as the relation between the uranium price and the cost of reprocessing. Figure 13.5 shows the results of the breakeven price of uranium (vertical) as a function of the reprocessing costs (horizontal). The light bar at the bottom indicates that at a uranium price of $50/kg, the equilibrium cost of reprocessing would need to decrease to US-$420, a 58% reduction.

13.3.3.2 Introducing Construction Cost Differences

Until now, I have ignored the capital costs of the reactors, i.e., the LWR, and the fast neutron liquid metal reactor. Bunn et al. (2003, 104) consider a linear approximation for the breakeven uranium price as a function of the difference in capital costs

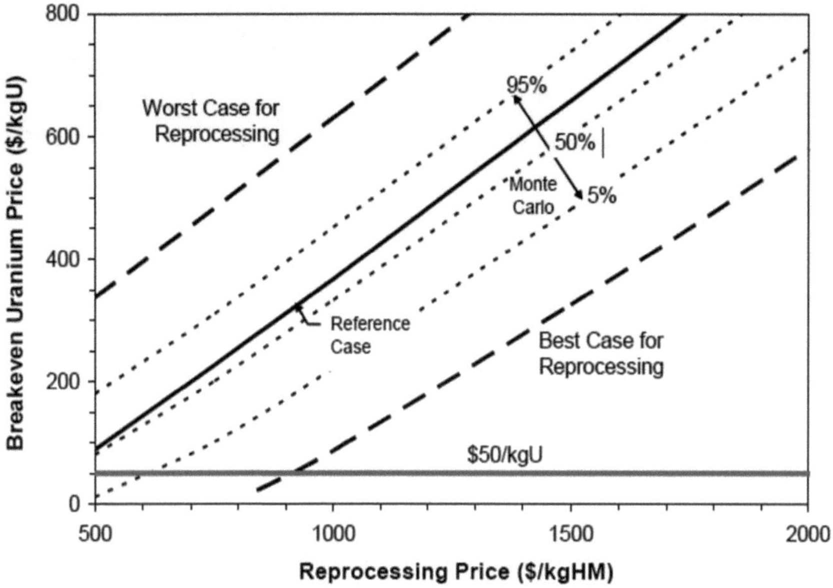

Fig. 13.5 Breakeven uranium price as a function of the cost of reprocessing (*Source* Bunn et al. [2003, 18])

(ΔC_{cap}). In equilibrium, the difference of the (per unit) capital cost between the light water and the fast reactor route (fixed costs) would equal the difference of the (per unit) variable fuel costs (Bunn et al. 2003, 104):

$$\frac{\Delta C_{cap} * F}{8776\eta} = C_{LWR}fuel(C_u) - C_{LMR}fuel$$

where

- ΔC_{cap} is the difference in the overnight construction costs (US-\$/kWe) between the light-water reactor ("once through") and the liquid metal fast reactor (hypothetical).
- F is a simplified term for interest during construction.
- η is the thermal efficiency of the plant.
- 8766 is the average number of hours per mean year.
- $C_{LWR}fuel(C_u)$ and $C_{LMR}fuel$ are the respective fuel costs.[13]

Given the cost disadvantage of plutonium reprocessing (the "negative value" of plutonium), the capital costs of the fast reactor would have to be lower than the capital cost of the light water reactor, in order to break even on the levelized cost of

[13] In this specification, only $C_{LWR}fuel(C_u)$ depends on the costs of the fresh uranium. Theoretically, C_{LMR}fuel is a function of Cu, too, but due to the high capital costs for reprocessing, this can be neglected.

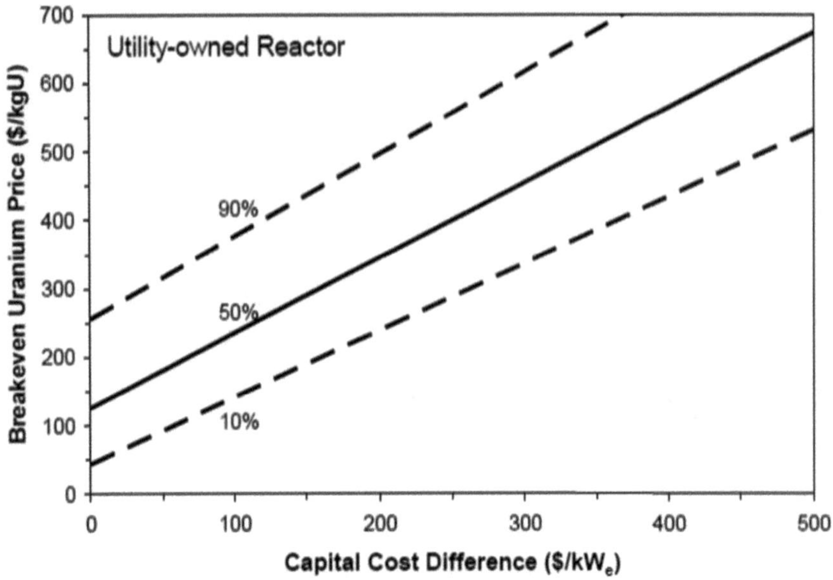

Fig. 13.6 Breakeven uranium price for utility-owned reactor as a function of the capital cost difference between a fast reactor and a light-water reactor (*Source* Bunn et al. 2003, 70 [Fig. 3.2])[14]

electricity. However, as discussed above, the opposite was the case, and still is. The construction costs of fast reactors are significantly higher than those of light water reactors (Bunn et al. 2003, 70).

In that case, one can once again determine a break-even price of uranium that would equalize the difference in capital costs (Fig. 13.6). Note that the estimated difference at the time (early 2000s) were considered to be significantly lower than today, so the values are only indicative. They confirm, however, that a very significant increase in uranium prices would be required for the plutonium route to become economic. As an approximation, I retain that the price of uranium had to be in the 4-digit range (> US-$999/kg) to become relevant.

13.3.3.3 Cost of Nuclear Waste Disposal

Finally, yet importantly, Eq. (13.1) also indicates that the different routes imply different costs of disposal of spent fuel (once-through process) and wastes (plutonium process). Direct disposal of spent fuel is more expensive than the disposal of "only" the radioactive waste after the multiple reprocessing (Bunn et al. 2003, 52 sq.). In that case, reducing the volume of the high-level waste, and/or reducing the specific substances that cause high disposal costs (e.g. because of heat development) could reduce storage costs. This is particularly important in cases where the available

[14] Assumptions in this specification differ slightly from the above in Fig. 13.5.

volume of the depository is constrained, such as in the US Yucca Mountain site (located within a mountain, not underground).

13.3.4 Intermediate Conclusion

After decades of unsuccessful attempts to use plutonium for electricity generation, have resource and energy economic conditions changed sufficiently to reverse this trend? In this section, I have reported the development of the plutonium route through different historical phases. Even though neither the technological nor the economic success expected have materialized, one observes a "renaissance" of breeder (and burner) research and the idea of a plutonium economic in some countries. The analytical framework reduces the economic analysis to a comparison between the plutonium route and the light-water once-through route. It allows identifying three topics to explore: (1) potential constraints on the availability of fresh uranium; (2) the perspective of construction costs for new designs, e.g. reactors of low capacity (e.g., SMR); and (3) the option to abate plutonium through burning. These topics will be explored in the next three sections.

13.4 Can Plutonium Benefit from Binding Constraints on Uranium Supply?

If, as shown in the last section, the plutonium fast reactor route is much more expensive than the traditional once-through cycle, could constraints on the latter, e.g. unavailability of uranium or extremely high prices, change the relation in favor of plutonium?[15] Due to the high capital intensity of the industry, in particular the plutonium route including reprocessing, the cost of the resource plays only a minor role in total costs. In this section, I assess whether uranium supply could become a binding constraint and open a case for the plutonium route.[16]

13.4.1 Historical Perspective

"For decades, consideration of reprocessing, recycling, and breeding plutonium has been driven in significant part by concerns that resources of uranium would not be

---

[15] Theoretically, thorium (90) can also be used for chain reactions, and it is more abundant than uranium, yet it is more complex to use and has not been developed at scale thus far (Nelson 2012).

[16] The significant environmental effects of uranium mining are not described here, but should be part of an environmental assessment of the overall uranium market, see (Neles and Pistner 2012, 152 sq.).

sufficient to support a growing nuclear energy system operating on a once-through cycle for long" (Bunn et al. 2003, 105). The initial hypothesis of the success of the plutonium breeder reactor was the danger of a shortage of uranium available to produce nuclear energy. This fear was not only expressed by physicists and engineers (such as Szilard (1945, 1947)),[17] but also shared by economists fearful of the combination of low availability of uranium and the high costs of enriching the uranium oxide to usable levels (Salin 1955; Baade 1958, 114). Advocates of reprocessing and breeding continued to argue that available resources of low-cost uranium are quite limited, making breeding and reprocessing essential in the relative near term (Bunn et al. 2003, 105).[18]

The lack of international trade of uranium in the early period of nuclear power further contributed to the effect and favored the fast-reactor plutonium route. In fact, only the United States, Canada, the Soviet Union and India had uranium (and/or thorium) resources available, while all others, including the UK and Germany, did not (Baade 1958, 121 sq). Thus, for countries not equipped with domestic resources, the contribution of nuclear energy independence was very low at best.

However, by the late 1970s, the situation had shifted, both on the uranium supply side and on the demand side. On the supply side, much more uranium was found, and known resources of uranium had increased manifold (Bunn et al. 2003, 4). On the demand side, less nuclear power plants than expected were built. Holdren (1975) was among the first to make the case that "uranium resources were sufficient at that time to delay deployment of breeder reactors."[19] It seems that this has become the mainstream assessment in the literature (Hall and Coleman 2013; Rooney et al. 2015; von Hippel et al. 2019).

Nonetheless, the issue of tight uranium supply is maintained in some of the literature. Time frames and expected dynamics of prices and reserves vary in this strand of the literature, in particular with respect to the forecasted demand. Liebert and Englert (2015) and Muellner et al. (2021) argue that with very high demand for nuclear energy, uranium supply could not be able to follow suit, both due to an inability "to produce enough uranium within the expansion phase of the growth scenario," and because of the insufficient "overall amount of uranium available for the total operating time of current and future plants" (Muellner et al. 2021, 6). Subsequently, it is concluded that "limited uranium-235 supply inhibits substantial expansion scenarios with the current nuclear technology" (Muellner et al. 2021, abstract). Based on a uranium market model, Monet et al. (2017) conclude that with rapidly growing demand, the uranium market may prove to be under stress in some periods

[17] "In a talk on nuclear energy in 1947, Szilard assumed it would be feasible to import 400 tons of natural uranium per year, an amount that would be sufficient to fuel only two 1,000 MWel light water reactors today." (von Hippel et al. 2019, 10).

[18] Reference is made to a rather pressimistic document by the US Department of Energy (2003, I–4), stating that uranium "is not an infinite resource. Expert organizations such as the World Nucler Association project that between 2050 and 2080, nuclear power plnats worldwide will encounter a seriours shortage of uranium needed to produce nuclear fuel."

[19] Cited according to (Bunn et al. 2003, 105, footnote 212).

of the twenty-first century. The textbook of Tietenberg (Tietenberg 2005, 148) also makes this argument.[20]

The assessment of the resource situation by the nuclear industry itself is split between optimism regarding resources and pessimism about low prices and investment incentives. On the one hand, the industry regularly assures stakeholders and the public that "sufficient uranium resources exist to support continued use of nuclear power and significant growth in nuclear capacity for low-carbon electricity generation and other uses … in the long term. Identified recoverable resources … are sufficient for over 135 years" (IAEA and NEA 2020, 14). But on the other hand, it insists on the high investment requirements to maintain the ample resource base, particularly in times of rather prices.[21]

13.4.2 Short-Term Price Volatility …

Given the complexity of technology, the secondary fuels available (e.g. recoverable resources from weapons, and highly enriched uranium), and other idiosyncrasies, the modeling of the uranium market is more complex than for simpler metals or energy fuels, such as copper or coal. There is no standard model that has emerged in the literature. On the supply side, the amount of uranium recoverable depends on resources, technology, and prices. The demand for uranium depends on the amount of nuclear power generation, and the forecasts thereof.

Different streams in the literature all suggest a high short-term price volatility. Owen (1983) highlights the role of inventories that lead to a constant difference between current production and current consumption, leading to significant price fluctuations. Trieu et al. (1994) confirm this by distinguishing spot market prices from long-term prices. Monnet et al. (2017) identify dynamic constraints and the role of market competition as major drivers of prices, suggesting that "long-term

20

> Resource availability is a problem with uranium as long as we depend on conventional reactors. However, if countries move to a new generation of breeder reactors, which can use a wider range of fuels, availability will cease to be an important issue. For the United States, for example, on a heat-equivalent basis, if they are used in conventional reactors, domestic uranium resources are 4.2 times as great as domestic oil and gas resources. With breeder reactors, however, the U.S. uranium base is 252 times the size of its oil and gas base. Tietenberg (2005, 148)

21

> However, considerable exploration, innovative techniques and timely investment will be required to turn these resouces into refined uranum ready for nuclear fuel production and to facilitate the deployment of prmising nuclear technologies. (IAEA and NEA 2020, 14)

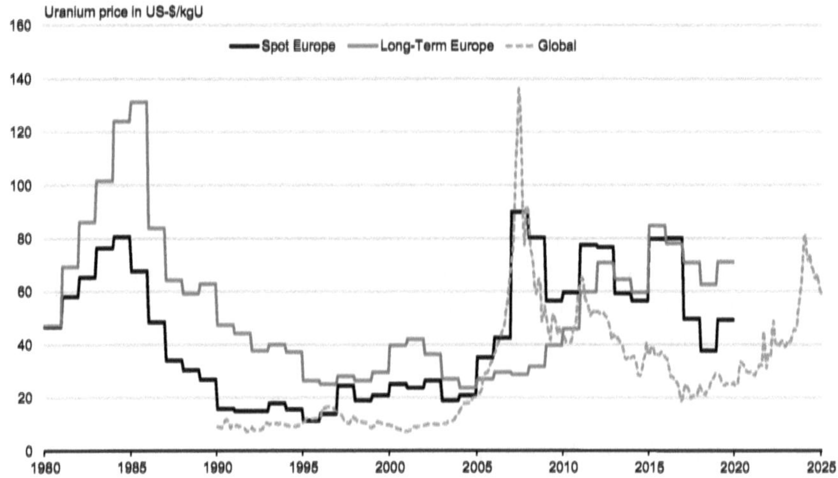

Fig. 13.7 Uranium prices and key events since the 1950s (*Source* Own depiction with data taken from the Euratom Supply Agency [European Prices] and the Federal Reserve Bank of St. Louis [Global])[22]

availability of uranium depends on demand scenarios and more on market dynamics than on ultimate resources." Some recent literature explores how turning points could be better identified, e.g. price increases, or the role of technical innovation (Pedregal 2020; Landajo et al. 2021): stationarity tests, etc.

Figure 13.7 shows price developments since the early years of nuclear power. In periods of high expectancy and limited supply, prices have gone up (early 1980s, 2000s). However, they have also come down, both as a result of changing demand expectations, and rising mining supply. Contrary to expectations, the supply–demand balance for uranium has not shifted significantly over the last decade. At the end of the 2010s, prices were lower than in 1980.

13.4.3 … and Long-Term Resource Availability

While short-run trends are quite diverse, evidence on the longer-term resource availability is somewhat clearer. In this subsection, I identify the longer-term trends and explain that from an economic perspective, the resource availability seems to be assured. Reserve and resource availability are large and can react flexibly to potential shortages. A comprehensive discussion of issues is provided by Bunn et al. (2003, 109 sq), which is still valid today; see also a critical assessment in (Hall and Coleman 2013), and a dynamic model of the global uranium market and the nuclear fuel cycle (Rooney et al. 2015).

[22] https://euratom-supply.ec.europa.eu/eu-natural-uranium-price-esa-indices-1980_en.

13.4.3.1 Market Structure

Although the market is somewhat concentrated, the probability of oligopolistic behavior by a high concentration of supply is modest. About 15 countries or more are active in the market. The countries with the highest shares of identified resources are Australia (28% of resources), Kazakhstan (15%), Canada (9%), Russia (8%), Namibia (7%), South Africa, Brazil (5%), and Niger (4%) (IAEA and NEA 2020, 16) (Fig. 13.8). The Hirschmann-Herfindahl Index (HHI, sum of squared market shares) is relatively high for uranium production in 2019 (2,196). There is a relatively high degree of vertical integration in the industry, including long-term contracts (Fig. 13.9).

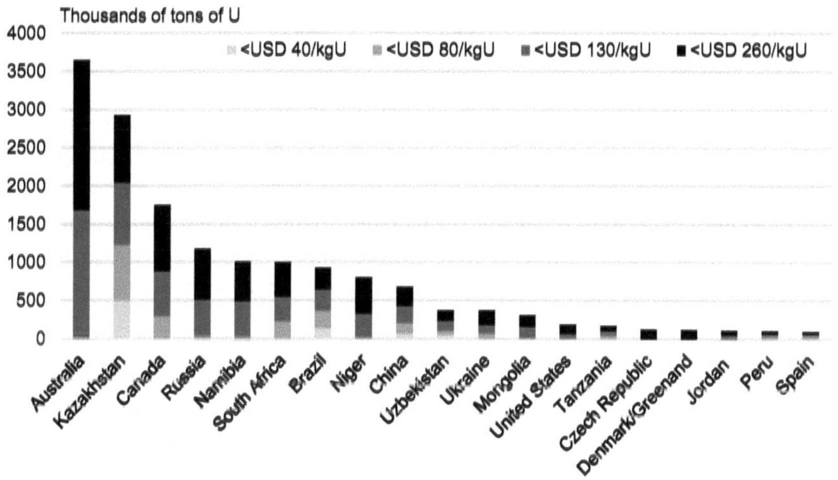

Fig. 13.8 Distribution of reasonably assured resources among countries with a significant share of resources (*Source* [IAEA and NEA 2020, 17])

Fig. 13.9 Uranium production worldwide (2018) (*Source* [IAEA and NEA 2020, 57])

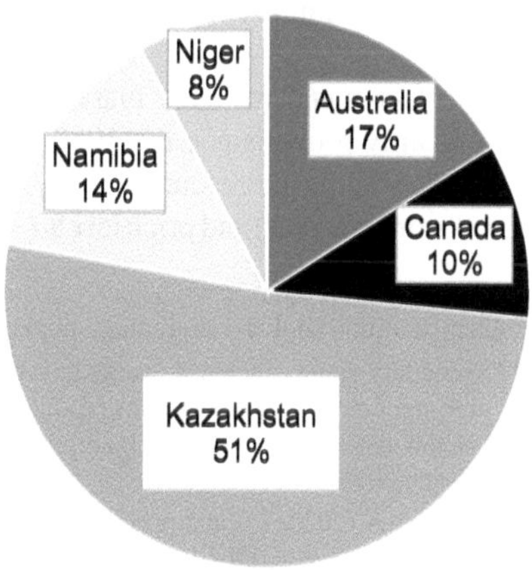

However, contrary to other resource markets, e.g. oil, the large producers are not those with the highest resources. Thus, Australia has by far the largest resources, but they are more expensive to mine than others. The US, currently producing hardly any uranium, could become a producer with rising prices, or by imposing import quotas (Considine 2019). The concentration with respect to resources (mined below US-$130/kg) is significantly lower with an HHI of 1,478 (own calculation based on IAEA and NEA (2020).[23]

13.4.3.2 Current Reserve and Resource Estimates

The available reserves ("economically minable") and resources ("physical potential") depend on market prices. The industry provides staggered reserve and resource estimates, currently split in four cost categories: < $40/kg, < $80/kg, < $130/kg, and, since 2010, < $260/kg. Within these categories, current uranium availability is as follows (IAEA and NEA 2020):

~ Recoverable resources at costs below US-$260/kg U are 8,070 kt, of which 4,723 kt are "reasonable assured resources" (Fig. 13.8), and the rest are inferred resources.

~ In addition, undiscovered resources include prognosticated resources and speculative resources. Together, they amount to 7,200 kt, consisting of 5,300 kt prognosticated resources (below US-$260/kg U) and another 1,878 kt of speculative resources (with an unassigned cost range).

Beyond conventional resources, uranium can be extracted from phosphate resources, if this was economically profitable. Also, it is theoretically possible to derive uranium from seawater (3–4 parts per billion), but this is technically unproven outside the laboratory setting, and elusively expensive (IAEA and NEA 2020, 38). Last, secondary resources are available, and were used for 10% of world reactor requirements in 2019 (IAEA and NEA 2020, 13).[24]

13.4.3.3 Dynamics of Resource Availability

As in other metal resource markets (like copper), the physical resource base has significantly grown since the 1960s, from about 1.5 Mt in 1965 to about 8 Mt of assured and inferred resources today (Fig. 13.10).[25] At current production levels (~ 60.000 tons per year) and price levels (~ US-$ 40–60/t), it is not profitable to invest and

[23] Chap. 8 of this book discusses these market concentrations in detail.

[24] "Secondary supply includes excess government and commercial inventories, spent fuel reprocessing, underfeeding and uranium produced by the re-enrichment of depltted uranium tails, as well as low-enriched uranium (LEU) produced by blending down highly enriched uranium (HEU). (IAEA and NEA 2020, 13).

[25] IAEA-NEA (2020, 15) and previous editions of the "red book" on uranium resources, production, and demand, e.g. (OECD NEA 2006; NEA/IAEA 2018; OECD NEA and IAEA 2014). Cost categories are not inflation adjusted.

obtain better resource estimates, or invest in additional exploration. Nonetheless, a further increase in the resource base is likely once prices would increase significantly.

In essence, the question boils down to the supply elasticity of uranium, i.e., the long-term resource potential as a function of costs or prices. Bunn et al. (2003, 111) suggested the following relation between resources and prices:

$$R = 2.1 * \left(\frac{p}{40}\right)^{\varepsilon} \tag{13.7}$$

where R is the total uranium resource (in Mt U) recoverable at price p (US-\$/kgU) and ε is the long-term price elasticity of supply. The estimated supply elasticities at the time (the early 2000) of 2.3–3.2 are certainly on the optimistic side, suggesting resources between 34 and 105 Mt at prices below US-\$130/kg. However, even with

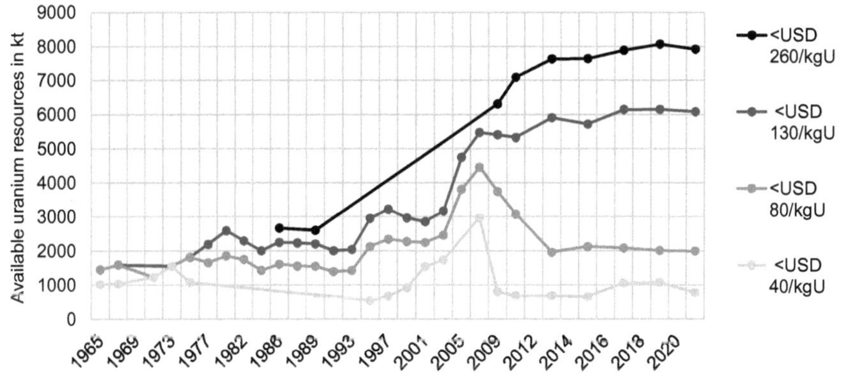

Fig. 13.10 Uranium reserve development from 1965–2021 (kt) (*Source* IAEA-NEA Uranium yearbooks, various issues)

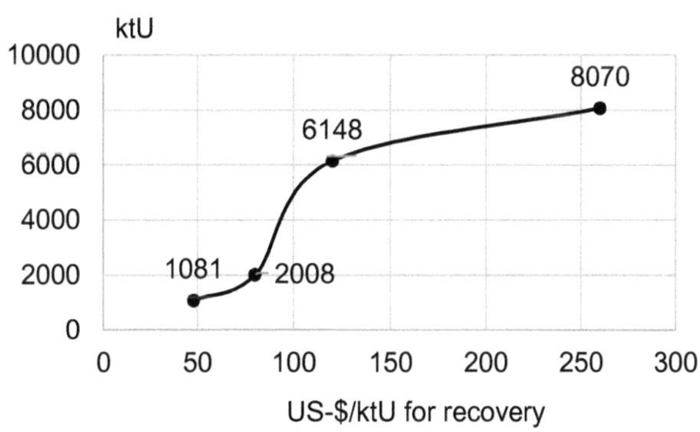

Fig. 13.11 Recoverable uranium resources by costb category (2019) (*Source* Own calculation, based on [IAEA and NEA 2020, 15])

significantly lower elasticities, it is reasonable to assume a significantly expanded resource base beyond the current cost categories.

Figure 13.11 shows the resource-price relation for 2019. There are increases in resource availability to be expected for prices beyond US-$260/kg, increasing the amount of recoverable resources, i.e., turning undiscovered resources into reasonably assured or inferred resources. Also, it is likely that within the reasonable resources, volumes move from higher-cost to lower-cost categories. Given this evidence, it is unlikely that the uranium supply will be constrained, even in the longer-term future, such that the plutonium route becomes economic, say, with uranium prices in the four-digit range.

13.4.4 Intermediate Conclusion

The uranium market is characterized by shorter-term price fluctuations, but longer-term stability at relatively low prices. Estimates of the recoverable resource base have gone up from 1.5 Mt (1965) to 8 Mt (2019) at prices below US-$260/kg (in September 2025, the US spot price was in the US-$ 60/kg range). In addition, about 7.2 Mt undiscovered resources remain to be mined, should prices go up significantly.

13.5 Can Reactors Become Competitive Through Standardized Mass Production?

In this section, I assess the potential for cost reduction for fast neutron reactors, and more generally, other types of reactors of small power output, through standardized production. This is an expectation that some industry observers connect with so-called SMR reactors ("small modular reactors"), i.e., of relatively low capacity (< 300 MWel, see Chap. 2 for details). Fast reactors are an element of the SMR-reactors, such as the Chinese CEFR (Chinese Experimental Fast Breeder), but other countries are also pursuing R&D in the field (Russia, Korea, USA), see Table 13.4.

13.5.1 Non-Conventional SMR-Reactors

Closely related to the SNR activities is an R&D stream on so-called "SMR-reactors," that has emerged over the last one or two decades. Some industry observers consider this to become a competitive source of nuclear energy in the medium-term future (Locatelli et al. 2014; Rothwell 2016; Lloyd et al. 2020; WNA 2020b). In fact, the International Atomic Energy Agency (IAEA) has developed a new research and development field dedicated to SMR-concepts, accompanied by bi-annual reviews

Table 13.4 Fast neutron designs with low power ratings (< 300 MWel) according to the World Nuclear Association

Reactor	Type, coolant	Power (Thermal or Electric) in MW	Fuel (Future)	Country	Expected operation
PRISM	Demonstration, pool, sodium	840/311	Metal	US	From 2020s
ARC-100	Prototype, pool, sodium	260/100	Metal	US	n.a
FMR	Demonstration, helium, HTR	50	?	US	2035
EM2	Helium HTR	500/240	Oxide?	US	n.a
Westinghouse LFR	Pool, lead	950/450	LEU, Oxide/ Silicide	US	n.a
Moltex SSR-U	MSR	750/300 (for 8 modules)	Pu + U Chloride	UK	n.a
Astrid	Prototype, pool, sodium	100–200	Oxide?	France & Japan	Delayed, after 2050
SVBR-100	Demonstration, pool, Pb-Bi	280/100	Oxide?	Russia	Cancelled
Gen4 module	Lead-Bi	70/25	LEU nitride	US	n.a
Sealer	Lead	3–10 MWe	LEU, Oxide/ Nitride	Sweden	By 2025
Aurora	Heatpipe	4/1.5	U–Zr Metal	US	n.a
eVinci	Heatpipe	0.2–5 MWe	Various	US	n.a

Source (WNA 2021)

of the most advanced concepts from the IAEA database on "Advanced Reactors Information System" (ARIS) (IAEA 2018; 2020).

Even though there is no commonly agreed definition, SMRs are often defined as "reactors in which a single reactor has an electrical power output of less than 300 MW_{el} (or a thermal power output of less than 1000 MW_{th})" (Pistner et al. 2021, 24). The modularity of production, and the possibility of standardized, industrial production of individual reactor modules, plays a particular role. SMR research takes as point of inception that traditional nuclear power plants with high electric power (up to 1,600 MW_{el}) have become too expensive, due to their long construction times, and lack of standardization (Lloyd et al. 2020; Boarin et al. 2021). Moving towards highly standardized, mass-produced modules should reverse this trend.

Pistner et al. (2021) provide a very detailed overview of SMR concepts. In many cases, these date back to developments in the 1950s, in particular the attempt to use nuclear power as a propulsion technology for military submarines. Today, a wide variety of theoretical concepts and demonstrators for SMR are discussed, but the majority only exist at the concept level ("paper reactors"). Very few SMRs are

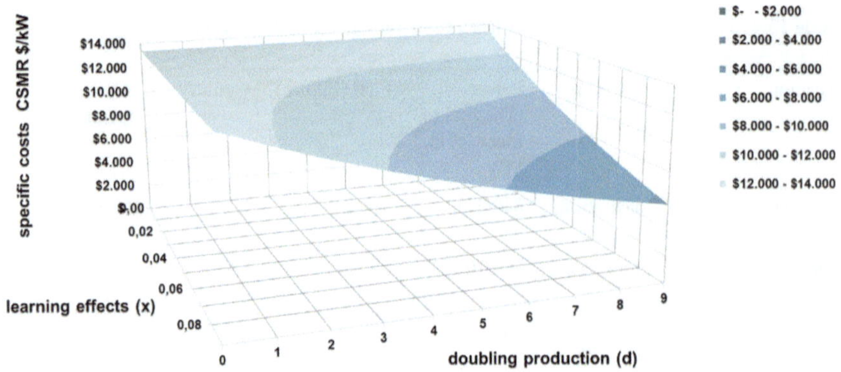

Fig. 13.12 Costs of an SMR taking into account learning, and doubling of production (*Source* Own calculations, based on Steigerwald et al. [2023])

already in operation, among them two Russian floating nuclear power plant types (KLT-40, and RITM).

There is a controversial discussion about the potential effects of SMR concepts in future energy systems (Locatelli et al. 2014; Lloyd et al. 2020; Pistner et al. 2021). In addition, it seems that the SMR optimism focuses mainly on light-water moderated, more traditional technologies, which have a certain established record of development and implementation (Ramana and Ahmad 2016; Pistner et al. 2021).

13.5.2 Some Indicative Evidence

Can SMR-reactors become cheaper than current reactors of high power? That depends on the learning and the mass production that can be generated. Given the absence of existing empirical evidence, I have to make simplifying assumptions to determine, for example, how many SMRs need to be produced to be competitive with the reactor of higher capacity. Figure 13.12 shows the development of construction costs as a function of the learning factor (x) and the number of units produced (represented by the doubling factor d). Clearly, a large number of units need to be produced to achieve significant cost reduction.

I use the example of an existing Westinghouse light water reactor with high capacity (AP1000, 1,117 MWel) and unit costs of US-$6,000/kWel and a (planned) Westinghouse SMR (225 MWel). Assuming a scale effect of 0.55 and an (optimistic) learning rate of 0.6, about 3,000 SMR with capacities of about 225 MWel would have to be built to compensate for the size penalty. If, in addition, one assumes that the higher reactor also achieves learning, this number would still go up.[26]

[26] Using an optimistic variant (specific costs: 6.000 USD/kW, b = 0,6, x = 0,1) would yield a value of 65, still above what Westinghouse was counting on. On the contrary, using less optimistic

It is even more difficult to benchmark fast reactors, due to a lack of representative models and reliable cost data. Table 13.4 shows a range of fast reactors characterized as SMRs by the World Nuclear Association. Currently, the Chinese Experimental Fast Reactor (CEFR) is the only fast SMR already in operation. Its power output is low (20 MWel), but it is designed in the tradition of sodium fast reactors (SFRs), eventually using plutonium–uranium-oxide fuels (MOX).[27] Like other SMRs, the CEFR faced significant cost escalations. From the initial estimate of US-$8,000/kWel, these have increased to US-$23.000/kWel (Pistner et al. 2021, 73). Other fast reactors SMR concepts are under development, such as the PRISM (Power Reactor Innovative Small Module, 311 MWel), and the ARC-100 (Advanced Reactor Concept, 100 MWel).

A direct benchmark with fast neutron reactors of high capacity is not useful. However, transferring the approach from the light-water reactor suggests that a very large number of fast neutron reactors would need to be built to benefit from learning effects.[28]

13.5.3 *Intermediate Conclusion*

High hopes are placed by some actors in the development of reactors of small capacity ("SMRs"), expected to become cost competitive after a period of learning and mass production. Assuming some basic production theory and parameters, this proposition is not convincing. This is shown using some production economics for the representative (Westinghouse) light water reactors. By analogy, there are indications that similar results hold for fast reactors under development. The perspective of fast SMRs to become operational and economic is slim. Thus, I conclude that expectations of fast reactors becoming substantially more competitive in the medium-term future are unlikely to be fulfilled.

---

values gives unbelievably high values, e.g. specific costs of 10.000 USD/kW, b = 0,5, x = 0,02: 666 billion.

[27] The CEFR went critical for the first time in 2011, but has had major interruptions since. Data on real utilization is not consistent between different sources. The CEFR has been developed further over the last years (Wang et al. 2013; Pistner and Englert 2017; Sokolski 2021).

[28] As a thought experiment, one could combine the costs of the CEFR with the power rating of the PRISM, benchmarked against the (optimistic) cost assumptionso of the Superphoenix demonstrator fast reactor ($ 4,580/kWel, Nature, Vol. 385, 9. Jan 1997, p. 104); assuming b = 0,5 and x = 5%, 9,168 would need to be produced.

13.6 Can Plutonium Be Efficiently Abated?

If, as the last two sections suggested, plutonium is unlikely to become a valuable resource, can it be abated efficiently? Environmental economics assumes that harmful outputs should be abated as long as this is less costly than the harm they generate. Therefore, options need to be explored to reduce the amount of harmful plutonium, and, if possible, to provide economic analysis thereof, i.e., to check whether the aggregated costs of reducing the plutonium are lower than the avoided damage. In essence, there are two options to reduce the danger from plutonium: (1) Deep geological depository, and (2) abatement. Attempts to place plutonium (and other radioactive waste) in long-term geological depositories is under way, though this is a very lasting and complex socio-technical process (Brunnengräber et al. 2015; Brunnengräber and Di Nucci 2019). If there were a way to get rid of plutonium more easily, this would be worthwhile exploring. In this section, I survey the literature on plutonium abatement, and discuss whether it could be done efficiently, i.e., at low cost and without creating other harm. The discussion focuses on fast metal cooled reactors, thus leaving aside other, still more hypothetical pathways.[29] This section does not cover individual research reactors currently under development in the Generation IV and/or SMR programs that use similar processes of waste burning.[30]

13.6.1 Partitioning and Transmutation (P&T)

Plutonium abatement ("burning") has been addressed by physicists from the very beginning, e.g. Leo Szilard (Szilard 1947),[31] and has become part of the narrative of plutonium as a large-scale source of energy (Gates 2021). Physically, it is indeed possible to reduce the amount of plutonium in the final waste by a process called partitioning and transmutation (P&T). Research interest in P&T has grown again in the context of the "Generation IV" and SMR-reactor developments (National Research Council 1996; IAEA 2004; Kirchner et al. 2015; Pistner and Englert 2017).

---

[29] Such as accelerator-driven systems (ADS) and molten salt reactors (MSR). While ADS-technology is not yet available (a demonstration reactor (MYRRAH) is planned in Belgium for the mid-2030s), the MSR requires pyrochemical separation that is far from being available, see (Frieß et al. 2021).

[30] For details on some of these projects, refer to (Locatelli et al. 2013; Pistner and Englert 2017; IAEA 2020).

[31] "If plutonium is allowed to disintegrate, or let us again say if it is burned, heat is produced in much the same way as in the case of uranium 235. Heat is produced however not only when we burn plutonium, but also when we manufacture it. As a matter of fact more heat is produced in the process of making plutonium than in the process of burning it. So if we consider the atomic fuel plutonium for purposes of power production we must keep in mind that we produce heat as a by-product at the time we manufacture plutonium and that we also produce heat again later when we decide to burn a certain quantity of it." (Szilard 1947, 3).

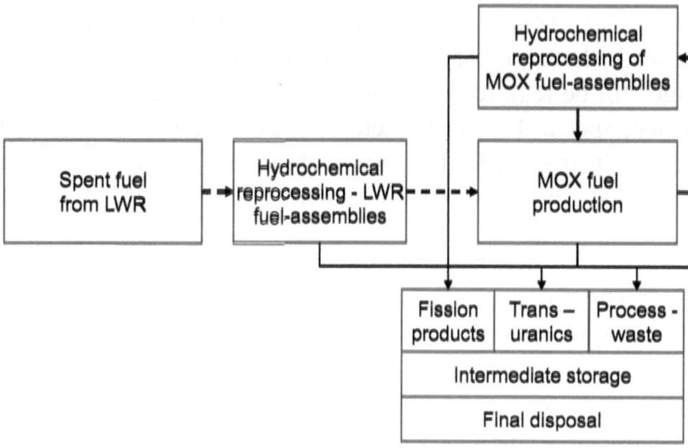

Fig. 13.13 Partitioning and transmutation using a fast reactor (*Source* Translated from [Frieß et al. 2021, 190])

Transmutation relies heavily on the availability of fast reactors at large scale,[32] and the reprocessing plutonium route in general (IAEA 2004).[33] A fast reactor can not only breed plutonium, but it can also burn it to reduce the burden of safe disposal and have lower safe disposal requirements than the parent nuclides, in particular less plutonium. In this event, some nuclides can even be reused as nuclear fuel. Figure 13.13 summarizes the process. A more detailled account of P&T is provided by Englert and Pistner (2025) in this book.

In a first step, radioactive waste is partitioned into different wave streams ("partitioning"), the objective being to reduce the amount of transuranic waste, not only the plutonium, but also other minor actinides such as neptunium, americium, and curium. Partitioning involves at least the separation of transuranic elements into plutonium (so-called major actinides) and the minor actinides. Currently, the only established technology is hydrochemical separation, relying on a repetitive process of separating the heavy metals by solution, extraction, precipitation, or adsorption in aqueous and organic solutions. The so-called hydrochemical PUREX process (plutonium–uranium–recovery by extraction) is used for fast neutron fuels (Frieß et al. 2021).

The second step, called transmutation, involves the conversion of radionuclides into other nuclides through nuclear physical transformations, particularly nuclear

[32] "Recycling of plutonium in LWR MOX reactors is an intermediate strategy to reduce separated plutonium stocks and to partially use its fissile content. However, full recycling and burning of plutonium is only possible when FRs become operational on the industrial scale." (IAEA 2004).

[33] "P&T is an alternative waste management strategy that aims to reduce the very long term radiological burden of nuclear energy. It relies on the nearly quantitative recycling of long lived and highly radiotoxic nuclides. It is therefore very important to stress the crucial role of reprocessing technologies in any further P&T development. For P&T to be a viable option it will be necessary for reprocessing expertise to be preserved either in existing industrial plants or by keeping R&D projects in this field alive." (IAEA 2004).

fission (Frieß et al. 2021, 25). Although transmutation is understood in theory, it has not been applied outside the laboratory context. Significant research and industrial development would still be required. Even though the principle of fast metal-cooled reactors is applied, special forms of reactors would have to be developed, among others, significantly increasing the proportion of minor actinides in the fuel. This requires significant modification to reactor designs, and longer R&D periods (Kirchner et al. 2015).

Further challenges of P&T are the choice of the appropriate fuel, and the related danger of proliferation. For P&T with fast metal cooed reactors, plutonium–uranium mixed oxide (MOX) fuels are the most obvious choice, though they would require a different composition than those used for light-water reactors, in particular with respect to the inclusion of minor actinides (Frieß et al. 2021, 29). Proliferation would be facilitated, because large amounts of weapon grade material would become available, without any radiation barrier.

13.6.2 Trade-Offs in Large-Scale P&T Scenarios

Research on "plutonium abatement" indicates trade-offs: the amounts of plutonium can indeed be reduced, but other products, mainly fission products, are generated in large quantities, so that the overall balance is unclear. Clearly this process is not "Pareto-efficient," because while some plutonium problems diminish (though they do not disappear), other problems appear.

Frieß et al. (2021) provide one of the most detailed analysis to-date of large-scale P&T scenarios, using the fast metal-cooled reactor, applied to the volumes available in Germany. On the one hand, they show that abating plutonium is possible. Provided technical solutions for MOX-elements with minor actinides, and a new fast reactor, are found in the coming decades, the scenario leads to a significant reduction of plutonium, i.e., for Germany, from 130 tons to about 17 tons (Fig. 13.14).[34] However, this would not only be very expensive and take a long time, but it would also generate significant amounts of additional waste:

~ A total of 23 new reactors would have to be constructed and operated over a period of about 300 years, to exhaust the possibility of transuranium reduction (Frieß et al. 2021, 34). In addition, a reprocessing facility for LWR-fuel and 18 reprocessing plants for FR-MOX fuel would be required, plus seven fuel fabrication plants (Table 13.5).

~ The amount of three long-lived fission products, technecium-99, iodine-129, and cesium-135, which serve as guide isotopes for evaluating the long-term safety of repositories, would increase significantly (48–71%).

---

[34] Americium would also be reduced (by over 50%), while neptunium would stay more or less constant. About 7 tons of Curim would be produced, which decay relatively quickly (in weeks) to another 6.5 tons of plutonium.

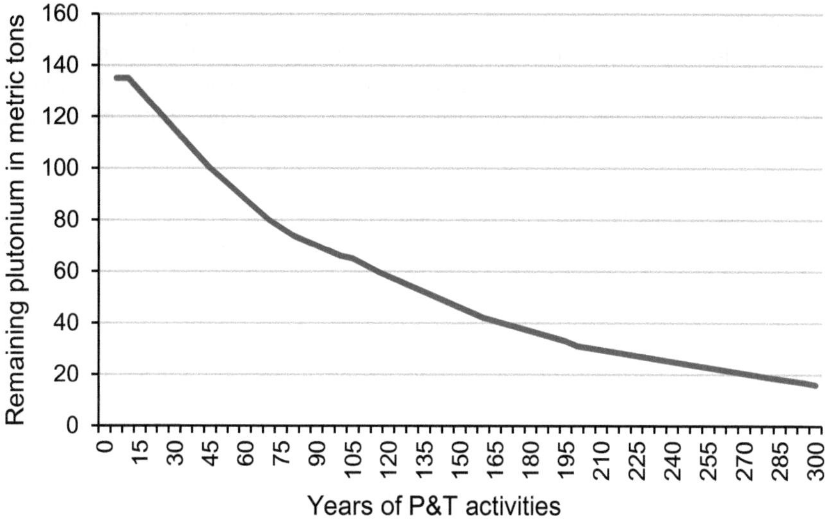

Fig. 13.14 Abatement of plutonium through transmutation in fast reactors (*Source* Translated from [Frieß et al. 2021, 197])

Table 13.5 Infrastructure requirements for the P&T scenario with fast reactors

Reactors	Amount	Comment	Total
Decommissioning	23 installations	5,000 m^3/installation	115,000 m^3
Operation	23 installations, 42 years each	42 years * 45 m^3/year	43,500 m^3
Reprocessing plants for light-water fuel			
Decommissioning	1	36,000 m^3	36,000 m^3
Operation	10,113 t HM	6.59 m^3/t HM	66,600 m^3
Reprocessing plants (reactor MOX-fuel)			
Decommissioning	18	1,500 m^3	27,000 m^3
Operation	2,710 t HM	6.59 m^3/t HM	17,900 m^3
Fuel fabrication			
Decommissioning	7	1,000 m^3	7,000 m^3
Operation	2,945 t HM	1.2 m^3/t HM	3,500 m^3
Total			**316,500 m^3**

Source (Frieß et al. 2021, 201)

~ In the course of the process, large quantities of low-and intermediate-level waste would be generated, mainly by the reprocessing and the dismantling of all the facilities (Frieß et al. 2021, 34) (Table 13.5). The volume would exceed 300,000 m3, which is the size of the Konrad depository in Germany for medium- and low-level waste.[35]

The figure shows the initial amount minus the transmuted amount of plutonium in a fast neutron reactor scenario. At the end of the scenario (after 300 years), at least 21.8 tons of plutonium remain, of which 2.3 t are in reprocessing and fuel fabrication.

Clearly abating plutonium comes at very high costs. Even without having precise figures, it seems that the benefit-cost-ratio of a hypothetical large-scale abatement of plutonium by partitioning and transmutation is low. In addition, P&T for a couple of centuries is at odds with the institutional setting in many countries.[36] Last but not least, abatement of plutonium in spent fuels would not resolve the requirement to develop a long-term depository: first, because not all plutonium can be abated, and, second, because about 40% of highly-radioactive waste has been already vitrified and cannot be further treated (Frieß et al. 2021, 237).

13.6.3 Intermediate Conclusion

Theoretical research on laboratory and pilot scale applications indicate that partitioning and transmutation can succeed in reducing the amount of plutonium, but that this requires significant amounts of new reactors, and generates additional amounts of other radioactive elements, mainly fission products and large amounts of medium- and low-level waste. No precise economic quantification is available on the cost and the benefits of abating plutonium, but the available evidence suggests that partitioning and transmutation are neither promising nor a secure route to support the plutonium economy. P&T programs require a large number of nuclear facilities, and a very long-term operation of up to several centuries (including the decades of R&D before a possible realization of the P&T program). The nuclear facilities needed for P&T are not available at a commercial scale, and it is highly unclear whether it will be technically possible. Not only would the amount of low-and intermediate-level secondary waste increase significantly, but the amount of high-level long-lived fission products (that are guide isotopes for the analysis of the repository) would also increase significantly.

[35] Planning of this mine started in the 1970s, it is currently expected to open in the late 2020s.

[36] In Germany, the 13th Amendment of the Nuclear Power Law (Atomgesetz) defined an end for commercial nuclear power in 2022, and the final depository act (Standortauswahlgesetz), obliged the government to site a final depository until 2031, and then to proceed with filling this depository rapidly.

13.7 Conclusion

In this chapter, I have highlighted specific economic aspects of plutonium. Interest in the topic is increasing recently due to the lack of competitiveness of traditional nuclear power, and various attempts to re-focus on the plutonium route using fast reactors that were once expected to become the foundation of commercial nuclear energy. Due to its high radioactivity and potential for military use, the treatment of plutonium is highly regulated and there is no "market" to define its prices. In addition, models of optimal resource depletion are not available to provide guidance on estimating the value of plutonium. Therefore, we develop a present analytical frameworks for different sub-questions having to do with resources, construction costs, and plutonium abatement.

I find no evidence that the historical trend, i.e., a negative value of plutonium, is likely to be reversed in the future. In the long term, uranium resources are sufficiently abundant to prevent extremely high uranium prices. Current research on SMR-reactors is unlikely to generate a new generation of fast reactors that would be cheaper. Abating plutonium is not an economically efficient way to deal with the problem of waste.

Further research should update the available analysis, some of which is quite outdated. Arguments also need to be developed as to whether research in fast reactor technologies should continue, even after seven unsuccessful decades. A full environmental economic assessment of the risks of fast reactors and plutonium, including proliferation, is still lacking in the literature. Last but not least, one needs to better understand the implication of these findings for the process of identifying depositories for nuclear waste.

References

Acheson-Lilienthal Report. 1946. "A Report on the International Control of Atomic Energy." Prepared for the Secretary of State's Committee on Atomic Energy. Washington, D.C. http://fissilematerials.org/library/ach46.pdf.

Baade, Fritz. 1958. *Welt-Energiewirtschaft: Atomenergie - Sofortprogramm oder Zukunftsplanung*. Hamburg: Rowohlt.

Besnard, Manon, Marcos Buser, Ian Fairlie, Gordon MacKerron, Allison Macfarlane, Eszter Matyas, Yves Marignac, Edvard Sequens, Johan Swahn, and Ben Wealer. 2019. "The World Nuclear Waste Report. Focus Europe." Berlin, Brussels. https://worldnuclearwastereport.org/.

Boarin, S., M. Mancini, M. Ricotti, and G. Locatelli. 2021. "Economics and Financing of Small Modular Reactors (SMRs)." In *Handbook of Small Modular Reactors*, 2nd ed., 241–78. Kidlington (UK) and Cambridge (USA): Elsevier.

Böse, Fanny, Alexander Wimmers, Björn Steigerwald, and Christian von Hirschhausen. 2024. "Questioning Nuclear Scale-up Propositions: Availability and Economic Prospects of Light Water, Small Modular and Advanced Reactor Technologies." *Energy Research & Social Science* 110.

Brunnengräber, Achim, and Maria Rosaria Di Nucci. 2019. *Conflicts, Participation and Acceptability in Nuclear Waste Governance: An International Comparison. Volume III*. Wiesbaden: Springer VS. https://link.springer.com/book/https://doi.org/10.1007/978-3-658-27107-7.

Brunnengräber, Achim, Maria Rosaria Di Nucci, Ana María Isidoro Losada, Lutz Mez, and Miranda A. Schreurs, eds. 2015. *Nuclear Waste Governance. An International Comparison.* Wiesbaden: Springer VS.

Bunn, Matthew, Steve Fetter, John P. Holdren, and Bob van der Zwaan. 2003. "The Economics of Reprocessing Versus Direct Disposal of Spent Nuclear Fuel." Project on Managing the Atom. Belfer Center for Science and International Affairs, Harvard Kennedy School.

Commissariat Général du Plan. 2000. "Report to the Prime Minister [of France]: Economic Forecast Study of the Nuclear Power Option." Paris, France. https://fire.pppl.gov/eu_fr_fission_plan.pdf.

Considine, Timothy J. 2019. "The Market Impacts of US Uranium Import Quotas." *Resources Policy* 63 (October):101445. https://doi.org/10.1016/j.resourpol.2019.101445.

Davis, Lucas W. 2012. "Prospects for Nuclear Power." *Journal of Economic Perspectives* 26 (1): 49–66. https://doi.org/10.1257/jep.26.1.49.

De Roo, Guillaume, and John E. Parsons. 2011. "A Methodology for Calculating the Levelized Cost of Electricity in Nuclear Power Systems with Fuel Recycling." *Energy Economics* 33 (5): 826–39. https://doi.org/10.1016/j.eneco.2011.01.008.

Englert, Matthias, and Christoph Pistner. 2025. "Partionioning and Transmutation - Technical Overview and Scenarios for Waste Disposal." In *Nuclear Power: Technology, Geopolitics, and Economics*, edited by Alexander Wimmers, Fanny Böse, Christian Von Hirschhausen, and Claudia Kemfert. Springer Fachmedien Wiesbaden.

Frieß, Friederike, Nikolaus Arnold, Wolfgang Liebert, and Nikolaus Müllner. 2021. "Sicherheitstechnische Analyse und Risikobewertung von Konzepten zu Partitionierungs- und Transmutationsanlagen für hochradioaktive Abfälle." Wissenschaftliches Gutachten im Auftrag des Bundesamtes für die Sicherheit der nuklearen Entsorgung (BASE) urn:nbn:de:0221–2021030826033. Berlin; Wien: Institut für Sicherheits- und Risikowissenschaften Universität für Bodenkultur (BOKU).

Gates, Bill. 2021. *How to Avoid a Climate Disaster - The Solutions We Have and the Breakthroughs We Need*. New York, NY: Alfred A. Knopf.

GIF. 2018. "GIF Annual Report 2017." GEN IV International Forum. https://www.gen-4.org/gif/upload/docs/application/pdf/2018-09/gif_annual_report_2017_210918.pdf.

Hall, S., and M. Coleman. 2013. "Critical Analysis of World Uranium Resources." Scientific Investigations Report 2012–5239. US Geologicsl Survey, prepared in cooperation with the U.S. Energy Information Administration. http://pubs.usgs.gov/sir/2012/5239/sir2012-5239.pdf.

Hippel, Frank von, Masafumi Takubo, and Jungmin Kang. 2019. *Plutonium: How Nuclear Power's Dream Fuel Became a Nightmare*. Singapore: Springer. http://ezproxy.st-andrews.ac.uk/login?url=http://link.springer.com/https://doi.org/10.1007/978-981-13-9901-5.

Holdren, John P. 1975. "Uranium Availability and the Breeder Decision." *Energy Systems and Policy* 1 (3).

IAEA. 2004. "Implications of Partitioning and Transmutation in Radioactive Waste Management." Technical Report Series 435. Vienna, Austria: IAEA. https://www-pub.iaea.org/MTCD/Publications/PDF/TRS435_web.pdf.

IAEA. 2017. "Nuclear Power for a Clean-Energy Future." IAEA Bulletin. https://www.iaea.org/sites/default/files/publications/magazines/bulletin/bull/bull584-nov2017_corr.pdf.

IAEA. 2018. "Advances in Small Modular Reactor Technology Developments. A Supplement to: IAEA Advanced Reactors Information System (ARIS)." Vienna, Austria: International Atomic Energy Agency.

IAEA. 2020. "Advances in Small Modular Reactor Developments. A Supplement to: IAEA Advanced Reactors Information System (ARIS)." IAEA Booklet. Vienna, Austria: International Atomic Energy Agency. https://aris.iaea.org/Publications/SMR_Book_2020.pdf.

IAEA and NEA. 2020. "Uranium 2020: Resources, Production and Demand." Paris: Nuclear Energy Agency / Organisation for Economic Co-operation and Development.

JAEC. 2011. "Estimation of Nuclear Fuel Cycle Cost." Tokyo, Japan: Technical Subcommittee on Nuclear Power, Nuclear Fuel Cycle, etc. http://www.aec.go.jp/jicst/NC/about/kettei/seimei/111110_e.pdf.

Jukes, J.A. 1956. "The Cost of Power and the Value of Plutonium from Early Nuclear Power Stations." In *The Economics of Nuclear Power*, 1:185 ff. London.

Kirchner, Gerald, Matthias Englert, Christoph Pistner, Beate Kallenbach-Herbert, and Julia Mareike Neles. 2015. "Gutachten „Transmutation"." K-MAT 48. Kommission Lagerung hoch radioaktiver Abfallstoffe. Hamburg, Darmstadt: Öko-Institut, Universität Hamburg, ZNF. https://www.bundestag.de/resource/blob/400438/f54e3da4bbe76395bce2e40721212379/kmat_48-data.pdf.

Landajo, Manuel, María José Presno, and Paula Fernández González. 2021. "Stationarity in the Prices of Energy Commodities. A Nonparametric Approach." *Energies* 14 (11): 3324. https://doi.org/10.3390/en14113324.

Lévêque, François. 2015. *The Economics and Uncertainties of Nuclear Power*. Cambridge, United Kingdom: Cambridge University Press.

Liebert, Wolfgang, and Matthias Englert. 2015. "Risiken der Uranproduktion und der Urannutzung - Risiken für die Nutzung der Kernenergie." Wien, Austria.

Lloyd, Clara, Robbie Lyons, and Tony Roulstone. 2020. "Expanding Nuclear's Contribution to Climate Change with SMRs." *Nuclear Future*, November, 39.

Löbl, O. 1956. "Wirtschaftlichkeit der Atomenergie." *Elektrotechnische Zeitschrift (ETZ)* 77 (13): 460 ff.

Locatelli, Giorgio, Chris Bingham, and Mauro Mancini. 2014. "Small Modular Reactors: A Comprehensive Overview of Their Economics and Strategic Aspects." *Progress in Nuclear Energy* 73 (May):75–85. https://doi.org/10.1016/j.pnucene.2014.01.010.

Locatelli, Giorgio, Mauro Mancini, and Nicola Todeschini. 2013. "Generation IV Nuclear Reactors: Current Status and Future Prospects." *Energy Policy* 61 (October):1503–20. https://doi.org/10.1016/j.enpol.2013.06.101.

Midttun, Atle, and Thomas Baumgartner. 1986. "Negotiating Energy Futures: The Politics of Energy Forecasting." *Energy Policy* 14 (3): 219–41. https://doi.org/10.1016/0301-4215(86)90145-X.

MIT. 2003. "The Future of Nuclear Power." Cambridge, MA: Massachusetts Institute of Technology. http://web.mit.edu/nuclearpower/pdf/nuclearpower-full.pdf.

MIT. 2018. "The Future of Nuclear Energy in a Carbon-Constrained World." Cambridge, MA, USA: Massachusetts Institute of Technology. http://energy.mit.edu/wp-content/uploads/2018/09/The-Future-of-Nuclear-Energy-in-a-Carbon-Constrained-World.pdf.

Monnet, Antoine, Sophie Gabriel, and Jacques Percebois. 2017. "Analysis of the Long-Term Availability of Uranium: The Influence of Dynamic Constraints and Market Competition." *Energy Policy* 105 (June):98–107. https://doi.org/10.1016/j.enpol.2017.02.010.

Morss, Lester R., Norman M. Edelstein, J. Fuger, and Joseph J. Katz, eds. 2006. *The Chemistry of the Actinide and Transactinide Elements*. 3rd ed. Dordrecht: Springer.

Muellner, Nikolaus, Nikolaus Arnold, Klaus Gufler, Wolfgang Kromp, Wolfgang Renneberg, and Wolfgang Liebert. 2021. "Nuclear Energy - The Solution to Climate Change?" *Energy Policy* 155 (August):112363. https://doi.org/10.1016/j.enpol.2021.112363.

Murakami, Tomoko. 2021. "A Historical Review and Analysis on the Selection of Nuclear Reactor Types and Implications to Development Programs for Advanced Reactors; A Japanese Study." *Energy Reports* 7 (November):3428–36. https://doi.org/10.1016/j.egyr.2021.05.049.

National Research Council. 1996. *Nuclear Wastes - Technologies for Separations and Transmutation*. Washington, DC: The National Academies Press. https://doi.org/10.17226/4912.

NEA/IAEA. 2018. "Uranium 2018: Resources, Production and Demand." NEA No. 7413.

Neles, Julia Mareike, and Christoph Pistner, eds. 2012. *Kernenergie: Eine Technik für die Zukunft?* Technik im Fokus. Berlin: Springer Vieweg.

Nelson, Andrew T. 2012. "Thorium: Not a Near-Term Commercial Nuclear Fuel." *Bulletin of the Atomic Scientists* 68 (5): 33–44. https://doi.org/10.1177/0096340212459125.

OECD NEA. 2006. "Forty Years of Uranium Resources, Production and Demand in Perspective." Paris. https://www.oecd-nea.org/ndd/pubs/2006/6096-40-years-uranium.pdf.

OECD NEA, and IAEA. 2014. *Uranium 2014: Resources, Production and Demand*. NEA 7209. Paris, Vienna: Organisation for Economic Co-operation and Development (OECD). https://www.oecd-nea.org/ndd/pubs/2014/7209-uranium-2014.pdf.

Owen, A.D. 1983. "The Economics of Uranium Demand." *Resource Policy* 9 (2): 110–21. https://doi.org/10.1016/0301-4207(83)90052-1.

Pedregal, Diego J. 2020. "Forecasting Uranium Prices: Some Empirical Results." *Nuclear Engineering and Technology* 52 (6): 1334–39. https://doi.org/10.1016/j.net.2019.11.028.

Pistner, Christoph, and Matthias Englert. 2017. "Neue Reaktorkonzepte. Eine Analyse des aktuellen Forschungsstands." Darmstadt: Öko - Institut e.V. https://www.oeko.de/fileadmin/oekodoc/Neue-Reaktorkonzepte.pdf.

Pistner, Christoph, Matthias Englert, Christian Küppers, Ben Wealer, Björn Steigerwald, Christian von Hirschhausen, and Richard Donderer. 2021. "Sicherheitstechnische Analyse und Risikobewertung einer Anwendung von SMR-Konzepten (Small Modular Reactors)." Vorhaben 4720F50500. Forschungsberichte zur Sicherheit der nuklearen Entsorgung. Berlin: Bundesamt für die Sicherheit der nuklearen Entsorgung. https://www.base.bund.de/SharedDocs/Downloads/BASE/DE/berichte/kt/gutachten-small-modular-reactors.pdf?__blob=publicationFile&v=6.

Radkau, Joachim. 1983. *Aufstieg und Krise der deutschen Atomwirtschaft 1945-1975: Verdrängte Alternativen in der Kerntechnik und der Ursprung der nuklearen Kontroverse*. Reinbek bei Hamburg, Germany: Rowohlt.

Ramana, M.V., and Ali Ahmad. 2016. "Wishful Thinking and Real Problems: Small Modular Reactors, Planning Constraints, and Nuclear Power in Jordan." *Energy Policy* 93 (June):236–45. https://doi.org/10.1016/j.enpol.2016.03.012.

Ramana, M.V., and J.Y. Suchitra. 2007. "Costing Plutonium: Economics of Reprocessing in India." *International Journal of Global Energy Issues* 27 (4): 454. https://doi.org/10.1504/IJGEI.2007.014867.

Rooney, Matthew, William J. Nuttall, and Nikolaos Kazantzis. 2015. "A Dynamic Model of the Global Uranium Market and the Nuclear Fuel Cycle." *Resources Policy* 43 (March):50–60. https://doi.org/10.1016/j.resourpol.2014.11.003.

Rothwell, Geoffrey. 2016. *Economics of Nuclear Power*. London, UK: Routledge.

Salin, Edgar. 1955. *Ökonomik der Atomkraft—Vor einer neuen Etappe der industriellen Revolution*. Cologne, Germany: Sigillum-Verlag.

Seaborg, Glenn T. 1970. "The Plutonium Economy of the Future." Santa Fe, New Mexico. http://fissilematerials.org/library/aec70.pdf.

Sokolski, Henry D. 2021. "China's Civil Nuclear Sector: Plowshares to Swords?" Occasional Paper 2102. Arlington, VA: Nonproliferation Policy Education Cente.

Steigerwald, Björn, Jens Weibezahn, Martin Slowik, and Christian Von Hirschhausen. 2023. "Uncertainties in Estimating Production Costs of Future Nuclear Technologies: A Model-Based Analysis of Small Modular Reactors." *Energy* 281 (October):128204. https://doi.org/10.1016/j.energy.2023.128204.

Strauss, Lewis. 1954. "Remarks Prepared by Lewis. L. Strauss, Chairman, United States Atomic Energy Commission, For Delivery at the Founders' Day Dinner, National Association of Science Writers, on Thursday, September 16, 1954, New York, New York." Washington D.C.: Atomic Energy Commission. https://www.nrc.gov/docs/ML1613/ML16131A120.pdf.

Suchitra, J.Y., and M.V. Ramana. 2011. "The Costs of Power: Plutonium and the Economics of India's Prototype Fast Breeder Reactor." *International Journal of Global Energy Issues* 35 (1): 1. https://doi.org/10.1504/IJGEI.2011.039982.

Szilard, Leo. 1945. "Liquid Metal Cooled Fast Neutron Breeders." In *The Collected Works of Leo Szilard, (Vol. 1), Scientific Papers (6 March 1945)*, edited by Bernard T. Feld and Gertrud Weiss-Szilard, 369–75. Cambridge MA (USA): MIT Press.

Szilard, Leo. 1947. "Atomic Energy, a Source of Power or a Source of Trouble (Speech given April 23 in Spokane, WA)." Leo Szilard Papers. MSS 32. Special Collections & Archives, UC San Diego Library, Spokane, WA. https://library.ucsd.edu/dc/object/bb43701801.

Tietenberg, Thomas H. 2005. *Environmental and Natural Resource Economics*. 7 edition. Boston: Addison Wesley.
Tietenberg, Thomas H., and Lynne Y. Lewis. 2016. *Environmental and Natural Resource Economics*. 11th edition, International student edition. London New York.
Trieu, Luan Ho, Eric Savage, and Gavan Dwyer. 1994. "A Model of the World Uranium Market." *Energy Policy* 22 (4): 317–29. https://doi.org/10.1016/0301-4215(94)90006-X.
U.S. Department of Energy. 2003. "Report to Congress on Advanced Fuel Cycle Initiative: The Future Path for Advanced Spent Fuel Treatment and Transmutation Research." Washington DC, USA: Office of Nuclear Energy, Science, and Technology. http://www.nuclear.gov/reports/AFCI_CongRpt2003.pdf).
Wang, Bo, Fudong Liu, Chunming Zhang, Yahua Qiao, Yongye Liu, Yang Yang, Yichuan Wang, Zhaorong Shang, and Qiang Shi. 2013. "Site Selection Considerations of Spent Nuclear Fuel Reprocessing Plant in China." *Energy Procedia*, The 3rd International Conference on Asian Nuclear Prospects (ANUP2012), 39 (January):382–86. https://doi.org/10.1016/j.egypro.2013.07.226.
Wealer, Ben, Simon Bauer, Christian von Hirschhausen, Claudia Kemfert, and Leonard Göke. 2021. "Investing into Third Generation Nuclear Power Plants - Review of Recent Trends and Analysis of Future Investments Using Monte Carlo Simulation." *Renewable and Sustainable Energy Reviews* 143. https://doi.org/10.1016/j.rser.2021.110836.
WNA. 2020a. "Generation IV Nuclear Reactors (Update December 2020)." https://world-nuclear.org/information-library/nuclear-fuel-cycle/nuclear-power-reactors/generation-iv-nuclear-reactors.aspx; https://world-nuclear.org/information-library/nuclear-fuel-cycle/nuclear-power-reactors/generation-iv-nuclear-reactors.aspx.
WNA. 2020b. "Small Nuclear Power Reactors." 2020. https://www.world-nuclear.org/information-library/nuclear-fuel-cycle/nuclear-power-reactors/small-nuclear-power-reactors.aspx.
WNA. 2021. "Fast Neutron Reactors (Update June 2021)." https://world-nuclear.org/information-library/current-and-future-generation/fast-neutron-reactors.aspx. https://world-nuclear.org/information-library/current-and-future-generation/fast-neutron-reactors.aspx.
Zhang, Hui. 2020. "China is Speeding up its Plutonium Recycling Programs." *Bulletin of the Atomic Scientists* 76 (4): 210–16. https://doi.org/10.1080/00963402.2020.1778372.

Chapter 14
New Ventures in Fusion Energy: Analyzing a Dynamically Emerging Market

Björn Steigerwald, Johanna Krauß, and Jens Weibezahn

14.1 Introduction

Recent advancements in research and development by private companies have rekindled both media and parliamentary interest in the potential application of fusion energy, with many discussions focusing on its prospective use for electricity generation in the near future.[1] Although still at an experimental stage, fusion is envisioned to provide a nearly limitless source of energy for human purposes by replicating the process that powers the sun (Takeda et al. 2023). There are a variety of potential fusion concepts under development, including magnetic and inertial confinement approaches, and more exotic concepts, all planned to run on different kinds of fuel, e.g., deuterium-deuterium, deuterium–tritium or proton-boron (Nuttall 2020). As of today, despite decades of research efforts, no commercial fusion power plant

[1] https://wirtschaftsrat.de/de/presse/medienresonanzen/cdu-macht-sich-f%C3%BCr-kerninfusion-stark/, https://www.gruene-hessen.de/landtag/pressemitteilungen/20-millionen-euro-fuer-ein-luftschloss/, last accessed on 06–03-2025. This date holds for all other online references unless otherwise indicated.

The authors thank the participants of the 14th International Conference on Energy Economics (IEWT) held at the Technical University of Vienna (TU Wien), Austria, in February 2025, for their helpful suggestions on an earlier version of this work.

B. Steigerwald (✉) · J. Krauß
Workgroup for Infrastructure Policy (WIP), Technical University (TU) Berlin, Berlin, Germany
e-mail: bs@wip.tu-berlin.de

B. Steigerwald
German Institute for Economic Research (DIW Berlin), Berlin, Germany

J. Weibezahn
Copenhagen School of Energy Infrastructure (CSEI), Copenhagen Business School, Frederiksberg, Denmark

A. Wimmers et al. (eds.), *Nuclear Power*, Studies in Energy, Resource and Environmental Economics, https://doi.org/10.1007/978-3-031-99894-2_14

exists, and many technical challenges remain (Grünwald 2024).[2] Moreover, research was dominated by large public research institutes and projects, with the focus on magnet confinement. Since the mid-1980s, the International Thermonuclear Experimental Reactor (ITER) Project is envisaged to enhance the magnetic confinement approach by achieving a burning plasma and demonstrate tritium breeding. However, the project faces severe delays and cost increases and is scheduled to fully operate by 2039 only. This places serious doubt on whether the planned demonstration project DEMO will ever be followed through (Wimmers et al., 2025).

In contrast, many new ventures with private co-financing have emerged in recent years and have raised substantial amounts of capital from private investors while claiming to commercialize fusion within the next twenty years. These claimed time horizons challenge the current dominance of public projects conducted by major research institutes. However, as major technical hurdles still need to be overcome, claims of rapid commercialization are questionable. This chapter aims to analyze the intensifying private investor support, which may suggest technological advances and the availability of a final product. We collected a comprehensive dataset to provide a global overview of new ventures that aim to develop fusion reactors.[3] The analysis is based on an adapted framework to incorporate the so-called life cycle theory into assessing the respective new ventures' funding series stages. This provides insights into the expectations and the actual progress observed within the emerging private sector of fusion power.

The remainder of this chapter is structured as follows: Sect. 14.2 provides the theoretical background. Section 14.3 introduces the dataset, while Sect. 14.4 provides the analysis. Section 14.5 concludes.

14.2 Theoretical Framework of New Venture Development

New ventures can be defined as new entities aiming to specifically develop, market, and produce a new product or service (Normann 1979). A fundamental objective for many technology-based new ventures is the growth of the venture, as this facilitates their ability to establish legitimacy, secure resources, attract investments, engage partners and stakeholders, and define technological standards (Nuscheler et al. 2019; Sanasi 2025). Thus, the advancement of fusion energy, like any other technological venture, relies on the growth and development of the underlying organizational structures.

A useful evaluation framework for the growth of new ventures is the so-called contingent life-cycle model that conceptualizes four stages of the development of new ventures (Kazanjian and Drazin 1990). Moreover, Kazanjian (1984; 1988) shows empirically that a technology-based new venture faces sequential problems from

[2] Refer to Chap. 4 of this book for an overview of the organizational and technological challenges faced by current fusion devices.

[3] Access to the continuously updated dataset will be provided upon request.

which the author defines stages that the venture must pass through to become a viable entity.

After the initial establishment, a new venture focuses on developing a product or technology, including business ideation and prototyping in Stage 1. In Stage 2, the commercialization stage, with initial funding secured, efforts shift to refining the product, solving engineering challenges, and preparing for market entry. The growth stage (Stage 3) involves scaling production, achieving market acceptance, and managing operational complexities. Finally, in Stage 4, the stability phase, the firm consolidates its market position, formalizes structures, and may develop next-generation products, often transitioning to professional management (Table 14.1).

Emerging, technology-based new ventures face unique challenges during their growth process. These challenges include the availability of general resources like material for building a product, skilled personnel and the challenges of attracting sufficient investment capital, engaging partners and other stakeholders, and setting technological standards (e.g., for inertial confinement fusion, this relates to production standards of fuel targets) (Coviello et al. 2024).

Table 14.1 Venture maturation stage model

Stage	Focus	Description
Stage 1	Concept and development	In this initial stage, new ventures concentrate on developing their business ideas, prototyping a product, and establishing a market presence. Furthermore, the companies still have technical issues, and the functional structure is usually still very minimal
Stage 2	Commercialization	In the second stage, companies produce the first editions of the product, aim to test the marketability of their product, and accelerate growth. This requires initial financial backing, which is invested in various aspects of the company's development with a focus on learning how to produce the product effectively. Functional areas start to emerge, but the structure is still informal
Stage 3	Growth	This stage is characterized by rapid expansion and the need for more formalized structures and processes. The organization must manage increased complexity, including scaling operations and enhancing decision-making capabilities. Often ventures face first "shocks" and "crises" which they must learn to overcome by building up working and efficient systems
Stage 4	Stability	The final and fourth stage is of critical long-term importance for technology-based ventures, as it represents a shift in focus to maintaining growth momentum and market position. There are numerous challenges to overcome when transforming from a new venture to a more structured and operating model. Often, in this stage, the top management team's experience is brought into the company. In this stage, second-generation products get developed and enter the market. For the company, it is important to balance innovation with economic efficiency to succeed

Source Own compilation based on Kazanjian and Drazin (1990)

One of the main problems during a venture's lifecycle is a lack of funding, as these companies operate in new and disruptive markets that carry high risks for the banking sector. Each venture will have to analyze the ideal and appropriate moment to seek external funding from a bank or from specialized investors based on its objectives and development stages, and this will require careful planning and preparation (Lassala and Ribeiro-Navarrete 2022). Thus, during the development of the new venture, it needs several funding rounds to gather sufficient financial resources for its products and its development. Each of these funding rounds has characteristics for the maturity of the new venture (Table 14.2).

A typical development process of a new venture with corresponding funding rounds is shown in Fig. 14.1. This process is summarized as follows: A venture begins in the concept and development phase, relying on internal funds or pre-seed investment to validate its business model and establish initial structures. Failure at this stage may lead to the "valley of death", where high risks and uncertain returns threaten the venture overall success. If successful, seed funding supports early scalability and product validation, bridging the gap to commercialization. In the growth phase, ventures secure Series A, B, and later-stage investments to scale production, expand market presence, and drive profitability. This funding supports aggressive expansion, leading towards IPO preparation. Near the end of this phase, Mezzanine financing can enhance liquidity and financial leverage, ensuring long-term stability and growth (Silbernagel and Vaitkunas 2012).

Table 14.2 Investment phases and typical characteristics

Investment phase	Characteristics
Pre-Seed	• Validation of the business model • Configuration of first organizational structures (i.e., the promoter team) • Problem solution • Stage with the greatest risk/return potential
Seed	• Recurring sales • Validation phase • Organic month-on-month increase in sales • First signs of the power of scalability
Series A	• Scalability • Expansion • Phase where profit generation begins
Series B, C, D…	• Leadership in a given market • Preparation for aggressive expansion and positioning as a benchmark in the sector • Growth with maximum liquidity • Pre-IPO phase
Mezzanine	• Substitution of ventures equity • Senior debt capacity has been maximized, and a company's cash flow has sufficient capacity for additional long-term borrowings

Source Own depiction based on Lassala and Ribeiro-Navarrete (2022) and Silbernagel and Vaitkunas (2012)

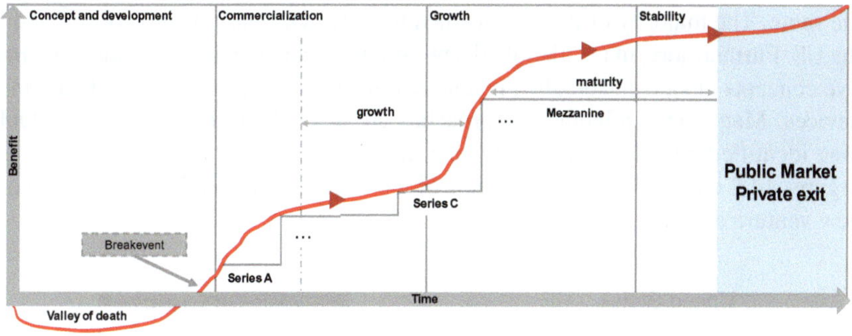

Fig. 14.1 Development of investment rounds (*Source* Own depiction based on Lassala and Ribero-Navarrete [2022, 5] and Kazanjian and Drazin [1990, 140f])

14.3 Global Overview of New Ventures and Fusion Devices

We collected a unique dataset of more than 80 new ventures that are publicly claiming to develop commercial nuclear fusion reactors for energy generation. The company-related data were gathered from CrunchBase, Traxcen, Dealroom, LinkedIn, Google News, and other sources, which were then cross-referenced with the International Atomic Energy Agency's (IAEA) Fusion Device Information System (FusDIS) to match the identified new ventures in the dataset with specific devices.[4]

The new ventures are spread across 15 countries and have accumulated private funding amounting to US-$10.2 billion,[5] with an additional US-$23 million of nondilutive funding, i.e., such not linked to equity purchases, mostly from publicly funded projects. The majority of the new ventures (47) is located in the United States (US), while eleven of these no longer exist. The US ventures have an average age of 10.7 years. Eight ventures are based in the United Kingdom with an average age of 8.4 years. Four each are from Germany, Japan, and China, and three are located in Canada.

When combining both datasets, i.e., the IAEA's FusDis with our new venture dataset, one finds that only 40% of identified new ventures can be linked to an actual device, either planned or under construction. Overall, the FusDIS lists 168 devices in 29 countries using various approaches. Note that each of these devices is an experimental device and the FusDIS removes devices that have been closed, such as the Joint European Torus that was closed in December 2023.[6]

For some devices, the initial location of the new venture and that of the device itself do not match. For example, German company Marvel Fusion is currently constructing its first experimental device in the US, and its competitor Focused Energy is doing

---

[4] The FusDIS was accessed on the 12-01-2025 and can be accessed here https://nucleus.iaea.org/sites/fusionportal/Pages/FusDIS.aspx.

[5] Unless otherwise indicated, each monetary value in given in 2020 values.

[6] https://www.science.org/content/article/european-fusion-reactor-goes-out-bang.

the same. The majority of devices planned to be (or currently being) constructed in the US. Furthermore, many identified new ventures seem to pursue so-called alternative concepts (19), e.g., so-called magnetic mirror machines or reversed field pinch devices. Magnetic confinement approaches are also often planned for (17). Only nine identified new devices are pursuing laser or inertial confinement approaches. Figures 14.2 and 14.3 give an overview of the devices that could be matched to our new venture dataset.

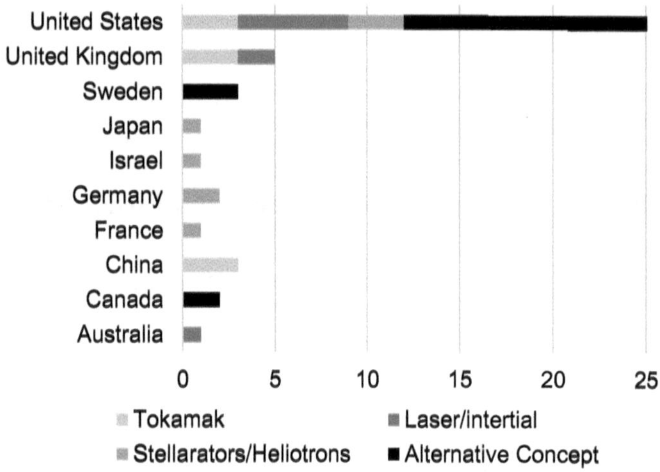

Fig. 14.2 Device distribution of new ventures with actual plans (according to FusDIS) (*Source* Own depiction from own dataset and IAEA FusDIS)

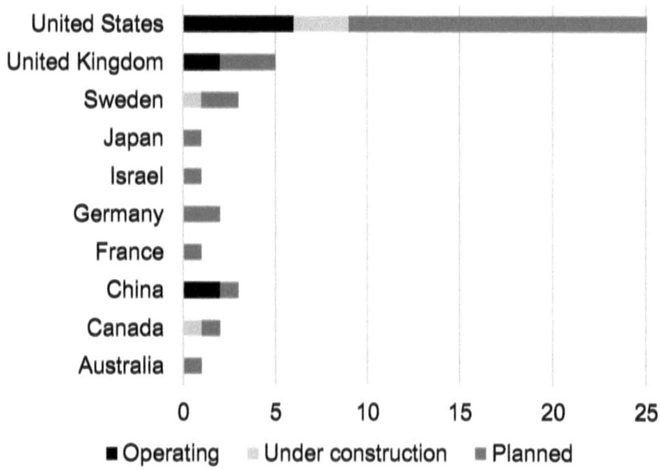

Fig. 14.3 Current status of new ventures' fusion device activities (according to FusDIS) (*Source* Own depiction from own dataset and IAEA FusDIS)

The mismatch between both datasets regarding the low number of new ventures actually pursuing the development of fusion device prototypes raises the question of whether the identified companies are actually pursuing the development of nuclear fusion for energy production or whether there might be other goals, such as magnet or laser development.

14.4 Analysis of New Venture Development and Funding from Entrepreneurial Perspective

At the time of writing in January 2025, according to the IAEA's FusDIS, there were ten private and 91 public experimental research facilities dedicated to basic and applied research of nuclear fusion. While initial development successes have been achieved in recent years, although no (commercial) prototype of a fusion reactor exists (Grünwald 2024). Consequentially, following the definitions provided in Table 14.1, each identified venture can be categorized in the initial stage of "Concept and Development".

Subdividing our dataset according to their announced and verifiable plans (according to the FusDIS) delivers two groups that are influenced by different actors with varying interactions. The first group consists of 32 ventures that can be tied to a concrete fusion device listed in the FusDIS (either "in operation", "planned" or "in construction", see Sect. 14.3) and corresponds to a total private funding of US-$7.5 billion. The second group comprises 48 companies officially planning to build a commercial reactor but not listed in the FusDIS, with a total funding volume of US-$2.7 billion. It seems that ventures whose device is known have collected more investments than those whose device has not yet been built or registered with the IAEA's FusDIS. The remaining devices listed in the FusDis comprise 71 public institutions or private universities with a known testing facility in "operation", "planned" or "in construction" and are not further discussed here.

Figure 14.4 shows the development of new ventures and corresponding funding. While the first ventures were founded in the second half of the twentieth century, the number of new ventures has substantially increased over the last few years, reaching a maximum of eleven newly founded companies in 2021. Funding volumes have also substantially increased, maxing at US-$3 billion in 2021, see the detailed assessment below. Notably, the average age of when new ventures received their first funding has been declining over the last 25 years, from about 15 years to only two.

Usually, newly funded companies must either move to a public test facility (in Germany, this would be, e.g., Garching or Greifswald) or to other countries for their experimental tests, which increases their development costs. There are obvious differences between the stage of a first laboratory design and a real production facility, which leads to further difficulties in development on the path to a potential marketable stage for a potential fusion power plant (Daniel and Alves 2020). Such scenarios are examples of critical vulnerabilities and failure risks especially in the development

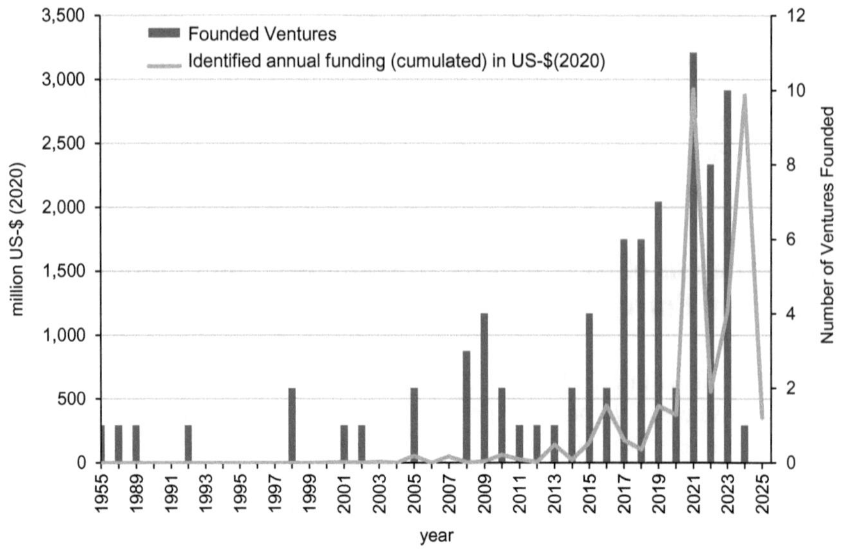

Fig. 14.4 Development of founding new ventures and identified funding (*Source* Own depiction from own dataset)

process of a potential fusion power plant with its ever-growing development costs. This often occurs in emerging technology companies and thus warrants a closer look at funding streams for fusion new ventures (Munoz and Terzidis 2025).

Out of 80 ventures, a total of 21 ventures have so far received no investor funding. Twenty-six companies have received funding between up to US-$10 million, and another 30 ventures between US-$10 million to 1 billion. Two companies have managed to attract funding volumes of more than US-$1 billion, one in China (Neo Fusion) and one in the US (Commonwealth Fusion Systems, CFS), see Fig. 14.5. However, it can be assumed that new private ventures will require further substantial funding requirements; for comparison, current cost estimates for ITER are based at over US-$22 billion (Wimmers et al. 2025).

Moreover, the distribution of the funding among funding series shows that funding is widely spread from early grants and seed funding to series G funding. A total of 51 early-stage investments (i.e., Pre–Seed, Seed) received funding volumes of US-$459 million and a total of 89 investments for later stages (i.e., Series A, Series B, etc.), totaling US-$6.3 billion. Twenty-two instances without temporal specification, corresponding to US-$1.8 billion, were identified (Fig. 14.6).

The two companies with the highest funding (each exceeding US-$1 billion) both received a series B funding. From new venture theory, this series would suggest the availability of a commercial product. But as we have shown, there are no functioning commercial fusion devices. This represents a deviation from the previously introduced theory of the introduced life-cycle contingency phases, as the companies have collected a high volume of investments in relatively late investment phases but

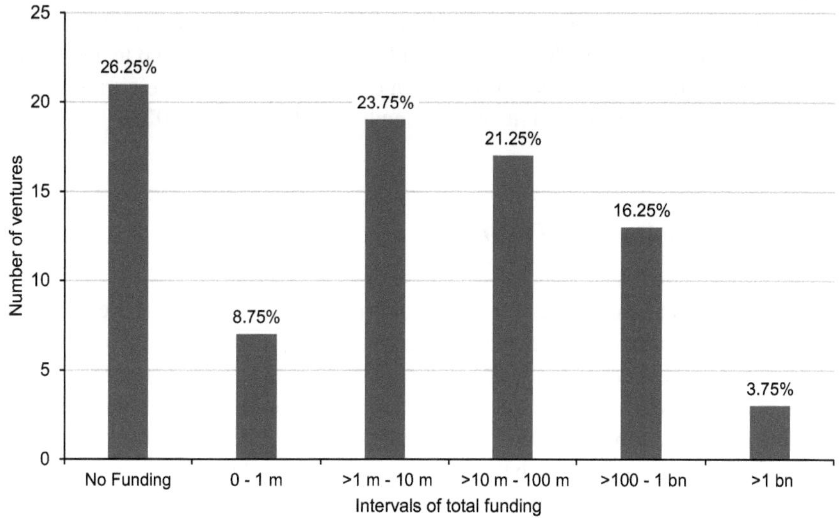

Fig. 14.5 Histogram of total funding (*Source* own depiction from own dataset)

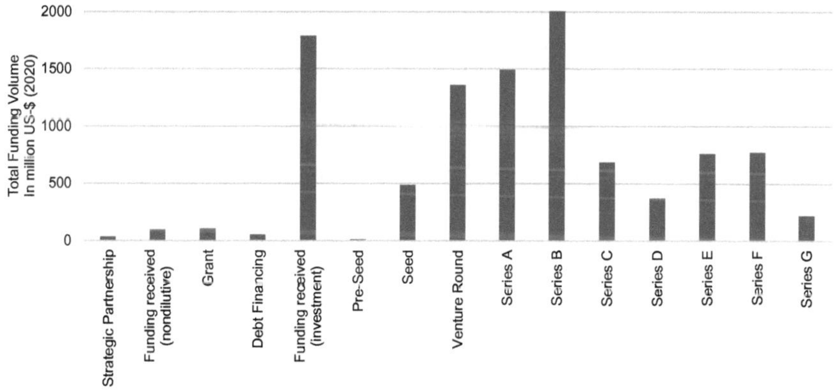

Fig. 14.6 Sum of funding per stage (*Source* Own depiction from own dataset)

have so far failed to scale (yet develop) their potential main product. Since typical literature refers to generating an initial profit or its scalability, this indicates a potential overvaluation of the market and a potential primary development of secondary products.

This hints towards potential spillover-effects, i.e., the development of sub-products necessary for the delivery of the main product (commercial nuclear fusion). For the identified fusion new ventures, such spillover products could be advancements in superconductors, laser technology, and plasma physics applications, to generate interim revenue streams (Bruns et al. 2012).[7] However, this can also indicate that ambitious promises are made which cannot hold up with the required technological progress to secure future funding, raising the question whether the current fusion market might be characterized by an ongoing hype (Bareis et al. 2023).

14.5 Conclusion and Outlook

In this chapter, we assess the development of private new technology-based ventures aiming to develop fusion power plants based on existing entrepreneurship literature. The first ventures to develop a fusion power plant have been known since the 1950s. From our initial assessments, the fusion market appears to be overvalued, with a possible tendency towards actual development efforts not aiming for commercial fusion for energy generation, but rather for spin-off products, such as laser technologies, magnets or other high-tech equipment to be sold on the long path towards establishing actually operating large-scale fusion demonstration devices.

From the perspective of entrepreneurial and technology venture theory, the funding trajectory of fusion ventures represents a mismatch. The life-cycle model for technological ventures emphasizes a clear progression from early-stage development, where product-market fit is established, to growth and scaling phases. The significant influx of capital into fusion ventures that remain in the "Concept and Development" stage (Table 14.1) challenges this structured approach.

This suggests a need to refine theoretical models for high-risk, deep-tech ventures that operate on longer innovation timelines. Unlike software startups, which may reach a minimum viable product (MVP) within months, fusion ventures require decades of R&D before a commercial application can emerge. The significant increase in funding for fusion ventures and ambitious promises may also indicate a hype cycle where expectations outpace the actual technological progress.

[7] For example, the company Quaise is planning to use a gyrotron, developed to heat plasma in tokamak devices, to melt boreholes to foster geothermal energy, see https://www.quaise.energy/.

Appendix

Table 14.3 List of nuclear fusion ventures their location and age

Company Name	Age (base year 2024)	Founded Year	Country
AGNI	7	2017	United States
Applied Fusion Systems / Pulsar Fusion	10	2014	United Kingdom
Avalanche	6	2018	United States
Cold Nuclear	9	2015	United States
Convergent Scientific	14	2010	United States
CrossFire Fusion Reactor	16	2008	Unknown
CTFusion	9	2015	United States
Deneum	6	2018	Estonia
Energy Solutions	26	1998	United States
Fusion Reactors	5	2019	United Kingdom
ITER	23	2001	France
Princeton Fusion Systems	7	2017	United States
Progressive Fusion Solutions	7	2017	Canada
Proton Scientific	12	2012	United States
Alpha Ring	9	2015	United States
Brillouin Energy	15	2009	United States
Canyon Magnet Energy	1	2023	United States
Commonwealth Fusion - CFS	6	2018	United States
Cortex Fusion Systems	3	2021	United States
Deuterium Energetics	9	2015	United States
EX-Fusion	3	2021	Japan
First Light Fusion	13	2011	United Kingdom
Focused Energy	3	2021	Germany
Fuse energy	6	2018	Canada
General Fusion	22	2002	Canada
HB11 Energy	7	2017	Australia
Helion	11	2013	United States
Horne Technologies	16	2008	United States
ISA Industries	8	2016	United States
Kyoto Fusioneering	5	2019	Japan
Lawrenceville Plasma Physics	50	1974	United States
LIS Technologies	1	2023	United States
Marvel Fusion	5	2019	Germany
MIFTI	15	2009	United States

(continued)

(continued)

Company Name	Age (base year 2024)	Founded Year	Country
Novatron Fusion Group	5	2019	Sweden
nT-Tao	8	2016	Israel
OpenStar Technologies	3	2021	New Zealand
Phoenix	19	2005	United States
Qdot	6	2018	United Kingdom
Renaissance Fusion	4	2020	France
SHINE Technologies	19	2005	United States
TAE Technologies	26	1998	United States
Tokamak Energy	15	2009	United Kingdom
Xcimer Energy	2	2022	United States
Zap Energy	7	2017	United States
Firefly Fusion	1	2023	Switzerland
Astral Systems	3	2021	United Kingdom
Blue Laser Fusion	2	2022	United States
Energy Singularity	3	2021	China
ENN Systems	35	1989	China
Gauss Fusion	3	2021	Germany
Laserfusion X	2	2022	United States
Next Step Fusion	1	2023	Luxembourg
Proxima Fusion	1	2023	Germany
Realta Fusion	2	2022	United States
Stellarex	2	2022	United States
Type One Energy Group	5	2019	United States
Helical Fusion	3	2021	Japan
Deutelio	2	2022	Italy
Nk Labs	16	2008	United States
Crossfield Fusion	5	2019	United Kingdom
Electric Fusion Systems	4	2020	United States
Near Star Fusion	3	2021	United States
Pulsar Fusion	10	2014	United Kingdom
Longview Fusion Energy	3	2021	United States
Thea Energy	2	2022	United States
LPPFusion	21	2003	United States
Princeton Sattelite Systems	32	1992	United States
General Atomics	69	1955	United States
HyperJet Fusion Corporation	7	2017	United States
Helicity Space	6	2018	United States

(continued)

(continued)

Company Name	Age (base year 2024)	Founded Year	Country
Neo Fusion	1	2023	China
Acceleron Fusion	1	2023	United States
Innoven Energy	14	2010	United States
Jupiter Volter	2	2022	United States
Linea Innovation	1	2023	Japan
Startorus	3	2021	China
Terra Fusion	1	2023	United States
Xtus Energy	1	2024	United States
Bridge 12	15	2009	United States
Sapientai	5	2019	United States
Polymath Research	26	1998	United States
Pacific Fusion	1	2023	United States

Source Own data collection

References

Bareis, Jascha, Maximilian Roßmann, and Frédérique Bordignon. 2023. "Technology Hype: Dealing with Bold Expectations and Overpromising." *TATuP - Zeitschrift Für Technikfolgenabschätzung in Theorie Und Praxis* 32 (3): 10–71. https://doi.org/10.14512/tatup.32.3.10.

Bruns, Peter, Minh Quang Tran, Daniel Kunz, Heinz Mueller, and Christian Soltmann. 2012. "Spillover Benefits from Controlled Nuclear Fusion Technology – A Patent Analysis." *World Patent Information* 34 (4): 271–78. https://doi.org/10.1016/j.wpi.2012.06.001.

Coviello, Nicole, Erkko Autio, Satish Nambisan, Holger Patzelt, and Llewellyn D.W. Thomas. 2024. "Organizational Scaling, Scalability, and Scale-up: Definitional Harmonization and a Research Agenda." *Journal of Business Venturing* 39 (5): 106419. https://doi.org/10.1016/j.jbusvent.2024.106419.

Daniel, Ana Dias, and Liliana Alves. 2020. "University-Industry Technology Transfer: The Commercialization of University's Patents." *Knowledge Management Research & Practice* 18 (3): 276–96. https://doi.org/10.1080/14778238.2019.1638741.

Grünwald, Reinhard. 2024. "Auf dem Weg zu einem möglichen Kernfusionskraftwerk. Wissenslücken und Forschungsbedarfe aus Sicht der Technikfolgenabschätzung." PDF. Büro für Technikfolgen-Abschätzung beim Deutschen Bundestag (TAB). https://doi.org/10.5445/IR/1000177720.

Kazanjian, R. K. 1984. "The Organizational Evolution of High Technology Ventures; The Impact of Stage of Growth on the Nature of Structure and Planning Processes." Unpublished doctoral dissertation, Wharton School of Business Administration, Philadelphia, PA

Kazanjian, R. K. 1988. "Relation of Dominant Problems to Stages Growth in Technology - Based New Ventures." *Academy of Management Journal* 31 (2): 257–79. https://doi.org/10.2307/256548.

Kazanjian, Robert K., and Robert Drazin. 1990. "A Stage-Contingent Model of Design and Growth for Technology Based New Ventures." *Journal of Business Venturing* 5 (3): 137–50. https://doi.org/10.1016/0883-9026(90)90028-R.

Lassala, Carlos, and Samuel Ribeiro-Navarrete, eds. 2022. *Financing Startups: Understanding Strategic Risks, Funding Sources, and the Impact of Emerging Technologies*. Future of Business and Finance. Cham: Springer International Publishing. https://doi.org/10.1007/978-3-030-94058-4.

Munoz, J. Mark, and Orestis Terzidis, eds. 2025. Technology Entrepreneurship in Theory and Practice. Perspectives in Science and Engineering. Routledge.

Normann, Richard. 1979. *Management for Growth*. Reprinted. A Wiley-Interscience Publication. Chichester: Wiley.

Nuscheler, Daniela, Andreas Engelen, and Shaker A. Zahra. 2019. "The Role of Top Management Teams in Transforming Technology-Based New Ventures' Product Introductions into Growth." *Journal of Business Venturing* 34 (1): 122–40. https://doi.org/10.1016/j.jbusvent.2018.05.009.

Nuttall, William J. 2020. "Commercial Opportunities for Nuclear Fusion." In *Commercialising Fusion Energy - How Small Businesses Are Transforming Big Science*, 1–21. Bristol, UK: IOP Publishing.

Sanasi, Silvia. 2025. "Structuring Experimentation: Implementing Growth Hacking in New Ventures." *Journal of Business Research* 188 (February):115084. https://doi.org/10.1016/j.jbusres.2024.115084.

Silbernagel, C., and D. Vaitkunas. 2012. "Mezzanine Financing." Bond Capital. https://www.bondcapital.ca/wp-content/uploads/pdfs/what-is-mezzanine-finance.pdf.

Takeda, Shutaro, Alexander Ryota Keeley, and Shunsuke Managi. 2023. "How Many Years Away Is Fusion Energy? A Review." *Journal of Fusion Energy* 42 (1): 16, s10894-023-00361-z. https://doi.org/10.1007/s10894-023-00361-z.

Wimmers, Alexander, Björn Steigerwald, Fanny Böse, Alexander Buschner, Claudia Kemfert, and Julia Rechlitz. 2025. "Power Generation from Nuclear Fusion Not Expected in the Foreseeable Future: Applied Research Developing Dynamically". DIW Weekly Report 15: S. 8995. https://doi.org/10.18723/DIW_DWR:2025-14-1

Part IV
The Nuclear Back-end

Chapter 15
State Involvement in Nuclear Decommissioning—Case Studies from the US and the UK

Niklas Wagner

15.1 Introduction

After closure, nuclear power plants (NPP) and other related nuclear facilities must be decommissioned. Aside from the utilization of land for other purposes, a closed NPP requires continued service and maintenance, as well as regulatory oversight to prevent the release of dangerous radiation that has accumulated during the facility's operation. The decommissioning of NPPs is no routine task. Experience is limited to individual, highly specific projects: as of mid-2024, of the 213 closed reactors, only 23 have completed the technical decommissioning process, geographically limited to Germany, Japan, Spain, and the United States of America (US) (Schneider et al. 2024). Despite early euphoria over the potential of civilian nuclear applications, the unavoidable task of decommissioning nuclear facilities has been neglected historically and represents a challenging and expensive legacy for current and future generations. Today, decommissioning incurs very high costs, and projects can last several decades. Additionally, projects regularly experience cost and schedule overruns (Bärenbold et al. 2024).

In general, the challenges regarding the provision of nuclear decommissioning can be assessed from two perspectives. First, it poses a technological problem. Engineers must develop suitable technologies and procedures for efficient and safe decommissioning at different reactors with heterogeneous operating histories. Second, organizational challenges arise. Thousands of people from different organizations must

This work was first presented at the 2023 and 2024 AT-OM Day Conferences in Berlin and at the IAEE 45th International Conference in Istanbul in June 2024. Gratitude to the particpants of all three sessions for their constructive comments on this work. Also, thanks to Christian von Hirschhausen and Alexander Wimmers for their helpful suggestions on earlier drafts of this chapter.

N. Wagner (✉)
TU Berlin, Berlin, Germany
e-mail: atom@wip.tu-berlin.de

A. Wimmers et al. (eds.), *Nuclear Power*, Studies in Energy, Resource and Environmental Economics, https://doi.org/10.1007/978-3-031-99894-2_15

work in a coordinated way, often over decades, to accomplish safe and efficient decommissioning. The coordination spaces between regulators, management, and workers along the value chain must be well designed due to the complexity and size of NPPs, all while ensuring radiation exposure minimization. This shows the connection between organizational and technological issues. False incentives, lack of liability, and insufficient oversight risk project success, leading to cost and time overruns and, in the worst case, to environmental harm and safety issues.

Worldwide, the role of the state in nuclear decommissioning varies substantially (Bärenbold et al. 2024). This raises the question of under what circumstances the state should take on which role and why. Economic analyses in this regard are absent. Therefore, this analysis explores the role of state involvement in nuclear decommissioning in two case studies while focusing on commercial power plants and individual associated facilities, such as the large Sellafield complex in the United Kingdom (UK). The UK is a case in which the state has now taken on a dominant role by directly conducting and contracting decommissioning work while also supervising the overall strategy. The US represents an opposite configuration, where the state mainly assumes the role of the regulator, and the private sector often takes the initiative to execute decommissioning activities. The comparison of these contrary cases is interesting, as both the US and the UK were pioneers of nuclear power utilization.

The approach of this study is a theory-informed explanatory case study with an institutional analysis. The goal is to explain the rationale behind the mode of state involvement in the two cases and explore the potential prerequisites, strengths, and weaknesses of these governance modes. Theory, especially transaction costs economics, helps to analyze the governance modes from an economic efficiency viewpoint and reduces the problem to analyzable units. The analysis starts with introducing the theory and contextualizes nuclear decommissioning within the theoretical approach. Thereafter, both cases are analyzed sequentially. Finally, the cases and results are jointly discussed.

15.2 Theoretical Framework and Methodology

From a theoretical viewpoint, the two domains are decisive for efficient and effective project execution. First, projects require certain resources and capabilities, which range from specialized machinery to tacit knowledge. Second, actors possessing these capacities and resources must be coordinated through governance structures that enable and constrain their behavior. With fast, safe, and efficient nuclear decommissioning as an overarching goal, a well-suited governance structure limits uncertainty and opportunism between actors.

15.2.1 State Involvement and Governance Structure

Two dimensions of a governance structure are the extent of vertical integration and state involvement. The first term refers to the extent to which tasks are integrated within a single organization. The second investigates the ratio of public and private task sharing. The latter is the focus of this analysis.

Effective governance structures[1] address the fundamental characteristics of human behavior, which are bounded rationality and opportunism. Bounded rationality refers to the inability of human beings to gather and process all the information needed for perfectly informed actions. Gathering and processing information is costly due to cognitive limitations, hidden information, and uncertainty. Opportunism is the potential for actors to use this lack of knowledge of their opponent to their advantage and act self-beneficially. However, it must be noted that even without opportunistic motivations, actors can act contrarily to their opponents' wishes due to incomplete contracts or the variability of certain goods (e.g., reduced material availability) (Furubotn and Richter 2005).

Transaction costs are a crucial metric for the success and efficiency of governance structures. They can be described as the costs of economic interactions: individuals must gather information, bargain and form agreements with other actors, and make decisions; they must also monitor the outcome and enforce the result of an agreement. Thus, transaction costs are the resources needed to establish, maintain, and transfer property rights. Property rights can be described as "one's ability to exercise choices over a good" (Allen 1991, 2). By providing orientation and reducing opportunism, effective governance structures can reduce transaction costs; their effect is thus twofold: (1) restricting opportunistic behavior, and (2) thereby enabling actors to efficiently adapt to changing situations (regarding the transaction of property rights).

Whether a governance structure effectively reduces transaction costs depends on the characteristics of the transaction. Williamson (1985) presents three fundamental dimensions that influence institutional effectiveness and, subsequently, transaction costs.

- Asset specificity describes to what extent investments in a transaction could be utilized for other applications. High specificity binds the actors together, which gives rise to potential opportunism. In this case, neither could the supplier find a buyer with an equally high valuation for their good or service, nor could the buyer simply divert to another supplier because an unspecialized supply would come at greater costs for the buyer (Williamson 1979). Asset specificity is a necessary but insufficient condition for the rise of transaction costs. In a world of perfect information, parties locked in a bilateral monopoly (due to their transaction-specific investments) could easily arrange a complete contract covering all contingencies ex-ante.

[1] Governance structures are institutions: "Institutions are the humanly devised constraints that structure political, economic, and social interaction" (North 1991, 97).

- However, such contracts are impossible to form due to environmental and behavioral uncertainty and the costly measurement of tasks.[2] In short, higher uncertainty makes negotiating, renegotiating, and enforcing a contract more expensive because heterogeneous and unknown situations must be negotiated for or hedged against (ex-ante), or disputes must be managed when unforeseen events occur (ex-post). Uncertainty can be reduced through monitoring during task execution, but this increases transaction costs even more.
- Higher frequency, meaning recurrent transactions, allows for economies of scale or learning of specialized governance modes that might come at a high initial cost but become cheaper over time.

Extending these commercial–private and voluntary interactions to include issues relating to involuntary exchange, like theft or environmental externalities, necessitates the addition of measurability as another dimension (Barzel 1985, 2005). This dimension emphasizes the costliness of information, which could vary depending on the good and the given institutions. Information on the characteristics of the good and the contractual partner is a prerequisite for writing contracts. This dimension is also important for the post-contractual enforcement of an agreement. In the absence of perfect information, undetected cheating is possible, and property rights are disputed. Actors spend resources during the monitoring process (e.g., quality control of goods), which could also include credible signaling of private information to other actors. Alternatively, resources can be spent on reducing the measurement costs, e.g., through standardization.

Transactions with low asset specificity and uncertainty (or low measurement costs) can be efficiently conducted on the market. The possibility of turning to a competitor reduces opportunism. Furthermore, adverse behavior can easily be detected. In this configuration, market prices account for necessary information and drive the efficient allocation of resources. In the case of asset-specific investments, however, switching suppliers or customers is costly. Therefore, the possibility of a buyer turning to a competitor is a minor deterrent. Additionally, high uncertainty and high monitoring costs complicate the detection of opportunistic actors because, first, not all situations and associated behavioral consequences can be anticipated and contracted for, and second, it is challenging to differentiate and identify root causes of noncompliance, i.e., environmental reasons or opportunistic behavior. In such cases, specialized governance structures are needed to safeguard the transaction. These, however, generate transaction costs for setup and maintenance.

If asset specificity and uncertainty are high, a transaction should be integrated into the firm hierarchy. In this case, the vertically integrated firm represents a structure of various incomplete contracts. In contrast to market competition and prices, the (management) hierarchy, related management practices of control, and the implicit social networks between workers and management confine opportunism and enable adaptation (Williamson 1999).

[2] The same is true vice versa: A transaction with low specificity and high uncertainty would pose little problems transaction costs wise. The actors could engage in short-term contracts, each adapted to new contingencies; there are no sunk costs which bind the actors together.

A similar rationale applies to state involvement. First, state involvement (like grades of vertical integration) is a continuum ranging from simple guarantees for contractual security between private actors over regulation to public (state-owned) enterprises. Therefore, the interesting question for this issue is when state involvement is needed beyond merely guaranteeing the enforcement of property rights, and what exactly this involvement might be. The following paragraphs represent examples of necessary state involvement.

Negative environmental externalities directly result from transaction costs (Krutilla 2011). In a world with zero transaction costs, each citizen could directly bargain with a polluter to attain the social optimum. However, property rights are usually not solely assigned to the polluter or the affected entities. Furthermore, transaction costs for private bargaining would be high, and each citizen would have to accurately monitor and quantify the externality to find the liable producer and be able to enforce a contract. Further, citizens cannot ban polluters from using the (common) environmental good. The state can safeguard environmental property rights more efficiently through its territorial power. This can be done through regulatory measures like establishing markets for environmental goods or by command-and-control instruments. Alternatively, the state can directly control the polluter via nationalization. Naturally, this incurs transaction costs for the state that depend on the chosen governance mode. The extent of state involvement can be explained by the amount of measurement costs, e.g., if attained transparency is too low via regulation (e.g., via third-party audits), it might be reasonable to integrate the polluter directly into the hierarchy of the state to gain deeper insights.

Other reasons for state involvement are natural monopolies and other situations in which market dynamics lead to a concentration of market power. Here, economies of scale or network effects favor fewer market participants, conversely increasing transaction costs on the market due to the lack of a disciplining effect of competition. The extensive transaction costs of regulating a monopolist can be a reason for state integration. For example, monitoring and skimming monopoly profits require deep insights into the company.

Another function of state involvement can be the absorption of uncertainty. Private actors might be reluctant to engage in (necessary) tasks because of high uncertainty in conjunction with transaction costs, making hedging impossible. In such a case, resources are provided by the state (e.g., via subsidies or public–private partnerships). The creation of artificial markets or direct contracting with the state might again be limited due to high uncertainty, making direct integration into the state hierarchy a viable option. Similar reasoning applies to knowledge spillovers or first-mover disadvantages.

Williamson (1999) introduces "probity hazards" of transactions, which demand a high level of integrity and professionalism in an interaction. To address these, the author suggests that effective governance should involve: (1) very low-powered incentives, (2) extensive administrative controls and procedures, and (3) a well-trained, socially conditioned staff with secure positions. As public bureaucracies stand outside competition and are less oriented toward profit-making, they are better suited to meet these criteria than private ones.

Another complementary stream of theories is focused on resources and capabilities that entities hold or control. Most prominently, this includes the theories of resource-based view, knowledge-based view, and resource dependency. These resource-based theories emphasize the specific value and complement different resources and property rights. To prevail, entities must control both the property rights and the resources that serve the specific purpose. This explains why entities seek to control certain resources and why the associated trade-offs, e.g., contracting, might increase transaction costs. Self-production, however, implies efforts for learning or a less efficient production function[3] (Foss and Foss 2005; Jacobides and Winter 2005; Williamson 1988). Considering these various factors, the question arises for which goods and under which circumstances actors, including the state, might prefer vertical integration and internal learning over procuring capabilities from other actors.

15.2.2 Nuclear Decommissioning Governance and Capabilities

The main reason for the high technical complexity of nuclear decommissioning is the radiological contamination of NPPs (or the uncertainty regarding this contamination). This means that personnel on site must have specific knowledge regarding handling radioactive material. Decontamination skills are necessary to reduce the volume of highly contaminated waste. Radiological contamination of every part or material is measured to identify the appropriate waste stream for conventional disposal (and recycling) when radiation levels are sufficiently low or for final storage. In addition to the specific know-how, specific physical assets are necessary. Special equipment is required for dismantling the reactor and its adjacent parts and subsequent decontamination and monitoring activities. Subsequently, transportation, storage, and disposal assets, such as casks for radioactive materials, are needed (Wimmers and von Hirschhausen 2024).

Regarding governance, both knowledge and assets can be regarded as specific assets. Furthermore, the dismantling of nuclear components is a lengthy process with high complexity and uncertainty, leading to process specificity (Wimmers and von Hirschhausen 2024). It is costly to replace (management) contractors as the tasks would have to be halted, the site's status assessed, and a new procurement process restarted. Another specificity that can be encountered is site specificity: Although some nuclear fleets are relatively homogeneous, standardized techniques may not apply to all plants, necessitating specialized tools with limited reusability (Bärenbold et al. 2024).

Decommissioning projects are prone to uncertainty due to three main characteristics of nuclear power plants: (1) the overall complexity and size of the facilities pose a challenge to assess the status of the facility comprehensively, (2) most projects are

[3] If capabilities and resources can be imitated at all.

first-of-a-kind projects and actors lack experience, and (3) the extent of radiological contamination is unknown.

Aside from efficient structuring of coordination to attain cost-efficient decommissioning, governance must prevent radiological contamination of the environment. It must also prevent the abandonment of the facility and ensure that actors are liable for decommissioning. Abandoned NPPs would threaten human health and the environment, as structural deterioration could result in the release of hazardous radiation. Additionally, blocking the sites with unused power plants incurs opportunity costs, as these sites might be suitable for other (non-nuclear) uses. However, decommissioning shifts the focus away from profit-making. The operator is no longer concerned with balancing regulatory compliance costs against profits, as the facility now represents a financial burden rather than a source of income. This change in motivation challenges regulators, who must ensure sufficient resources are allocated for decommissioning, and that safety and security standards are upheld (Bärenbold et al. 2024).

15.2.3 Methodology

The approach of this analysis is a theory-informed explanatory case study and institutional analysis. Case studies can be a fruitful approach for an in-depth understanding of institutional configurations, especially if the operationalization is difficult and observations are limited. The primary objective is to explain why different types of state involvement exist for nuclear decommissioning and what potential prerequisites are needed for certain governance configurations.

Beyond a direct causal analysis, this institutional analysis also identifies advantages, disadvantages, and potential hazards for governance success, which could serve as a focal point for further research on transaction costs in this sector.

The basis for the theory-driven institutional analysis is the identification of participating actors and their motives, as well as the institutional and technical environments. This serves as the basis for identifying "action situations" (Ostrom 2011) related to the question of state involvement to understand the prerequisites, strengths, and weaknesses of different governance configurations.

The selection of cases is based on previous research by Wimmers et al. (2023), Bärenbold et al. (2024), and Wimmers and von Hirschhausen (2024), on whose work the technical and organizational overviews of the US and UK nuclear decommissioning approaches rely.

15.3 Decommissioning in the United States

15.3.1 Technical Overview

Regarding the commercial reactor fleet, nuclear decommissioning governance in the US can be characterized as a prototypical open-market model. Private actors bear the main liability and responsibility. Although the state primarily acts as a regulator, it maintains substantial influence over nuclear decommissioning projects, e.g., for fund availability.

The US commercial reactor fleet mainly consists of pressurized water reactors (PWR) and boiling water reactors (BWR). Ninety-four reactors are in operation, and 21 are closed. The latter are in various stages of decommissioning (Schneider et al. 2024). Additionally, the US operates 15 research, weapons, and fuel sites across the country that are undergoing decommissioning, but are out of this analysis' scope. The US has four commercial near-surface Low-level Radioactive Waste (LLW) disposal facilities. Furthermore, the Department of Energy (DOE) operates six near-surface LLW repositories that are reserved for its own waste.

Currently, there is no disposal site for Greater than Class C (GTCC) waste. GTCC waste is LLW that exceeds the concentration of long-lived radionuclides according to 10 CFR 61.55. In the case of decommissioning, this especially concerns components such as irradiated reactor parts. Several options for GTCC waste are under consideration. Principally, the US Nuclear Regulatory Commission (NRC) could allow near-surface disposal and has proposed modifying the regulation to facilitate this option. In 2016, the project for the identification of a final repository for High Level Radioactive Waste (HLW), including spent nuclear fuel from commercial NPPs, failed. Responsibility for HLW waste disposal was supposed to have been transferred to the Federal Government, but this was on hold in the mid-2020s, with spent nuclear fuel being stored in decentralized interim storage facilities across the country. Currently, there are plans by private actors to provide long-term consolidated interim storage facilities.[4]

15.3.2 Overview of Organizations

The NRC is an independent federal agency concerned with regulating nuclear activities, including licensing nuclear facilities. For decommissioning, this includes safety regulations, but also enabling best practices at decommissioning sites. Furthermore, the NRC is responsible for the terms of accumulation and reimbursement from Nuclear Decommissioning Trust Funds (DTF).

---

[4] The Waste Isolation Pilot Plant (WIPP) is a deep geological repository operated by the DOE in New Mexico that only accepts transuranic radioactive waste (TRU) from nuclear weapons research and production see https://www.wipp.energy.gov/. Last accessed on 02-09-2024. This date holds for all other references in this chapter, unless otherwise indicated.

To permanently close an NPP, the nuclear license holder must inform the NRC 30 days in advance. The NRC must be notified again when the facility is defueled. The licensee must submit a "Post-Shutdown Decommissioning Activities Report" (PSDAR) two years after the closure, at the latest. The PSDAR outlines the schedule of the decommissioning approach and includes cost estimates for decommissioning and spent fuel management. The NRC reviews the PSDAR, but it must not be explicitly approved. The licensee can start decommissioning 90 days after filing the documents. The licensee must file a "License Termination Plan" (LTP) at least two years before the planned completion of decommissioning. The LTP contains details regarding the status of the site, the remaining work, cost estimates, and subsequent use of the site. The NRC must approve it. Finally, the licensee must provide a "Final Status Survey Report" (FSSR) after completing decommissioning, in which they report on the final status of the site. They must also request license termination or modification (in case spent fuel remains at the site in an Independent Spent Fuel Storage Installation (ISFSI)). The NRC terminates the license if the status of the site complies with the LTP, the radiological hazards are removed (10 CFR 50.82), or the license is modified according to 10 CFR 72 for (interim) spent fuel storage.

The DOE directly owns various commercial facilities, nuclear weapons production facilities, and governmental nuclear energy research sites. The Office of Environmental Management is directly involved in decommissioning these legacy sites. Furthermore, the DOE operates the WIPP for TRU stemming from its own facilities. It is also the leading actor in finding and developing an HLW disposal site. These decommissioning efforts are not part of this analysis. The Environmental Protection Agency is involved in decommissioning as the regulator for public health and radiation safety standards, as mandated by federal regulation 40 CFR Part 190.

Most states assume direct regulatory control over nuclear facilities as the so-called Agreement States (according to Sect. 274 of the Atomic Energy Act). Nevertheless, the NRC's specifications and regulations are decisive and represent the minimum requirements. According to the Radioactive Waste Policy Act, states must dispose of their LLW domestically but are allowed to form so-called "compacts" to establish a joint solution for LLW disposal.

Commercial NPPs can be owned either by private investors and/or by public entities. Especially,in the case of investor-owned NPPs, it is a widespread practice that investors are shareholders of a Limited Liability Company (LLC), which possesses the assets, obtains the license, and operates the NPP. With these LLC schemes come concerns of final responsibility, as parent companies might not be financially liable (Lordan-Perret et al. 2021).

The decommissioning efforts for commercial NPPs are funded via DTFs. These must be accumulated mainly during the operation of the NPP by the operator. This can be done via prepayment, external sinking funds, statement of intent, surety methods, or a combination of the above (10 CFR 50.75(e)). The funding assurance of the DTF must correspond to a predefined formula representing the minimum threshold for decommissioning costs or "Site-specific Decommissioning Cost Estimates" (DCE) (10 CFR 50.75). The terms of providing the assurance are more favorable if a DCE is provided (10 CFR 50.75(e)). The licensee must report the status of the DTF to the NRC every two years; five years before the planned closure, annual reports are required (10 CFR 50.75(f)(2)). Compared with the decommissioning costs estimates, a shortfall during the accumulation phase can be addressed by providing proof of a parent company guarantee, a volume increase in the trust fund, or additional (future) contributions to the trust fund. Similarly, if a shortfall is identified during decommissioning, the licensee is required to provide additional financial guarantees to cover the estimated decommissioning costs (10 CFR 50.82(a)(8)(vi)).

Multiple decommissioning governance structures can be observed in the US. Conventional organizational models are being challenged by recently emerged models (Fig. 15.1). Conventionally, incumbent operators retain ownership over the to-be-decommissioned NPP and the license. Then, they can follow one of two approaches. The first option is "self-performance." Here, the owner (and licensee) assumes the strategic and management position (Tier 1)[5] and eventually also performs specific dismantling tasks (Tier 2), whereby these tasks can also be contracted to other market participants. In the second option, the operation contractor model, the incumbent owner contracts also for the Tier 1 position. This includes the development of the decommissioning strategy and site management.

Within the last decade, new models have appeared. In the license acquisition model, the ownership of the NPP is transferred from the incumbent owner to a decommissioning specialist, including all assets, liabilities, and the DTF. The license stewardship model differs in two key characteristics. First, while the operating license is transferred to the contractor, the original licensee retains ownership of the site, spent fuel, and other assets. Along with the operational license, the contractor gains access to the DTF, which includes associated financial liabilities and responsibilities. Second, once the decommissioning tasks are completed, the license is returned to the original owner.

The license transfers must be permitted by the NRC (10 CFR 50.80(a)). Within both schemes, the liabilities of the incumbent owner are minimized. Decommissioning contractors vary regarding their vertical integration. Usually, they contract with other firms (Tier 2) to fulfill individual decommissioning tasks. The most prominent organizations appearing as Tier 1 contractors within the new models are Holtec, ADP (both License Acquisition), and EnergySolutions (License Stewardship).

Holtec is a privately owned energy technology provider focused on nuclear technology. The company has experience in modifying and developing NPP assemblies and manufactures its own dry nuclear waste container system. It plans to construct a consolidated interim storage facility for spent nuclear fuel in New Mexico. Currently, Holtec (via its subsidiary Holtec Decommissioning International) is decommissioning the NPPs at Oyster Creek, Pilgrim-1, Indian Point, and Palisades within the License Acquisition scheme.[6]

[5] Throughout this text, Tier 1 refers to strategic management tasks covering the whole project, Tier 2 tasks are of limited scope and depend on decisions of the Tier 1 level. This includes all executive works like dismantling, engineering, or construction services.

[6] https://holtecinternational.com/company/divisions/hdi/our-fleet/.

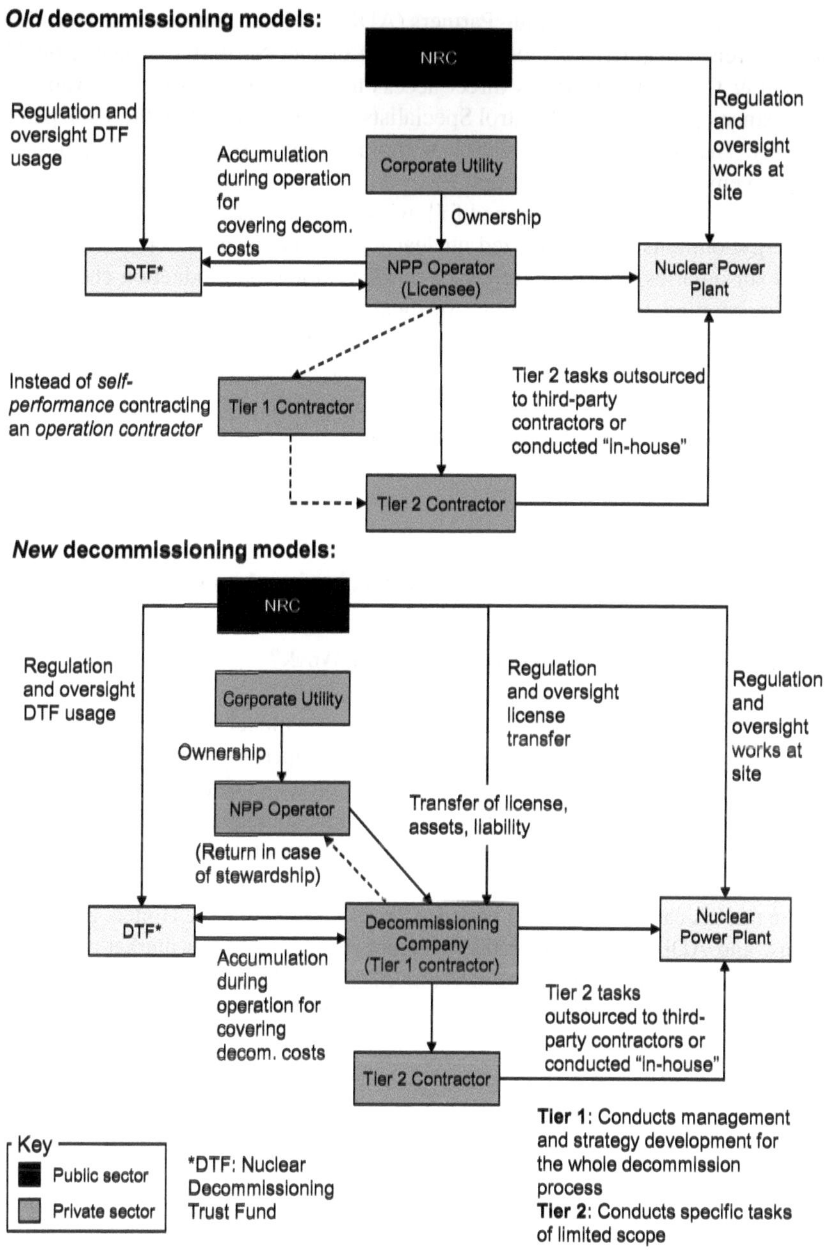

Fig. 15.1 Stylized organizational models for nuclear decommissioning in the US. (*Source* Own depiction based on Wimmers and von Hirschhausen [2024])

Accelerated Decommissioning Partners (ADP) is a joint venture of the NorthStar and the French majority-state-owned company Orano. NorthStar is owned by J.F. Lehman & Company. ADP has direct access to the LLW repository in Andrews, Texas, since operator Waste Control Specialists (WCS) is owned by the same parent company. The experience of NorthStar within nuclear decommissioning is limited, but complemented by Orano. Crystal River-3 and Vermont Yankee are ADP's main projects, both to be decommissioned following the license acquisition model.

EnergySolutions is a specialized nuclear waste services provider. The company and its subsidiaries provide a variety of services along the production chain, from decommissioning to transportation, verification, recycling, and treatment up to disposal. EnergySolutions owns and operates the LLW facility in Clive, Utah, and the Barnwell LLW facility in South Carolina. The company pioneered the License Stewardship model for decommissioning the Zion and La Crosse NPPs. Kewaunee and Unit 2 of the Three Miles Island NPP shall also be decommissioned under a License Acquisition scheme.[7]

15.3.3 Analysis of the US Decommissioning Markets

15.3.3.1 Does the Decommissioning Market Work?

The main concerns of the open market decommission model are whether the market is functional, able to provide capabilities, and offers superior governance by allowing autonomous adaptation of actors through competition and resulting prices. Companies like EnergySolutions and Holtec certainly have a unique knowledge and asset base for decommissioning and, therefore, can offer capabilities that utilities do not have.

The market can be segmented between Tier 1 contractors like Holtec, EnergySolutions, and ADP that dominate this segment, implying a high market concentration, and heterogeneous Tier 2 contractors not focused upon here. The question is whether market concentration also translates into effective market power that impairs efficiency.

Tier 1 Contractors

Effective pre-contractual competition relies on the availability of many competitors and bidder parity between the incumbent and potentially incoming firms. Due to the necessity to possess a certain set of resources and capabilities, the initial number of potential competitors is limited. Most physical and managerial decommissioning tasks are unique to the nuclear sector; therefore, other experiences can only be transferred to some extent. Especially on the project management level, unique nuclear

[7] https://www.energysolutions.com/decommissioning-and-decontamination/.

decommissioning knowledge is necessary, reducing the number of eligible competitors on the market. This is reinforced by the high need for funding assurance during decommissioning, implying that the contractor must have a certain size and financial security. They must also possess the capacity to deal with the regulatory effort. Additionally, privileged access to waste facilities can be a major advantage for decommissioning contractors. Therefore, owning or operating waste facilities (e.g., ADP and EnergySolutions) might be advantageous. Building and licensing new waste facilities is a burdensome and risky process, which might deter potential entrants to the segment.[8] Furthermore, if assets, knowledge, sites, or processes are specific to a transaction, bidder parity is endangered. Potentially, the advantage of an incumbent could be neutralized if assets, knowledge, etc., are transferred to an alternative firm after winning a tender. But the measurements, as well as bargaining and adaption efforts, can imply prohibitively high transaction costs. Additionally, establishing a tender, defining award criteria, and evaluating bids (including assessing the reliability of information) are transaction cost-intensive processes, particularly for complex tasks under uncertainty (Williamson 1976).

Post-contractual market barriers imply that the disciplining effect of the market, i.e., the mitigation of hold-up or moral hazard issues, is limited due to process, site, asset, and knowledge specificity. Even if physical assets can be transferred to a new contractor, site-specific knowledge might not be transferable, resulting in advantages for the incumbent contractor. The resulting lock-in implies that the threat of an engagement cancellation is low for the incumbent, which could favor opportunistic behavior. Additionally, nuclear decommissioning projects involve high complexity and uncertainty, complicating monitoring efforts by contractual partners or regulators to detect malperformance or misbehavior. This can manifest in the fact that it is impossible to distinguish whether underperformance results from unexpected conditions or from negligent, even opportunistic, behavior of the contractor. These properties imply a bilateral monopoly once a contractual relation is created, potentially limiting the disciplining effect of market competition.

Tier 2 Contractors

These considerations are less plausible for Tier 2 contractors as they execute specific non-management tasks. There are many more market participants, especially regarding non-nuclear-specific decommissioning tasks, e.g., the demolition of auxiliary (non-contaminated) buildings. Certainly, specific tasks also require specific knowledge, especially in the context of working in a radioactive environment. However, these skills can be learned off-site, as the context of the entire site is not needed for learning, and tasks are executed in a limited scope. Further, Tier 2

[8] NRC. 2007. 'History and Framework of Commercial Low-Level Radioactive Waste Management in the United States'. NUREG-1853. Washington, D.C.: Nuclear Regulatory Commission. https://www.philrutherford.com/CalRadForum/NUREG-1853.pdf.

contractors can be present at multiple sites and transfer their experiences. Additionally, tasks of smaller scope and uncertainty are expected to be easier to tender from the perspective of transaction costs.

It is also reasonable to assume lower post-contractual market barriers for Tier 2 activities. Due to the limited scope and uncertainty, more complete contracts can be designed to cover most contingencies, reducing the possibility of costly bargaining. This also eases the detection of misbehavior and malperformance. Furthermore, because the process and asset specificities are limited, the case of replacing a contractor is less costly because the likelihood of the complete decommissioning project being terminated is limited. Therefore, the lock-in between the parties is limited. Nevertheless, this cannot be generalized without restriction. Tier 2 tasks can also be associated with high uncertainty, for example, if first-of-their-kind tasks are delegated. Furthermore, if Tier 2 tasks cover a bottleneck, this can be reflected in higher process specificity.

Deficits of markets and contractual arrangements between private actors must not directly imply stronger state engagement. The transition toward the emerging decommissioning models in the US can be interpreted as market-based organizational innovation based on two rationales. First, contrary to "Self-performance," outsourcing Tier 2 decommissioning allows incumbent operators to leverage the capabilities of expert organizations and focus on their core capabilities (electricity generation). Second, in contrast to conventional models, the emerging models mitigate most contractual hazards of Tier 1 tasks by vertically integrating the function and aligning ownership (liabilities) and execution (Moloney and Hess 2013; Stenger et al. 2019). However, the quasicontractual relationship between the NRC and the decommissioning contractor persists, especially regarding the accessibility of DTF funds. To make this interaction feasible, uncertainty and probity hazards must be limited.

15.3.3.2 Regulatory Challenges for the NRC

Table 15.1 summarizes the most important regulatory tasks of the NRC, divided between the domains of ownership, production, and funding. For all these regulatory tasks, the NRC needs highly project-specific information. The NRC staff must acquire detailed knowledge about the status of the sites and have in-depth (technical) knowledge regarding decommissioning to assess the feasibility of plans (including their funding feasibility) and observe the execution. The main reason for this regulatory burden is that it cannot rely on easily observable (or measurable) outcomes, and nuclear decommissioning is mostly an unstandardized site-specific task. Additionally, the licensee has less interest in retaining the license after the operations have ceased. With this, profit generation has ended, reducing the licensee's motivation.

In the case of the safety of the workforce and exposure of the environment to radiation, all regulatory means are focused on preventing significant harm. Damage caused by exposure to radiation or contamination with radioactive materials is irreversible or extremely costly to remediate (e.g., soil excavation and disposal). Regulators with

Table 15.1 Regulatory tasks of the NRC

Ownership	Production	Funding
Regulation and approval of license transfer (if necessary)	Regulation and oversight (via inspections) regarding safety and best practice	Regulation and oversight over DTF accumulation
Review and approval of the request for license termination		Regulation and oversight (via inspections) regarding withdrawals from DTF
	Review and approval of the decommissioning plan and cost estimates	
		Review and approval of cost estimates during decommissioning

Sources Own depiction, based on NRC information

a focus on prevention must know the contingencies of processes and the associated risks to assess whether the regulated practices are appropriate. Furthermore, the regulator must employ some extent of foresight, since decommissioning is not a continuous process, but a sequence of steps, making the effect of a potential (non-) action relevant, not only for the present, but also for the future.

Ensuring that the so-called "polluter pays principle" is upheld is a multi-level process. In an ideal world, the mere obligation to cover all liabilities produced by operating an NPP would result in private actors ensuring sufficient funding and efficient decommissioning.[9] However, having only the obligation (and insufficient monitoring) offers the potential for actors to evade these costs. The actor can act myopically (Ridge et al. 2014), and the assessment of liabilities might be subject to uncertainties, leading to the provision of insufficient funds. This alone would not be a major social problem, as a private actor would incur unexpected decommissioning expenses. But if liability regimes are incomplete, liabilities are at risk of falling to the public sector, i.e., being backstopped by taxpayer money.

It is reasonable to assume conflicting interests between the NRC and profit-maximizing decommissioning firms because the NRC is interested in efficient and safe decommissioning. The decommissioning contractor is mostly interested in profit maximization, which is not necessarily congruent with the maximization of safety. In most instances, the licensee may retain excess DTF resources after finalizing a decommissioning project. In a hypothetical situation where any cost overrun or damage would be borne by a rational decommissioning contractor, this contractor would minimize expected costs and mitigate cost overruns. In the case of limited liability of the contractor (or myopia), they could engage in risky cost-saving behavior since the expected costs of failure would not have to be completely borne. The favored ownership structure for NPPs is that of the LLC, which limits the liabilities of the parent company and could lead to increased public liability instead (Lordan-Perret et al. 2021).

[9] Meaning that the externality of abandonment is perfectly priced by society.

Therefore, the NRC has the important task of regulating and safeguarding the DTFs. The decommissioning firm has only limited property rights over the DTF (despite accumulating or inheriting it). Significant control over these property rights is located with the NRC, which regulates for which purposes DTF resources are permitted to be used. Therefore, the decommissioning firm and the NRC are in a quasi-contractual relationship in which the NRC acts as the principal and the decommissioning firm as the agent. The focus of the NRC is on the prevention of misuse of funds or malperformance; missing resources cannot be replaced, likely shifting the liability to the public budget. This issue concerns the initial accumulation of the funds and, therefore, the assessment of cost estimates, but especially the regulation of reimbursement and the development of the DTF during decommissioning. License transfers involve special regulatory tasks because the NRC must ensure sufficient financial assurance and capabilities of the potential new licensee.

Additionally, the regulatory interaction regarding the accumulation of sufficient funds for decommissioning is exposed to several uncertainties: (1) the remaining operational lifetime (e.g., regarding increased energy market competition), (2) the risk of underestimation of actual on-site costs, and (3) the risk of market downturns decreasing DTF investment returns (Lordan-Perret et al. 2021).

Thus, NRC regulation is motivated mainly by the necessity to enforce the liability of the operators and decommissioning contractors. Other motivations lie in direct concerns for safety and environmental harm, rules for DTF accumulation and funding assurance, ensuring efficient use of resources during decommissioning by approving plans, observing actions on site, and controlling withdrawals from the funds. The prerequisites for the application of DTFs are sufficiently low uncertainty and limited monitoring costs. This might be challenging in the case of heterogeneous reactor designs (contrary to the US fleet) or accident sites like Three Mile Island. Furthermore, it also requires the regulator to assess the executing organizations and their capabilities before a license transfer is carried out.

15.3.3.3 Wrong Incentives in the Case of Former Rate-Regulated Utilities

In the case of former rate-regulated public utilities, the licensee (e.g., a decommissioning contractor) is not allowed to retain excess DTF cash after decommissioning is completed. 18 CFR 35.32(a)(7) states that these excess funds must be returned to the ratepayers. Therefore, the only way to generate profit is via production-related revenues. The decommissioning contractor (in the emerging schemes) can principally add a markup to actual production costs for a certain service. For formerly rate-regulated utilities, this favors slow and inefficient decommissioning, as the contractor is incentivized to prolong work and generate further production-related revenues. The NRC regulates excess spending by reviewing site-specific cost estimates, which are the basis for the permitted spending of the contractor. These regulatory tasks require site-specific and extensive technical knowledge to assess whether the initial estimates or potential changes are appropriate. These circumstances result in additional transaction costs for all actors. Public utilities and the contractual partners must negotiate and implement agreements that regulate the final distribution of the surplus funds while preserving the incentives of the contractor.[10]

15.3.3.4 Increased Uncertainty Due to Lack of Suitable Waste Repositories

A special case regarding the oversight over the DTFs is when decommissioning firms request funds beyond the initially envisioned purpose of the DTF. Generally, the use of funds is only permitted for radiological decontamination,[11] which explicitly excludes spent fuel management or land restoration activities. The NRC may grant exemptions, but must first assess the site-specific uncertainties of radiological decommissioning and of other tasks. A prominent case is the request by Holtec to use DTF funds for spent fuel management at the Pilgrim NPP.

Based on the cost estimates for all activities and expected returns on the DTF, the NRC granted Holtec's request. Generally, the DOE is obligated to assume the spent fuel liabilities. As no final or interim DOE repository for spent fuel is available, this introduces further uncertainties to fuel management costs. It is unclear when (and if) the DOE will take possession of the waste and whether and what will compensate for long-term interim fuel storage.[12] Currently, storage costs are funded via the DTFs, the Judgment Fund, and the Nuclear Waste Fund. The Judgment Fund allows for the disbursement of government funds without needing specific Congressional approval, bypassing the typical political processes. This mechanism may contribute to the continued problem of spent fuel and other HLW being stranded at interim storage sites. In contrast, accessing the Nuclear Waste Fund for waste relocation and constructing a permanent repository requires Congressional approval and could increase the federal deficit. This requirement could be a major reason for the ongoing lack of progress in advancing the siting process for a final repository (Wimmers and von Hirschhausen 2024).

15.3.4 Summary

The US nuclear decommissioning market is dominated by a few specialized Tier 1 contractors, creating high entry barriers and potential efficiency concerns. The NRC plays a critical role in overseeing license transfers, cost estimates, and DTF

[10] An exemplary case is the transfer of the Vermont Yankee station (State of Vermont 2002).

[11] According to 10 CFR 50.82(a)(8)(i)(A) and 10 CFR 50.75(h)(1)(iv) in conjunction with the definition for decommissioning 10 CFR 50.2.

[12] The risk of a funding shortfall due to the exemption is stressed by the NGO Pilgrim Watch (Pilgrim Watch 2019).

allocations to ensure safety and financial accountability. Former rate-regulated utilities face incentives for inefficiency, while the lack of permanent waste repositories increases long-term uncertainties—financial, organizational, and safety-related. Despite a seemingly structured market, regulatory and financial challenges persist, especially regarding the hedging of environmental and health-related risks against profit-maximizing entities.

15.4 Decommissioning in the United Kingdom

The governance of nuclear decommissioning in the UK can be characterized as a prototypical state-dominated model. Public entities control the entire decommissioning production chain, and the state assumes most of the responsibilities (and liabilities). Private actors are only employed for tasks of limited scale and scope. The chapter will also briefly touch upon the historical development of the current governance scheme, due to the UK's decommissioning governance undergoing significant changes in the last decade. This change of governance is an interesting point of investigation as it discloses the shortfalls of the abolished semi-privatized decommissioning scheme.

15.4.1 Technical Overview

In the UK, 36 reactors are permanently closed, two reactors are under construction, and nine reactors remain operational (Schneider et al. 2024). One of the sites, housing several closed reactors, is the Sellafield site in Cumbria. It is one of the largest multipurpose nuclear sites in Europe. Here, decommissioning work entails first-of-its-kind reactors (intended for plutonium production), reprocessing plants, and storage facilities. The overall size and complexity of the site, unique facility designs, historically neglected decommissioning, and complex retrieval of radioactive legacy waste are the main challenges here. For example, 22 so-called "Magnox Swarf Storage Silos" are located at the site, each with a volume of about 720 m^3. Their content is not precisely known: Most silos are filled with swarf of reprocessed Magnox fuel (mostly fuel rod cladding made from magnesium) that has corroded and turned into sludge over time. This process has generated hydrogen gas, which is highly flammable and makes the emptying process even more challenging.[13]

Additionally, 26 gas-cooled and graphite-moderated Magnox reactors are undergoing decommissioning.[14] Several design aspects complicate the decommissioning of these reactors. One notable challenge is the use of graphite as a moderator that,

[13] https://www.gov.uk/government/case-studies/what-is-the-magnox-swarf-storage-silo.

[14] In the following, gas-cooled reactors will be abbreviated as GCR. This technology line includes Magnox reactors, French UNGG reactors, and British AGRs, see Chap. 2 of this book.

in its natural form, is not radioactive but can be activated through neutron radiation in the reactor core.[15] Irradiated graphite accounts for one of the largest volumes of radioactive waste sources from decommissioning in the UK, accounting for up to 45,600 m^3 with an approximate weight of 57,000 metric tons. There is no final solution for proceeding with this activated material (Pick 2010).

Other nuclear facilities are the Dounreay research site, which hosted the fast breeder research program (two reactors) and a naval reactor test site (both state-owned); the eight operating and four closed Advanced gas-cooled reactors (AGRs); the operational PWR at Sizewell; and the Hinkley Point C construction project, all owned and operated by EDF Energy, a subsidiary of French-state-owned utility Électricité de France (EDF).

The UK operates two LLW disposal sites. The first is the Low-Level Waste Repository (LLWR) in Cumbria, which receives waste from most decommissioning sites and other commercial nuclear waste. The second site, located at the Dounreay research site, will house waste from the breeder decommissioning and the Vulcan Naval Reactor Test Establishment.[16]

Currently, there is no permanent disposal option for Higher Activity Waste (HAW). HAW includes HLW, Intermediate Level Waste (ILW), and certain LLW materials not suitable for the LLWR (e.g., graphite cores). Plans for a Geological Disposal Facility (GDF) are under development.[17] The main purpose of this project is to dispose of HLW. ILW will be disposed of in the GDF or near service disposal sites, which are still to be sited (DESNZ 2024).

15.4.2 *Overview of Organizations*

The Nuclear Decommissioning Authority (NDA) is a non-departmental, so-called "arms-length" public body. It bears the main responsibility for the decommissioning of all nuclear facilities in the UK (with the exception of some military instances). The NDA develops strategic programs and is the parent of several management and operational subsidiaries. The most important are:

- Sellafield Ltd is the Site Licensing Company (SLC) for the decommissioning of the Sellafield nuclear complex.
- Nuclear Restoration Services (NRS) is the SLC responsible for the decommissioning of all other nuclear sites under NDA supervision, including the commercial Magnox fleet, the Dounreay research site and reactors, as well as the AGRs. NRS was formed from a merger of SLCs Magnox Ltd and Dounreay Ltd in 2023.[18]

[15] https://www.grs.de/de/aktuelles/eins-aus-10-billiarden-radioaktive-abfaelle-mit-beschleuniger-massenspektrometrie-noch.

[16] https://ukinventory.nda.gov.uk/about-radioactive-waste/how-do-we-manage-radioactive-waste/

[17] https://www.gov.uk/guidance/gdf-geological-disposal-facility.

[18] https://www.gov.uk/government/news/dounreay-has-successfully-joined-with-magnox-limited.

- Nuclear Waste Services (NWS) provides services around nuclear waste management, including operating the LLWR in Cumbria and developing a GDF for HAW.
- Nuclear Transport Solutions (NTS) offers transportation services for radioactive material, including appropriate packaging solutions.

Therefore, the NDA is the main strategic management entity for nuclear decommissioning in the UK. The SLCs mainly assume decommissioning project management and procurement functions (Tier 1), while actual decommissioning tasks are mostly procured on the market (Tier 2). This is framed as the so-called "market-enhanced" SLC (right side of Fig. 15.2). In the cases of NWS and NTS, a substantial share of Tier 2 tasks is directly integrated into the respective organization. For example, NTS owns trains and ships, and employs corresponding staff.

Until 2021, the NDA employed the Parent Body Organization (PBO) scheme (left side of Fig. 15.2). The ownership and strategic management position of the SLC were tendered to private actors. Therefore, a significant share of responsibility for decommissioning was borne by market actors, on behalf of the NDA, which was still ultimately responsible for project success (Wimmers and von Hirschhausen 2024).

In 2009, when EDF Energy acquired the ownership of the AGR fleet, it was agreed that EDF Energy would manage spent fuel and operational waste generated at the AGRs after January 15, 2005, until these materials could be transferred to the Sellafield reprocessing facility (now closed). Following a 2021 amendment to

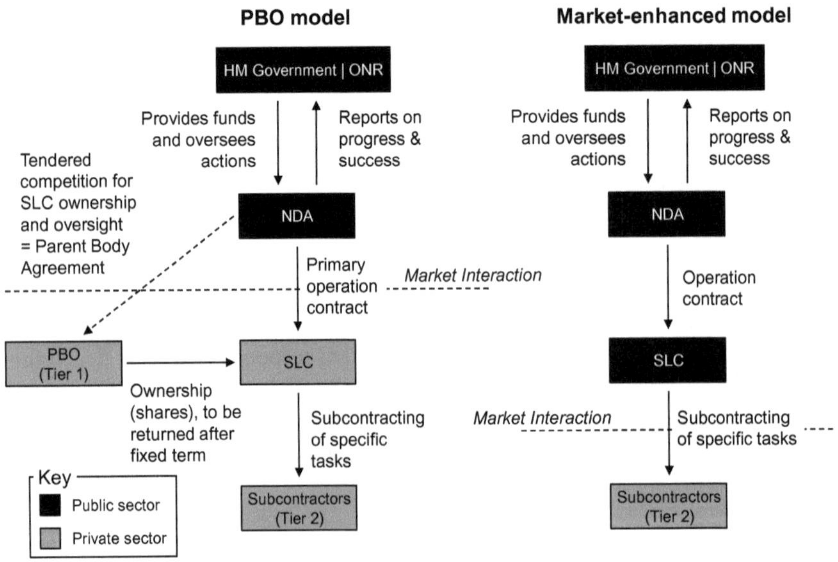

Fig. 15.2 Discontinued PBO model (left) and currently used Market-enhanced model (right) for decommissioning. (*Source* Taken from [Wimmers and von Hirschhausen 2024, 6])

this contract, the AGRs will remain in EDF's ownership until they are defueled, after which they will be transferred to the NDA for further decommissioning. EDF Energy will handle the decommissioning of the PWRs at Sizewell and Hinkley Point, while the NDA will assume responsibility for the spent fuel and waste.

As a statutory corporation (Energy Act 2013, Part 3), the Office for Nuclear Regulation (ONR) assumes regulatory oversight and sets out the regulatory framework for safe and efficient decommissioning while also granting licenses for nuclear facilities. The license holder bears full liability for the facility. A new nuclear facility must provide decommissioning plans when it applies for a license. If ownership changes occur, the license must be renewed. This is the case when the respective SLCs obtain ownership from the incumbent operators. Furthermore, the ONR must grant permission before decommissioning work can begin. The permission process includes an environmental impact assessment and approval of the decommissioning and waste management strategy, including the corresponding program, which is funded by the Department for Energy Security & Net Zero (DESNZ). The ONR monitors the respective licensee, and it has the authority to release a site from its nuclear license if all radiological hazards have been remediated. Additionally, the Environment Agency, the Scottish Environment Protection Agency, and Natural Resources Wales are involved with environmental permits and radioactive materials. The NDA must report to the DESNZ[19] annually via business plans and via an updated strategy report every five years. The Department and its minister are ultimately accountable for the performance of the NDA.

Funding for decommissioning work is provided by public funds, commercial revenue of the NDA, and the Nuclear Liabilities Fund (NLF) for the future decommissioning of the AGRs and the operational PWR at Sizewell. The NLF is fed by provisions of the commercial operators of NPPs, namely, EDF, through the public budget and, ideally, from investment returns. The decommissioning of Hinkley Point C shall be paid for by a currently non-existent "Funded Decommissioning Program" modeled after the NLF. The private sector is engaged as Tier 2 contractors.[20]

15.4.3 Analysis of the Decommissioning Organization in the United Kingdom

15.4.3.1 Path Dependencies and Failed Privatization

Although the British energy sector underwent privatization in the 1990s, the nuclear industry largely deviated from this trend due to resulting waste management liabilities that no private actor was willing to assume. The AGR plants and the PWR were privatized in 1996 via the sale of British Energy, which was later acquired by EDF.

[19] This function was formerly part of the former Department for Business, Energy & Industrial Strategy (BEIS).

[20] https://www.gov.uk/government/organisations/nuclear-restoration-services/about/procurement.

Dounreay and Sellafield were not considered for privatization (Wimmers and von Hirschhausen 2024). The failure to pass on decommissioning and waste liabilities can briefly be interpreted as an institutional lock-in because of path-dependencies, creating high transaction costs of institutional change.

Before the privatization efforts, the development of the nuclear industry was mainly driven by the state and strongly influenced by military interests and certain expectations regarding the development of nuclear technology. The UK is one of the few countries that opted for GCR designs (the early Magnox facilities and later AGRs). These are generally more expensive to decommission. The three main reasons for choosing this technological route were:

- military interest in plutonium production;
- a strong bet on the development of breeder reactors, which could utilize GCR waste as fuel, as well as using spent fuel through reprocessing; and
- the interest of the UK Atomic Energy Authority (UKAEA) to push their domestic technology, therefore limiting the competition with alternative (foreign) light-water technologies (MacKerron 2012).

Further, planning for decommissioning, including the estimation of costs and the set-up of an institutional framework, was of no or only minor consideration—there were no provisions for decommissioning. Instead, a focus on revenue-generating activities persisted for decades. Decommissioning moved into focus with the privatization efforts, as this demanded an assessment of the liabilities connected to the plants. The cost estimates made in this context quickly exceeded previous valuations. As a direct consequence, escalating liabilities increased the uncertainty of the potential transaction.

Furthermore, decommissioning cost estimates rendered aging Magnox plants worthless. Negative values of assets do not prohibit ownership transfers to the private sector per se; however, instead of receiving money, the state would have to pay for the private sector to take on the assets, which would, in a situation of high uncertainty, incur extensive transaction costs. The liabilities for decommissioning must be either accounted for precisely (ex-ante), or the state may provide a backstop for these liabilities, which in turn requires a tailored governance structure to assess and manage claims by private actors. Presumably, neither private actors nor the state would have accepted these high transaction costs or the high residual uncertainty remaining from an incomplete contract.

This means that past decisions (on strategies, governance, and, therefore, property rights structures and technologies) prevented the transition toward different governance structures because of high transaction costs for the reconfiguration of ownership. Subsequently, this led to the creation of the NDA to coordinate the decommissioning efforts across all British sites. In this role, the NDA fulfills several properties of a mission-oriented agency (Mazzucato 2018). As a coordinating commercial contractor, the NDA effectively "creates" a market for nuclear decommissioning. In addition, it engages with private and public actors (e.g., universities) to build capacities for its mission. Crucial elements of the NDA strategy are the provision of absorptive capacity and institutional learning.

15.4.3.2 Depth of State Involvement: The Failed PBO Scheme and the Market-Enhanced SLCs

The rationale behind the failed PBO scheme was to introduce competition to drive performance and innovation, and to utilize the capabilities of private actors for domestic decommission efforts. The scheme was later abolished, first at Sellafield and later at the other sites. Subsequently, the NDA has assumed full strategic management authority over the SLCs. The developments and challenges that led to the failure of the PBO scheme are outlined below.

Pre-Contractual Difficulties of the PBO Scheme

The PBO contracts were to be distributed via tenders. In the initial tender, the NDA was bound to a limited number of consortiums that had the necessary skills. The last Magnox tender of 2012, in which four consortiums competed over the PBO position for Magnox Ltd, the SLC responsible for the decommissioning of all British Magnox reactors and the sites at Harwell and Winfrith, was valued at £6.2 billion over 14 years. Both sides, i.e., the NDA and the competitors, made meaningful specific investments during the competition. On the supplier side, the competitors were specifically set up as diverse consortia for this task and had to provide information regarding their capabilities and approaches (signaling). On the demand side, the NDA had to evaluate the different proposals (screening). Due to the specificities of the sites and of required skills, the Statement of Response Requirements (SORR) of the Magnox tender counted over 350 pages and about 700 criteria (Holliday 2021). In transaction cost terms, this vast number of criteria could be interpreted as the measurement cost to assess the potential abilities of candidates. Another reason for transaction costs was the uncertain status of the decommissioning sites. As a basis for cost estimates and the award of the contract, the NDA had to present a baseline condition of the decommissioning site and measure the capabilities of the bidders against this. Uncertainty regarding the radiological content of the various sites, constructive peculiarities, the actual dismantling approach, and tasks already completed by predecessors complicated measurement activities. Therefore, it was impossible to create complete contracts covering all possible contingencies, especially over a period of 14 years.

In the case of actors bound to each other, this opens the space for post-contractual opportunism.[21] For the Magnox contract, the Holliday Inquiry (2021) attested to the NDA's lacking capabilities, resources, and specific knowledge to reliably execute this overly complex procurement. These failures led to a faulty award procedure and, subsequently, to the termination of the contract. This was related to the creation of SORR evaluation criteria and the handling of the evaluation itself, during which the NDA deviated from the original evaluation scheme to presumably prevent the

[21] If situations occur that were not contracted upon, entities will likely behave in a self-beneficial way if there are no safeguards in place that align the interests of the actors.

exclusion of most competitors due to non-material criteria. In summary, the transaction costs were too high under the PBO governance structure to reliably safeguard the NDA's property (the SLC plus the decommissioning site), i.e., allocating it in a way that was aligned with the NDA mission. Furthermore, the lack of pre-contractual safeguarding induced post-contractual transaction costs in the form of contract consolidation and court litigation.

Post-Contractual Difficulties of the PBO Scheme

Post-contractual asset-specific investments were probable, especially regarding the learning effects of the PBO personnel and alignment with the SLC workforce. This may have been one motivation for the long duration of the PBO contracts. The longevity of the contract should have incentivized the PBO to make long-term investments in its knowledge and in the SLC, for example, in training and strategy building.

This led to a bilateral monopoly situation because bidder-parity at contract renewal phases, or after a contractor replacement, could not be assured due to outsiders facing higher costs and risk than the incumbents. Imagining competition between an incumbent PBO and a potential PBO candidate, and assuming that the site-specific knowledge of the SLC staff is not transferable, the incumbent could potentially make better-informed decisions that are more cost-effective and face fewer risks regarding the uncertainties of the site. This likelihood would be reflected in the bid price. In contrast, the outsider must take the costly on-site learning process into account and hedge against unknown contingencies of the facility. Although the SLC, including the knowledge and skills of its workforce, is transferred to a new PBO, the specific (informal) alignment between the incumbent PBO and the SLC is not. This change, the lack of alignment, and resulting uncertainty can reduce the performance of the workforce (Bordia et al. 2003). The outsider must invest in realignment (e.g., time to build trust) and/or costly bargaining with the workforce to ensure a productive work relationship. Due to the incompleteness or impossibility of such bargaining, the outsider faces a higher risk regarding the behavior of the SLC workforce if unexpected contingencies occur (Williamson et al. 1975).

Additionally, the NDA was locked to the incumbent PBO because the transfer of the SLC and the decommissioning site to another supplier is costly. These costs stem from the necessity to determine the state of the site as a basis for a new tender. Additionally, due to the described oligopoly structure of the decommissioning market, turning to competitors as punishment for the failure of an incumbent was a limited deterrent (Williamson 1973). Also, possible attitudes of governmental agencies could have prevented the contract from breaking, as the displacement of a PBO could have been considered as a failure of the agency. An indication regarding this could be observed during the contract consolidation with the winner of the Magnox contract in 2014. The NDA "conditionally approved" all change control requests. In general, the NDA was reluctant to replace the PBOs despite significant cost and schedule overruns, and described the continuation of the contracts as the best option at present (Holliday 2021).

It cannot be ruled out that the cost and schedule overruns, as well as post-contractual consolidation claims, were not at least partly caused by opportunistic behavior. In an uncertain environment with high measurement costs (and incomplete contracts), the NDA was not able to reliably differentiate between justified overruns and avoidable costs. This primarily benefited the PBO because the NDA lacked knowledge of the sites, representing a large contingency in the transaction (Holliday 2021; KPMG 2013). The NDA was only indirectly present at the site through the SLCs, and so, this scheme limited oversight and increased measurement costs through complicated communication (KPMG 2013). Also, the reliability of the information provided by the SLC to the NDA, was questionable because the PBO and its secondees could have influenced communication to their advantage.

Specifically in the case of the Magnox tender, false assumptions regarding the baseline for the tender competition and contract development led to the contract termination due to material post-contractual variations in the contract consolidation phase, which also posed a legal issue. It could not be effectively assessed whether post-contractual claims were justified, and whether the material deviation rendered the initial competition for the best-suited candidate as inefficient. Additionally, this posed the risk of violating procurement law (Holliday 2021; MacKerron 2012; Wimmers and von Hirschhausen 2024). The lack of oversight also created an opportunity for opportunistic behavior of the incumbent firm (e.g., regarding the accuracy of reported task fulfilment).

As a result of uncertainty, the PBO contracts were limited in their ability to achieve long-term goals. The incentive mechanisms were focused on short-term accomplishments, which discouraged investments in long-term strategy, knowledge, and leadership at the SLC, highlighted by the Legacy Ponds and Silos (LP&S) projects at Sellafield. This was fundamentally rooted in the different interests of the PBO and the NDA. The PBO needed to maximize shareholder value, which is not consistent with the NDA's decommissioning mandate unless adequate safeguards and incentives are in place.

Summarized, the remaining probity hazards and the associated disutility led to the termination of the PBO model and to a higher state integration. This comes at the cost that the NDA expands capacities for contracting and strategic management at the sites.

15.4.3.3 Economizing on Transaction Costs and Capabilities—Transfer of AGRs

The AGR fleet and the Sizewell PWR were privatized when British Energy was transformed from a public to a stock company in 1996. In 2002, British Energy received its first government bailout. The following year, in 2003, a restructuring agreement was reached, under which the Department of Trade and Industry assumed £1.7 billion of liabilities and became responsible for the remaining £3.9 billion in liabilities not covered by the NLF. The restructuring was completed in 2005 and included the assumption of all nuclear liabilities. In 2009, however, British Energy was sold to EDF after having filed for bankruptcy. Following this sale, liabilities exceeding the NLF remained with the taxpayer because the governmental focus was to establish a stronger nuclear operator, also with nuclear new build plans in mind. Decommissioning was not in focus (and other arrangements might have deterred EDF) (MacKerron 2012).

Subsequently, an agreement between the NDA and EDF was reached to transfer the AGRs to NDA ownership after closure and completed defueling. From a governance perspective, it might be easier to control the decommissioning management and strategy of an integrated entity like the NDA instead of creating a dedicated governance structure to safeguard resources provided by the state if EDF conducted the decommissioning by themselves (probity hazards). From a capabilities perspective, it can be argued that the NDA has notable experience (therefore capabilities) with the decommissioning of Magnox reactors, which could be applied to the decommissioning of the AGR fleet (both concepts are GCRs). EDF follows a strategy of deferred dismantling for the GCRs in France, and consequently, the decommissioning experience for these reactor types is limited (Bärenbold et al. 2024). Furthermore, AGRs and Magnox reactors share sites, offering the possibility to leverage site-specific knowledge and NDA assets. As fuel transfers are part of the operational routine tasks at NPPs, it is expected that EDF can leverage its experiences. As an incentive for efficient defueling, EDF can earn a fee of up to approximately £100 million, but in the case of malperformance, it might be charged the same amount.

The arrangement between the BEIS (today's DESNZ) and EDF is still prone to uncertainties and potential conflicts, some of which are:

- Specific terms of ownership transfer of the sites
- Limited access and information upon site transfer
- Unclear baseline for "efficient" defueling performance
- Coordination during defueling (EDF is responsible for defueling, and NDA for transportation to Sellafield and subsequent treatment)

15.4.3.4 Controlling Public Hierarchies—Governance of the NDA

The NDA's position is not subject to market competition and its disciplining effects. However, public bureaucracies can still be prone to opportunism and rent-seeking without appropriate governance, too. Radiological safety and environmental issues are regulated by the ONR, the Environment Agency, the Scottish Environment Protection Agency, and Natural Resources Wales (NAO 2018). Additionally, the NDA is controlled by the Scottish (for Dounreay) and the UK Parliaments which must approve the annual NDA budgets. The minister of the Department (DESNZ) is ultimately accountable for the performance and actions of the NDA. On the UK level, two additional actors are of importance: UK Government Investments (UKGI), and the Treasury. UKGI is a government-owned company under the Treasury's oversight that oversees the NDA on the Department's behalf on a day-to-day basis. The

Department approves strategies and annual plans and monitors the performance of the NDA, while the Treasury approves spending. Furthermore, both entities approve business cases submitted by the NDA for projects exceeding £100 million or when significant changes or cancellations to an approved plan or project are necessary (BEIS 2021; NAO 2018; NDA 2021).

This represents a quasi-contractual situation prone to transaction costs. The main concern in these interactions is to ensure "value for money."[22] Possible methods to exercise control are transparency requirements, review and approval loops, performance assessments, and budgetary control (NDA 2021). However, as noted, higher uncertainty and measurement costs can complicate regulation. Furthermore, the governance of the NDA was criticized as being overcomplicated with unclear responsibilities, leading to ineffective control (NAO 2018).

The main issue is the proprietary information that the NDA and the SLC hold regarding the status of the sites, the assessment of strategic approaches, and other issues. This establishes a position of power for the NDA (Weber 1946). Disclosure requirements (e.g., accounts, strategy) can create transparency (Bendor et al. 1987), but this depends on the ability of the authorities to interpret this information. Even if a transparent chain of accountability is established, enforcement is reliant on the detection of malperformance.

The value of delivery by the NDA is hampered by conflicting demands of regulators and politics. The focus of the Department is on value for money. Cost minimization would imply quickly decommissioning sites with low hazards to cut fixed costs for maintenance. Meanwhile, the ONR demands to focus on high-hazard sites (like fuel ponds) to minimize radiological risks (NAO 2018).

A critical issue is the integration of budgetary discipline while parallelly ensuring flexibility to counter unforeseen events. The NDA must improve efficiency and allow for experimentation and innovation to be able to accomplish its tasks in time and to ensure the best value for money. High budgetary constraints can lead to a lower value for money due to high fixed costs. Furthermore, the budgets annually approved by Parliament reduce the planning security for long-term tasks (also for contractors). Additionally, the limitation of carrying budgets across years might hamper efficiency in improving flexibility. Transitioning toward external funding via the NLF as a steady source of funding could ease this issue (NAO 2018), albeit risking future funding shortfalls, as discussed in Chap. 17. of this book. Furthermore, the agency's budget must not be prone to political opportunism, e.g., unjustified budgetary cuts or interventions in day-to-day activities. This is mainly addressed through the introduction of the NDA as an "arm's length" agency (Christensen and Lægreid 2006). These issues are also linked to the extent to which probity regarding the mission is successfully anchored as an intrinsic goal in the organization, e.g., through the management or recruitment, and renumeration strategy.

---

[22] Notably, this has been the NDA's mission since its creation in 2004, and it was also the motivation for setting up the PBO scheme, see Wimmers and von Hirschhausen (2024).

15.4.4 Summary

The UK 's nuclear decommissioning sector faced path dependencies and privatization failures stemming from technological complexities of to-be-decommissioned facilities, historical neglect, unresolved liabilities, and high procedural complexities. The NDA was established as an "arm's length" agency to coordinate decommissioning. The PBO scheme, which hoped to introduce market competition to decommissioning, failed due to high transaction costs and governance issues. The AGR fleet transfer process to the NDA is driven by efficiency and capability considerations, but remains complex, and the exact organization thereof remains unclear. Overall, historical NDA governance was constrained by bureaucratic inefficiencies, conflicting priorities, and budgetary challenges, requiring improved oversight and funding mechanisms. Whether the NDA can overcome these challenges, remains uncertain—decommissioning activities in the UK will continue for many decades, regardless.

15.5 Discussion

In this work, two distinct cases for decommissioning governance were analyzed (see Table 15.2). In the US, the decommissioning of commercial NPPs is mainly driven by private enterprises. These enterprises have changed the decommissioning business both technically and organizationally. This includes contracts for decommissioning, in which entire NPPs, including their license, are transferred to the decommissioning contractor. As the main supervising state body, the NRC ensures the safety and funding for decommissioning. The predominant entity in the UK is the NDA, a public agency, which assumes strategic and managerial authority through its subsidiaries across the decommissioning sites. Private actors are included as contractors for tasks of limited scope.

The main promise of introducing private actors is two-fold: (1) utilizing existing capabilities within the market, and (2) achieving efficient governance through competition and profit-seeking. In the US and Europe, numerous firms are engaged in nuclear and decommissioning-related activities, which could provide value for decommissioning. Additionally, the profit-seeking motive of market participants incentivizes them to innovate technology- and governance-wise. Innovation of the latter can lead to lower transaction costs (emerging models in the US). Limitations of this promise are possible for Tier 1 contractors (strategic management position). Such contractors must have specialized skills and experience tailored toward nuclear decommissioning management. This might reduce the number of possible actors and manifest advantages of incumbents. This limitation can be magnified if the management entity must bear the liability, as this increases the need for financial assurance of the entity (e.g., emerging models in the US). Furthermore, competition can be hampered by limited access to waste facilities. During the decommissioning process,

Table 15.2 Summary of findings on the US and UK decommissioning models

	Open market/state as regulator	State dominated
Strength	• Leverage resources and capabilities in the market • Disciplining effect of competition • Efficiency incentives through profit-seeking motive • Spontaneous reconfigurations of governance	• "Probity" due to different organizational structures and renumeration mechanisms increases safety and efficiency (mission orientation instead of profit seeking) • Direct involvement in decommissioning reduces measurement costs and enforcement costs • Clear liability
Weakness	• Remote position of the state from decommissioning work increases the information costs for regulation - Ensure financial assurance - Safeguard safety and budget - Detect misbehavior/ malperformance	• Need to build (additional) capabilities/capacities • Lacking external market incentives for efficiency and innovation • Risk of dispersion of responsibility • Depending on politics • Easier access to public budget (rent-seeking)
Technological environment factors favoring the model	• Limited scale, complexity, and uncertainty - Standardizable - E.g., western PWR and BWR reactors	• High decommissioning liabilities in conjunction with high uncertainties • GCRs, Sellafield
Social environment factors favoring the model	• Restricted market competition (Tier 1) • Competitive market (Tier 2) • Sufficient market size	• Competitive market (Tier 2) • Path dependencies led to the assumption of liabilities by the state - Technological decision -Absent political focus on decommissioning

(continued)

Table 15.2 (continued)

	Open market/state as regulator	State dominated
Focus points for ensuring functioning	• Ensure a strong liability regime incl. financial assurance • Reduce external uncertainties (e.g., spent fuel storage question) • Ensure regulatory competence • Ensure a sufficient flow of information from the decommissioning contractor • Ensure appropriate approaches for decommissioning cost estimates • Prevent monopolization • Ensure access to waste facilities	• Only contract for tasks of limited scope (Tier 2) • Employ a strong mission-oriented culture through recruitment and career pathways • Ensure oversight (in the absence of competition) via other state organs (e.g., Department of Treasury) • Shield the agency from political influence to allow flexibility • Enable strong control of the agency for better performance
Open questions	• Are new decommissioning models consistently successful? • How competitive is the Tier 1 market? • Can the NRC cope with the additional regulatory effort of new decommissioning models? • What are effective arrangements to return excess funds to rate payers in case of former rate-regulated utilities?	• Do NDAs internal structures and external government controls incentivize efficient decommissioning? • Are probity and experimentation/innovation opposites? What prevails in the NDA? • Is the new Market-enhanced approach successful?

Source Own compilation

the disciplining forces of competition are weakened due to the costs of replacing an incumbent and high information costs for detecting and pursuing malperformance.

An essential concern for efficient decommissioning governance is the scale, complexity, and uncertainty of the decommissioning site. High uncertainty regarding the status, remediation costs, best practices, etc., increases transaction costs for interactions on the market, with regulators, and within hierarchies.

The basic premise for the presence of the state is the existence of externalities of environmental and health risks due to inadequate decommissioning or the potential abandonment of a nuclear facility. A particular problem is the latter, which aims to cover liabilities for otherwise worthless assets. In this case, the state takes two distinct roles within the two countries. In the US, DTFs were established, which must be accumulated during the lifetime of the NPP. The NRC ensures that the responsible entity can finally meet its liability by setting up requirements for the DTF (including prescribing minimum cost estimates for decommissioning). During decommissioning, the NRC intensively regulates actions at the site to prevent direct health and environmental hazards, ensure sufficient resources, and avoid DTF shortfalls. These regulatory actions require in-depth knowledge of the site and practices. Therefore, the uncertainty must be limited to enable effective regulation. This concerns uncertainty about the technologies, actors, and interacting interfaces (e.g., issues of spent fuel management). The institutional environment in the US poses an additional challenge for ensuring the coverage of liabilities. Parent companies can be shielded from LLCs that are license holders. If no other revenue-generating activities exist for the LLC, the NRC must focus on preventing bankruptcy and subsequent abandonment of the NPP.

While the nuclear industry in the US was open for private actors from 1954 onward, the UK efforts for privatization, including the nuclear liabilities, started only in the 1990s. Military and political choices lead to high liabilities (e.g., choice of GCR technology, waste storage at Sellafield, etc.) in conjunction with high uncertainties regarding these liabilities. These path dependencies prevented the transfer of liabilities to the private sector. Therefore, the state assumed a role beyond that of the regulator. Today, the NDA now assumes the strategic oversight as a commercial contractor and promotes the UK's decommissioning mission.

The development of the role of the NDA after its introduction sheds light on the question of how intensive government involvement should be. The NDA contracted for management positions via the PBO model. Especially high uncertainty and lock-ins with private actors prevented incentive-compatible arrangements and ultimately led to the abandonment of the PBO scheme. In the current market-enhanced model, the NDA focuses on contracting tasks of limited scope and uncertainty. This promises to circumvent the main contractual difficulties of the PBO model, but it also means that the NDA and its subsidiaries must build additional capacities for assuming Tier 1 positions. This governance shift illustrates the importance of finding a sweet spot between developing commercial contracting capacities (transaction costs) and the effort to build one's own capacities for production. Subsequently, the NDA will also decommission the newer AGR fleet. Aside from the difficulties of contracting, it was argued that the NDA possesses favorable capabilities and experience for decommissioning for this task.

Furthermore, the NDA's governance poses challenges, and it remains unclear whether the NDA will live up to its role as an entrepreneurial, mission-oriented entity aiming for the best value for money. Transparency and control mechanisms, and a strong mission culture, are essential for ensuring efficient decommissioning. However, the analysis showed that alternative governance modes, like full privatization or the PBO model, were not feasible in the UK.

15.6 Conclusion

The decommissioning and dismantling of nuclear power plants is a complex operation that involves high asset specificity, long-term planning, and multiple safety concerns. Therefore, it cannot be organized by "market competition" itself but requires complex organizational models for financing and physical delivery. In this chapter, we have laid out a conceptual framework for such models, based on institutional economics. The focus is on different experiences from two interesting and complex country case studies, one on the United States, and the other one, on the United Kingdom.

The US is certainly the most interesting case study, even though absolute figures on decommissioning and dismantling are low. The market for the decommissioning of nuclear power plants is dominated by a small number of specialized companies. This makes market entry difficult and can lead to inefficiencies. The NRC oversees license transfers, cost estimates, and funding allocations to ensure safety and financial accountability. Tier 1 companies are oligopolists, and try to integrate vertically. At the same time, former rate-regulated utilities often have little incentive for efficiency. There is a lack of repositories for radioactive waste, which creates long-term uncertainties—financially, organizationally, and in terms of safety. Despite a structured market, regulatory and financial challenges remain, particularly when it comes to protecting the environment and health from profit-oriented companies.

In the UK, it is very difficult to determine whether the respective governance modes are the best possible fit. This is because implementation is still in its infancy, and it takes a long time for a decommissioning project to be completed. Sub-questions related to this are a deeper investigation of the NRC's regulatory efforts, decommissioning cost assessments, and an investigation of the market environment and competition for Tier 1 contractors. Regarding the NDA, an investigation of the internal governance and incentivization of private actors and their interaction with the government would be of interest. Expanding the gathered knowledge to other countries, and applying both approaches, the US' market-based and the UK's market-enhanced models, could bring potential benefits to countries still facing the challenge of nuclear decommissioning.

The chapter also hints at a larger debate on comparative institutional approaches, including other national experiences. Potential candidate countries would be Germany, Japan, and France.

Annex

See Table 15.3.

Table 15.3 Commercial LLW facilities in the US (National Academies of Sciences, Engineering, and Medicine 2017)

	Barnwell, SC	Clive, UT	Richland, WA	Andrews, TX
Operator	EnergySolutions	EnergySolutions	US Ecology (subsidiary of Republic Services)	Waste Control Specialists (WCS)
Types of LLW accepted*	A, B, C	A	A, B, C	A, B, C
Available to	Atlantic Compact: South Carolina, Connecticut, and New Jersey	All states	Northwest Compact (Alaska, Hawaii, Idaho, Montana, Oregon, Utah, Washington, and Wyoming) and Rocky Mountain Compact (Colorado, Nevada, and New Mexico)	Texas Compact (Texas and Vermont) and other states on a case-by-case basis
Accepts DOE LLW		yes		yes

* The classification is based on a multidimensional classification scheme of radio nuclide concentration according to 10 CFR 61.55. Simplified, the harmfulness of waste due to radionuclide concentration increases from A (lowest) to C (highest)

References

Allen, Douglas W. 1991. 'What Are Transaction Costs?' *Rsch. in L. & Econ.* 14:1.

Bärenbold, Rebekka, Muhammad Maladoh Bah, Rebecca Lordan-Perret, Björn Steigerwald, Christian von Hirschhausen, Ben Wealer, Hannes Weigt, and Alexander Wimmers. 2024. 'Decommissioning of Commercial Nuclear Power Plants: Insights from a Multiple-Case Study'. *Renewable and Sustainable Energy Reviews* 201:114621. https://doi.org/10.1016/j.rser.2024.114621

Barzel, Yoram. 1985. 'Transaction Costs: Are They Just Costs?' *Zeitschrift Journal of Institutional and Theoretical Economics* 141 (1): 4–16.

Barzel, Yoram. 2005. 'Organizational Forms and Measurement Costs'. *Journal of Institutional and Theoretical Economics* 161 (3): 357–73.

BEIS, Department for Business, Energy & Industrial Strategy. 2021. 'Departmental Review of the Nuclear Decommissioning Authority'. https://assets.publishing.service.gov.uk/media/60d07781d3bf7f4bd9814f82/nda-departmental-review-2021.pdf.

Bendor, Jonathan, Serge Taylor, and Roland Van Gaalen. 1987. 'Politicians, Bureaucrats, and Asymmetric Information'. *American Journal of Political Science* 31 (4): 796–828. https://doi.org/10.2307/2111225.

Bordia, Prashant, Elizabeth Hobman, Elizabeth Jones, Cindy Gallois, and Victor J. Callan. 2003. 'Uncertainty During Organizational Change: Types, Consequences, and Management Strategies'. *Journal of Business and Psychology* 18 (4): 507–32. https://doi.org/10.1023/B:JOBU.0000028449.99127.f7.

Christensen, Tom, and Per Lægreid. 2006. 'Agencification and Regulatory Reforms'. In *Autonomy and Regulation*, edited by Tom Christensen and Per Lægreid, 8–50. Social and Political Science Archive. Cheltenham, UK: Edward Elgar Publishing. https://doi.org/10.4337/9781781956229.

DESNZ. 2024. 'Managing Radioactive Substances and Nuclear Decommissioning Policy Consultation: Government Response'. London: Department for Energy Security & Net Zero. https://www.gov.uk/government/consultations/managing-radioactive-substances-and-nuclear-decommissioning.

Foss, Kirsten, and Nicolai J. Foss. 2005. 'Resources and Transaction Costs: How Property Rights Economics Furthers the Resource-Based View'. *Strategic Management Journal* 26 (6): 541–53. https://doi.org/10.1002/smj.465.

Furubotn, Eirik, and Rudolf Richter, eds. 2005. *Institutions and Economic Theory: The Contribution of the New Institutional Economics*. 2nd ed. Economics Cognition and Society. Ann Arbor: University of Michigan Press. https://doi.org/10.3998/mpub.6715.

GAO. 1995. 'Radioactive Waste - Status of Commercial Low-Level Waste Facilities'. GAO/RCED-95–67. Washington, D.C.: United States General Accounting Office. https://www.gao.gov/assets/rced-95-67.pdf

Holliday, Steve. 2021. 'Report of the Holliday Inquiry: Inquiry into Award of the Magnox Decommissioning Contract by the Nuclear Decommissioning Authority, Related Litigation and Its Subsequent Termination'. London: Dandy Booksellers Ltd. https://assets.publishing.service.gov.uk/government/uploads/system/uploads/attachment_data/file/966572/The_Holliday_Inquiry.pdf.

Jacobides, Michael G., and Sidney G. Winter. 2005. 'The Co-Evolution of Capabilities and Transaction Costs: Explaining the Institutional Structure of Production'. *Strategic Management Journal* 26 (5): 395–413. https://doi.org/10.1002/smj.460.

KPMG. 2013. 'The PBO Model at Sellafield: Performance to 31st May 2013'. https://assets.publishing.service.gov.uk/government/uploads/system/uploads/attachment_data/file/534449/KPMG_report_on_Sellafield_Performance_to_31_May_2013.pdf.

Krutilla, Kerry. 2011. 'Transaction Costs and Environmental Policy: An Assessment Framework and Literature Review'. *International Review of Environmental and Resource Economics* 4 (3–4): 261–354. https://doi.org/10.1561/101.00000035.

Lordan-Perret, Rebecca, Robert D. Sloan, and Robert Rosner. 2021. 'Decommissioning the U.S. Nuclear Fleet: Financial Assurance, Corporate Structures, and Bankruptcy'. *Energy Policy* 154 (July):112280. https://doi.org/10.1016/j.enpol.2021.112280.

MacKerron, Gordon. 2012. 'Evaluation of Nuclear Decommissioning and Waste Management'. Sussex: Science and Technology Policy Research (SPRU). https://sussex.figshare.com/articles/report/Evaluation_of_nuclear_decommissioning_and_waste_management/23386643/1/files/41117729.pdf.

Mazzucato, Mariana. 2018. 'Mission-Oriented Innovation Policies: Challenges and Opportunities'. *Industrial and Corporate Change* 27 (5): 803–15. https://doi.org/10.1093/icc/dty034.

Moloney, B. P., and J. Hess. 2013. 'Progress on the Decommissioning of Zion Nuclear Generating Station'. In *Volume 2: Facility Decontamination and Decommissioning; Environmental Remediation; Environmental Management/Public Involvement/Crosscutting Issues/Global Partnering*. Brussels: ASME. https://doi.org/10.1115/ICEM2013-96321.

NAO. 2017. 'The Nuclear Decommissioning Authority's Magnox Contract'. London: National Audit Office. https://www.nao.org.uk/reports/the-nuclear-decommissioning-authoritys-magnox-contract/

NAO. 2018. 'The Nuclear Decommissioning Authority: Progress With Reducing Risk at Sellafield'. HC 1126. London: National Audit Office. https://www.nao.org.uk/wp-content/uploads/2018/06/The-Nuclear-Decommissioning-Authority-progress-with-reducing-risk-at-Sellafield.pdf.

NAO. 2022. "The Decommissioning of the AGR Nuclear Power Stations." HC 1017. London: National Audit Office. https://www.nao.org.uk/wp-content/uploads/2022/01/The-decommissioning-of-the-AGR-nuclear-power-stations.pdf
National Academies of Sciences, Engineering, and Medicine. 2017. *Low-Level Radioactive Waste Management and Disposition: Proceedings of a Workshop*. Washington, D.C.: National Academies Press. https://doi.org/10.17226/24715.
NDA. 2021. 'Nuclear Decommissioning Authority: Framework Document'. Nuclear Decommissioning Authority. https://assets.publishing.service.gov.uk/media/60c8cd1b8fa8f57ceec3c944/20210526_FINAL_NDA_Framework_Document.pdf.
NDA, and ONR. 2023. "Memorandum of Understanding Between the Office for Nuclear Regulation and The Nuclear Decommissioning Authority." Nuclear Decommissioning Authority and Office for Nuclear Regulation. https://www.onr.org.uk/media/lq0mlers/mou-onr-nda.pdf
North, Douglass C. 1991. 'Institutions'. *The Journal of Economic Perspectives* 5 (1): 97–112.
NRC. 2016. 'Agreement State Program Policy Statement'. NRC–2016–0094. Nuclear Regulatory Commission. https://www.govinfo.gov/content/pkg/FR-2017-10-18/pdf/2017-22514.pdf
NRC 2019. 'One NDA - The Handbook'. Nuclear Decommissioning Authority. https://microcontrib.hays.com/documents/4706216/4758784/One+NDA+Handbook+2019.pdf/fdbce3da-bdf7-83ca-4fa4-169297bfe094?t=1580726745690
NRC. 2024. "Integrated Low-Level Radioactive Waste Disposal; Proposed Rule Draft Regulatory Analysis for Comment." Nuclear Regulatory Commission. https://www.nrc.gov/docs/ML2324/ML23242A259.pdf
Ostrom, Elinor. 2011. 'Background on the Institutional Analysis and Development Framework'. *Policy Studies Journal* 39 (1): 7–27. https://doi.org/10.1111/j.1541-0072.2010.00394.x.
Pick, M. E. 2010. 'Magnox Graphite Core Decommissioning and Disposal Issues'. Magnox South, Gloucester, United Kingdom: International Atomic Energy Agency (IAEA). https://www-pub.iaea.org/mtcd/publications/pdf/te_1647_cd/pdf/tecdoc_1647.pdf.
Pilgrim Watch. 2019. 'Pilgrim Watch Motion Under 10 CFR §2.323 to Stay Staff Order of August 22, 2019 Granting Exemption'. No. 50–293 & 72–1044 LT. https://www.mass.gov/doc/september-3-2019-pilgrim-watch-motion-under-10-cfr-ss-2232-to-stay-staff-order-of-august-22-0/download.
Ridge, Jason W., Dave Kern, and Margaret A. White. 2014. 'The Influence of Managerial Myopia on Firm Strategy'. *Management Decision* 52 (3): 602–23. https://doi.org/10.1108/MD-01-2013-0037.
Schneider, Mycle, Antony Froggatt, Julie Hazemann, Özgür Gürbüz, Paul Jobin, Phil Johnstone, Timothy Judson, et al. 2024. 'The World Nuclear Industry Status Report 2024'. Paris, France: Mycle Schneider Consulting. https://www.worldnuclearreport.org/IMG/pdf/wnisr2024-v2.pdf.
State of Vermont. 2002. 'Memorandum of Understanding Among Entergy Nuclear Vermont Yankee, Llc, Vermont Yankee Nuclear Power Corporation, Central Vermont Public Service Corporation, Green Mountain Power Corporation, And The Vermont Department Of Public Service'. Docket No. 6545. https://www.google.com/url?sa=t&rct=j&q=&esrc=s&source=web&cd=&ved=2ahUKEwjGmb30saaIAxXNS_EDHWJxH4IQFnoECBIQAQ&url=https%3A%2F%2Fepuc.vermont.gov%2F%3Fq%3Ddownloadfile%2F232535%2F27332&usg=AOvVaw2Kz_NdHvArBw-TyuhWQB_S&opi=89978449.
Stenger, Daniel F., Amy C. Roma, and Sachin S. Desai. 2019. 'Innovations in Decommissioning and Their Application Abroad'. *Nuclear News*, no. July, 45–52.
Weber, Max. 1946. 'Bureaucracy'. In *From Max Weber: Essays in Sociology*, edited by H. H. Gerth and C. Wright Mills. New York: Oxford University Press.
Williamson, Oliver E. 1973. 'Markets and Hierarchies: Some Elementary Considerations'. *The American Economic Review* 63 (2): 316–25.
Williamson, Oliver. 1976. 'Franchise Bidding for Natural Monopolies-in General and with Respect to CATV'. *The Bell Journal of Economics* 7 (1): 73. https://doi.org/10.2307/3003191.
Williamson, Oliver. 1979. 'Transaction-Cost Economics: The Governance of Contractual Relations'. *The Journal of Law & Economics* 22 (2): 233–61.

Williamson, Oliver. 1985. *The Economic Institutions of Capitalism: Firms, Markets, Relational Contracting*. New York: Free Press.
Williamson, Oliver. 1988. 'Technology and Transaction Cost Economics: A Reply'. *Journal of Economic Behavior & Organization* 10 (3): 355–63. https://doi.org/10.1016/0167-2681(88)90055-8.
Williamson, Oliver. 1999. 'Public and Private Bureaucracies: A Transaction Cost Economics Perspective'. *Journal of Law, Economics, & Organization* 15 (1): 306–42.
Williamson, Oliver E., Michael L. Wachter, and Jeffrey E. Harris. 1975. 'Understanding the Employment Relation: The Analysis of Idiosyncratic Exchange'. *The Bell Journal of Economics* 6 (1): 250–78. https://doi.org/10.2307/3003224.
Wimmers, Alexander, Rebekka Bärenbold, Muhammad Maladoh Bah, Rebecca Lordan-Perret, Björn Steigerwald, Christian von Hirschhausen, Hannes Weigt, and Ben Wealer. 2023. 'Decommissioning of Nuclear Power Plants: Regulation, Financing, and Production'. 104. DIW Data Documentation. Berlin, Basel: Deutsches Institut für Wirtschaftsforschung (DIW). https://www.econstor.eu/handle/10419/268719.
Wimmers, Alexander, and Christian von Hirschhausen. 2024. 'Organizational Models for the Decommissioning of Nuclear Power Plants: Lessons from the United Kingdom and the United States'. *Utilities Policy* 91 (December):101843. https://doi.org/10.1016/j.jup.2024.101843.

Chapter 16
Disposal of High-Level Nuclear Waste: Global Status and German Experience with the Unresolved Liability of Nuclear Energy

Jochen Ahlswede, Jan Graefje, and Hendrik Schopmans

16.1 Introduction

We have lived through a century of radioactive waste. The discovery and subsequent theoretical description of nuclear fission in 1938/39 marked not only the beginning of the age of nuclear weapons and nuclear energy generation (both for military and civilian purposes) but also the era of radioactive waste. Although minor volumes of radioactive waste had been generated earlier—for example, as a by-product from the mining and processing of pitchblende (uraninite) to extract radium and produce pigments—the significant production of radioactive waste began with the use of nuclear fission in the 1940s. Initially, this waste originated from military programs that focused on the production of highly enriched uranium and fissile plutonium for fission (and later fusion) weapons. Shortly afterward, the first reactors for civilian nuclear energy production were built. In the following decades, the world's reactor fleet continued to grow,[1] leading to a continuous increase in global radioactive waste inventories. Today, civilian nuclear energy production accounts for most of the waste, both in terms of volume and activity levels. In addition, civilian and military research activities continue to produce waste, as well as—in comparatively smaller quantities than energy production and research—industrial and medical activities. Nuclear weapons testing and catastrophic accidents at nuclear facilities have also contributed to the accumulation of radioactive waste. With the expansion of nuclear power across numerous countries, radioactive waste is now present in most regions of the world. Due to the enormous risks that high-level radioactive waste (HLW) poses for human

[1] IAEA Power Reactor Information System PRIS, https://pris.iaea.org/pris/home.aspx, last accessed on 27–02-2025. This date holds for all other online references unless otherwise indicated.

J. Ahlswede (✉) · J. Graefje · H. Schopmans
Federal Office for the Safety of Nuclear Waste Management (BASE), Berlin, Germany
e-mail: jochen.ahlswede@base.bund.de

A. Wimmers et al. (eds.), *Nuclear Power*, Studies in Energy, Resource and Environmental Economics, https://doi.org/10.1007/978-3-031-99894-2_16

health and the environment, its safe disposal represents a global challenge requiring the concerted efforts of national actors and international collaboration.

Radioactivity is the decay of atomic nuclei accompanied by the emission of particles or electromagnetic radiation. It is governed by physical laws and thus cannot be altered through biological or chemical processes (see Chap. 2 of this book). Consequently, there are only three conceivable principal options for "managing" radioactive waste:

- Transmutation: Transforming radioactive nuclei to stable nuclei through targeted irradiation with neutrons.
- Dilution: Spatial distribution of waste to reduce activity concentrations to levels that no longer significantly increase natural environmental radioactivity.
- Isolation: Permanent isolation of waste from the biosphere and shielding of radiation through artificial or natural barriers.

All existing waste management strategies or technologies rely on one of these options or a combination. Targeted transmutation for waste management at an industrial scale is neither possible nor practical: Although physical methods for transmutation of waste are conceivable and partially applied, they involve extreme effort in terms of financial resources and time, and cannot convert all radioactive substances into stable forms without residue.[2] Dilution is feasible only under very limited conditions and is not an option for most waste because of its composition, aggregate state, or radioactivity concentrations. Nevertheless, it is practiced regularly—for example, through the release of radioactive gases from nuclear facilities (for the European Union, see, e.g., EC 2024) or the discharge of treated but still slightly radioactive cooling water from the damaged Fukushima nuclear plant.[3] This leaves isolation, i.e., the storage of waste in artificial facilities or using geological barriers in a repository, as the only feasible option for most radioactive waste streams.

HLW from nuclear reactors is composed of different radioactive nuclei with half-lives from fractions of a second to hundreds of thousands of years and beyond. Once produced, radioactive waste remains a persistent reality, necessitating long-term management based on its potential risks to humans and the environment. In this sense, radioactive waste aligns closely with the concept of the Anthropocene—a proposed geological epoch characterized by the significant influence of human activities on Planet Earth. Radioactive waste embodies humanity's ability to intervene deeply in natural processes through technology, creating substances in forms and concentrations previously nonexistent in nature (Richardson et al. 2023). It serves as a representative example of "technofossils," artificially created materials that characterize the Anthropocene (Dibley 2018) or can be classified as "novel entity" with regard to the concept of planetary boundaries (Böse et al. 2024).

This chapter provides an overview of established practices and the international state of nuclear waste disposal. It also addresses the challenges associated with the

[2] See for details Frieß et al. (2021), Pistner et al. (2024), and Chap. 18 of this book.

[3] https://www.base.bund.de/en/nuclear-safety/nuclear-accidents/fukushima/discharge-radioactive-water.html.

disposal of radioactive waste, especially that stemming from commercial electricity generation from fission reactors. The focus will be on HLW, as these materials pose the greatest challenges and long-term risks to humans and the environment due to their high levels of radioactivity and heat generation. HLW incorporates spent nuclear fuel (SNF) that contains fissile material that could be used for nuclear weapons, raising concerns about proliferation risks (Marka 1993). By compiling data on the global disposal situation, it becomes clear that the claim that HLW management is a "solved problem" is not tenable. However, this very assertion plays a crucial role in international debates on climate protection and future energy scenarios, as it is used as evidence that nuclear energy is a sustainable form of energy production. A prominent example of this discussion was the EU taxonomy legislation, especially the interpretation of the so-called "do no significant harm" criteria (see JRC (2021) and BASE 2021).

The chapter closes with an exploration of Germany's history of nuclear waste disposal, an example of a country that, despite a long history of nuclear energy use and ample financial as well as scientific resources, has not yet satisfactorily resolved the issue of radioactive waste disposal.[4]

16.2 Forms of Radioactive Waste and Their Treatment

While radioactivity itself can be precisely defined in physical terms, what qualifies as (radioactive) "waste" depends on subjective categories, societal conventions, and legal regulations. Consequently, the internationally accepted definition of radioactive waste is abstract. The International Atomic Energy Agency (IAEA) (2019, 186) describes it as material "for which no further use is foreseen that contains, or is contaminated with, radionuclides at activity concentrations greater than clearance levels as established by the regulatory body." This variability complicates universal statements about the management of radioactive waste.

The differentiation of radioactive waste types is not internationally uniform. However, the IAEA (2019) categorization provides a useful framework:

- Low-level waste (LLW) is waste that exceeds clearance levels for conventional disposal because it contains limited amounts of long-lived radionuclides. It requires robust isolation and containment, typically for up to a few hundred years. This category covers a broad range of waste, including contaminated protective clothing, tools, filter materials, and concrete. Shielding requirements for such waste are generally minimal. It is often stored in cement containers or steel plate drums.

[4] Further useful references include the World Nuclear Waste Report (2019), the IAEA Publication "Status and Trends in Spent Fuel and Radioactive Waste Management", and the European Commission's "Questions and Answers on the EU Taxonomy Complementary Climate Delegated Act covering certain nuclear and gas activities".

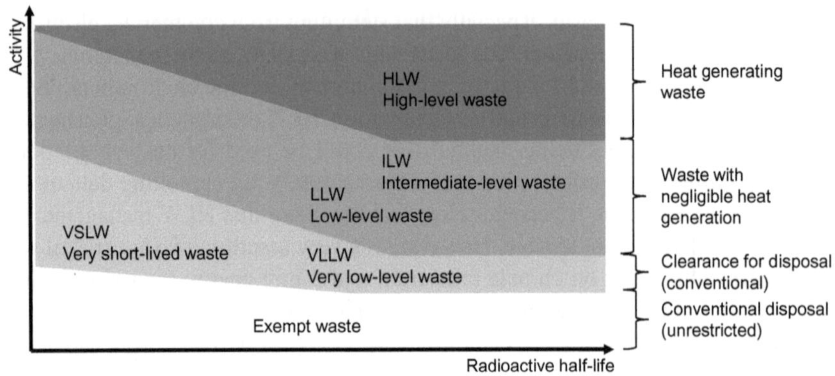

Fig. 16.1 Simplified waste classification and disposal routes. *Source* Taken from Bärenbold et al. (2024, Fig. 6)

- Intermediate-level waste (ILW) has higher levels of long-lived radionuclides that require additional containment and isolation compared to LLW. ILW may contain alpha-emitting radionuclides and typically needs no or limited heat dissipation measures during storage or disposal. Examples include ion exchange resins, activated products, and dismantled reactor components.
- High-level waste (HLW) comprises activity concentrations high enough to generate significant heat through radioactive decay and typically contains large amounts of long-lived radionuclides. This category includes SNF from commercial, research, and military reactors, as well as waste from reprocessing activities.

There are also other categories of waste, like very short-lived waste (VSLW, for example, medical isotopes, which can be cleared from regulatory control after a few years), very low-level waste (VLLW, suitable for disposal in near-surface landfill-type facilities), and exempt waste (EW, waste that meets the criteria for clearance of radiation protection). Figure 16.1 shows the different waste types and their properties regarding radioactivity and radioactive half-life.

Table 16.1 provides an overview of the world's inventory of radioactive waste as of 2016, excluding liquid and military waste. The disposal of highly radioactive waste has not happened so far. Despite having relatively low volumes compared to other waste types, HLW from nuclear energy production for civilian purposes represents by far the largest source of radioactive waste in terms of activity (IAEA 2022).

16.2.1 Disposal of Low- and Intermediate-Level Radioactive Waste

The disposal strategies for LLW and ILW vary depending on the radiation intensity, chemical composition, and half-life of the radionuclides they contain. The half-lives

Table 16.1 Status of the global solid radioactive waste inventory (as of 2016, military waste not included)

Waste Type	Storage	Disposal	Total
VLLW (m^3)	2,918,000	11,842,000	14,760,000
LLW (m^3)	1,471,000	18,499,000	19,970,000
ILW (m^3)	2,739,000	133,000	2,872,000
HLW (m^3)	29,000	0	29,000
	Storage	Reprocessed	Total
SNF from NPP (tHM)	263,000	127,000	390,000

Source Own compilation with data taken from IAEA (2022, 55)

of radionuclides of these waste types range from a few years to several thousand years. In general, the disposal process for LLW and ILW involves the following steps (compare, for example, IAEA 2010):

- Collection and sorting: LLW/ILW is collected at the point of origin, classified, and sorted according to its activity and chemical and physical properties. The goal is to separate waste that can be recycled or decontaminated from waste that might require long-term disposal.
- Conditioning: LLW/ILW for which long-term storage has been deemed necessary is treated to enhance its stability and reduce its volume before final disposal. Common procedures include solidification, compaction, and embedding in matrix materials such as concrete, bitumen, or glass. These methods ensure radionuclides are permanently fixed to prevent their spread.
- Interim storage: The waste is stored in interim facilities. These facilities shield the environment from radiation ("safety") and, in turn, protect waste forms from outside influences ("security") until final repositories become available. Interim storage facilities are typically designed to last a few decades.
- Final disposal: LILW, can be disposed of in near-surface underground repositories if it consists mostly of short-lived waste with half-lives of a few decades. Deep geological repositories are used for longer-lived LLW/ILW, especially when secure isolation over centuries is necessary. Some countries, such as Germany, plan to place all LLW/ILW in deep geological repositories.

Disposal of waste from naturally occurring radioactive materials (NORM) is based on their specific characteristics and associated radiation risks. NORM waste arises when materials from the Earth's crust contain radioactive isotopes such as radium-226, radon-222, thorium-232, and uranium-238, or when industrial processes concentrate these isotopes. Such waste is generated in various industries, including oil and gas production, mining, metal processing, and the use of phosphate fertilizers. Although NORM waste exhibits much lower radiation levels than waste from fission reactors (typically classified as LLW or VLLW), it still requires management to protect the environment and human health (IAEA 2013).

Storage and disposal facilities must ensure the security of the waste. While LLW and ILW are not of interest for constructing nuclear weapons due to the non-existent or very limited fissile material content, some of the materials could be used to build so-called "dirty bombs" by combining conventional explosives and radioactive materials, for example, discarded radioactive sources (Levi and Kelly 2002). These bombs would disperse the radioactive material into the environment, contaminating an extended area and thus creating radiological exposure for the affected population, as well as psychological uncertainty or panic.

16.2.2 Disposal of High-Level Radioactive Waste

The safe and secure disposal of HLW represents the greatest challenge of nuclear waste management. While alternative disposal methods (e.g., disposal in outer space, deep-sea burial, or disposal in the "eternal ice" of Antarctica) have been considered since the 1950s (National Research Council 1957), they have been consistently dismissed due to comparatively higher safety risks than disposal in deep geological formations. Unlike earlier hopes of a fully closed fuel cycle, it has become evident that SNF cannot be fully recycled. Reprocessing methods, as practiced in countries like France and Russia, can recover certain materials (e.g., uranium or plutonium) but still leave behind significant quantities of high-level radioactive waste requiring final disposal. Thus, a closed fuel cycle remains a theoretical possibility but is not feasible in the foreseeable future (Frieß et al. 2021).

The internationally recognized standard for HLW disposal is to permanently isolate this waste in deep geological formations. This method has gained acceptance due to its scientific and technical advantages; but it remains a topic of intense debate: There are states that pursue a de facto "wait and see" approach (IAEA 2009), opting for long-term interim storage to potentially benefit from future technological advancements in waste treatment. These speculative solutions can be transmutation technologies and/or establishing a plutonium economy using breeder reactors.[5] It is sometimes even argued from industry representatives that, in such cases, final disposal relying on long-term deep geological containment might no longer be necessary—a claim which is not valid according to independent analysis (e.g., Englert et al. 2024).

Deep geological repositories, at depths of several hundred meters, should offer a long-term safe solution based on two main arguments:

- Predictability of geological structures: With current scientific knowledge, the behavior of well-studied geological formations, such as salt domes, clay, or granite rock, is predictable over timeframes of hundreds of thousands to millions of years. These formations, often combined with engineered barriers, can safely contain wastes for very long periods of many millennia and isolate them from groundwater.

[5] https://world-nuclear.org/information-library/nuclear-fuel-cycle/fuel-recycling/processing-of-used-nuclear-fuel.

- Technical feasibility with acceptable risks: The construction and operation of deep geological repositories are considered technically feasible. Risks associated with such facilities are manageable when strict safety standards are adhered to. Multilayered barriers—comprising both engineered components (containers) and natural geological protections—minimize the risk of radioactive substance release.

Until a final repository becomes available, HLW is stored in spent nuclear fuel pools at the reactor site or specialized interim facilities. These facilities use either dry storage (waste is sealed in containers placed in ventilated halls or concrete casks/chambers to allow heat dissipation) or wet storage (waste is submerged in deep pools filled with water, which serves to shield radiation and remove heat). While wet storage facilities enable easier accessibility to the fuel, dry storage facilities provide enhanced safety because they rely on robust casks and passive cooling mechanisms without risk of coolant loss.

SNF contains fissile material, especially plutonium-239, which can be reprocessed and repurposed for military applications. Although plutonium extracted from civilian SNF is not ideal for fission bombs (due to contamination with other plutonium isotopes), it still represents a fundamental risk (Mark 1993).

For decades, the disposal of radioactive waste, especially HLW, has played a critical role in discussions about the risks and benefits of nuclear energy. The complexity of safe disposal has been historically underestimated, as evidenced by the multi-decade delays in establishing facilities like deep geological repositories in many countries. Despite significant progress in a few nations, the situation remains largely unresolved globally.

Adding to this challenge is the potential future implementation of "alternative" reactor concepts—such as molten salt reactors and sodium-cooled fast reactors—that could further qualitatively complicate the waste disposal issue. Krall and Macfarlane (2018) point out that the chemical composition of radioactive waste produced in such reactors differs significantly from the uranium oxide-based fuel used in currently widely deployed light water reactors, introducing new requirements for deep geological disposal.

16.3 International Experiences in HLW Management

There is no comprehensive and reliable global inventory of radioactive waste. Although the IAEA works to publish aggregated reports from member states, they are neither up to date nor complete. The IAEA has provided the "Spent Fuel and Radioactive Waste Information System" to collect data under the umbrella of the Joint Convention on the Safety of Spent Fuel Management and on the Safety of Radioactive

Waste Management.[6] Still, the data is incomplete and partially outdated. Nonetheless, a rough estimate of the scale of accumulated high-level radioactive waste is possible. Between 1954 and 2016, nuclear power plants (NPPs) worldwide generated approximately 390,000 tons of heavy metal (tHM) of SNF (IAEA 2022). About one-third has been reprocessed, while the rest remains in interim storage. Of the latter, 70% is stored in wet storage facilities.

16.3.1 International Regulation

As the use of nuclear energy expanded, the issue of managing HLW increasingly became the subject of international regulatory regimes. A multilayered system of conventions, agreements, and standards has been established to create globally uniform minimum safety standards. At their core, these regulations emphasize national responsibility for the safe disposal of domestically generated radioactive waste.

A central element of this framework is the "Joint Convention on the Safety of Spent Fuel Management and on the Safety of Radioactive Waste Management" of 1997 (IAEA 1997). This agreement sets international standards for safely handling, storing, and disposing of radioactive materials. Member states commit to developing and implementing national waste management programs. Regular review conferences foster transparency and the exchange of experiences among states.

The London Convention from 1972, supplemented by the 1996 Protocol, prohibits the disposal of radioactive waste in the seas. This ban was a response to the uncontrolled dumping of nuclear waste in the oceans during the 1950s and 1960s. Additionally, the United Nations Convention on the Law of the Sea (UNCLOS) from 1982 regulates the general protection of the marine environment, ensuring that activities with potentially harmful effects on marine ecosystems are monitored. Although not directly aimed at waste disposal, other international agreements, such as the Comprehensive Nuclear-Test-Ban Treaty (CTBT), indirectly contribute to reducing radioactive releases into the environment by prohibiting nuclear tests. The Bamako Convention from 1998, a regional treaty, bans the import of hazardous and radioactive waste to African countries to prevent illegal dumping.

At the European level, nuclear waste management is supplemented by a binding directive for all European Union (EU) member states. The Directive 2011/70/Euratom establishes a legal framework for safely disposing of radioactive waste and SNF. Its main goal is to create uniform safety standards within the EU. The directive requires each member state to develop a national program covering the entire waste management chain from waste generation to final disposal. These programs must define responsibilities, include timelines and technical concepts, and are subject to regular review by the European Commission. A key principle of the directive is that the responsibility for waste disposal lies fundamentally with the country where the

[6] https://sris.iaea.org/home.

waste was generated. The export of waste is not excluded but permitted only under strict conditions and must not undermine EU safety standards. Member states are encouraged to identify suitable sites for deep geological repositories for HLW and to promote societal acceptance of such projects through transparency and public participation. The directive also mandates the independence of the responsible regulatory authority from all other entities involved in generating and managing radioactive waste. Beyond the directive, additional efforts, such as linking financial incentives under the EU taxonomy to nuclear waste disposal requirements, aim to motivate member states to make progress in nuclear waste management (EC 2022). Whether these measures will have any practical effect remains to be seen.

Below the level of international conventions, the work of the IAEA and the OECD Nuclear Energy Agency (NEA) plays a key role in developing uniform minimum standards for the safe disposal of radioactive waste. The IAEA Safety Standards cover all aspects of waste management, including geological disposal, interim storage, and the decommissioning of nuclear facilities. IAEA and NEA promote knowledge sharing among member states through working groups, conferences, databases, and other measures. However, both organizations are mandated not only to ensure safe operations and waste disposal, but also to promote the civilian use of nuclear energy.[7] In practice, representatives from reactor developers and the nuclear industry are closely involved in their work and exert influence on the formulation of safety standards. As a result, these organizations do not follow the principle of independence to the same degree as required, for instance, by Directive 2011/70/Euratom at the European level. Consequently, the outcomes of safety-related work by the IAEA and NEA risk being biased toward the promotion of nuclear energy, especially compared to regulatory bodies that are functionally separated from nuclear operators.

Remarkably, the safe management of high-level radioactive waste holds little significance in international sustainability discussions. As Böse et al. (2024) demonstrate, while the safe disposal of radioactive waste was included in earlier frameworks like the Agenda 21 of 1992, it was omitted in the later formulation of the United Nations' Sustainable Development Goals.

16.3.2 Nuclear Waste Management in Nuclear Energy-Producing Countries

Although the civilian use of nuclear energy dates back to the early 1950s, at the time of writing in February 2025, no fully operational repository for high-level radioactive waste has been established anywhere in the world.[8] Table 16.2 (in the Annex of this

[7] Refer to the Statute of the International Atomic Energy Agency. 26/10/1956. UNTS 276 (p. 3). Entry into Force 29 July 1957 and that of the Nuclear Energy Agency. 1995. "NEA Statute". Paris: OECD Publishing.

[8] "Fully operational" is understood as the beginning of disposition of HLW in a completely licensed facility.

chapter) provides an overview of the waste management situation in all countries that produce nuclear energy for commercial electricity, including countries that have closed all of their commercial power reactors (Germany, Italy, and Lithuania). Information about the status of interim and final storage of HLW is compiled from diverse sources, including national waste management programs, reports to international conventions, scientific publications, and news items.

16.3.3 Expected Milestones for HLW Repositories: 2025–2050

Our compilation provides a broad overview and does not account for the historical, social, and technical complexities characterizing nuclear waste management in each national context. Yet, it allows for a cross-country comparison, based on which several general observations about the global state of nuclear waste management can be made.

First, a closer look at the situation in individual countries reveals significant differences regarding the plans for and progress toward final repositories (Figs. 16.2 and 16.3). A few countries are relatively advanced in the planning and construction of deep geological repositories. Finland has identified a site, approved a repository, and constructed an underground facility. Currently, it is conducting test runs, and the first waste is expected to be disposed of in the late 2020s. Sweden has identified a site and recently begun construction, which is expected to be completed in a decade. Operator companies in Canada and Switzerland have recently proposed specific sites following extensive selection procedures. The outcome of the licensing processes, however, is still open, and the actual start of disposal operations is decades away.

Another group of countries has identified potential sites for further investigation, but is also estimated to be several decades away from initiating disposal operations. These include China, France, and Russia, all of which have established underground laboratories at their designated repository sites. In Belgium, Germany, Japan, and the United Kingdom, processes to identify repository sites for HLW are underway but still in the early stages.

16.3.4 Expected Milestones for HLW Repositories: 2050–2150

Even more states aspire to construct deep geological repositories but have yet to develop or initiate a site selection procedure. Once again, the reasons are varied. As previously mentioned, some countries deliberately opt for a wait-and-see approach, hoping that technological progress might bring a solution for waste disposal that would make disposal in deep geological repositories redundant. In the Netherlands,

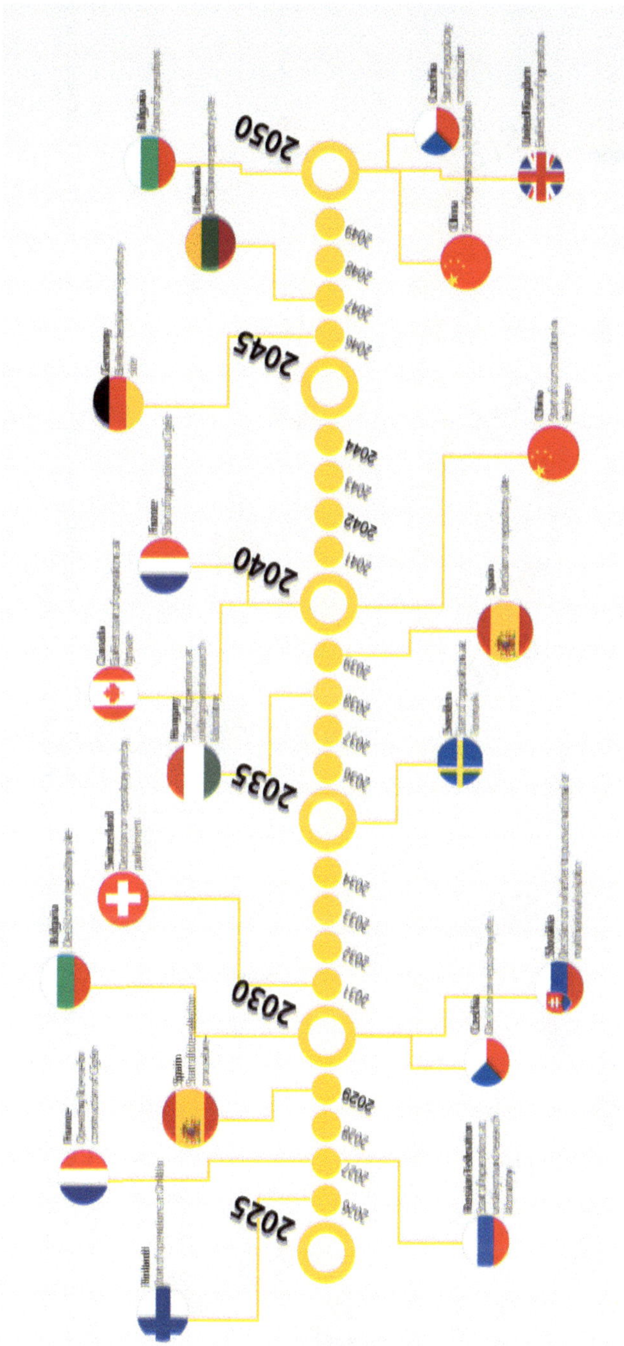

Fig. 16.2 Expected milestones for HLW repositories across the world (2025–2050). (*Source* Official national documents [see Annex 16.2]. Indicated milestones reflect rough estimates communicated by official national bodies and are subject to change)

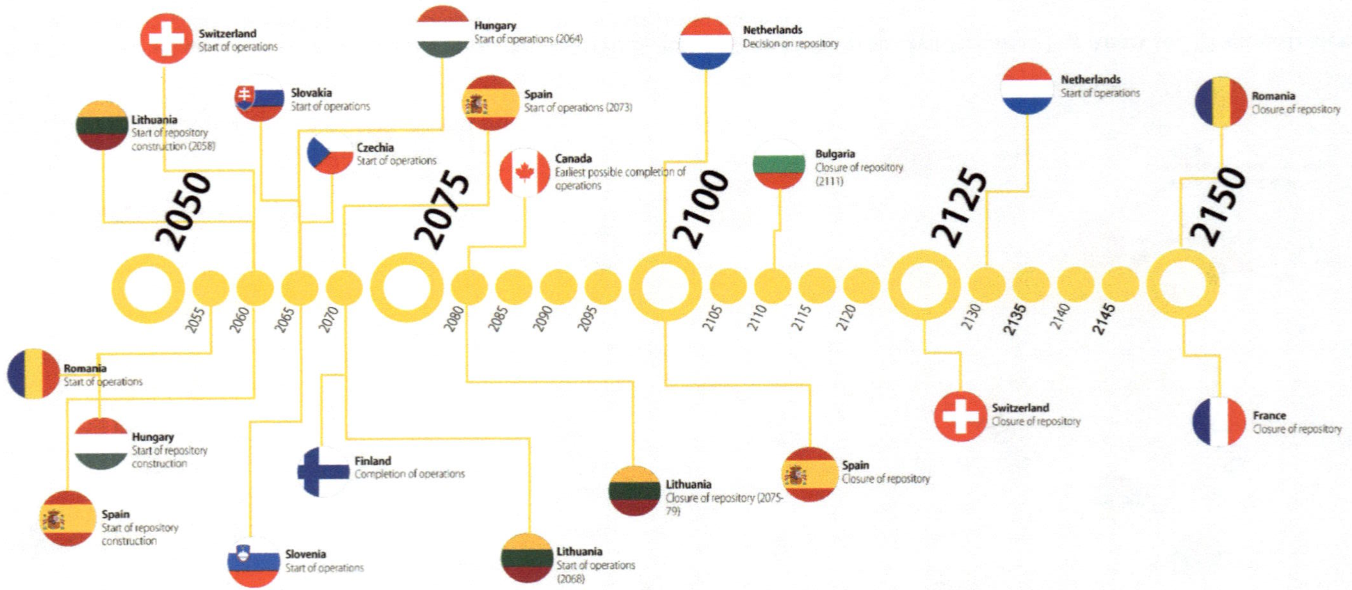

Fig. 16.3 Expected milestones for HLW repositories across the world (2050–2150) (*Source* Official national documents [see Annex 16.2]. Indicated milestones reflect rough estimates communicated by official national bodies and are subject to change)

for instance, a decision about final disposal is envisioned to be taken in 2100. Furthermore, some states possessing only a small nuclear inventory, such as Slovakia or Slovenia, are pursuing dual-track policies; in other words, they still must decide whether to dispose of HLW in a national repository or whether to establish a multinational repository in cooperation with other states. In the United States, the country with the largest inventory of spent nuclear fuel, Yucca Mountain in Nevada was selected as the only potential repository site as early as 1987. Yet, a perceived lack of scientific consideration behind the selection of Yucca Mountain and resulting public and scientific opposition have brought the site's evaluation for final disposal to an indefinite halt. Therefore, the future road to a final repository in the US remains unclear. In the last few years, efforts have concentrated on establishing centralized long-term interim storage facilities.

A second observation is that among countries that have initiated a site selection procedure, processes vary significantly in their procedural steps, objectives, and governance structures. A key distinguishing feature of site selection approaches is the question of voluntariness: In consent-based approaches, a site is chosen based on the active approval of the population or municipality that would be directly affected in the selected region. By contrast, in comparative approaches, the site selection is guided by the principle of maximizing safety and minimizing risk. As the German example discussed below demonstrates, the public can actively participate in the site selection. However, the eventual proposal for a site is still based on explicit criteria determining what site would offer the best possible safety.

Third, comparing countries' progress toward deep geological repositories demonstrates that, without exception, a significant time lag exists between the beginning of HLW production and the yet-to-be-completed development of operational and safe final repository solutions. Even in Finland, which is expected to be the first country to dispose of commercial HLW in a deep geological repository, half a century will have passed until the first HLW is finally disposed of. However, Finland marks an exception rather than the rule. The time gap for the two largest nuclear energy-producing countries, the United States and France, is already nearing 70 years. Additional decades will pass before a repository becomes operational, is back-filled, and eventually sealed. Only after the successful final closure of the repository can waste disposal truly be considered complete. In total, the time required for the entire process ranges between one and two centuries. These time horizons, however, constitute a safety risk in their own right. In most countries, HLW is stored above the surface, either in spent fuel pools or overground interim storage facilities, which are often neither designed nor licensed to operate beyond a couple of decades. So far, the international regulations outlined in Sect. 3.1 have failed to urge countries to provide operational and safe national infrastructures for the deep geological disposal of HLW.

Overall, as nuclear inventories are growing, interim storage facilities are filling up, and operational licenses for interim storage are being consistently extended, a fully operational deep geological repository does not yet exist anywhere in the world. Regardless, actors from government, the private sector, and the scientific community continue to portray the safe disposal of HLW for all future generations as a solved issue, downplaying the remaining challenges. Moreover, proponents of nuclear energy point to the existing—still empirically unproven—repository projects to argue that the production of electricity from fission reactors is sustainable. These proponents include international organizations such as the IAEA[9] and OECD-NEA,[10] the Joint Research Center of the European Union (JRC 2021), and the World Nuclear Association.[11]

Against the backdrop of this global situation—the discrepancy between the production of highly hazardous waste and the actual availability of a long-term, secure disposal solution—it can be stated that nuclear energy has violated and continues to violate the precautionary principle. This principle stipulates that technologies should not be used if their impacts could endanger the public or the environment and if there is no scientific consensus on the issue.[12] Even if one disregards the inevitably remaining residual risks of the few already licensed HLW repositories, it is certainly not possible to speak of a scientific consensus regarding non-existent disposal facilities.

The preceding analysis also raises questions about the reasons behind the temporal divergence between waste production and disposal. Even though recurring patterns and interests can be identified, an in-depth, country-specific comparative analysis would far exceed the scope of this chapter. Instead, we will exemplarily focus on the disposal of HLW in Germany.

16.4 Nuclear Waste Disposal in Germany

Germany[13] is an interesting case because it could have become a role model for safe and effective nuclear waste disposal given its starting conditions: an early entry into (commercial) nuclear energy, a highly developed industrial economy, high state capacity, a strong scientific system, stable political conditions, and diverse and favorable geological conditions for a deep geological disposal facility (GDF). While the country was internationally leading with its nuclear waste disposal program in the 1970s (Möller 2009), it is now lagging behind several of its European neighbors after multiple setbacks.

---

[9] https://www.iaea.org/newscenter/news/finlands-spent-fuel-repository-A-game-changer-for-the-nuclear-industry-director-general-grossi-says.

[10] https://www.oecd-nea.org/jcms/pl_97502/moving-towards-the-construction-of-A-safe-deep-geological-repository.

[11] https://world-nuclear.org/information-library/nuclear-fuel-cycle/nuclear-waste/radioactive-wastes-myths-and-realities.

[12] https://eur-lex.europa.eu/legal-content/EN/TXT/?uri=LEGISSUM:precautionary_principle.

[13] This chapter focusses on the history of the Federal Republic of Germany, and excludes the activities of the German Democratic Republic.

16.4.1 Historical Nuclear Energy Production and Government Measures to Address the Problem of HLW

The history of nuclear waste disposal in Germany is closely tied to the development of nuclear energy and the associated political, economic, and societal challenges. The civilian use of nuclear energy in West Germany began in 1955, following the lifting of the post-World War II ban on nuclear research. From then on, Chancellor Konrad Adenauer's government actively promoted nuclear energy to ensure that the expected growth of Germany's electricity demand could be satisfied (Armand et al. 1957). Research and development were state-directed, and the first research reactors were commissioned in the late 1950s. The commercial use of nuclear energy began with the Kahl prototype reactor in 1961 (Müller 1990), marking the start of large-scale production of HLW.

By the 1960s, the necessity of final disposal of radioactive waste had been recognized, but concrete solutions were not available at that time.[14] Between 1957 and 1963, Germany waited for results on the suitability of salt domes in the USA for the disposal of radioactive waste. Germany had a long tradition of salt mining, and many salt domes were available in the northern part of the country, so it was considered a possible host rock early on. In 1965, the about-to-be-abandoned Asse II salt mine was acquired by the Federal Government with the intent of turning it into a repository for radioactive waste, including HLW. What seems remarkable from today's perspective is that geological disposal was considered an economical alternative to storage in constructed aboveground structures—the low acquisition cost of Asse II was perceived not to endanger the economic prospects of nuclear power.

Problems with the stability of the mine and water ingress were recognized before the acquisition, but opinions within government institutions differed about the severity of these issues. Asse II was first used as an experimental facility for geological disposal of small amounts of radioactive waste, and experiments with HLW were also envisioned. Subsequently, the usage of the Asse mine increasingly shifted from a research facility to a disposal facility for significant amounts of LLW and ILW. Asse II never underwent a licensing procedure as required for a nuclear repository by the 1976 amendment of the Atomic Energy Act ("Atomgesetz"). The deposition of waste stopped when this law came into force. Since research activities continued until 1992, it was officially referred to in public as a "research mine" up until 2008. Over time, significant issues arose, including water ingress that threatened the safety of the stored waste. Public criticism of Asse undermined trust in German waste disposal policies. Despite early intentions and preparatory work, HLW was never placed into the Asse II mine. Instead, a reorientation of the waste management strategy began in the 1970s; this introduced the concept of the so-called integrated disposal center (Möller 2009).

---

[14] The following description of the history of nuclear waste management in Germany between the 1960s to 1970s is largely based on the work by Möller (2009, Chaps. 2–4).

In the 1960s, LLW and ILW from research institutions and hospitals, exacerbated by the commissioning of a nuclear reprocessing facility in Karlsruhe in 1971, were Germany's primary waste management challenge. While government institutions were aware of the necessity for and the challenges of HLW disposal, it did not yet pose an immediate problem: Due to significant heat generation, spent fuel must spend several years in wet storage, usually in a pool right next to the reactor, before it can be handled. Reprocessing was preferred in the early years as part of the so-called "closed fuel cycle," where fissile materials like uranium-235 and plutonium were recovered from spent fuel for reuse. The remaining radioisotopes would require disposal. However, this approach faced growing criticism due to its complexity, high cost, and the generation of significant amounts of LILW and radioactive emissions. German utilities sent SNF to reprocessing facilities abroad in La Hague in France and Sellafield in the United Kingdom, but were also obliged to take back remaining HLW.

With the expansion of nuclear power plant capacities in the 1970s, the volume of HLW grew significantly. In 1977, the concept of a German integrated waste management center was announced, to be located in Gorleben near the border to the German Democratic Republic (Möller 2009). This facility was intended to include a reprocessing plant, a GDF, and a fuel fabrication plant. However, the selection of the Gorleben site lacked transparent scientific criteria and triggered widespread protests. Plans for a reprocessing plant were abandoned, leaving Gorleben as a site for interim storage and a potential final repository. An underground research laboratory in Gorleben was built in the early 1980s to gather data for the license application for a later GDF.

The time lag, described in Sect. 16.3, between the commercial utilization of nuclear power and the provision of operational options for the safe disposal of nuclear waste was already an important topic in Germany back in the 1970s. The planned GDF in Gorleben was intertwined with a legal instrument introduced in the 1976 amendment to the Atomic Energy Act: the "Proof of Provision for Waste Disposal" (German: "Entsorgungsvorsorgenachweis"). This measure was designed to link nuclear waste production in power plants to a solution for its disposal by requiring operators to lay out "realistic plans" of how they would safely manage and dispose of their spent fuel and waste. This proof had to be furnished in the licensing processes for new nuclear power plants and, in the absence of an operational GDF, had to be continually updated[15] with evidence of "progress made in realization" of a disposal solution.[16] The goal was to ensure that all parts of the integrated disposal concepts interlock[17] and to legally bind operators to the long-term consequences of nuclear energy use. While the legal requirement was a step forward, it was a formal, vaguely worded hurdle that did not lead to actual disposal solutions—the Proof of

[15] During operation of a reactor, updates were required every three years.

[16] Bekanntmachung der Grundsätze zur Entsorgungsvorsorge für Kernkraftwerke, 1980, Bundesanzeiger Nr. 58.

[17] Bundestags-Drucksache 7/4794, S. 7, Entwurf eines Vierten Gesetzes zur Änderung des Atomgesetzes (AtG), 24.02.1976.

Provision for Waste Disposal never had the status of a strict legal requirement.[18] Initially, operators were required to reprocess spent fuel or outline plans for its direct disposal, if reprocessing was not possible for technical reasons or economically justifiable. However, in practice, compliance with the Waste Management Proof of Provision did not deliver a breakthrough in waste management. Instead, sending SNF abroad to reprocessing facilities, as required until 1994, merely deferred the issue of domestic final disposal until the resulting HLW had to be repatriated.

The legal framework for nuclear disposal evolved over the following decades.[19] The 1994 amendment of the Atomic Energy Act equally allowed direct disposal and reprocessing as waste management strategies, allowing operators to focus on final disposal without the need for reprocessing. The 2002 amendment went further, banning reprocessing and requiring operators to store spent fuel in interim storage facilities before eventually disposing of it in a GDF.[20] This increasing commitment to direct disposal focused attention on advancing the Gorleben project—since the commissioning of new nuclear power plants was linked to showing "progress made in realization" of a GDF in the sense of the Proof of Provision for Waste Disposal. This link also involved nuclear waste disposal in a broader societal conflict over the use of nuclear power in Germany, linking a rejection of the use of nuclear power with a rejection of Gorleben, further eroding the public acceptance of the site (Syrovatka 2016).

Overall, the Proof of Provision for Waste Disposal did not lead to the preparation of a timely GDF solution for HLW. The framework gave the impression that disposal was—if not resolved—at least addressed, even though the actual implementation of a secure repository lagged far behind. This approach led to delays, increased public skepticism, and political disputes, which eroded confidence in nuclear waste management. Today, the Proof of Provision for Waste Disposal can be seen as an

---

[18] Hofmann, Hasso. 1991. "Zum Stand des Rechts der atomaren Entsorgung." In: Wissenschaftskolleg: "Jahrbuch 1989/90." Berlin: Nicolaische Verlagsbuchhandlung Beuermann GmbH.

[19] In the course of reunification in 1990, the FRG also inherited GDR radioactive waste in various forms: While spent fuel from GDR power reactors was returned to the UdSSR until the middle of the 1980s (Bundestags-Drucksache 11/7661, S. 2, Lagerung radioaktiver Abfälle der DDR und Konsequenzen aus den atomrechtlichen Bestimmungen des Umweltrahmengesetzes, 20.08.1990), about 580 tHM of spent fuel in wet storage and reactors (EWN 2020; BMUV 2024) became the responsibility of reunified Germany along with LLW/ILW from subsequent decommissioning of East German NPPs. The management of some small-volume "special" inventories e.g., from Rossendorf Research Reactor (https://www.vkta.de/erinnerungsschrift-zur-uebergabe-der-rossendorfer-brennelemente-an-den-bund/) or an unshielded plutonium-berrylium source in Berlin (https://www.base.bund.de/shareddocs/kurzmeldungen/de/2018/3101-strahlenquelle.html) have been active topics up until the present day. Conversely, the Morsleben LLW/ILW repository, whose operating license was grandfathered in under the German Reunification Treaty, has been used for the disposal of waste from Western Germany until 1998 (https://www.einblicke.de/was-war-und-was-ist-daten-und-fakten-uber-das-endlager-morsleben). Overall, legacy HLW from the GDR (with its low total volume and activity in comparison to HLW from Western Germany) has been integrated into German waste management without influencing it decisively at the national level.

[20] Hennenhöfer, Gerald. 2021. "AtG § 9 Rn. 5." In: Hennenhöfer, Gerald, Thomas Mann, Norbert Pelzer and Dieter Sellner (Hrsg.): "Atomgesetz, Mit Pariser Atomhaftungs-Übereinkommen: Kommentar." München: C.H. Beck.

example of the challenges in addressing complex technical issues through regulatory measures with presumably good intentions, but without sufficient practical solutions. It illustrates how political and legal frameworks were used to buy time without fundamentally resolving the central issue of HLW disposal.

16.4.2 Nuclear Phase-Out, Current Situation, and Outlook

The election of a new government in 1998 marked a turning point for Germany's nuclear energy policy and its HLW management. The 2002 Nuclear Phase-Out Act initiated a gradual exit from nuclear energy and reorganized the search for a GDF site. The exploration of Gorleben was paused, and the Committee on Site Selection Procedures for Repository Sites (German: "Arbeitskreis Auswahlverfahren Endlagerstandorte (AkEnd)") developed criteria for a new, transparent site selection process. Despite this, little progress was made until 2010, as a lack of political will and consensus continued to impede the search (Brunnengräber 2016).

Shortly after implementing a lifetime extension policy for NPPs in 2010, the Fukushima reactor disaster in 2011 brought about a definitive decision on the Nuclear Phase-Out in Germany carried by a broad political consensus. This decision also broke the political gridlock regarding HLW disposal: In 2013, the Site Selection Act was passed by the Federal Parliament (Bundestag), terminating the Gorleben exploration and initiating a restart of the search for a GDF site (Sellner et al. 2021).

The new process aims to identify the site with the best possible safety for a GDF that can safely isolate HLW from the biosphere for at least one million years. Transparency, scientific selection criteria, and public participation are central to this process. However, a critical weakness of this ambitious site selection process was recently uncovered: If carried out as stipulated by law, the selection and decision making process will take at least until 2046–2068 (according to the operator company, BGE 2022) and the 2070s (according to a study commissioned by the supervisory authority, Krohn et al. 2024). This timeframe is significantly longer than envisioned by the Site Selection Act, which seeks a siting decision by the federal parliament in 2031. The decision would be followed by an approval process and a construction phase, with the commissioning of the GDF perhaps not to be expected before the end of the twenty-first century, with closure likely occurring in the mid-twenty-second century. As a result, the supervisory authority has called for an evaluation and acceleration of the site selection process.[21] Whether and how this will be achieved is subject to decisions made at the political level.

[21] https://www.base.bund.de/shareddocs/interviews/de/2023-24-07-zeitindex-koenig.html.

To conclude, Germany had good prerequisites for the safe and rapid construction of a GDF for HLW. In addition, the problem was recognized at an early stage and addressed in regulatory terms. Nevertheless, the time lag between HLW production and the availability of a safe disposal situation is not only growing, but also appears very uncertain at present. Rather than a single reason, a mixture of reasons is responsible, among them the underestimation of the technical and social complexity of the task, economic interests, and non-transparent decision-making processes.

16.5 Conclusion

After nearly seven decades of using nuclear fission to generate (mainly) electricity, the global situation regarding the disposal of HLW remains problematic. While it is encouraging that a few countries, such as Finland and Sweden, have completed approval processes and are coming close to commissioning operational geological repositories, these cases are exceptions. They represent only a fraction of the global high-level radioactive waste inventory, leaving the vast majority of accumulated waste without a clear disposal solution.

International conventions and organizations, such as the Joint Convention on the Safety of Spent Fuel Management, have laid the groundwork for addressing this issue on an intergovernmental level. Further Initiatives like the London Convention, the Bamako Convention, and European regulations aim to establish global or regional safety standards. However, the past decades have revealed a stark disconnect between announcements and meaningful progress. Safe implementation of reliable disposal infrastructure has failed due to political conflicts, technical obstacles, societal resistance, and the sheer complexity of the challenge.

The history of German HLW disposal efforts amplifies these challenges. Despite being an industrial and technological leader with advanced nuclear expertise, the country has yet to develop an operational repository for HLW.[22] The problem of HLW disposal was recognized early, and Germany even introduced legal instruments such as the Waste Management Proof of Precaution to ensure waste disposal was integrated into nuclear energy policy. However, these instruments were often circumvented or poorly enforced, resulting in delays and a reliance on temporary and insufficient measures like interim storage or overseas reprocessing. The controversial Gorleben site and the broader societal disputes surrounding nuclear energy and waste management further highlight the difficulties of balancing technical, political, and social considerations.

[22] For LLW/ILW, the Konrad repository is currently under construction. It was licensed in 2002 under nuclear law for a waste volume of 303,000 m^3. For additional waste, the possibility has been created to store it at the site of the future HLW repository (separately from the HLW) if there are no safety-related disadvantages.

Despite these issues, some international organizations, scientific institutions, and industry representatives continue to present the HLW disposal problem as "solved." This misrepresentation risks fostering complacency and delaying the critical infrastructure development needed to seriously and persistently address the issue. Adding to this is the lack of a comprehensive and up-to-date global inventory of radioactive waste. This deficiency complicates full assessments of the global scale of the problem and hinders effective international solutions and coordination.

A critical re-evaluation of current strategies is urgently needed. If the precautionary principle is taken seriously, an operational and safe disposal infrastructure must be established before further increases in radioactive waste production take place. While some countries have demonstrated progress, the timelines required to develop and implement geological repositories—spanning decades to over a century—underscore the enormity of the challenge. The fact that it often takes up to 100–200 years to fully finish HLW disposal highlights the inadequacy of current efforts to align waste management solutions with the ongoing production of radioactive materials.

To truly address this global challenge, a paradigm shift is required: from reactive, incremental measures to proactive and coordinated action. This means prioritizing waste disposal solutions not as an afterthought but as a fundamental prerequisite for the continued or expanded use of nuclear energy. Without such a commitment, the promise of nuclear energy as a sustainable and safe power source remains deeply compromised.

16.6 Annex 16.1: International Disposal Strategies

See Table 16.2.

Table 16.2 Status of HLW disposal in 34 nuclear energy-producing countries, sorted by installed nuclear net electrical capacity as of 2024

Country	Total nuclear net Electrical Capacity From in 2024 (MW)	Beginning of nuclear electricity production[23]	Status Interim Storage	Status Final Disposal	Host Rock
United States of America	96,952	1957	Spent fuel pools and dry cask storage; siting process for centralized interim storage facility ongoing	Evaluation of the site at Yucca Mountain halted, no ongoing site selection	n.a
France	61,370	1959	Waste is reprocessed, storage facilities at La Hague and Marcoule	Site selected: CIGEO in Bure; licensing procedure ongoing, expected for 2035–2040	Clay
China	54,152	1991	Spent fuel pools and dry cask storage; reprocessing planned	Underground research laboratory in Beishan, completion of the repository planned for 2050	Crystalline
Russia	26,802	1954	Spent fuel pools and dry cask storage at reactor sites; siting process for centralized interim storage facility ongoing	Site proposed in Yeniseisky, an underground research laboratory planned	Crystalline

(continued)

[23] First grid connection of a nuclear reactor according to IAEA PRIS database (https://pris.iaea.org/pris/home.aspx).

Table 16.2 (continued)

Country	Total nuclear net Electrical Capacity From in 2024 (MW)	Beginning of nuclear electricity production	Status Interim Storage	Status Final Disposal	Host Rock
South Korea	25,825	1977	Spent fuel pools at reactor sites; centralized interim storage planned	Site selection ongoing; plan to construct an underground research laboratory, followed by a permanent disposal facility; 37 years planned between initiation of site selection procedure and construction of permanent disposal facility	n.a
Canada	13,699	1962	Interim storage in wet or dry storage facilities	Selected Ignace (Wabigoon Lake Ojibway Nation) as site for repository; start of construction planned for 2033, operation in 2040s	Crystalline
Ukraine	13,107	1977	Spent fuel pools at reactor sites; centralized interim dry storage facility	Site selection ongoing in Chernobyl Exclusion Zone	Likely crystalline
Japan	12,631	1983	Spent fuel pool at reactor sites	Site selection ongoing	n.a
Spain	7123	1968	Spent fuel pool and dry cask storage at reactor sites; Storage in the Centralised Temporary Storage facility (CTS) planned	Start of site selection procedure in 2029, decision expected for 2039	Crystalline or clay

(continued)

Table 16.2 (continued)

Country	Total nuclear net Electrical Capacity From in 2024 (MW)	Beginning of nuclear electricity production	Status Interim Storage	Status Final Disposal	Host Rock
Sweden	6944	1964	Centralized spent fuel pool ("Clab ") at Oskarshamn NPP	Forsmark site selected and licensed, construction started, start of operation planned in the 2030s	Crystalline
India	6920	1969	Reprocessing of spent fuel; storage in the Solid Storage Surveillance Facility at Tarapur	Geological disposal facility envisioned, no information on specific project	n.a
United Kingdom	5883	1956	Storage at reactor sites and interim storage at Sellafield	Site selection ongoing, areas of focus identified; proposal for site planned in the 2030s, earliest operation in the 2050s	No explicit preference
United Arab Emirates	5321	2020	Spent fuel pool at reactor site, dry cask storage planned	unclear	n.a
Finland	4394	1977	On-site storage at NPPs	Trial runs for the repository at Onkalo started in 2024	Crystalline
Czechia	3934	1985	Dry storage at reactor sites	Site selection process since 1992, decision planned for 2030, start of operation in 2065	Crystalline

(continued)

Table 16.2 (continued)

Country	Total nuclear net Electrical Capacity From in 2024 (MW)	Beginning of nuclear electricity production	Status Interim Storage	Status Final Disposal	Host Rock
Belgium	3908	1962	Interim storage at reactor sites	Site selection process in preparation	Clay
Pakistan	3262	1971	Wet and dry storage at reactor sites	unclear	n.a
Switzerland	2973	1968	Centralised dry storage	Proposal for site "Nördlich Lägern," start of operation planned in the 2060s	Clay
Slovakia	2308	1972	Storage at reactor sites	Site selection process since 1998, site selection planned for 2030, operation planned for 2060s; considers participation in shared international repository project	n.a
Belarus	2220	2020	Wet storage at the reactor site, construction of a storage facility planned by 2030; reprocessing in Russia	Unclear	n.a
Bulgaria	2006	1974	Reprocessing in Russia halted; wet and dry storage at reactor site; plans to construct an interim storage facility at Kozloduy NPP	Draft strategy indicates planned construction of the repository by 2050	n.a

(continued)

Table 16.2 (continued)

Country	Total nuclear net Electrical Capacity From in 2024 (MW)	Beginning of nuclear electricity production	Status Interim Storage	Status Final Disposal	Host Rock
Hungary	1916	1982	On-site storage at Paks NPP	Planned construction of an underground research facility in the Boda region after 2032; construction start of the repository in 2055, commissioning in 2064	Clay is being investigated as a potential host rock
Brazil	1884	1982	Wet and dry storage on-site at Angra	No information on the specific project	n.a
South Africa	1854	1984	Wet and dry storage on-site at Koeberg; off-site Centralised Interim Storage Facility planned	No information on the specific project	n.a
Argentina	1641	1974	Wet and dry storage at reactor sites	Site selection process in planning	n.a
Mexico	1552	1989	Wet storage and Independent Spent Fuel Storage Installation (ISFSI) at Laguna Verde site	Deep geological repository favored, no information on specific project	n.a
Romania	1300	1996	Interim spent fuel storage facility at Cernavoda	Plans to commission a deep geological repository by 2055	n.a

(continued)

Table 16.2 (continued)

Country	Total nuclear net Electrical Capacity From in 2024 (MW)	Beginning of nuclear electricity production	Status Interim Storage	Status Final Disposal	Host Rock
Iran	915	2011	Spent fuel pools at reactor site; Joint Comprehensive Plan of Action intended all spent fuel to be shipped out of Iran, status unclear	Unclear	n.a
Slovenia	688	1981	Wet spent fuel pools and dry storage at the Krško plant	Dual track approach, decision for either national or multinational Deep geological repository; if national, start of Deep geological repository operation planned for 2065	n.a
Netherlands	482	1968	Centralised interim storage facility at Borselen, operational since 2003, and commissioned for 100 years	Decision on geological disposal planned for 2100, operational repository by 2130	Clay
Armenia	416	1976	Wet storage on-site and long-term interim dry storage at ANPP	No information on the specific project	n.a
Germany	n.a	1961	Interim storage at sites of former NPP and centralized storage facilities	Site selection procedure ongoing; decision expected between 2046 and 2070s	Clay, crystalline or salt rock

(continued)

Table 16.2 (continued)

Country	Total nuclear net Electrical Capacity From in 2024 (MW)	Beginning of nuclear electricity production	Status Interim Storage	Status Final Disposal	Host Rock
Italy	n.a	1963	Most spent fuel is sent abroad for reprocessing; wet storage at former NPP sites	Ongoing siting process for national repository	n.a
Lithuania	n.a	1983	On-site interim storage at Ignalina NPP	Decision for repository site expected in 2047, completion of construction and start of operations in 2068	Clay

Source Own compilation compiled from various sources, see Annex 16.2

16.7 Annex 16.2: Sources on International Disposal Strategies

United States

- National Academies of Sciences, Engineering, and Medicine. Merits and Viability of Different Nuclear Fuel Cycles and Technology Options and the Waste Aspects of Advanced Nuclear Reactors. Washington, DC: The National Academies Press. https://doi.org/https://doi.org/10.17226/26500. (2023)
- US Department of Energy. Consent-based siting for consolidated interim storage. https://eedgis.pnnl.gov/portal/apps/storymaps/stories/34462804fe664a5980e93fc4b6026f42 (2024)

France

- OECD-NEA, "Radioactive Waste Management Programmes: France" https://www.oecd-nea.org/rwm/profiles/France.pdf
- BASE, "Search for a repository in France: Application for construction licence submitted" https://www.base.bund.de/shareddocs/kurzmeldungen/en/2023/0118-repository-search-france.html (2023)

China

- World Nuclear Association, "China's Nuclear Fuel Cycle" https://world-nuclear.org/information-library/country-profiles/countries-a-f/china-nuclear-fuel-cycle (2024)
- IAEA, "China begins construction of its first underground research laboratory for high level waste disposal" https://www.iaea.org/newscenter/news/china-begins-construction-of-its-first-underground-research-laboratory-for-high-level-waste-disposal
- China Atomic Energy Authority, "Nuclear Fuel Cycle", https://www.caea.gov.cn/english/n6759365/n6759368/c6793479/content.html
- IAEA Bulletin, "Coping with Growth: China's Spent Fuel Management Strategy" (2019)

Russia

- 6th National Report of the Russian Federation on compliance with the obligations of the Joint Convention on the Safety of Spent Fuel Management and on the Safety of Radioactive Waste Management (2020)
- NWMO, "What other countries are doing" https://www.nwmo.ca/canadas-plan/what-other-countries-are-doing
- World Nuclear Association, "Russia's Nuclear Fuel Cycle", https://world-nuclear.org/information-library/country-profiles/countries-o-s/russia-nuclear-fuel-cycle#HLWdisposal

Republic of Korea

- KORAD https://www.korad.or.kr/resources/ebooks/brochure2023/korad_brochure_eng_2023.pdf (2023)

Canada

- Nuclear Waste Management Organization (NWMO), "How is it stored today?" https://www.nwmo.ca/Canadas-used-nuclear-fuel/How-is-it-stored-today
- NWMO, "The NWMO updates the timeline for site selection to 2024," https://www.nwmo.ca/news/the-nwmo-updates-the-timeline-for-site-selection-to-2024
- NWMO "The NWMO selected site for Canada's deep geological repository" https://www.nwmo.ca/news/the-nuclear-waste-management-organization-selects-site-for-canadas-deep-geological-repository

Ukraine

- Report on Nuclear and Radiation Safety in Ukraine for 2023
- ENERG, "Radioactive waste disposal in geological formations in Ukraine," https://energnet.eu/ukrainian-news/radioactive-waste-disposal-in-geological-formations-in-ukraine/#:~:text=According%20to%20the%20results%20of,deep%20geological%20formations%20is%20ongoing (2022)

Japan

- Nuclear Regulation Authority/NRA Japan, "Status of the spent fuel management in Japan," https://www.nra.go.jp/data/000473313.pdf (2024)
- Fujiyama & Kaku, „Current status of the geological disposal programme and an overview of the safety case at the pre-siting stage in Japan," Rock Mechanics Bulletin, Vol. 2 (3) https://www.sciencedirect.com/science/article/pii/S2773230423000355 (2023)

Spain

- IAEA SRIS, National Framework for the management of radioactive waste and spent fuel, https://sris.iaea.org/country-overview/introduction/ES/Spain

Sweden

- Sweden's Eighth National Report under the Joint Convention on the safety of spent fuel management and on the safety of radioactive waste management (2024)
- Power Magazine, "Sweden begins construction on world's second deep geological repository for spent nuclear fuel," https://www.powermag.com/sweden-begins-construction-on-worlds-second-deep-geological-repository-for-spent-nuclear-fuel/ (2025)

India

- Bhabha Atomic Research Centre, "Reprocessing and Nuclear Waste Management," https://www.barc.gov.in/randd/rwm.html
- Stimson Center, "Spent Nuclear Fuel in India," https://www.stimson.org/2020/spent-nuclear-fuel-india/

United Kingdom

- The United Kingdom's Ninth National Report on Compliance with the Convention on Nuclear Safety (2022)
- Nuclear Waste Services, "GDF progress as Areas of Focus Identified," https://www.nuclearwasteservices.uk/2025/01/gdf-progress-as-areas-of-focus-identified/ (2025)

United Arab Emirates

- Spence & Ahmad, Risks to Persian Gulf Cities from Spent Fuel Fires at the Barakah and Bushehr Nuclear Power Plants. Science & Global Security, Vol. 29, no. 2 (2021).
- Emirates Nuclear Energy Corporation, "The UAE's Radioactive Waste Management Strategy" https://www.enec.gov.ae/doc/factsheet-the-uaes-radioactive-waste-management-strategy-eng-5f195cd79948b.pdf

Finland

- 8th Finnish National Report to the Joint Convention on the Safety of Spent Fuel Management and on the Safety of Radioactive Waste Management, https://www.julkari.fi/handle/10024/149620 (2024)
- Posiva, "First stage of encapsulation completed successfully in Posiva's Trial Run of Final Disposal" (2024)
- https://www.posiva.fi/en/index/news/pressreleasesstockexchangereleases/2024/firststageofencapsulationcompletedsuccessfullyinposiva8217strialrunoffinaldisposal.html

Czech Republic

- Radioactive Waste Repository Authority/SURAO, "Selection of a site for the deep geological repository" https://www.surao.cz/wp-content/uploads/2021/05/presskit_20210504_EN_selection.pdf (2021)
- ENSREG, "Czechia," https://www.ensreg.eu/country-profile/Czech%20Republic (2021)

Belgium

- FANK, "Dialog über den Umgang mit radioaktivem Abfall" https://fank.fgov.be/de/dossiers/radioaktive-abfaelle/management-radioaktiver-abfaelle/endlagerung/geologische-endlagerung-0
- FANK, "Tiefenlagerung radioaktiver Abfälle," https://fank.fgov.be/de/dossiers/radioaktive-abfaelle/management-radioaktiver-abfaelle/endlagerung/tiefenlagerung
- Stimson Center, "Spent Nuclear Fuel in Belgium," https://www.stimson.org/2020/spent-nuclear-fuel-belgium/

Pakistan

- Mahmood et al., "The role of nuclear energy in the correction of environmental pollution: Evidence from Pakistan," Nuclear Engineering and Technology, Vol. 52, 6 (2020) https://www.sciencedirect.com/science/article/pii/S1738573319308150

Switzerland

- Nagra, "Wie läuft die Suche nach dem besten Tiefenlagerstandort ab?" https://nagra.ch/wissensforum/standortsuche/

Slovakia

- Mrskova, Deep Geological Repository and Ideas on Shared Solution in Slovak National Program https://www.djs.si/proc/nene2020/pdf/NENE2020_2003.pdf (2020)
- BASE, Ländersteckbrief Slowakei (2024)

Belarus

- World Nuclear Association, https://world-nuclear.org/information-library/country-profiles/countries-a-f/belarus#waste-management

Bulgaria

- Draft of updated strategy for the management of spent nuclear fuel and radioactive waste in Bulgaria (2023)

Hungary

- Seventh Report prepared within the Framework of the Joint Convention on the Safety of Spent Fuel Management and on the Safety of Radioactive Waste Management (2020)
- IAEA Country Nuclear Power Profile—Hungary

Brazil

- National Report of Brazil for the 7th Review Meeting of the Joint Convention on the Safety of Spent Fuel Management and on the Safety of Radioactive Waste Management (2020)
- Nuclear Engineering International, "Brazilian dry storage facility begins operation," https://www.neimagazine.com/news/brazilian-dry-storage-facility-begins-operation-8725959/

South Africa

- ESKOM—Solid Radioactive Waste Management Plan for Koeberg Nuclear Power Station (2023) https://nnr.co.za/wp-content/uploads/2024/03/8.-Koeberg-Solid-Radioactive-Waste-Management-Plan-Rev-1.pdf
- National Radioactive Waste Disposal Institute, Centralised Interim Storage Facility https://www.nrwdi.org.za/solutions/cisf/

Argentina

- 8th National Report to the Joint Convention on Spent Fuel Management and on the Safety of Nuclear Waste Management (2023)

Mexico

- National Report Presented by the United Mexican States to Mett the Requirements of the Convention on Nuclear Safety (2022)

Romania

- IAEA Mission Finds Romania Committed to Safe Management of Radioactive Waste, Encourages Improvements in Long Term Planning https://www.iaea.org/newscenter/pressreleases/iaea-mission-finds-romania-committed-to-safe-management-of-radioactive-waste-encourages-improvements-in-long-term-planning (2022)

- Energy Industry Review, "Romania to build new radioactive waste repositories" https://energyindustryreview.com/power/romania-to-build-new-radioactive-waste-repositories/

Iran

- Spence & Ahmad, Risks to Persian Gulf Cities from Spent Fuel Fires at the Barakah and Bushehr Nuclear Power Plants. Science & Global Security, Vol. 29, no. 2 (2021).
- Joint Comprehensive Plan of Action (2015).

Slovenia

- ARAO, "Slovenian Approach to Strategy and Planning for High Level Waste and Spent Fuel Deep Geological Disposal" https://arhiv.djs.si/proc/nene2016/html/pdf/NENE2016_1003.pdf
- BASE, Ländersteckbrief Slowenien (2023)

Netherlands

- Report of the ARTEMIS Mission to the Netherlands, https://www.iaea.org/sites/default/files/documents/review-missions/final-artemis-report-netherlands.pdf (2023)
- COVRA, "Deep Geological Disposal," https://www.covra.nl/en/radioactive-waste/deep-geological-disposal/

Armenia

- Keshishyan & Petrosyan, Spent Fuel and Radioactive Waste Management in Armenia https://inis.iaea.org/records/pkppr-ytt10 (2020)

Italy

- Report of the IRRS Mission to Italy https://www.iaea.org/sites/default/files/documents/review-missions/final_report_artemis_italy.pdf (2023)

Lithuania

- State Nuclear Power Safety Inspectorate (VATESI), "Spent nuclear fuel and long-lived radioactive waste treatment facilities," https://vatesi.lrv.lt/en/nuclear-facilities/radioactive-waste-management-facilities/spent-nuclear-fuel-and-long-lived-radioactive-waste-treatment-facilities/

Germany

- Website of the Federal Office for the Safety of Nuclear Waste Management, https://www.base.bund.de/shareddocs/kurzmeldungen/de/2024/forschung-pasta.html

References

Armand, Louis, Franz Etzel, Francesco Giordani, and European Atomic Energy Community. 1957. "A Target for EURATOM: Report." Brussels: Euratom. https://ec-europa-finder.primo.exlibrisgroup.com/discovery/fulldisplay?context=L&vid=32EUC_INST:VU1&search_scope=MyInstitution&tab=LibraryCatalog&docid=alma990004516120104061.

Bärenbold, Rebekka, Muhammad Maladoh Bah, Rebecca Lordan-Perret, Björn Steigerwald, Christian Von Hirschhausen, Ben Wealer, Hannes Weigt, and Alexander Wimmers. 2024. "Decommissioning of Commercial Nuclear Power Plants: Insights from a Multiple-Case Study." *Renewable and Sustainable Energy Reviews* 201:114621. https://doi.org/10.1016/j.rser.2024.114621

BASE. 2021. "Expert Response to the Report by the EU Commission's Joint Research Centre 'Technical Assessment of Nuclear Energy with Respect to the "Do No Significant Harm" Criteria in Regulation (EU) 2020/852, the "Taxonomy Regulation"'. Particularly Considering the Suitability of Criteria for Including Nuclear Energy in EU Taxonomy." Berlin: Federal Office for the Safety of Nuclear Waste Management with support from the Federal Office for Radiation Protection. https://www.base.bund.de/SharedDocs/Downloads/BASE/EN/reports/2021-06-30_base-expert-response-jrc-report.pdf.pdf?__blob=publicationFile&v=9.

BGE. 2022. "Zeitliche Betrachtung des Standortauswahlverfahrens aus Sicht der BGE". *Peine: Federal Company for Radioactive Waste Disposal*. https://www.bge.de/fileadmin/user_upload/Standortsuche/Wesentliche_Unterlagen/05_-_Meilensteine/Zeitliche_Betrachtung_des_Standortauswahlverfahrens_2022/20221216_Zeitliche_Betrachtung_StandAW-48_barrierefrei.pdf.

BMUV. 2024. "Verzeichnis radioaktiver Abfälle (Bestand zum 31. Dezember 2023 und Prognose)." Bonn: Federal Ministry for the Environment, Nature Conservation, Nuclear Safety and Consumer Protection. https://www.bmuv.de/fileadmin/Daten_BMU/Download_PDF/Nukleare_Sicherheit/verzeichnis_radioaktive_abfaelle_2023_bf.pdf.

Böse, Fanny, Alexander Wimmers, Julia Neugebauer, Theresa Lösel, Timo Hermes, Jasmin Beppler, Marie-Sophie Nickel, Pauline Morawe, Maximilian Weber, and Christian von Hirschhausen. 2024. "Putting Radioactive Materials on the Sustainability Agenda: A Report from a Workshop on the Sustainability of Human-Made Radioactive Materials Held at the safeND Research Symposium 2023." *Radiation and Environmental Biophysics*, February. https://doi.org/10.1007/s00411-024-01061-8.

Brunnengräber, Achim. 2016. "Die atompolitische Wende Paradigmenwechsel, alte und neue Narrative und Kräfteverschiebungen im Umgang mit radioaktiven Abfällen." In *Problemfalle Endlager*, edited by Achim Brunnengräber, 11–32. Baden-Baden: Nomos Verlagsgesellschaft mbH & Co. KG. https://doi.org/10.5771/9783845278131-10.

Dibley, Ben. 2018. "Technofossil: A 'Memento Mori.'" *Journal of Contemporary Archaeology* 5 (1): 44–52. https://doi.org/10.1558/jca.33380.

EC. 2022. "Commission Delegated Regulation 2022/1214 of 9 March 2022 Amending Delegated Regulation (EU) 2021/2139 as Regards Economic Activities in Certain Energy Sectors and Delegated Regulation (EU) 2021/2178 as Regards Specific Public Disclosures for Those Economic Activities." L 188/1. Offical Journal of the European Union. European Commission. http://data.europa.eu/eli/reg_del/2022/1214/oj.

Englert, Matthias, Simone Mohr, Saleem Chaudry, Stephan Kurth, and Florian Krob. 2024. "Verfolgung und Aufbereitung des Standes von Wissenschaft und Technik bei alternativen Entsorgungsoptionen für hochradioaktive Abfälle (altEr)." BASE-029/24. Berlin: Federal Office for the Safety of Nuclear Waste Management. https://doris.bfs.de/jspui/handle/urn:nbn:de:0221-2024052844041.

EC. 2024. *Study on Monitoring of Radioactive Discharges from Nuclear Facilities in the EU: Final Report*. LU: European Commission Publications Office. https://data.europa.eu/doi/https://doi.org/10.2833/253956.

EWN. 2020. "Castoren bei EWN. Bauarten und Inventare." Entsorgungswerk für Nuklearanlagen. https://www.ewn-gmbh.de/fileadmin/user_upload/EWN/Projekte/ESTRAL/Dokumente/Castoren_bei_EWN_Web.pdf.

Frieß, Friederike, Nikolaus Arnold, Wolfgang Liebert, and Nikolaus Müllner. 2021. "Sicherheitstechnische Analyse und Risikobewertung von Konzepten zu Partitionierungs- und Transmutationsanlagen für hochradioaktive Abfälle." Wissenschaftliches Gutachten im Auftrag des Bundesamtes für die Sicherheit der nuklearen Entsorgung (BASE) urn:nbn:de:0221–2021030826033. Berlin; Wien: Institut für Sicherheits- und Risikowissenschaften Universität für Bodenkultur (BOKU).

IAEA. 1997. "Joint Convention on the Safety of Spent Fuel Management and on the Safety of Radioactive Waste Management." Joint Convention INFCIRC/546. International Atomic Energy Agency. https://www.iaea.org/sites/default/files/infcirc546.pdf.

IAEA. 2009. "Policies and Strategies for Radioactive Waste Management." NW-G-1.1. IAEA Nuclear Energy Series. Vienna: International Atomic Energy Agency. https://www-pub.iaea.org/MTCD/Publications/PDF/Pub1396_web.pdf.

IAEA. 2010. "Disposal Approaches for Long Lived Low and Intermediate Level Radioative Waste." NW-T-1.20. IAEA Nuclear Energy Series. Vienna: International Atomic Energy Agency. https://www-pub.iaea.org/MTCD/Publications/PDF/Pub1407_web.pdf.

IAEA. 2013. "Management of NORM Residues." IAEA-TECDOC-1712. Vienna: International Atomic Energy Agency. https://www-pub.iaea.org/MTCD/Publications/PDF/TE-1712_web.pdf.

IAEA. 2019. "IAEA Safety Glossary. 2018 Edition." Vienna: International Atomic Energy Agency. https://www-pub.iaea.org/MTCD/Publications/PDF/PUB1830_web.pdf.

IAEA. 2022. *Status and Trends in Spent Fuel and Radioactive Waste Management*. IAEA Nuclear Energy Series, NW-T-1.14 (Rev. 1). Vienna: IAEA. https://www.iaea.org/publications/14739/status-and-trends-in-spent-fuel-and-radioactive-waste-management.

JRC. 2021. "Technical Assessment of Nuclear Energy with Respect to the 'Do No Significant Harm' Criteria of Regulation (EU) 2020/852 ('Taxonomy Regulation')." JRC Science for Policy Report 124193. Brussels, Belgium: Joint Research Centre, European Commission. https://www.politico.eu/wp-content/uploads/2021/03/26/JRC-report_March-2021-clean-Copy-printed.pdf.

Krall, Lindsay, and Allison Macfarlane. 2018. "Burning Waste or Playing with Fire? Waste Management Considerations for Non-Traditional Reactors." *Bulletin of the Atomic Scientists* 74 (5): 326–34. https://doi.org/10.1080/00963402.2018.1507791.

Krohn, Judith, Angelika Spieth-Achtnich, Silvia Schütte, Mbah Melanie, Alexandra Lampke, Katja Hünecke, Dörte Fouquet, Jörg Kuhbier, and Tobias Reinhardt. 2024. "Unterstützung des BASE bei der Prozessanalyse des Standortauswahlverfahrens (PaSta)." Abschlussbericht FKZ 4718F10001. Bundesamt für die Sicherheit der nuklearen Entsorgung. https://www.base.bund.de/SharedDocs/Downloads/BASE/DE/fachinfo/fa/pasta_abschlussbericht.pdf?__blob=publicationFile&v=3.

Levi, Michael A., and Henry C. Kelly. 2002. "Weapons of Mass Disruption." *Scientific American* 287 (5): 76–81. https://doi.org/10.1038/scientificamerican1102-76.

Mark, J. Carson. 1993. "Explosive Properties of Reactor-grade Plutonium." *Science & Global Security* 4 (1): 111–28. https://doi.org/10.1080/08929889308426394.

Möller, Detlev. 2009. *Endlagerung Radioaktiver Abfälle in der Bundesrepublik Deutschland*. Frankfurt am Main: Peter Lang GmbH.

Müller, Wolfgang D. 1990. *Geschichte der Kernenergie in der Bundesrepublik Deutschland. Anfänge und Weichenstellungen*. Stuttgart: Schäffer Verlag für Wirtschaft und Steuern.

National Research Council. 1957. *The Disposal of Radioactive Waste on Land*. Washington, D.C.: National Academies Press. https://doi.org/10.17226/10294.

Pistner, Christoph, Matthias Englert, Christian von Hirschhausen, Fanny Böse, Björn Steigerwald, Lukas Gast, and Richard Donderer. 2024. "Analysis and Evaluation of the Development Status, Safety and Regulatory Framework for So-Called Novel Reactor Concepts." BASE – Research reports on the safety of nuclear waste disposal BASE-031/24. Berlin: Federal Office for the

Safety of Nuclear Waste Management. https://www.base.bund.de/shareddocs/downloads/en/expert-info/f/final-report-novel-reactor-concepts.pdf?__blob=publicationFile&v=1.

Richardson, Katherine, Will Steffen, Wolfgang Lucht, Jørgen Bendtsen, Sarah E. Cornell, Jonathan F. Donges, Markus Drüke, et al. 2023. "Earth beyond Six of Nine Planetary Boundaries." *Science Advances* 9 (37): eadh2458. https://doi.org/10.1126/sciadv.adh2458.

Sellner, Dieter, Ulrich Waas, Norbert Pelzer, and Ulrich Karpenstein. 2021. "Einführung." In *Atomgesetz: mit Pariser Atomhaftungs-Übereinkommen: Kommentar*, edited by Gerald Hennenhöfer, Thomas Mann, Norbert Pelzer, and Dieter Sellner, 1. Auflage, 1–82. Beck-online Bücher. München: C.H.Beck.

Syrovatka, Felix. 2016. "Zwischen Konfrontation und Kooperation die Anti-Atom-Bewegung bei der Suche nach einem Endlagerstandort." In *Problemfalle Endlager*, edited by Achim Brunnengräber, 211–34. Nomos Verlagsgesellschaft mbH & Co. KG. https://doi.org/10.5771/9783845278131-210.

Chapter 17
Very Long-Term Financing of Nuclear Waste Management Under Uncertainty: Overview and Model-Based Case Study of Germany

Mahdi Awawda and Alexander Wimmers

17.1 Introduction

Managing nuclear waste is a multifaceted, long-term responsibility that entails considerable financial, technical, and ethical difficulties (Yim 2022). In Germany, the Site Selection Act of 2017 ("Standortauswahlgesetz," StandAG) envisions the selection of a suitable site by 2031.[1] From then, the plan envisages several years of construction works until the waste is transferred to the repository sometime in the second half of the twenty-first century. However, in 2022, the Federal Company for Radioactive Waste Disposal (BGE for German Bundesgesellschaft für Endlagerung) announced a delay in the site selection process, now expected to be completed sometime between 2046 and 2068. In 2024, another report, commissioned by the overseeing Federal Office for the Safety of Nuclear Waste Management (BASE for

[1] Gesetz zur Suche und Auswahl eines Standortes für ein Endlager für hochradioaktive Abfälle. https://www.gesetze-im-internet.de/standag_2017/BJNR107410017.html, last accessed on 7 March 2025. If not otherwise indicated, this date holds for all references in footnotes.

Preliminary results from this work were presented at the 2024 ATOM Day Conference in Berlin, and further discussed at the concluding workshop of the TRANSENS Project in Hannover in September 2024. The authors thank conference participants for their helpful comments that improved the quality of this work. Furthermore, the authors extend their gratitude to Doug Koplow, Christian von Hirschhausen, Fabian Präger, Wolfgang Irrek and Fanny Böse for fruitful discussions on this work. A draft version of this chapter was published to ArXiv: https://doi.org/10.48550/arXiv.2412.16126.

M. Awawda (✉) · A. Wimmers
Workgroup for Infrastructure Policy (WIP), Technical University (TU) Berlin, Berlin, Germany
e-mail: maw@wip.tu-berlin.de

A. Wimmers
German Institute for Economic Research (DIW Berlin), Berlin, Germany

A. Wimmers et al. (eds.), *Nuclear Power*, Studies in Energy, Resource and Environmental Economics, https://doi.org/10.1007/978-3-031-99894-2_17

German Bundesamt für die Sicherheit der nuklearen Entsorgung), placed the estimated year of the site selection even later, to 2074 (Krohn et al. 2024). This delay has implications regarding the preservation of knowledge and the affectedness on the German population. It also raises technical concerns regarding the re-organization and refurbishment of interim storage sites and, as discussed in this work, it poses future funding challenges (Ott et al. 2024).

Apart from legacy activities, such as the clean-up of the Asse II mine or of former GDR sites, most nuclear waste management activities are funded by the German Nuclear Waste Disposal Fund (KENFO), an external segregated fund which was created in 2017 with an initial fund volume of €24.1 billion (Bärenbold et al. 2024). The goal of the fund is to accumulate sufficient return on investment (ROI) to ensure that future taxpayers are not obliged to pay for future nuclear waste management activities in Germany. However, with the delay of the site selection process, and historic cost underestimation of waste disposal activities in Germany, the question arises whether the fund can fulfill its purpose (Hirschhausen and Wimmers 2023; Brunnengräber 2024, Wimmers and Vorwerk 2025).

This work provides a first investigation into the fund adequacy of KENFO. We first introduce the challenges of nuclear waste management funding and provide several examples of external segregated funding approaches for nuclear back-end activities. We then describe the German approach in detail before providing a quantitative assessment of the adequacy of current fund volumes for several potential site selection dates in the second half of this century and beyond.

17.2 Background

17.2.1 Nuclear Waste Management

The internationally accepted classification of radioactive waste differentiates different types of waste depending on their (average) half-life and their radioactivity. The six waste categories are exempt waste (EW), very short-lived waste (VLSW), very low-level waste (VLLW), low-level waste (LLW), intermediate-level waste (ILW), and high-level waste (HLW). HLW consists mainly of spent nuclear fuel (SNF), while other waste categories are generated mostly during decommissioning of nuclear facilities when activated and contaminated components are dismantled. There is a global consensus that nuclear waste shall be disposed of in storage sites. Due to their long half-life, HLW shall be stored in deep geological facilities for millennia. This necessitates the construction of durable, safe, and secure facilities that prevent harmful radiation from dissipating into the environment. While many countries have implemented solutions for LLW and ILW, e.g., near-surface disposal sites in Sweden or the US, currently, no deep geological disposal facility (GDF) for HLW is fully operational anywhere in the world.[2] Consequently, while countries are in different process stages of implementing GDFs, HLW is stored in interim storage facilities in decentralized facilities, often located near NPPs, or in consolidated storage sites (Bärenbold et al. 2024).

17.2.2 *Funding of Nuclear Waste Management*

17.2.2.1 The Organization of Long-Term Funding Provision in the Nuclear Back-End

The main challenges in providing funds for nuclear back-end activities stem from the uncertainty in the technological, organizational, and time dimensions. Each of these dimensions can induce actual and transaction costs that are unlikely to be accounted for when initial funds are set up. In this regard, the provision of a final repository, including other back-end activities, can be regarded as another form of large infrastructure provision, known for cost escalations and delays (Johnstone and Stirling 2020; Vorwerk 2024). For the remainder of this chapter, we assume that a GDF will be used for the final storage of nuclear wastes.

Uncertainty about the provision of repository infrastructure originates from uncertain geological compositions at a given site, requiring substantial discovery efforts, potential planning issues regarding the design and size of a repository, and the design of the storage casks, among others. Organizational challenges arise from the potential lack of final liability assurances or undefined responsibilities. The time dimension differentiates nuclear waste management (NWM) activities from other large infrastructure projects in that, in most states, the siting and construction processes, as well as the waste transfer and the operation of a GDF, will last many decades, even centuries. From an organizational perspective, this raises concerns regarding the provision of knowledge, the availability of skilled personnel, and the existence of regulatory agencies and competent oversight.

Grouped together, these uncertainties complicate accurate ex-ante assessments of costs. With unexpected developments, potential delays, and to-be-expected cost overruns, the risk of allocated funds being exhausted is substantial (Lordan-Perret et al. 2023). Depending on the organization of these funds, see below, the risks of costs being socialized are substantial (Brunnengräber 2024).

Funding must be ensured for long-term projects that are NWM activities. Given the high complexity and uncertainty of these processes, it is challenging to accurately determine costs, which in turn provides challenges regarding the provision of sufficient funds. Several efforts have been made to estimate the costs of NWM activities, as outlined below. However, given the historic underestimation of costs in the whole

[2] At the Finnish repository Onkalo, trial runs with empty casks were started mid-2024, see https://world-nuclear-news.org/articles/successful-start-to-trial-run-at-finnish-repository, last accessed on 20 December 2024.

nuclear sector and the lack of independent (ex-post) assessments (Schneider et al. 2023), these estimations are to be taken with a grain of salt.

Currently, the organization of funding for nuclear back-end activities differs among, and sometimes even within, countries. Options are internally segregated and non-segregated funds, from which operators finance waste management and decommissioning activities; external segregated funds, fully separated, independently operating funds out of reach of operators or other actors; state funds, i.e., taxpayer money; and a combination of IOUs and guarantees (Bärenbold et al. 2024). For this chapter, we limit our assessment to external segregated funds to provide a comparison to the German approach demonstrated in Sect. 17.3.

17.2.2.2 International Experience

In the US, decommissioning activities are funded mostly through so-called Decommissioning Trust Funds that are set up for a specific nuclear plant. In rate-regulated markets, the funds are mostly accumulated through rates charged to electricity consumers. The cost for decommissioning is calculated through a standardized formula or site-specifically. Regardless of the selected option, the calculated costs have been estimated as too low, but calls for a reassessment of the formulas have gone unheard, and final liability challenges in the case of fund shortfalls remain uncertain (Lordan-Perret et al. 2023). NWM is supposed to be funded through the Nuclear Waste Fund (US-$43.5 billion as of 2019), accumulated through a fee paid by the nuclear operators. Since the failure of the Yucca Mountain project and the subsequent standstill of siting activities in the US, fee collection has been halted, and the federal government has paid compensations totaling approximately US-$9 billion to NPP operators for interim storage of spent fuel and other wastes. Releasing money from the Nuclear Waste Fund requires congressional approval. So, current interim storage activities hope to be funded through the separate Judgment Fund (Wimmers and Von Hirschhausen 2024). Thus, despite the external organization of funds, the (probable) underestimation of cost and organizational failures will likely lead to US taxpayers having to fund both decommissioning and waste management activities in the long run.

In the United Kingdom (UK), the Nuclear Liabilities Fund (NLF) was initiated in 1996 to cover the decommissioning costs for the advanced gas-cooled reactors and the only currently operational pressurized water reactor, Sizewell B. The initial endowment was £228 million. Through several cash sweeps during extensive sector restructuring, including the final transfer of spent fuel management liabilities to the UK government, the fund grew by an additional £2.6 billion by 2007. The goal of the NLF was to cover future costs through returns on invested capital. However, it has been underperforming for years, resulting in government cash injections totaling £10 billion in 2020 and 2021. The bad performance of the NLF highlights the risks of external funding, especially if investment strategies underperform over long periods. Once funds are exhausted, British taxpayers will have to step in (Wimmers and Von Hirschhausen 2024).

The UK and US cases highlight the risk of cost underestimation and poor fund performance. Through good regulation, these risks for taxpayers can be mitigated and imposed on the operators. This is the approach of the Swiss fund STENFO. It consists of two funds, one for decommissioning (since 1985) and one for nuclear waste management (since 2001). Fees are collected from the operators of nuclear facilities. Most notably, the Swiss fund provision includes an assessment of costs and required fees that is conducted every five years. The operators provide cost estimates, which are then fed into a cost study conducted by Swissnuclear, the association of Swiss NPP operators, and presented to STENFO and the Swiss Ministry of Environment, Transport, Energy, and Communications. Once approved, rates are adjusted to new assessments. This way, the risk of fund shortfalls is reduced. The final liability lies with the Swiss state, albeit before this occurs, all funds of STENFO must have been used up, and all nuclear operators must have gone bankrupt (Wimmers et al. 2023; Brunnengräber 2024). As a result, the funding organization in Switzerland appears to be the most sophisticated, as it includes measures for cost evaluations that are enforced. As will be discussed, the German funding organization has no such adaptation mechanism.

17.2.2.3 Cost Estimations for Nuclear Waste Management

As mentioned above, NWM cost estimations remain ex-ante in nature until projects have been completed. Nevertheless, several attempts have been made to estimate the costs of NWM projects, most of which focus on the storage of SNF. For example, the IAEA (2009a) introduces cost categories and components along with cost estimations, but limits the assessment to a non-descript exemplary comparison of storage options over roughly twenty years.

Different methods can be used to estimate the cost of NWM activities. The decision depends on the specific phase of the NWM process, whether it is before the operation, during storage and handling, or in the post-closure phase of a repository. For instance, the Analogy Method is generally employed in the early stages of disposal programs, utilizing cost data from similar past projects and adjusting for scale, complexity, and material differences. While this method offers quick initial cost estimates, its accuracy is limited due to the variability in disposal costs. Another example is the Engineering Buildup Method, or Bottom-up Method, which becomes more relevant as the disposal plan matures. It breaks down the program into components within a Work Breakdown Structure (WBS) to estimate individual costs. As the plan evolves, the WBS becomes more detailed, improving cost accuracy. Alternatively, the Parametric Method establishes statistical relationships between plan parameters and costs using historical data, though its applicability in radioactive waste disposal is limited due to the scarcity of completed programs (IAEA 2007).

Regarding NWM modeling, multi-criteria analysis methods like PROMETHEE introduce an effective tool for navigating the complexity of NWM financing by considering time scenarios, disposal location, and funding mechanisms (Briggs et al. 1990). Further, Rothwell (2021) examines the size-cost relationship of storage alternatives and presents models estimating the economic impact of wet or dry and onsite and offsite options, respectively. The assessment compares centralized interim storage facilities (CISF) with decentralized onsite storage. Furthermore, the OECD (2010) provides a global survey of cost components, estimation techniques, and documentation obligations related to decommissioning NPPs. In any case, given that no empirical evidence exists of completed final repositories for HLW, the accuracy of these estimations cannot be assessed. Cost experience from other sections of the nuclear production chain, such as new build or decommissioning, shows that cost estimations in this sector are usually overly optimistic (Bärenbold et al. 2024).

17.3 The German Case

The following section provides an overview of Germany's legal, organizational, and operational framework for managing nuclear waste. It outlines the Site Selection Act of 2017, which envisions the selection of a safe and suitable location for long-term nuclear waste disposal sites. Moreover, the section introduces primary governmental entities responsible for interim storage and final disposal of radioactive waste. Additionally, relevant governmental actors, along with the German waste fund, KENFO, will be introduced.[3] Finally, potential paths regarding German NWM and its volumes are discussed to motivate the selection of cost scenarios introduced in Sect. 17.3.

17.3.1 Legal Framework

According to the act on the reorganization of responsibility in NWM (short VkENOG), the complete responsibility for both interim storage and final disposal of radioactive waste, as well as the clean-up of legacy facilities operated by the former German Democratic Republic, lies with the Federal State of Germany. The decommissioning of West German commercial NPPs is to be conducted by the former operators.[4] The primary objective of the StandAG is to choose the safest possible site within the country for the disposal of HLW, ensuring the protection of people and the environment from ionizing radiation for up to one million years.[5] The legislation also prohibits any agreements that would permit the disposal of radioactive waste outside of Germany, in line with Council Directive 2011/70/EURATOM.[6] The legislation identifies rock salt, clay, and crystalline rock as the key geological formations

[3] https://www.kenfo.de/en/about-us/about-kenfo.

[4] Gesetz zur Neuordnung der Verantwortung in der kerntechnischen Entsorgung. https://www.gesetze-im-internet.de/vkenog/BJNR011409017.html.

[5] Gesetz zur Suche und Auswahl eines Standortes für ein Endlager für hochradioaktive Abfälle. https://www.gesetze-im-internet.de/standag_2017/BJNR107410017.html.

[6] https://eur-lex.europa.eu/eli/dir/2011/70/oj/eng.

suitable for HLW storage. It requires selecting a site with deep geological layers that are appropriate for permanent closure, while also allowing for the retrievability of wastes during the operational phase and the possibility of recovery for up to 500 years after closure. The site selection process is set for completion by 2031. Furthermore, the legislation allows for the disposal of LLW and ILW at the same site. In addition to these efforts, the Konrad mine in Lower Saxony is being converted into Germany's first repository for LLW and ILW, with operations expected to begin in the early 2030s (Hirschhausen and Wimmers 2023).

The site selection process for an HLW repository shall be conducted in three distinct phases. The first phase consists of two steps. The first was completed in 2020 when an interim report on the so-called "sub-areas" was published.[7] Areas of Germany not included here are geographically unsuitable for a GFF. The second step is currently scheduled for completion in 2027. Here, potentially viable regions are to be selected for which overground discovery activities shall be conducted in the second phase. The duration of Phase 2 depends on the number of selected regions (e.g., six or ten), and the duration of regulatory oversight and control procedures to be conducted by BASE. In Phase 3, discovery shall be conducted underground. Here, too, the duration can vary. If boreholes are used, the phase is expected to require five to six years. However, if exploratory mines are constructed, this will add additional work requiring up to 18 years in this phase alone (BGE 2022).

17.3.2 Institutional Actors

This section provides a brief introduction to the governmental and institutional organizations and actors involved in the German NWM process, as shown in * MERGEFORMAT Fig. 17.1 (Hirschhausen and Wimmers 2023). The Federal Ministry for the Environment, Nature Conservation, Nuclear Safety, and Consumer Protection (BMUV) plays a significant role in safeguarding the public from environmental toxins and radiation, promoting resource efficiency, and advancing climate action, including biodiversity protection.[8] The BMUV oversees nuclear safety and radiation protection, including the authorization of NPPs, which are managed by state authorities. Subordinate bodies are the BASE and the Federal Office for Radiation Protection (BfS). In addition, there are several advisory bodies, such as the Reactor Safety Commission (RSK) for nuclear safety matters, the Commission on Radiological Protection (SSK) for radiation issues, and the Nuclear Waste Management Commission (ESK), for concerns related to NWM.[9]

The Federal Office for the Safety of Nuclear Waste Management (BASE) is responsible for overseeing the management and disposal of nuclear waste in

[7] https://www.bge.de/en/repository-search/site-selection-procedure/.

[8] https://www.bmuv.de/en/ministry/tasks-and-structure.

[9] https://www.bmuv.de/en/topics/nuclear-safety/overview-nuclear-safety/legal-provisions-and-technical-rules-and-regulations.

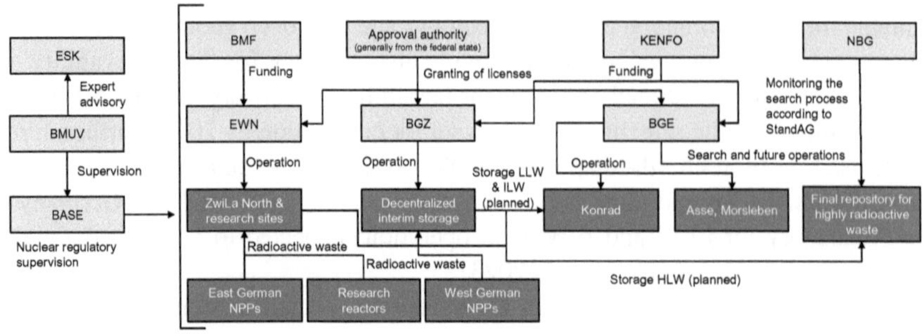

Fig. 17.1 Overview of the actors in German nuclear waste management. *Source* Adapted and translated from Hirschhausen and Wimmers (2023, 10)

Germany. It governs the site selection process for HAW disposal, ensures public involvement, and keeps the public informed. BASE also supervises administrative procedures related to mining, water, and nuclear laws, and ensures safety compliance during all repository phases of construction, operation, and decommissioning. It monitors the Morsleben repository, the Asse II mine, and the Konrad site. BASE also reviews permits for interim storage and nuclear fuel transport.[10] Following the closure of Germany's last nuclear plants in 2023, BASE continues to oversee safety during NPP decommissioning and participates in national and international safety committees, advising the federal government.[11]

The Nuclear Waste Management Commission (ESK) was established in 2008 to provide expert advice on nuclear safety. Supported by a secretariat within BASE, the ESK's advisory role can be requested by BMUV or initiated independently.[12]

The Federal Company for Radioactive Waste Disposal (BGE) was founded in July 2016 and is responsible for managing radioactive waste disposal in Germany.[13] Since April 2017, BGE has been conducting final disposal projects. They are constructing and operating final disposal sites, inspecting waste containers, retrieving waste from Asse II, operating the Konrad and Morsleben sites, and managing the closure and backfilling of Gorleben. BGE is also responsible for identifying a site for a final HLW repository.

The Federal Company for Interim Storage (BGZ) was established to ensure the safe and secure interim storage of radioactive waste (BGZ 2023). Operating independently under private legal status, BGZ is funded by the federal budget, with reimbursements from KENFO. Since 2020, BGZ has been managing twelve facilities for LILW and 14 for HLW at NPP sites transferred to its control (BGZ 2023).[14]

[10] https://www.base.bund.de/en/base/tasks/supervision/supervision-site-selection/supervision-site-selection.html and https://www.base.bund.de/en/base/tasks/tasks_content.html.

[11] https://www.base.bund.de/en/nuclear-safety/nuclear-phase-out/nuclear-phase-out_content.html.

[12] https://www.entsorgungskommission.de/en/our_work.

[13] https://www.bge.de/en/bge/organisation/.

[14] https://bgz.de/en/our-mandate/.

17.3.3 Nuclear Waste Volumes in Germany

In contrast to the international definition, radioactive waste in Germany is classified into two main categories: (i) Heat-generating radioactive waste, i.e., HLW from SNF and reprocessing; and (ii) radioactive waste with negligible heat generation, i.e., LLW and ILW. About 95% of the waste comes from research and the operation or decommissioning of NPPs, with the rest from industrial, medical, and research sources. Germany has approximately 27,000 m^3 of heat-generating waste and up to 620,000 m^3 of waste with negligible heat generation, which includes future waste from the Asse II mine and uranium enrichment. As of December 2022, over 130,000 m^3 of LLW and ILW are stored across Germany. This volume is expected to rise by the middle of the century, driven by the dismantling of NPPs. By 2060, this waste volume is projected to increase by 170,000 m^3, bringing the total to approximately 300,000 m^3, which will eventually be disposed of in the Konrad repository.[15]

Germany had generated around 15,000 tons of heavy metal (uranium and plutonium) of HLW from SNF by the end of 2016, including fuel sent to France and the UK for processing. Forecasts from waste producers help assess the required repository capacity, based on which the total predicted volume of HLW by 2080 will rise to around 10,500 tons, translating to an estimated 27,000 m^3 for final disposal. Special containers, including CASTOR containers, are used to store and transport this highly heat-generating waste.[16]

Additionally, other categories of waste, such as the 220,000 m^3 from the Asse mine and 100,000 m^3 of uranium enrichment residues, contribute to the overall volume. Future repository sites must be capable of accommodating both HLW and LILW, with container designs and repository adaptation based on the chosen host rock.

17.3.4 Funding

The reorganization of the German nuclear waste disposal process resulted in the establishment of the KENFO. In 2016, the nuclear operators that had previously been responsible for the provision of disposal funds agreed to transfer the full liability for NWM to the state by paying a fee of €24.1 billion, including a risk premium of €6.2 billion. KENFO was initiated to provide stable investment returns over the

[15] The former iron ore mine Konrad is being converted into a final repository for German LLW and ILW. Construction works began in 2007 and the completion date has been rescheduled multiple times. The most recent completion data of 2027 has been delayed indefinitely, see https://www.ndr.de/nachrichten/niedersachsen/braunschweig_harz_goettingen/Atommuell-Endlager-Schacht-Konrad-wird-noch-spaeter-fertig,schachtkonrad398.html and https://www.bge.de/de/konrad/. See also https://www.bge.de/en/radioactive-waste/current-inventory/ and https://www.bge.de/en/abfaelle/hintergruende/types-of-waste/ for information on German waste composition.

[16] https://www.bge.de/en/radioactive-waste/current-inventory/.

coming decades to ensure that future taxpayers need not pay for NWM procedures.[17] KENFO is overseen by a Board of Trustees comprising representatives from federal ministries and experts in finance, law, and energy. Additionally, BMUV supervises the foundation to ensure compliance with legal and financial obligations. KENFO's team of around 30 persons, of whom nearly half focus on investment and risk management, aims to function autonomously while maintaining transparency and accountability in managing Germany's nuclear waste disposal funds.[18] The investment strategies are legislated by the German Insurance Supervision Act (VAG), emphasizing security, profitability, liquidity, and a diversified portfolio.[19] The fund's strategic asset allocation ensures adequate capital reserves to cover nuclear waste disposal costs over 80 years, with a target ROI of 3.7% (KENFO 2024a). Note that KENFO was set up with the assumption that the last waste containers would be transferred into a completed repository by 2080. With the abovementioned delay in the site selection process, this schedule has become unachievable.

KENFO's portfolio is divided into liquid and illiquid assets. Liquid assets are traded in organized markets, while illiquid assets include corporate investments, loans, and non-tradable financial instruments. As of year-end 2023, publicly traded equities accounted for 47% of the portfolio, amounting to approximately €10.8 billion, which was invested in global equities and Real Estate Investment Trusts (REITs). These equities were diversified across developed and emerging markets to mitigate risks while enhancing potential returns. With 28%, or €6.4 billion, non-governmental bonds, invested in corporate and emerging market bonds, represented another significant portion of the portfolio. An additional 15%, or €3.4 billion, was allocated to governmental bonds primarily from industrialized nations. Illiquid assets, including private equity, infrastructure, and credit portfolios, made up 9% of the portfolio, equivalent to about €2.03 billion. Finally, KENFO held a cash reserve of €461 million (KENFO 2024a).

According to KENFO's investment strategy, the share of illiquid assets is expected to increase to 30% by 2028 and increase further in the decades after that. Until the shift is completed, the capital reserved for these illiquid investments is temporarily allocated to liquid assets with comparable risk profiles. This aims to enhance flexibility in managing the portfolio's risk-return dynamics while maintaining liquidity for fund payouts. Moreover, KENFO aims to invest about 35% of the portfolio in publicly traded equities, 25% in risky bonds, and 10% in low-risk government bonds (Mikus 2020).[20]

KENFO's gradual increase in illiquid assets aims to enhance portfolio diversification into non-public sectors, such as mid-sized enterprises and infrastructure. However, high real estate entry prices and rising interest rates limit broader real

[17] Gesetz zur Neuordnung der Verantwortung in der kerntechnischen Entsorgung. https://www.gesetze-im-internet.de/vkenog/BJNR011409017.html.

[18] https://www.kenfo.de/en/en-kapitalanlagen/how-we-invest.

[19] Gesetz über die Beaufsichtigung der Versicherungsunternehmen (Versicherungsaufsichtsgesetz—VAG). https://www.gesetze-im-internet.de/vag_2016/__124.html2.

[20] https://www.kenfo.de/en/en-kapitalanlagen/how-we-invest.

estate investments outside of REITs, reducing exposure to current market volatility. Additionally, KENFO's near-cash holdings, which are not ROI-generating, provide liquidity for capital commitments and will be progressively shifted into illiquid investments. This strategy balances immediate flexibility with long-term investment goals, adapting to current market conditions while advancing toward target allocations. Consequently, KENFO aims to reduce its cash and near-cash holdings from 8 to 0% in the foreseeable future. The reasoning behind this is the risk of having to pay negative interest on cash funds, while shares can be relatively easily monetized for fund payouts and earn returns (KENFO 2024b).

While the KENFO is designed to ensure that the potential costs are covered, there is a high uncertainty regarding the actual costs of NWM, especially in Germany. The only comprehensive and extensive study to provide some form of cost estimations was conducted by consulting firm Warth and Klein Grant Thornton in 2015 (WKGT 2015). The assumptions are based on provisions by the former nuclear operators and assume that the final repository will be closed by 2098. KENFO's volume is to some extent based on this study. It is unlikely that the cost estimates are accurate, given recent developments and the sector's tendency for cost escalation. First assessments show that costs for interim waste storage have been underestimated by €150–420 million per year (Irrek 2023). However, to the best of the authors' knowledge, there is no other (publicly available) study on the potential costs for the provision of a GDF in Germany, not to mention all other nuclear back-end activities.

17.3.5 Uncertainty in Germany's Waste Disposal Process

As mentioned in the introduction, delays of several decades for the site selection of a GDF for HLW have been announced by separate entities. With these potential delays comes the necessity to prepare for different so-called disposal pathways to ensure the safe and secure (interim) storage of nuclear waste in Germany. In this work, we limit our assessment to three central pathways that all have the final goal of constructing and operating a GDF for HLW, albeit with different time horizons. These paths are discussed by Scheer et al. (2024).

- The first path follows the legally mandated process outlined in the StandAG, which prioritizes the search for a permanent disposal site based on the current legal framework in Germany. Key elements of the StandAG path include decentralized interim storage leading up to the final site decision as early as 2031. The waste will be transported in specialized containers to a dedicated reception facility at the final disposal site. However, uncertainties remain regarding the timing of the closure phases and the final site decision.
- Second, the Extended Interim Storage path envisions continued use of existing decentralized storage facilities for 65–100 years, far beyond their initial 40-year licenses, due to delays in establishing a final disposal site. This path requires enhanced safety measures, including ongoing monitoring of transport and storage

containers for issues like sealing integrity and radiation containment. The long-term reliability of these containers is uncertain, especially as materials age.

- Finally, the Consolidated Interim Storage path involves constructing new, centralized storage facilities to replace the current decentralized ones. These newly built storage sites would consolidate HLW from multiple locations into a smaller number of technologically advanced, safety-optimized facilities. This approach ensures that the storage complies with modern safety and security standards, enabling easier maintenance and upgrades. However, while it requires significant investments and additional transportation of waste, the centralized approach might offer increased long-term safety compared to extending the use of older facilities.

Currently, every actor apart from the legislative authorities seems to have accepted the delay of the site selection process from 2031—the StandAG has not been adapted since the first official publication by BGE in 2022. Consequently, the already high degree of uncertainty characterizing NWM activities, see Sect. 17.2.2, is further increased by procedural and organizational uncertainty unique to the German situation in which, as of 2024, all three discussed options are under consideration. This also influences the risk of funding shortfalls and the consequential necessity of future generations to provide funding for projects that could have been dealt with many decades before, and for which sufficient funding could have been provided. Thus, the remainder of this work discusses the potential funding shortfalls and assesses the theoretically necessary ROI increases or near-term cash injections to ensure that shortfalls do not occur.

17.4 Method and Scenarios

We define the long-term financing of NWM activities from an external fund's perspective as a mathematical optimization problem, specifically a minimization problem. We then propose an algorithm to solve this non-linear problem. This method is useful for developing scenarios to explore the trajectories of NWM funding, focusing on helping the funding entity determine its target ROI, initial funding volume, or any additional subsidies needed. Note that this work does not include an assessment of KENFO's commitment to so-called ESG (environment, sustainability, governance) criteria in its investments. Refer to Brunnengräber and Denk (2024) for a critical assessment of these practices.

17.4.1 Problem Definition and Model

This section presents a model for a hypothetical external segregated fund, which includes an initial capital amount and an annual ROI to cover recurring yearly expenses. The model is based on detailed cost projections while also recognizing the inherent uncertainty and challenges in forecasting costs over long-term periods. This approach leverages the engineering buildup method, as outlined in Sect. 17.2.2.2, allowing for the design of a fund that can accommodate all or a selected subset of costs related to NWM projects from their inception to completion. Furthermore, the development of these cost projections relies on Rothwell (2021). For each type of interim waste storage (is), on-site and consolidated wet and dry storage, the following stylized cost formula was used:

$$C_{is} = a_{is} + \frac{b_{is}}{V} \tag{17.1}$$

These cost models estimate the total costs for interim storage C_{is} based on the volume V of nuclear waste (in metric tons of heavy metal, MTHM) using a constant term $a_{is,}$ representing fixed cost, and a variable term b_{is}, which scales with the inverse of waste volume. This inverse relationship implies economies of scale, meaning larger volumes reduce the cost per unit of nuclear waste.

The fund's investment portfolio is structured to comply with state regulations while effectively managing projected costs within an acceptable risk framework. Therefore, key financial assumptions, such as the target ROI, inflation rates, and risk management measures (e.g., the probability of event occurrence and its estimated impact over time), are integrated into the model. According to risk management principles, a higher target ROI is accompanied by greater risk, and conversely, a lower target ROI will result in lower financial risks (Bitsch et al. 2010). Hence, the higher the initial fund's value, the lower the target ROI could be, and vice versa.

Therefore, we define a so-called "constrained optimization problem" with two primary objectives: (i) determine the minimal ROI to fulfill future cost projections given a specific initial fund balance, and (ii) determine the minimal initial capital investment that meets future cost projections given a constant ROI. The capital injection for a given year can be derived from the defined minimization problem by subtracting the current fund balance from the variable being solved for (the minimal required fund balance). In Fig. 17.2, the optimization problem is solved for the lowest possible initial balance. Hence, the fund balance must exceed the annual projected costs for every single year. Moreover, regardless of whether the problem is solved for (i) or (ii), the balance in the final period must be zero.

We define a minimization problem with constraints to solve the first proposed constrained optimization problem (i), which could be reconstructed to solve the second problem (ii). The following mathematical definition presumes the presence of both cost projections and the initial required fund balance, with an unknown target ROI.[21]

$$\textit{Objective Function} : \textit{MinimizeROI} \tag{17.2}$$

[21] The minimization problem can be rewritten to minimize the fund balance at t = 1 instead of optimizing for ROI.

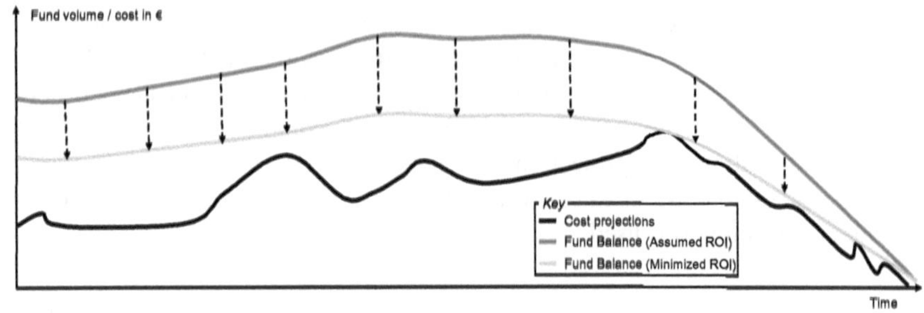

Fig. 17.2 Stylized description of the minimization problem. *Source* Own depiction

$$\text{s.t. } Costs_t \leq LiquidAssets_{t-1} \tag{17.3}$$

$$ROI \geq \left(\frac{1}{T}\sum_{t=1}^{T}\frac{FundBalance_t + Costs_t}{FundBalance_{t-1}}\right) - 1 \tag{17.4}$$

$$FundBalance_t = FundBalance_{t-1} * (1 + ROI) - Costs_t \tag{17.5}$$

$$FundBalance_t = LiquidAssets_t + IlliquidAssets_t \tag{17.6}$$

$$LiquidAssets_t = (CashPercentage_t + LiquidPercentage_t) * FundBalance_t. \tag{17.7}$$

$$\text{Liquid Percentage t} = \begin{cases} 0.70 \quad \text{if t} < \frac{T}{4}\text{(first quarter)} \\ 0.60 \text{ if } \frac{T}{4} \leq \text{t} < \frac{T}{2}\text{(second quarter)} \\ 0.50 \text{ if } \frac{T}{2} \leq \text{t} < \frac{3T}{4}\text{(third quarter)} \\ 0.40 \quad \text{if t} \geq \frac{3T}{4}\text{(fourth quarter)} \end{cases} \tag{17.8}$$

$$\text{CashPercentage}_t = \max\{\left(0.08 - \frac{0.08}{5} * (t-1)\right); 0\} \tag{17.9}$$

$$\text{Costs}_t \in \text{CostProjections} = \{\text{Costs}_{t=1}, \text{Costs}_{t=2}, \cdots, \text{Costs}_{t=T}\} \tag{17.10}$$

$$\text{Costs}_t = i\sum_s F_{t,s} + j\sum_s I_{t,s} + k\sum_s D_{t,s} \tag{17.11}$$

$$\begin{aligned} F_{t,s} = {} & \text{Capital}_{t,s} + \text{Operation}_{t,s} + \text{Transport}_{t,s} + \text{Disposal}_{t,s} \\ & + \text{Regulatory}_{t,s} + \text{Misc}_{t,s} \end{aligned} \tag{17.12}$$

$$\begin{aligned} I_{t,s} = {} & \text{Capital}_{t,s} \\ & + \text{Operation}_{t,s} + \text{Transport}_{t,s} + \text{Regulatory}_{t,s} + \text{Misc}_{t,s} \end{aligned} \tag{17.13}$$

$$D_{t,s} = \text{Decontamination}_{t,s} + \text{Demolition}_{t,s} + \text{Transport}_{t,s} + \text{SiteRestoration}_{t,s} + \text{Safeguarding}_{t,s} + \text{Misc}_{t,s} \quad (17.14)$$

$$t \in \{0, 1, \ldots, T\} \in \mathbb{N} \quad (17.15)$$

$$s \in \{\text{Site}_1, \ldots, \text{Site}_n\} \quad (17.16)$$

$$i, j, k \in \{0, 1\} \quad (17.17)$$

The average annual rate of return must cover the projected costs. ROI is the return on investment expressed as a percentage (Eqs. 17.3 and 17.4). FundBalance_t is the balance of the fund at the end of year t (Eqs. 17.5 and 17.6). The fund must cover costs until the project's end year T (Eq. 17.3). Costs_t are the projected costs for NWM services in year t (Eqs. 17.10 and 17.11). LiquidAssets_t is the portion of the fund that is liquid in year t, consisting of easily monetizable assets, such as government and corporate bonds, REITs, and cash. IlliquidAssets_t cannot be easily monetized, such as private equity, infrastructure investments, private debt, and real estate (Eq. 17.6). $\text{LiquidPercentage}_t$ gives the share of liquid assets in the total fund that changes according to the KENFO's targets discussed in Sect. 17.3.4 (Eqs. 17.7 and 17.8). KENFO plans to reduce its cash holdings from 8 to 0% within five years. Thus, there will be no generated yield for the respective percentage of the fund. Thus, cash holdings must be deducted from the ROI calculation. For this, a linear decrease is assumed, starting at 8% in $t = 1$, down to 0% by $t = 5$ (Eq. 17.9).

Depending on the country and its external segregated fund's regulation, the external segregated fund might have defrayed costs of the whole process, i.e., projects relating to NWM, such as a subset of those. Hence, subdividing the cost projection variable by formulating these into binary variables is particularly important.

$F_{t,s}$ are the final disposal cost projections for a given site s in year t. Analogously, $I_{t,s}$ and $D_{t,s}$ are the interim storage and decommissioning costs, respectively. Binary variables i, j, and k indicate whether the cost component is covered by the fund. For KENFO, $k = 0$. Definitions of the cost parameters are found in the Annex.[22]

17.4.2 Algorithm for the Minimization Problem

The binary search method, often called the bisection method, efficiently finds a target value within a specific range. Given an initial fund balance, this method can identify the minimum target ROI required to cover projected costs. The algorithm outlined below describes the steps for solving this minimization problem. The process begins

[22] The detailed cost structure of the de jure scenario is found in the folder "Cost Projections" of the GitHub project (https://github.com/Mahdi-Awawda/KENFO).

by establishing lower and upper bounds that encompass the expected ROI range. The gap between these bounds is gradually narrowed until a predefined tolerance is reached to ensure the accuracy of the result. In each iteration, the midpoint of the current bounds, referred to as the "assumed ROI," is calculated and assessed against the constraints of the model. If the fund balance turns negative at any point during the projection, it suggests that the assumed ROI is too low, prompting an upward adjustment of the lower bound. However, if the fund remains solvent throughout the projection, the upper bound is adjusted downward. This iterative process continues until the fund's balance at the end of the projection is a minimal positive value. More precisely, the algorithm terminates when the difference between the upper and lower bounds of the ROI is within a very small tolerance level (e.g., 10^{-12}). Find the detailed algorithms for both the minimum ROI and the required capital increase in the annex of this work.

17.4.3 Scenario Definition and Results

We assess KENFO's fund adequacy for different scenarios of the site selection process. The scenarios differ only in the date of the site selection, i.e., in delays in the different phases of the site selection process. From this date, all activities remain analogous to the schedule as outlined by ESK (2023). In the following, we describe the general approach and define the different scenarios.

17.4.3.1 Data Assessment

Cost projections originate primarily from the abovementioned report by WKGT (2015). Here, cost projections are divided into five categories: decommissioning and dismantling; interim storage; costs for containers, transports, and operational waste; the Konrad LLW and ILW repository; and the final HLW GDF.

Since the publication of the report in 2015, the funding organization in Germany has been reorganized to the extent that decommissioning and dismantling obligations are now funded by the former operators, or from the state budget for former GDR and research reactors. These costs are thus not included in the cost assessment for KENFO. Regarding the second category, cost estimates were derived from balance sheets and internal data from the utilities. In the various scenarios of this work, these estimates were standardized and extended beyond the 40-year operating licenses to account for final waste disposal by 2088. The data of the third category comprises unit costs for containers, including CASTOR and POLLUX containers for HLW, and MOSAIK containers for LLW and ILW, as well as costs for transport and costs for conditioning facilities that are necessary to package the waste into these containers. The individual utilities used these standardized data points provided by the cask manufacturer GNS.[23] They adjusted them according to their specific operational conditions and waste volumes, which were later adjusted in the study to derive cost estimates. The cost projections of the fourth category were based on cost calculations provided by the BfS and calculations of the utilities. Finally, the cost projections for the last category were primarily derived from past estimates for the Gorleben site,[24] inflation adjustments, and the cost-sharing responsibility of the utilities, with adjustments reflecting standardized recalculations.

Due to the lack of any other source, cost projections were taken from the outdated study by WKGT (2015), following the engineering buildup method by breaking down the NWM program into components in a WBS, estimating their costs, and summing them for an overall estimate. As the plan evolves, the WBS becomes more detailed, improving cost accuracy for later work where more accurate cost estimates can be substituted in the model.

The detailed cost projections were adjusted based on precise expenses provided in the annual federal budget reports, along with various reports from official stakeholders such as BMUV, BGE, and BGZ. See Wimmers et al. (2023) for an overview. In addition, the timeline for each scenario was adjusted by adapting project structures following the details outlined in ESK (2023) and BGE (2023). For instance, if the determination of the final disposal site is delayed by n years, the resulting contingent liabilities of interim storage activities, such as maintenance and site security, are increased, and potentially, replacement interim storage containers become necessary. Thus, overhead costs for CASTOR containers arise periodically (once every 40 years), with additional container costs of about €2 million. The costs of maintenance and site security are accounted for annually.

Furthermore, an inflation rate (IR) of 1.72%[25] and nuclear-specific cost increase (NSCI) of 1.97%[26] is assumed for n years (ceteris paribus) across all scenarios to enhance the comparability and clarity of the analysis by removing economic variability as a factor. This allows differences in projected outcomes to be attributed directly to project-specific uncertainties, such as delays or operational risks. In long-term projects of NWM, where the timeline spans several decades, maintaining uniform inflation and cost increases aligns with the expectation of stable macroeconomic conditions.

Another important detail for the development of the scenarios is that the time to repackage and transport wastes from interim facilities to the hypothetical final disposal site, and that allocated to final disposal operations, is assumed to be at least 30 years. This does not consider delays regarding the site selection process.

[23] Gesellschaft für Nuklear-Service mbH, see https://www.gns.de/, accessed on 14 November 2024.

[24] See https://www.bge.de/en/the-gorleben-mine for more information.

[25] The inflation rate was calculated based on the IR average of the years 1992–2021. We excluded 2022 and 2023, considering the pandemic and Ukrainian war as anomalous events. The assumption is that inflation will revert to the German central bank's target long-term rate of 1.7–1.9% (European Central Bank 2021).

[26] Analyses of past projects show that nuclear sector costs rise faster than in other industries, due to technical complexity and regulatory demands. This led to an estimated annual cost increase of 1.97% following WKGT (2015).

Consequently, the cost projections for the final disposal site have been extended to align with adjusted (extended) timelines, ensuring that these constraints are met.

17.4.3.2 Scenarios

The scenario development involved calculating time estimates for Phases 2 and 3 of the site selection process, assuming Phase 1 would conclude by the end of 2027, as mentioned in BGE (2023), see Sect. 17.3.2. These three phases comprise the final disposal site selection procedure (FDSP). Following BGE (2023), we also include an additional delay based on a risk assessment of 23 potential events. After a site has been selected, the transport from interim storage to final disposal sites begins, and the disposal operations commence. These activities are referred to collectively as "final storage and disposal activities" (FDSA). These activities will commence once the FDSP is completed, requiring a minimum of 49 years.

Table 17.1 provides an overview of the developed scenarios.[27] The first scenario represents the "de jure" situation outlined in the site selection process. Here, the site for the GDF is selected in 2031 as initially planned, and the repository itself will be closed in 2080.

The "de facto" scenarios illustrate the variation of site selection delay dates defined by BGE (2022). Scenario 2 assumes the earliest possible selection year of 2046. In Scenario 3, the site is chosen in 2057; in Scenario 4, we assume the most extended published delay of 2068. All three scenarios require additional funds for long-term interim storage and potential cask refurbishments.

The three remaining scenarios consider the potential pathway of consolidated interim storage as presented in Sect. 17.3.5. Here, we assume that wastes are consolidated into one of three possible centralized storage sites at Ahaus, Gorleben, or Lubmin instead of extending the operational lifetimes of existing decentralized interim storage facilities. Each consolidated facility requires an initial investment for capacity extension and refurbishment.

In the first four scenarios, cost projections were disaggregated into four categories: Interim Storage, Containers, Transports & Operational Waste, Konrad Mine Repository, and HAW Final Disposal Site. These categories represent the scope of obligations of KENFO. Two supplementary cost categories were introduced for the three remaining scenarios: "Consolidated Interim Storage" and "Transportation to Consolidated Interim Storage."

[27] The detailed cost structure of the scenarios is found in the folder "Cost Projections" of the GitHub project (https://github.com/Mahdi-Awawda/KENFO).

Table 17.1 Scenario overview

Scenario	Phase I completion	Phase II duration	Phase III duration	Risk management[28] delay	Determination date of FDSP	Delay according to CIS	Date of the completion of the FDSA
1	2028	10	5 (boreholes)	3	2031	0	2080
2	2028	10	5 (boreholes)	3	2046	0	2095
3	2028	10	13 (exploratory mines)	4	2057	0	2106
4	2028	12	23 (exploratory mines)	5	2068	0	2118
5	2028	10	5 (boreholes)	3	2070	15	2120
6	2028	10	13 (exploratory mines)	4	20xx[29]	20	2154
7	2028	12	23 (exploratory mines)	5	2xxx	25	2180

Abbreviations: FDSP = Final Disposal Site Procedure, FDSA = Final Disposal Site Activities, CIS = Consolidated Interim Storage

[28] Delays in transportation and construction of a repository can also occur in parallel.

[29] It is assumed that the completion of a repository requires at least 49 years after site selection, allowing for flexibility in determining the selection in the last two scenarios.

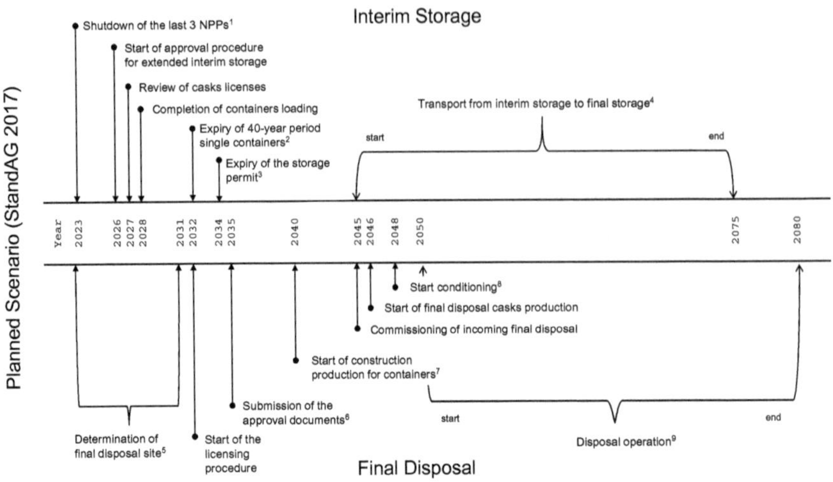

Fig. 17.3 Simplified timeline of interim storage and final disposal milestones. *Source* Own translation with modifications based on (ESK 2023)

Scenario 1: "De Jure"

The following is a description of the "de jure" scenario focusing on milestones derived from a recent report by ESK (ESK 2023). Figure 17.3 illustrates these milestones by distinguishing between the responsibilities of the state-owned companies (BGZ and BGE in blue and red, respectively). In 2026, BGZ must begin the extended interim storage approval process, which requires inspection and approval by BASE.[30] By 2027, several legal requirements from §§ 6 and 7 of the Atomic Energy Act will be enforced for facilities like the Ahaus Interim Storage Facility. Notably, nuclear fuel storage may exceed 40 years from the start of container storage, and extensions beyond this period are only allowed for compelling reasons after consultation with the German parliament.[31] All BGZ-operated interim storage facilities use dry storage for SNF, with loading set to be completed by 2028 (ESK 2023). From 2032 onwards, the first containers in various facilities will begin to reach their technical expiry date, including Ahaus (BZA), Northern Facility (ZLN) in Lubmin, and sites at former NPPs, e.g., in Philippsburg (KKK), Biblis (KWB), or Neckarwestheim (GKN). At all sites, extended storage approvals must begin by 2034 at the latest. Germany's current plan mandates the interim storage of irradiated fuel at NPP sites until a permanent disposal facility is available. Once such a facility has been established, nuclear waste transport could begin by 2045 and will last for 30 years, concluding all interim storage projects by 2075.

[30] https://www.base.bund.de/de/zwischenlager/transport-behaelter/zulassungen/zulassungen.html.

[31] Gesetz über die friedliche Verwendung der Kernenergie und den Schutz gegen ihre Gefahren (Atomgesetz). § 6 (5). https://www.gesetze-im-internet.de/atg/__6.html.

The first project of BGE is the determination of the final disposal site. Phase 1 of the site selection procedure involves identifying suitable subregions based on geological safety criteria, excluding unsuitable areas. Afterward, safety assessments are then conducted, and BGE proposes areas for surface exploration, with the legislator deciding which regions to explore. In Phase 2, BGE will conduct site-specific exploration, incorporating results into safety assessments and socio-economic analyses, leading to the identification of potential sites for underground exploration. BGE then conducts detailed underground exploration and site comparisons, followed by BASE's environmental impact assessment. In Phase 3, BASE reviews BGE's proposal, balancing private and public interests, and recommends the safest site to BMUV. The final decision rests with the legislator. Additionally, BGE develops final disposal container concepts for all host rocks as a separate project (BGE 2023).

We start with the base case (which is no longer realistic due to the delay). Following the completion of the site selection procedure in 2031, the approval processes for the GDF, the treatment facility, on-site storage infrastructure, and the final disposal containers will begin. The submission of final application documents is scheduled for 2035. By 2040, approvals for the construction of the GDF, on-site storage infrastructure, and treatment plant must be secured, with final disposal activities commissioning by 2045. Starting in 2046, the production of 1870 final repository containers (around 60 per year), 5,940 repository coquilles plus packaging (about 200 per year), and several containers for 3900 glass coquilles will begin. This will take 30 years. In 2048, the conditioning phase will commence, during which 340 Mg of spent fuel and 130 glass coquilles will be treated annually over a period of 30 years. A sufficient number of final repository containers must be ready when conditioning starts. The actual disposal process will run from 2050 to 2080, during which 60–200 repository containers and 130 glass coquilles will be disposed of annually (ESK 2023).

Scenarios 2, 3, 4: "De facto"

In the three "de facto" scenarios, we assumed extended or prolonged interim storage due to the delay in the site selection procedure, as communicated by BGE, along with adjustments to the detailed breakdown of the de jure scenario (see Sect. 17.4.3.2.1). The adjusted timeline was calculated following BGE (2023). After establishing the expected date for the final disposal site selection, all activities from the previous "de jure" scenario were integrated into the new timeline. These adjustments account for the delays of 15, 26, and 38 years in the three respective scenarios. The delays are relative to the timeline estimated by ESK and result in periodic additional interim storage costs, particularly due to the necessary replacements of CASTOR containers. For example, in the interim storage facilities, Biblis and Neckarwestheim, the first container replacement will be required in 2032 across all three scenarios. A second round of replacements is scheduled for 2072, and a third for 2112 (applicable only for Scenario 4). This schedule strongly depends on the project management technique used in the corresponding queue management. Potential optimization could involve prioritizing the transport of SNF from facilities with containers nearing expiry to the final disposal site before other facilities. In addition to this periodic overhead, continuous cost increases must be factored in for ongoing maintenance and security expenses. Furthermore, the cost projections for final disposal activities included additional overhead, depending on the approach utilized in Phase 3 of the site selection process, such as boreholes or mine sites. For example, the cost projections for boreholes, which would span five years across six to ten regions, are significantly lower than mine sites, which could extend for 13–23 years across six to twelve regions.

Scenarios 5, 6, 7: "Consolidated Interim Storage"

In the three remaining scenarios, we investigate the potential installation of CISFs as temporary solutions until the establishment of a final repository. CISF might offer higher safety standards compared to multiple dispersed sites, and might be more economical due to economies of scale. To assess the financial impact, we extended the three "de facto" scenarios by introducing two additional cost categories while keeping all other variables unchanged (ceteris paribus). These cost projections are calculated as follows:

The average cost (AC) per kilogram of heavy metal (kgHM) is expressed as a function of the facility's capacity in tons (see Eq. 17.1). In the following scenario, we assume the use of pool storage instead of dry casks, as the calculated maintenance and security costs are lower than the costs of replacing the containers, given the extended duration of NWM activities in comparison to the first four scenarios. Below is the cost estimation function for the off-site consolidated pool storage, transformed into € using a conversion rate of 1 US$ ≈ 0.93 €, and applied to the estimated total waste volume of $10.5*10^5$ kgHM. The average cost per kgHM for off-site pool storage is calculated as

$$AC_{off,pool} = \left(176US - \$ + \frac{370,000US - \$}{Capacity_{off,pool}}\right) per\ kgHM \tag{17.18}$$

After converting to euros and substituting the volume of $10.5*10^5$ kgHM, the value obtained is

$$AC_{off,pool} = \left(164€ + \frac{344,100€}{10,500}\right) = 196.78€ per\ kgHM \tag{17.19}$$

This gives the total costs of approx. €2.066 billion. These costs must then be allocated across the specified timelines and discounted accordingly for the three scenarios, incorporating the annual effect of NSCI of 1.97% and an inflation rate of 1.72% on the cost projections. Due to these investments, FDSA activities are completed at later (uncertain) dates, pushing GDF closure well into the twenty-second century.

17.5 Results and Discussion

17.5.1 Results

In the context of this work, we consider German NWM activities, which are to be funded by KENFO. The assessment of long-term funding adequacy in this chapter is based on seven scenarios that reflect possible trajectories of the so-called cost side. These cost projections are then incorporated into the proposed model to determine: (i) the average minimal ROI required to cover projected costs; and (ii) the initial fund needed for a given average ROI, from which a necessary capital injection as of the year 2024 can be derived. With the latter, we propose a reduction of generational inequality by adding additional funds paid for by today's generation to minimize the risks of funding shortfalls for future generations. To limit the already substantial uncertainty of fund development, this work does not consider KENFO's investment portfolio, i.e., it does not differentiate between different investment strategies. Furthermore, it does not assess the suitability of investment decisions regarding, e.g., ESG criteria. The scenario range for the completion of final disposal activities was from 2080 to 2180. Depending on the scenario, additional cost projections included the effect of such a long timeframe. We account for an NSCI of 1.97% and an IR of 1.72% and solve the minimization problem defined above.[32]

Cost projections range from €71 to 804 billion. The required ROI to cover these costs ranged between 5.56% and 6.63% in the seven scenarios. The influence of the inflation rate and the NSCI is large and could thus strongly influence the required target ROIs. Moreover, in no scenario will the current fund volume suffice for the target ROI of 3.7%. Required capital injections (made in 2024) range from €11.54 to 31.07 billion.

For Scenario 1, the discounted cost projection totals €71 billion. The minimum average ROI required to cover projected costs to 2080 is calculated to be 5.56%. Moreover, solving for the publicly communicated target ROI of 3.70% (assumed as constant over the project lifetime), we find that the fund's current value would require a one-time capital injection of €11.25 billion to cover cost projections by 2080. The minimum average ROI required to cover presumed cost projections is 5.95% by 2095 for the second scenario, 6.24% by 2106 for the third scenario, and 6.34% by 2118 for the fourth scenario. To meet KENFO's publicly stated target ROI of 3.70%, the fund's current size would need to be increased via one-time capital injections of €17.79 billion, €23.60 billion, and €28.43 billion, for Scenarios 2, 3, and 4, respectively. Finally, results show the minimum average ROI required to cover assumed cost projections is 6.63% to 2120, 5.67% to 2154, and 5.54% to 2180 for Scenarios 5, 6, and 7, respectively. To reach KENFO's target ROI of 3.70%, the fund's current value would need to be adjusted to €52.48, €44.99, and €46.02 billion. Notably, the last two scenarios require a lower capital balance this year and a

[32] The Java code is found in the folder "Code" of the GitHub project (https://github.com/Mahdi-Awawda/KENFO).

smaller capital infusion (€23.58 and €24.61 billion) compared to the fifth scenario, which requires €31.07 billion. These capital infusions cover cost projections to 2120, 2154, and 2180, in Scenarios 5, 6, and 7, respectively.

17.5.2 Discussion

A limitation of this work is linked to the challenge of accurately predicting the costs of such complex and long-term projects that are NWM activities. The actual costs can vary due to several factors, including the chosen methods for interim storage and final disposal, the type and quantity of waste, the regulatory environment, and the availability of suitable storage facilities (IAEA 2009b; OECD 2010). Furthermore, long-term macroeconomic assumptions, such as a uniform inflation rate, are highly influential and pose risks of misrepresentation. Because the proposed model assumes that the external fund maintains sufficient annual liquidity, which may not always reflect real-world conditions, it does not account for potential periodic increases in the fund, e.g., via subsidies. Additionally, we assume positive and constant ROIs over the entire time frame, even though actual ROI values can fluctuate and funds might generate negative returns due to changes in interest rates, market conditions, government policies, wars, and catastrophes, among others.

The most substantial limitation of this particular work, however, relates to the lack of reliable, accurate, and timely cost data for NWM activities. While there are several assessments on an international level, country-specific data is limited. For Germany, the only comprehensible cost study dates back to 2015 and is based on utility provisions. A reassessment of the potential cost of final disposal of nuclear waste in Germany is necessary. This future cost study would have to account for various potential project delays, CISF options, and the general underestimation of costs in the nuclear sector.

As demonstrated in Table 17.2, all scenarios indicate the necessity for additional capital. While it remains impossible to accurately forecast the nuclear-specific cost increase or inflation rate over the next 50–150 years, it is evident that BGE will not finalize the disposal site selection by 2031. Instead, this process is anticipated to extend for several more decades. Consequently, BGE and other key stakeholders must develop multiple project management scenarios, each accounting for varying timelines. These are critical for the financing body, as delaying such complex and expensive projects could be advantageous only if the fund's annual returns are appropriately aligned with the anticipated financial requirements over time, which in turn strongly depend on the timelines.

Table 17.2 Scenario comparison

Scenario	Timeframe (until)	Total cost projections* (billion)	Required yearly ROI (%)	Balance required (end of 2024)** (billion)	Required capital injection (end of 2024)** (billion)
1	2080	€70.97	5.56	€32.66	€11.25
2	2095	€118.78	5.95	€39.20	€17.79
3	2106	€178.06	6.24	€45.02	€23.60
4	2118	€263.43	6.34	€49.84	€28.43
5	2120	€273.63	6.63	€52.48	€31.07
6	2154	€530.46	5.67	€44.99	€23.58
7	2180	€803.13	5.54	€46.02	€24.61

Assumptions: *Initial total fund value of €21.41 billion, IR: 1.60%, and NSCI: 1.97%; **Average ROI of 3.70%

In Scenarios 6 and 7, one can observe a decline in the required ROI and, consequently, the required funds compared to Scenario 5. Although all three scenarios consider a CISF strategy, the break-even point for the same target ROI strongly depends on the cost structure. We conclude that the strategy may not be economically viable if the construction and transport activities occur too early or are significantly delayed. However, suppose the costs associated with CISF are adequately managed before establishing the final disposal site. In that case, the fund could gain the necessary time to generate sufficient returns, allowing for the coverage of prolonged final disposal costs.[33] This analysis highlights CISF construction as a possible strategy for the German site selection process. Future research should assess whether there might be an economic gain in constructing CISFs and delaying the final waste repository construction to allow the fund to grow. However, our analysis shows that the sooner the final repository is sealed, the cheaper the process becomes. Thus, potential CISF solutions should only be considered once it is impossible to complete the process before the turn of the next century.

Furthermore, future research should address the implications of capital injections for KENFO. While this would, in theory, reduce the risks of fund shortfalls, it might induce further process delays due to available funds and the certainty that they will be provided, and it remains uncertain how these injections would be financed. Debt financing could induce interest rates that might reduce the impact of the injection in the long run. Reinstating the former nuclear operators as final creditors could reduce potential future taxpayer liabilities, but might bring substantial legal implications.

[33] Note that this strategy has been observed in US nuclear decommissioning projects (Wimmers and Von Hirschhausen 2024).

17.6 Conclusion

Long-term financing of nuclear waste management is a very complex undertaking that is full of uncertainties. In the case of Germany, a shortfall of funds is expected. To ensure that funds collected in the KENFO are sufficient to cover all future costs of NWM in Germany, there are two general options. First, one could attempt to reduce the "cost" side of the equation. This would first require a detailed assessment of to-be-expected costs, and then merit an acceleration of the whole site selection process. Even if a site was chosen in 2031, current estimates indicate that the KENFO volume will not suffice. Thus, potential cost reductions, e.g., via international collaboration in cask manufacturing and repository construction, could reduce the pressure. The second option is to increase the "revenue" side of the equation. This can be achieved either via cash injections, funded by today's taxpayers or a fee, or via increased ROIs. However, on average, increasing ROI incurs increased investment risks.

Thus, from a current perspective, the long-term funding of German NWM remains uncertain, and future generations will likely have to pay for today's failures.

Additional future research should address the composition and risk distribution of KENFO's investments. While the KENFO is committed to ESG criteria, so-called "sustainable" or "green" investments usually garner limited returns. Thus, KENFO's portfolio includes investments in fossil fuel companies ("in transition"), potentially contradicting the self-proclaimed ESG goals. Furthermore, as the experience of the NLF has shown, externally segregated funds are always at the risk of too low returns.

Annex

Algorithm 1

Algorithm 1: Algorithm to Calculate Smallest Possible ROI

```
input:
    an array of double values costs representing annual cost projections,
    a double value initialBalance representing the initial fund balance
output:
    a double value targetROI representing the resulting ROI that ensures cost coverage over a given
    period
begin
initialization of variables:
    lowerBound := 0;
    upperBound := 100;
    tolerance := 0.000000000001;
while (upperBound - lowerBound) > tolerance do
begin
    assumedROI := (upperBound + lowerBound) / 2;
    balance := initialBalance;
    T := size(costs);
    for year := 1 to T, do
    begin
        liquidAssets := liquidPercentage(t) * balance;
        cashPercentage := max(0.08 - (0.08 / 5) * (year - 1), 0);
        totalLiquid := liquidAssets + cashPercentage * balance;
        if costs[year] > totalLiquid then
            break;
        balance := balance * (1 + assumedROI / 100) - Costs[year];
        if balance < 0 then
            break;
        end;
        if balance < 0 then
            lowerBound := assumedROI
        else
            upperBound := assumedROI;

    end;
end;
return (upperBound + lowerBound) / 2;
end;
```

Algorithm 2

The following is a brief introduction of Algorithm 2, which was used to solve for the minimum required capital injection (see ***b)*** of the constrained optimization problem in Sect. 17.4.2).

This algorithm is analogous to Algorithm 1, as both employ the bisection method to solve for a target value within a specified tolerance. However, while Algorithm 1 aims to find the smallest possible ROI, Algorithm 2 focuses on determining the minimum initial balance required for a given ROI. The primary difference lies in their objectives: Algorithm 1 iterates to adjust ROI, whereas Algorithm 2 iterates to adjust the initial fund balance.

Algorithm 2: Algorithm to Calculate Minimum Initial Balance

```
input:
    an array of double values costs representing annual cost projections,
    a double value targetROI representing the target return on investment (ROI)
output:
    a double value initialBalance representing the resulting minimum initial balance required

begin
initialization of variables:
    lowerBound := 0;
    upperBound := Double.MAX_VALUE;
    tolerance := 0.00001;
while (upperBound - lowerBound) > tolerance do
begin
    initialBalance := (upperBound + lowerBound) / 2;
    balance := initialBalance;
    T := size(costs);
    cashFlowPercent := 0.08;
    for year := 1 to T, do
    begin
            cost := costs[year];
            if year < 5 then
                balance := balance * (1 + targetROI / 100) * (1 - cashFlowPercent) - cost;
                cashFlowPercent := cashFlowPercent - (0.08 / 8);
            else
              balance := balance * (1 + targetROI / 100) - cost;
              if year < T / 4 then
                liquidAssets := balance * 0.70;
              else if T / 4 ≤ year < T / 2 then
                liquidAssets := balance * 0.60;
              else if T / 2 ≤ year < 3T / 4 then
                liquidAssets := balance * 0.50;
            else
                liquidAssets := balance * 0.40;
            totalLiquid := liquidAssets + cashFlowPercent * balance;
            if cost > totalLiquid then
                break;
            balance := balance * (1 + targetROI / 100) - cost;
            if balance < 0 then
                break;
    end;
    if balance < 0 then
            lowerBound := initialBalance;
    else
            upperBound := initialBalance;
end;
return (upperBound + lowerBound) / 2;
end;
```

Model Parameters Definition

The following is a brief description of the parameters introduced in the model as outlined in Sect. 17.4.1.

Capital $_{ts}$ = Costs associated with building the facility, including land, construction materials, equipment, and labor.

Operation $_{ts}$ = Operation and maintenance costs of the storage facilities and costs associated with environmental monitoring and surveillance of the storage facility and the surrounding area.

Transport $_{ts}$ = Costs associated with transporting the waste from the point of generation to the storage facility. This includes the cost of packaging and transporting the waste and any necessary infrastructure improvements to support transportation.

Disposal $_{ts}$ = Costs associated with the final disposal of the waste. This includes the cost of preparing the waste for disposal, the cost of disposal containers, and the cost of placing the waste in the final repository.

Regulatory $_{ts}$ = Costs associated with complying with regulatory requirements related to the storage and disposal of nuclear waste. This includes the cost of obtaining permits, monitoring and reporting data, and maintaining compliance with environmental and safety regulations.

Decontamination and Demolition $_{ts}$ = This can involve several methods, such as chemical or mechanical removal, in situ decontamination (like heating), or simply letting the radioactive material decay over time. Demolition, however, is the safe tearing down of the structures at the nuclear facility once they have been sufficiently decontaminated.

Site Restoration ts = After the nuclear facility has been decommissioned and demolished, the site often needs to be restored. This can involve various activities, including removing any remaining structures, remediation of contaminated soil or groundwater, and regrading or replanting to restore the site to a safe and usable condition.

Safeguarding $_{is}$ = Throughout the decommissioning and dismantling process; ongoing security measures are needed to protect against theft or sabotage of radioactive materials. Safeguarding also refers to the measures taken to ensure the safety of workers and the public during this process. It could include personal protective equipment, safety training, air and water monitoring, emergency response planning, and other safety measures.

Misc. $_{ts}$ = This could contain different types of special costs varying from insurance, security, research, and development costs for NWM tasks.

References

Bärenbold, Rebekka, Muhammad Maladoh Bah, Rebecca Lordan-Perret, Björn Steigerwald, Christian von Hirschhausen, Ben Wealer, Hannes Weigt, and Alexander Wimmers. 2024. "Decommissioning of Commercial Nuclear Power Plants: Insights from a Multiple-Case Study." *Renewable and Sustainable Energy Reviews* 201:114621. https://doi.org/10.1016/j.rser.2024.114621.

BGE. 2022. "Zeitliche Betrachtung Des Standortauswahlverfahrens." Bundesgesellschaft für Endlagerung. https://www.bge.de/fileadmin/user_upload/Standortsuche/Wesentliche_Unterlagen/05_-_Meilensteine/Zeitliche_Betrachtung_des_Standortauswahlverfahrens_2022/20221028_STA_Zusammenfassung_der_Rahmenterminplanung_Schritt_2_Phase_I_und_Zeitschaetzungen_Phase_II_und_III.pdf.

BGE. 2023. "Zeitliche Betrachtung des Standortauswahlverfahrens aus Sicht der BGE." Bundesgesellschaft für Endlagerung. https://www.bge.de/fileadmin/user_upload/Standortsuche/Wesentliche_Unterlagen/05_-_Meilensteine/Zeitliche_Betrachtung_des_Standortauswahlverfahrens_2022/20221216_Zeitliche_Betrachtung_StandAW-48_barrierefrei.pdf.

BGZ. 2023. "Sichere Zwischenlagerung." Bundesgesellschaft für Zwischenlagerung. https://bgz.de/wp-content/uploads/2023/07/BGZ_Broschuere_Sichere-Zwischenlagerung.pdf.

Bitsch, Florian, Axel Buchner, and Christoph Kaserer. 2010. "Risk, Return and Cash Flow Characteristics of Infrastructure Fund Investments." *EIB Papers* Vol. 15 (No. 1): 106–36. https://papers.ssrn.com/sol3/papers.cfm?abstract_id=1992961.

Briggs, Th., P.L. Kunsch, and B. Mareschal. 1990. "Nuclear Waste Management: An Application of the Multicriteria PROMETHEE Methods." *European Journal of Operational Research* 44 (1): 1–10. https://doi.org/10.1016/0377-2217(90)90308-X.

Brunnengräber, Achim. 2024. "Fondsmodelle Für Den Hochradioaktiven Atommüll. Aber Sind Die Kosten Für Die Entsorgung in Deutschland Und Der Schweiz Überhaupt Berechenbar?" *GWP – Gesellschaft, Wirtschaft, Politik* 73 (4–2024): 481–93. https://doi.org/10.3224/gwp.v73i4.09.

Brunnengräber, Achim, and Albert Denk. 2024. "Generationenkapital: Der Staat als Zocker." *Blätter für deutsche und internationale Politik* 4. https://www.blaetter.de/ausgabe/2024/april/generationenkapital-der-staat-als-zocker.

ESK. 2023. "Verlängerte Zwischenlagerung Bestrahlter Brennelemente Und Sonstiger Hochradioaktiver Abfälle in Abhängigkeit von Der Auswahl Des Endlagerstandorts." Entsorgungskommission. https://umweltfairaendern.de/wp-content/uploads/2023/05/ESK_Positionspapier_verlaengerte_ZL_40plus_ESK105_23032023.pdf.

European Central Bank. 2021. *Inflation Expectations and Their Role in Eurosystem Forecasting.* LU: Publications Office. https://doi.org/10.2866/878799.

Hirschhausen, Christian von, and Alexander Wimmers. 2023. "Rückbau von Kernkraftwerken und Entsorgung radioaktiver Abfälle in Deutschland: ordnungspolitischer Handlungsbedarf." *Perspektiven der Wirtschaftspolitik* 24 (3): 286–301. https://doi.org/10.1515/pwp-2023-0032.

IAEA. 2007. *Cost Considerations and Financing Mechanisms for the Disposal of Low and Intermediate Level Radioactive Waste.* IAEA-TECDOC 1552. Vienna: International Atomic Energy Agency. https://www-pub.iaea.org/MTCD/Publications/PDF/te_1552_web.pdf.

IAEA. 2009a. *Classification of Radioactive Waste: General Safety Guide.* Vienna: International Atomic Energy Agency. https://www-pub.iaea.org/MTCD/Publications/PDF/Pub1419_web.pdf.

IAEA. 2009b. *Costing of Spent Nuclear Fuel Storage.* IAEA Nuclear Energy Series, NF-T-3.5. Vienna: Internat. Atomic Energy Agency. https://www.iaea.org/sites/default/files/costingfuel0909.pdf.

Irrek, Wolfgang. 2023. "Kosten der Langzeit-Zwischenlagerung und verlängerten Standortsuche." Presented at the „Zwischen. Sicher? Ende?" - atommuellreport.de, Hannover, June 23. https://www.atommuellreport.de/fileadmin/Dateien/bilder/Projekte-Termine/20230623_AMR-Fachtagung_IrrekKosten.pdf.

Johnstone, Phil, and Andy Stirling. 2020. "Beyond and beneath Megaprojects: Exploring Submerged Drivers of Nuclear Infrastructures." *Journal of Mega Infrastructure & Sustainable Development* 2 (2): 220–41. https://doi.org/10.1080/24724718.2021.2012351.

KENFO. 2024a. "Digitale Pressemappe zum Pressehintergrundgespräch mit „unter 1"-Teil." Fonds zur Finanzierung der kerntechnischen Entsorgung. Berlin. https://www.kenfo.de/fileadmin/user_upload/pressemitteilungen/kenfo_digitale_pressemappe_2024.pdf.

KENFO. 2024b. "Geschäftsbericht zum 31.12.2023." Fonds zur Finanzierung der kerntechnischen Entsorgung. Berlin. https://www.kenfo.de/fileadmin/user_upload/geschaeftsberichte/kenfo_geschaeftsbericht_2023.pdf.

Krohn, Judith, Angelika Spieth-Achtnich, Silvia Schütte, Mbah Melanie, Alexandra Lampke, Katja Hünecke, Dörte Fouquet, Jörg Kuhbier, and Tobias Reinhardt. 2024. "Unterstützung des BASE bei der Prozessanalyse des Standortauswahlverfahrens (PaSta)." Abschlussbericht FKZ 4718F10001. Bundesamt für die Sicherheit der nuklearen Entsorgung. https://www.base.bund.de/SharedDocs/Downloads/BASE/DE/fachinfo/fa/pasta_abschlussbericht.pdf?__blob=publicationFile&v=3.

Lordan-Perret, Rebecca, Rebekka Bärenbold, Hannes Weigt, and Robert Rosner. 2023. "An Ex-Ante Method to Verify Commercial U.S. Nuclear Power Plant Decommissioning Cost Estimates." *Energy Economics* 127 (November):107023. https://doi.org/10.1016/j.eneco.2023.107023.

Mikus, Anja. 2020. "Wir investieren automatisch in Technologiewerte" Börsen-Zeitung, 7 November. https://www.kenfo.de/fileadmin/user_upload/dokumente/anja_mikus_boersenzeitung_kenfo.pdf.

OECD. 2010. *Cost Estimation for Decommissioning: An International Overview of Cost Elements, Estimation Practices and Reporting Requirements.* Radioactive Waste Management. OECD. https://doi.org/10.1787/9789264106864-en.

Ott, Konrad, Klaus-Jürgen Röhlig, Fabian Präger, and Christian von Hirschhausen. 2024. "Für mehr Tempo in der Endlagerung hochradioaktiver Abfälle." *Forschungsjournal Soziale Bewegungen Plus* 37 (4). https://forschungsjournal.de/fjsb/wp-content/uploads/fjsb-plus_2024-4_ott.pdf.

Rothwell, Geoffrey. 2021. "Spent Nuclear Fuel Storage: What Are the Relationships between Size and Cost of the Alternatives?" *Energy Policy* 150 (March):112126. https://doi.org/10.1016/j.enpol.2020.112126.

Scheer, Dirk, Frank Becker, Thomas Hassel, Peter Hocke, Thorsten Leusmann, and Volker Metz. 2024. "Trittsicherheit auf Zukunftspfaden? Ungewissheitsbewältigung bei der Entsorgung hochradioaktiver Abfälle." In *Entscheidungen in die weite Zukunft*, edited by Anne Eckhardt, Frank Becker, Volker Mintzlaff, Dirk Scheer, and Roman Seidl, 113–40. Energiepolitik und Klimaschutz. Energy Policy and Climate Protection. Wiesbaden: Springer Fachmedien Wiesbaden. https://doi.org/10.1007/978-3-658-42698-9_7.

Schneider, Mycle, Antony Froggatt, Julie Hazemann, Christian von Hirschhausen, M.V. Ramana, Alexander James Wimmers, Nina Schneider, et al. 2023. "World Nuclear Industry Status Report 2023." Paris: Mycle Schneider Consulting. https://www.worldnuclearreport.org/IMG/pdf/wnisr2023-v1-hr.pdf.

Vorwerk, Lukas. 2024. "Wirtschaftssystemdesign und die Bereitstellung und Finanzierung von Energieinfrastrukturen im Kontext der Transformation zur Klimaneutralität. Eine institutionenökonomische Analyse." Dissertation, Berlin: Technische Universität Berlin. https://doi.org/10.14279/depositonce-22321.

Wimmers, Alexander, Rebekka Bärenbold, Muhammad Maladoh Bah, Rebecca Lordan-Perret, Björn Steigerwald, Christian von Hirschhausen, Hannes Weigt, and Ben Wealer. 2023. "Decommissioning of Nuclear Power Plants: Regulation, Financing, and Production." DIW Data Documentation 104. Berlin: DIW Berlin, German Institute for Economic Research. https://www.diw.de/documents/publikationen/73/diw_01.c.864222.de/diw_datadoc_2023-104.pdf.

Wimmers, Alexander, Leonard Göke, Christian von Hirschhausen, and Claudia Kemfert. 2023. "Ökonomische Aspekte der Atomkraft." Kurzgutachten im Auftrag der Bundestagsfraktion Bündnis 90/Die Grünen. Berlin. https://www.gruene-bundestag.de/fileadmin/media/gruenebundestag_de/themen_az/atomausstieg/PDF/Studie_Oekonomische_Aspekte_der_Atomkraft.pdf.

Wimmers, Alexander, and Christian Von Hirschhausen. 2024. "Organizational Models for the Decommissioning of Nuclear Power Plants: Lessons from the United Kingdom and the United States." *Utilities Policy* 91 (December):101843. https://doi.org/10.1016/j.jup.2024.101843.

Wimmers, Alexander and Lukas Vorwerk. 2025. "Finanzierungsfragen vertagen? Ergebnisse eines Workshops zur Sicherstellung der langfristigen Finanzierung der kerntechnischen Entsorgung". *atw - International Journal for Nuclear Power* 70(5), 76–80. https://www.yumpu.com/en/document/view/70713952/atw-international-journal-for-nuclear-power-052025.

WKGT. 2015. "Gutachtliche Stellungnahme zur Bewertung der Rückstellungen im Kernenergiebereich." Berlin: Warth & Klein Grant Thornton AG Wirtschaftsprüfungsgesellschaft. https://www.bmwk.de/Redaktion/DE/Downloads/S-T/stresstestkernenergie.pdf?__blob=publicationFile&v=3.

Yim, Man-Sung. 2022. *Nuclear Waste Management: Science, Technology, and Policy*. Vol. 83. Lecture Notes in Energy. Dordrecht: Springer Netherlands. https://doi.org/10.1007/978-94-024-2106-4.

Chapter 18
Partitioning and Transmutation: Technical Overview and Discussion of Critical Issues

Matthias Englert and Christoph Pistner

18.1 Introduction

One of the main challenges in radioactive waste management pertains to the hazards posed by highly radioactive waste (HLW) contained in spent nuclear fuel (SNF), which contains transuranic (TRU) elements (plutonium and minor actinides (MA) such as americium and curium) and fission products with long half-lives that emit significant radioactivity and heat. Typically, depending on the time spent in the reactor, SNF contains approximately 1% plutonium, around 0.1–0.2% MA, and 4–6% fission products.

In contrast to the internationally accepted strategy of radioactive waste management to store SNF in deep geological facilities (see Chap. 16 of this book), the concept of a partitioning and transmutation (P&T) system is reemerging as a means to segregate the most problematic radionuclides from the bulk waste and transform them into shorter lived isotopes (Scheer et al. 2024). Proponents of P&T systems envision the ease of the constraints on repository designs and the shortening of the required isolation period by reducing both the radioactive inventory and the thermal load of the waste. A series of technical challenges must be overcome to achieve this, including the development of suitable separation processes, fuel and/or target production, and reactor systems capable of efficient transmutation. In this chapter, we provide a brief overview of such P&T systems. We introduce the technical approaches to P&T, including their potential advantages, while also discussing environmental,

This chapter summarizes Chap. 8 of a detailed report funded by the Federal Office for the Safety of Nuclear Waste Management (Pistner et al. 2024, 458–598). Further information on P&T is provided by Kirchner et al. (2015).

M. Englert (✉) · C. Pistner
Öko-Institut e.V., Darmstadt, Germany
e-mail: m.englert@oeko.de

A. Wimmers et al. (eds.), *Nuclear Power*, Studies in Energy, Resource and Environmental Economics, https://doi.org/10.1007/978-3-031-99894-2_18

proliferation-related, and economic challenges; and we provide guidance on the feasibility of current waste management programs.

18.2 Technical Background on P&T Systems

A P&T system requires substantial infrastructure provision and the availability of an array of processes. First, partitioning processes to achieve the chemical and physical separation of elements in SNF from today's commercial reactors and from dedicated transmutation reactors' SNF must work. Second, there must be a fuel production infrastructure to reuse the extracted plutonium. Third, a functioning transmutation process must be implemented to convert long-lived fission products into shorter lived or stable isotopes. Fourth, an established iterative process, i.e., the repetition of the first three steps, must be implemented (Fig. 18.1). Finally, as some long-lived waste will remain in every foreseeable P&T scenario, it must be transferred to a still-required final repository facility for long-term storage.

In practical terms, partitioning involves applying sophisticated chemical and physical separation processes that target specific isotopes. This is achieved by exploiting differences in the chemical behavior or physical properties of the radionuclides present in SNF. Once the most problematic elements are isolated, transmutation employs neutron-induced reactions—either in dedicated reactor systems or within modified conventional reactors—to transform these long-lived isotopes into forms that are either less radiotoxic or more amenable to subsequent decay.

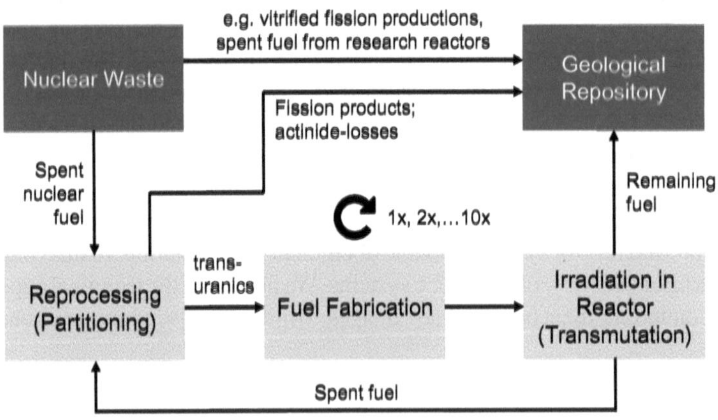

Fig. 18.1 Schematic diagram of a P&T system. *Source* Own depiction based on Pistner et al. (2024, fig. 8–1)

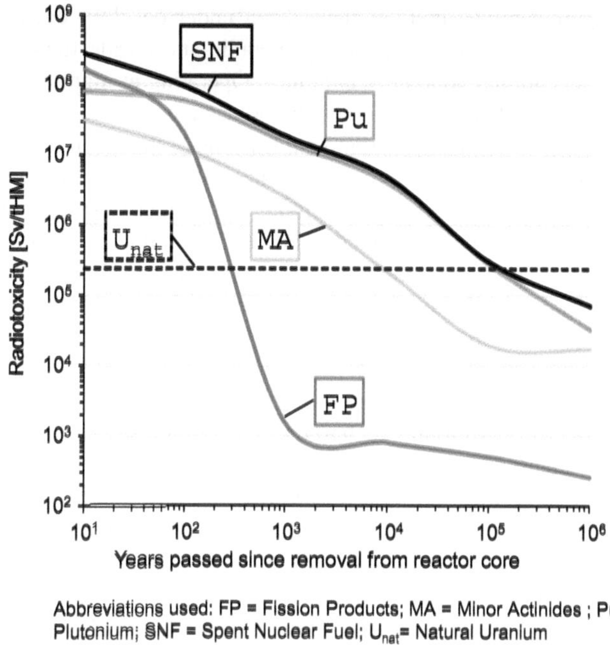

Fig. 18.2 Radiotoxicity in SNF. *Source* Own depiction based on Gompper et al. (2010, fig. 1)

18.2.1 Partitioning

The development of partitioning technology has been defined by a progressive refinement of both aqueous and pyrochemical processes developed to separate key radionuclides from SNF. Both approaches must distinguish and isolate plutonium and MA,[1] as well as select fission products whose long half-lives contribute disproportionately to the long-term radiotoxicity of nuclear waste (Fig. 18.2). Partitioning lays the groundwork for subsequent transmutation by focusing on these elements, offering the potential to convert long-lived isotopes into those that decay more rapidly.

Aqueous reprocessing has long been the main approach to nuclear fuel separation. Traditional methods, such as the Plutonium Uranium Redox Extraction (PUREX) process, are well-established and have been used successfully to recover uranium and plutonium from SNF. However, the PUREX process is not optimized for the selective removal of MAs, and modifications to its chemical framework are required to achieve the high separation factors needed for effective partitioning. Certain improvements in solvent extraction chemistry have led to the development of advanced processing schemes capable of differentiating between closely related elements. However, the processes are highly complex and are mostly done on a laboratory scale.

[1] Minor actinides are neptunium, americium, curium, berkelium, californium, einsteinium, and fermium.

In parallel, pyrochemical methods have been under development for decades, as they are particularly suited for handling metallic and nitride fuels that present challenges to conventional aqueous techniques. Operating at high temperatures in molten salt media, pyrochemical processes take advantage of the solubility properties and redox potentials of actinides and lanthanides. One of the theoretical advantages of these methods is their ability to process fuel in a more compact, potentially more efficient manner, thereby reducing the volume of secondary waste. Furthermore, pyrochemical processes can be designed to operate continuously, offering potential benefits regarding process throughput and overall efficiency. However, these high-temperature methods also bring significant technical challenges, including issues related to material corrosion, the control of process atmospheres, and the management of complex salt chemistries.

The interplay between chemical process design and materials science is critical in developing effective partitioning technologies. Current research focuses on novel solvents and extractants that provide high selectivity while minimizing the generation of secondary waste streams. Advances in membrane technology and solid-phase extraction are also being explored as complementary techniques to traditional liquid-to-liquid extraction processes. These developments are supported by experimental programs and computational modeling efforts that aim to optimize process conditions and ensure robust performance under varying operational scenarios.

In practical applications, the choice between aqueous and pyrochemical partitioning will depend on a variety of factors, including the specific composition of the spent fuel, the desired level of separation, and the economic and environmental trade-offs. However, except for plutonium separation, most of the more advanced separation technologies are still in an early phase of research despite large-scale investments and research going on for decades.

18.2.2 Fuel Fabrication

After separation, the TRU elements are used to produce fresh fuel, either in solid or liquid form. They can be embedded in a uranium or thorium matrix, similar to how plutonium is used in mixed-oxide (MOX) fuel for light-water reactors (LWR). Alternatively, so-called “Inert Matrix Fuels” (IMF) are used which do not contain fertile material. IMF fuels avoid the reproduction of plutonium and are therefore particularly suitable for a P&T strategy to treat an existing waste stock, as the TRU transmutation efficiency is increased compared to MOX fuels. So far, only laboratory-scale research and initial irradiation tests have been carried out for IMFs that are also being developed for accelerator-driven systems (ADS). Thorium fuels could also be conceivable for P&T systems, as no additional plutonium and MA are created from thorium. However, these fuels also produce long-term radiotoxicity. Thorium fuels are still under development.

While MOX production with plutonium has been demonstrated on an industrial scale, the production of fuels with MA has not. It is technically challenging due to the increased activity, especially when using curium. The volatility of americium also places unique demands on the process controls. Other fuel types, such as carbide, nitride, and metallic fuels (or composites like CERCER or CERMET) are also being developed, but these hardly differ in terms of their transmutation properties.

Since only a small fraction of TRU elements contained in fresh transmutation fuel is fissioned during reactor operations, the steps of separating the fuel, producing the fuel, and using it in the reactor must be repeated many times, depending on transmutation efficiency and their envisioned reduction.

The time that is required for partitioning and fuel production (processing time) also includes an interim storage period for SNF to allow for the decay of short-lived radionuclides. Thus, fuel production typically takes three to five years for solid fuels. Molten-salt reactors (MSR) could integrate pyrochemical reprocessing during operation and thus reduce this timeframe. Depending on the various fuel types (oxide, carbide, nitride, metallic, and CERMET), the fuel properties can differ in density and burn-up potential, with nitride fuels containing up to 40% more fissile material than oxide fuels, but also in their partitioning chemistry.

18.2.3 Transmutation

In this step, reactor technologies utilize neutron-induced fission reactions. In particular, two reactor technology lines could be suitable in this domain: reactors using a fast neutron spectrum in a critical operational mode or ADS, which are operated as fast reactors with a subcritical reactor core. Both technologies differ in design, operation, and the specific challenges they could address. The reactor concept itself influences final disposal requirements due to the materials and operating resources used, e.g., neutron activation and surface contamination of the reactor internals, as well as the creation of secondary waste. Thermal reactors are only suitable for transmutation to a limited extent due to the lower number of neutrons produced per fission, which are needed to sustain the chain reaction in the reactor and to transmute the added TRU content. Independent of the choice of reactor technology, different elements of the SNF can potentially be transmuted, as is briefly described in the following.

18.2.3.1 Plutonium Recycling

Plutonium recycling or reprocessing was originally envisaged in a plutonium economy based on the use of fast reactors, aiming at breeding additional fissile material and reaching a "closed fuel cycle" and thus providing new fissile material that was deemed scarce in the early years of commercial nuclear power production. This never materialized due to the sufficient availability of uranium resources and the high cost of SNF reprocessing, MOX fuel production, and fast breeder reactor technology.

Plutonium reprocessing and the one-time use in MOX fuels for LWRs is currently carried out on a large industrial scale only in France and has been conducted in a few other countries worldwide in the past, even though MOX fuel is significantly more expensive than uranium fuels. Thereby, SNF from LWRs is reprocessed, plutonium is separated, and then used to produce MOX-fuel for use in specially licensed LWRs.[2]

Currently, multi-recycling of spent MOX fuel is not carried out on an industrial scale. Handling of spent MOX fuel is challenging due to the high heat load and activity of the fuel. The overall amount of plutonium can be significantly reduced, but large amounts of MA are created. Moreover, the proportion of fissile Pu-239 in the total plutonium contained in the spent fuel decreases; thus, for multiple plutonium recycling procedures, there is a need to compensate for the lower fissile content of the plutonium by increasing plutonium concentrations or by adding other fissile materials, e.g., via increased uranium enrichment levels. According to Grambow (2021), multi-recycling does allow plutonium inventories to be stabilized and radioactive waste with a low plutonium content (99.8% separation efficiency) to be produced, but this requires total inventories of plutonium in the fuel cycle that are about a factor of 10 higher than today's inventories. However, the total reduction in radiotoxicity of the remaining wastes to be disposed of in a geologic repository is only threefold over a period of 100 years. This small reduction must be compared to the radiological risks posed by the overground inventories during this period.

18.2.3.2 Recycling of Minor Actinides

In addition to the transmutation of plutonium, there is the option to add MA to the transmutation system. If the MA are not treated in a P&T system, the effect of P&T on long-term storage remains very limited. Therefore, many reactor concepts and scenarios envisage the simultaneous use of plutonium and MA in the fuel.

There are fuel cycles with reactors that are specifically optimized for burning plutonium and MA, but also for breeding new fissile material (homogeneous vs. heterogeneous core). Fast reactors can be specifically optimized for this purpose. For example, MOX fuels with plutonium can be used in the center of the reactor core, and MOX fuels with plutonium and MA in the periphery. Estimating transmutation rates, transmutation fractions, and final waste generation depends on the reactor configuration and the entire fuel cycle design.

The higher the proportion of MA in the fuel, the more fissile material must be added to the fuel, for example, in the form of U-235 due to the lower plutonium content and the neutrons from fission lost for the transmutation of actinides. While the transmutation processes would reduce the levels of plutonium, americium, and neptunium, curium typically accumulates in the fuel, complicating handling during subsequent partitioning, fuel production, and transport processes due to the heat production and neutron radiation from curium isotopes. The fraction of MA in the fuel is limited to a few percent by the safety properties of the typical fast reactor

[2] Refer to Kuperman (2019) for an overview of global plutonium use.

concepts if the MA is mixed homogeneously into the fuel. In heterogeneous cores, the MA fraction in the outer area of the reactor core can be increased to up to 40%.

18.2.3.3 Transmutation of Fission Products

Long-lived fission products are especially relevant for the long-term safety of a geological repository due to their high mobility in typical geological environments.

Thus, it is necessary to include long-lived fission products in a transmutation scenario, in particular technetium-99, iodine-129, caesium-135, and carbon-14, to have a significant impact on the long-term safety of geologic repositories.

In theory, it is possible to separate and transmute long-lived fission products, but advanced partitioning methods and fuel designs are required. Additionally, for their transmutation, several challenges must be overcome. For example, the transmutation of fission products has a low cross-section of neutron capture compared to most TRU elements, especially in fast neutron environments. For example, selenium-70 and tin-126 have particularly low cross-sections, complicating their transmutation. Additionally, for some elements, long- and short-lived as well as stable isotopes are generated, requiring isotopic separation and an isotope-specific treatment.

The production of suitable target materials for transmutation remains a significant technical hurdle. According to NEA (2002), the need for isotope separation and the complex process of target fabrication are major obstacles. Currently, only technetium-99 and iodine-129 are considered as viable candidates for transmutation. However, only the feasibility of technetium-99 transmutation has been demonstrated. Iodine-129 molecules are thermally and chemically unstable under irradiation. In addition, the transmutation rates are low.

Due to the technical difficulties posed by partitioning and target production for irradiation, as well as the isotope separation that is sometimes necessary, fission product transmutation is technically limited to the nuclides technetium-99 and possibly iodine-129. In countries that have undertaken reprocessing, parts of the fission product waste have already been immobilized in glass. Re-partitioning this vitrified waste is considered to be very challenging. Overall, the transmutation of fission products is rarely pursued today.

18.2.3.4 Specifics of Fast Reactors and ADS

Fast reactor concepts are designed to operate with a neutron spectrum that is significantly higher in energy than that found in thermal reactors. This high-energy environment is particularly effective at inducing fission in TRU nuclides that are not fissioned but transmuted to heavier isotopes in conventional LWRs. In these systems, a significant fraction of the reactor's neutrons is dedicated to fissioning long-lived actinides, thereby continuously reducing the TRU inventory of the fuel.

ADS consists of a subcritical reactor core coupled with a high-intensity neutron source driven by a high-energy particle accelerator. In this configuration, the reactor is deliberately maintained below criticality. This means that the fission chain reaction is not self-sustaining. Instead, an external neutron source—provided by spallation reactions induced by the accelerator—drives the transmutation process. This design may be needed for safety reasons to achieve a high content of TRU elements in the fuel, which complicates reactor control in a critical fast reactor as it diminishes the fraction of delayed neutrons stemming from fission. An ADS can be rapidly shut down by turning off the accelerator, thereby reducing the risk of a runaway chain reaction. Additionally, ADS can accommodate a broader range of fuel compositions and waste streams.

Both fast reactors and ADS still face considerable technical challenges. For fast reactors, issues such as fuel fabrication, core design, and the management of high neutron fluxes demand advanced materials solutions. The high-energy environment within these reactors can accelerate material degradation, necessitating the development of new alloys and cladding materials that can withstand prolonged irradiation. For ADS, the integration of the accelerator with the subcritical core poses challenges related to beam-target interactions, heat removal, and the reliable operation of high-current accelerators in a radiation environment. Although being under development for more than 60 years with substantial funding, the operational history of fast reactor concepts, mainly sodium-cooled concepts, was mixed at best, and most countries have abandoned their development programs. Other concepts, e.g., with a liquid lead coolant, have never reached maturity. Currently, extensive computational modeling and nonradioactive experimental testing are ongoing to optimize these systems and ensure their safe and efficient operation. Still, very few concepts have reached the point of building a prototype or demonstration plant in recent decades. Refer to the abovementioned study by Pistner et al. (2024) for details on these concepts.

18.3 Effects of P&T on the Final Disposal of HLW

18.3.1 Residual Waste Streams

In currently planned geological repositories, TRU elements, long- and short-lived fission products, as well as the uranium from the fuel matrix of LWRs and (mono-recycled) MOX fuels, are to be disposed of together. The requirements for the repository are therefore determined by the properties of these waste groups. The idea behind P&T is to separate and convert the various material groups and thus reduce requirements for final repositories.

For TRU elements, a large part of the waste inventory is to be converted into fission products by fission processes. But even if separation technologies were available for all types of waste and corresponding technologies for transmutation were available for all relevant TRU elements, residual quantities would still remain in the waste streams. Since only a certain percentage of the respective inventory is transmuted per individual P&T cycle, and a fraction of the TRU is lost to the waste streams during reprocessing, a residual quantity of TRU elements always remains in the HLW. Thus, a geological repository is still necessary. The higher the separation efficiency, the less waste remains. To achieve a reduction factor of the TRU of 100, a separation efficiency of 99.9% is necessary with the burn-ups that can be achieved in today's fast reactor systems (NEA 2002).

The types and compositions of waste that remain at the end of a P&T treatment depend heavily on the initial inventories, the specifics of the technologies used, and the process schemes applied. In practice, it would be likely that certain types of waste are not treated via transmutation (e.g., in Germany, vitrified waste canisters from former LWR fuel reprocessing, special waste forms such as wastes from high temperature test reactors THTR/AVR, or other fuels from research reactors). After the transmutation process concludes, residual amounts of TRU elements remain in spent transmutation fuel elements. Finally, when the remaining waste is disposed of in a repository, the long-lived fission products continue to dominate the long-term safety assessment, as their transmutation capability is extremely limited.

Overall, in any foreseeable P&T scenario, a repository for high-level radioactive waste will still be needed, as residual amounts of TRU elements, as well as long-lived fission and activation products, remain in the waste streams. In addition, there would be significantly higher amounts of low- and medium-level radioactive waste to be disposed of from the operation and dismantling of the P&T plants.

18.3.2 Heat and Volume

The size of a final repository and thus the space required for storage and the associated costs are largely determined by the total volume of waste and its thermal output. HLW emits heat and thus requires a minimum distance between the waste containers so that the maximum permissible temperature in the final repository is not exceeded. Therefore, the volume needed in a final repository strongly depends on the thermal output of the waste (Swift and Sassani 2019).

In the first 100–200 years after unloading fuel from a reactor, fission products are particularly relevant for the heat production of highly radioactive waste and would thus determine the size of the repository. The separation of fission products such as strontium and cesium and their separate storage would significantly reduce the size of a repository. This also applies without any need for a transmutation of radionuclides, but would necessitate the safe storage of these fission products for a couple of hundred years. In contrast, actinides only amount to 10–20% of the heat production.

A volume reduction of HLW is achieved by reprocessing, mainly by separating uranium from the remaining components. But as uranium does not produce relevant decay heat, this volume reduction would change little in the required size of the repository, and the separated uranium would have to be disposed of separately.

The reduction in volume and mass of highly radioactive waste through the separation of uranium is shown as an advantage of P&T in many studies. Each time the uranium is reused as a fuel matrix, the amount of waste per unit of energy generated could be reduced (i.e., halved). This applies to scenarios with permanent recycling of uranium. The reprocessed uranium ultimately results in a different waste stream that contains medium-level radioactive waste, which must also be disposed of.

18.3.3 Long-Term Safety

The standard methodology for assessing the radiological risks to humans from underground repositories for SNF is a long-term safety analysis. This considers the conceivable release pathways for radioactive substances from a repository. The long-term safety analysis includes, among other things, geological and man-made barriers, dissolution rates of containers and waste, radionuclide retention in the subsurface, (non-)existence of groundwater pathways to the biosphere, radionuclide mobility during transport to and in the biosphere, and the accumulation of radionuclides in food chains.

Actinides are extremely immobile under liquid-saturated and chemically reducing conditions in the enclosing geological layers of typical geologic repositories. Even when the effects of glaciation on the repository are taken into account and different host rocks such as granite, clay, and salt are assumed, actinides do not contribute to possible future human exposure in most long-term safety analyses. The only exception is the American Yucca Mountain Tuff with its limited geological barriers: In contrast to other host rocks, water can penetrate into the repository.[3] Only in the case of flood-water penetration into the rock, can the potential dose contributions to humans from plutonium and neptunium be comparable to those from the long-lived mobile fission products.

P&T mainly deals with plutonium and other TRU elements. Therefore, the transmutation of TRU elements alone cannot reduce the required isolation periods in the repository. The reduction in long-term risks that can be achieved with P&T treatment of TRU elements is thus less than expected when simply considering the achievable reductions in the radiotoxicity of the wastes to be disposed of in a geologic repository. To make an effective contribution to long-term safety, P&T processes would therefore have to consider mobile, long-lived fission products. The transmutation of fission products is, however, even more difficult to achieve technically. Some long-lived fission products, which cannot be treated by any current form of P&T, contribute significantly to the long-term risks.

In addition, the expected doses from the fission and activation products in a carefully selected repository are, according to calculations, well below the permissible limits. Bollingerfehr et al. (2014) therefore assume that there will be no positive contribution to the dose expected in the long-term from the fluid-borne transport of radionuclides through P&T applications of actinides.

Additionally, it can be seen that in scenarios with a fixed initial inventory of radioactive waste, the total inventories of the radionuclides that determine the dose in

[3] The Yucca Mountain project has been on hold since 2016 with no alternative under consideration.

transport calculations for long-term safety, are not reduced by P&T, but are increased by the fission products created during transmutation (Bollingerfehr et al. 2014). But even in scenarios of longer term use of nuclear energy, fuel cycles with reprocessing have no significant advantage in long-term safety considerations over the current use of LWRs if all waste streams, including low- and medium-level radioactive waste, are taken into account.

18.3.4 Radiotoxicity

Radiotoxicity measures the radiological effects of radioactive substances which have negative health effects when they enter the human body directly, e.g., via eating, drinking, or breathing. Comparisons of the ingestion and inhalation toxicity of radioactive waste can be used as a proxy parameter for very unusual scenarios of failure of the repository environment or human intrusion. But they are not suitable to assess the long-term safety of a geologic repository. Despite this limitation, radiotoxicity is used in many P&T studies to compare transmutation scenarios, since actinides contribute significantly to radiotoxicity, and actinide reduction by P&T also greatly reduces radiotoxicity. In contrast, due to their immobility, actinides play no significant role in long-term safety in most repository environments.

The radiotoxicity measure simplifies risk assessments but ignores all processes between the geological location of the waste and human exposure, as well as the normal development of a repository. It is only suitable for determining risk in the case of very unlikely scenarios. As P&T scenarios would necessitate the handling of the actinides in reprocessing and fuel fabrication facilities, as well as reactors, for typically a couple of hundred years, the benefit of radiotoxicity reduction for unlikely events in a geological repository must be weighed against the risks posed by the radiotoxicity of the radionuclide inventory during a P&T scenario.

The reduction of TRU elements in HLW through P&T significantly reduces the thus defined radiotoxicity, because actinides have a high radiotoxicity that only gradually reduces over very long periods of time. The effectiveness of the radiotoxicity reduction depends largely on the separation efficiency of the partitioning step, e.g., how much residue remains in the waste. Plutonium makes the greatest contribution to radiotoxicity. So, plutonium transmutation would make the greatest contribution to radiotoxicity reduction. But only a transmutation of MAs such as americium, curium, and neptunium would reduce radiotoxicity by more than a desired factor of 100. Different fuel cycles differ only slightly from one another in terms of radiotoxicity reduction (factor 2). Even in a thorium fuel cycle, although very few new actinides are created and the TRU inventories are low, the long-term radiotoxicity is still comparatively high.

18.4 Further Challenges of P&T Implementation

18.4.1 Safety and Regulation

Establishing a P&T system induces several safety and regulatory challenges that must be addressed to ensure the protection of public health and the environment. The complexity of handling high-activity materials, coupled with the novel operational conditions introduced by P&T infrastructure (including alternative reactor concepts), demands a comprehensive approach to safety analysis, risk assessment, and regulatory oversight.

A critical component of this framework is the meticulous management of radioactive materials during both partitioning and transmutation. Partitioning processes involve the separation of highly radioactive isotopes from SNF, a task that requires precise control over chemical reactions and the maintenance of stringent contamination limits. The high radiation associated with these processes necessitates robust shielding, remote handling systems, and automated process controls to protect personnel and minimize exposure risks.

When considering reactor concepts designed for transmutation, the safety challenges become even more pronounced. Fast reactors, for example, operate under conditions of intense neutron flux, which can lead to significant material degradation. The high-energy neutrons not only facilitate the fission of long-lived isotopes but also impose stress on reactor components, requiring the development of advanced materials with enhanced radiation tolerance. Furthermore, the dynamic behavior of these reactors under transient conditions—such as rapid changes in power output—must be carefully modeled and controlled to prevent the onset of unsafe conditions. Historically, fast reactor prototypes have been prone to unreliable operations and even accidents.

ADS concepts introduce a potential layer of safety through their subcritical operation. This approach relies on an external neutron source provided by a particle accelerator, making it possible for the reactor to remain below criticality under normal operating conditions. However, the coupling of high-power accelerators with reactor cores also presents unique challenges, including the need to manage beam-induced heating and potential target degradation. Comprehensive safety analyses of ADS require the integration of accelerator physics, nuclear reactor engineering, and materials science to ensure that all potential failure modes are identified and adequately mitigated.

Regulatory frameworks must evolve in tandem with these technological advances. Traditional regulations, which have been developed primarily for conventional reactor designs and corresponding fuel cycles, may not fully encompass the complexities introduced by a P&T system. In parallel with technical measures, regulatory regimes must also incorporate robust oversight mechanisms to ensure compliance with established safety standards. Licensing procedures, periodic safety reviews, and continuous performance monitoring are essential components of this approach.

Safety and regulation challenges in partitioning and transmutation are not merely technical; they are also deeply related to public perception and political acceptability. Transparent communication of risk assessments, including demonstrable safety records from pilot projects and experimental facilities, is vital to building trust among stakeholders.

18.4.2 Waste Management and Environmental Impact

The integration of P&T into existing waste management infrastructures could offer the potential for enhanced resource recovery. By enabling the recycling of fissile and fertile materials, these technologies could support a more efficient use of nuclear fuel resources. This closed-fuel cycle approach would reduce the volume of waste requiring long-term disposal and increase fuel utilization. But today, better fuel utilization is not needed from a resource availability perspective or for economic reasons. Furthermore, relying on P&T systems to emerge could prevent further progress in ongoing disposal efforts, which, as of now, remain mostly nascent, see Chap. 16 of this book.

18.4.3 Economic and Non-proliferation Challenges

The deployment of P&T systems comes with economic and non-proliferation challenges. The implementation demands substantial financial investments and the careful navigation of complex political and security landscapes.

Economically, the development and deployment of P&T technologies require large-scale investments in research, pilot plant facilities, and eventually, large-scale infrastructure. As of today, these do not exist. The capital costs associated with establishing new chemical processing plants and suitable reactor concepts are substantial. Partly, these costs stem from the need for specialized materials, state-of-the-art control systems, and rigorous safety protocols to manage the highly radioactive materials involved in the processes.

From a technical perspective, ensuring the technical viability of these technologies requires continuous improvements in process efficiency, waste minimization, and operational reliability. Research programs have focused on developing advanced separation techniques that offer higher decontamination factors while minimizing secondary waste generation. Likewise, the design of transmutation systems—whether fast reactors or ADS—must optimize neutron flux, fuel utilization, and system stability to maximize performance and economic return. These technical challenges necessitate extensive computational modeling, experimental validation, and iterative design improvements, all of which contribute to the overall cost structure. Consequently, the widespread implementation of a P&T system lies far ahead from today's perspective.

Beyond the economic considerations, the P&T system presents significant non-proliferation challenges. The chemical separation processes required for partitioning will produce streams containing weapons-usable materials, particularly plutonium and certain minor actinides. This possibility has raised concerns about the diversion of nuclear materials or the clandestine use of these technologies for non-peaceful purposes. To mitigate these risks, strict safeguard measures must be implemented throughout the entire process—from initial fuel handling to the final disposition of separated materials.

The long-term use of P&T technologies will always carry the potential to be used for military purposes in the future, either clandestinely or openly, once a state withdraws from the international nonproliferation regime.

Ultimately, the implementation of P&T systems hinges on a technical innovation coupled with robust economic planning and stringent non-proliferation safeguards. From today's perspective, and an economic viewpoint, the establishment of such a system seems unlikely given the technological readiness and non-availability of the necessary technologies, the high upfront costs, and the to-be-expected high operational cost of a P&T system.

18.5 Conclusion

In conclusion, the advances in P&T research have yielded interesting insights into the potential for reducing the long-term hazards of nuclear waste. However, significant challenges remain that hinder their near-term implementation in nuclear waste management, be that technical, economic, regulatory, or non-proliferation related.

On the technical front, uncertainties persist regarding the long-term reliability and efficiency of reprocessing methods and transmutation systems despite huge development efforts over many decades. While laboratory-scale experiments and pilot projects have demonstrated proof-of-concept for some key technologies, scaling these processes to commercial levels introduces complex challenges, such as ensuring material durability under high neutron fluxes, managing high radiation fields, and maintaining process stability over extended operational periods. These issues necessitate extensive further experimental validation and a much more robust integration of advanced materials and process controls. Further, proposed reactor concepts remain unavailable as of today, and a widespread implementation in the next decade seems unlikely (Böse et al. 2024).

Economically, the high capital and operational costs associated with developing and deploying these technologies remain significant barriers. The initial investments and long-term operational expenses are likely to be prohibitive without substantial public support.

From a regulatory and non-proliferation perspective, the introduction of novel chemical separation processes and reactor systems requires an overhaul of existing frameworks. The current regulatory environment, designed for specific nuclear technologies like LWRs, is not fully equipped to address the specific challenges posed by P&T, leading to potential delays in licensing and implementation. Furthermore, the risks associated with the separation of weapons-usable materials remain a critical concern that must be rigorously managed, while once deployed, reprocessing technologies remain a possible acquisition path for weapons-grade material.

P&T has also little effect on the requirements for long-term storage of HLW and does not make the repository indispensable. The most comprehensive report on P&T by the US National Research Council of the National Academy of Sciences (NRC 1996), came to the same conclusion in 1996: it is highly unlikely, if not absolutely impossible, that a repository facility for HLW can be avoided. More recent comprehensive studies also show the same result, i.e., that P&T fuel cycles are unlikely to contribute to the elimination of the long-term challenges of nuclear waste disposal (Lensa et al. 2007; NEA 2011; Salvatores and Palmiotti 2011; Frieß et al. 2021). In contrast, P&T treatment of the waste inventory of a national nuclear energy system such as Germany would require at least ten (more likely dozens) of transmutation reactors and several reprocessing facilities operating consecutively over a period of at least a hundred (more likely hundreds of) years. The risks posed by the radioactive inventory and nuclear material flows of such a large nuclear infrastructure must be weighed against the promise of P&T to reduce disposal requirements.

Overall, while proponents of P&T systems promise a theoretical pathway to a reduced environmental footprint in nuclear waste management, a cautious and critical approach is warranted. The convergence of unresolved technical issues, high economic demands, and stringent regulatory requirements suggests that further research and a more comprehensive risk and benefits assessment are essential before these technologies can be considered viable options for high-level waste management. Consequentially, as of today, established and internationally recognized waste disposal routes should be followed through, to avoid a further postponement of waste management responsibilities for future generations to deal with.

References

Bollingerfehr, W., D. Buhmann, W. Filbert, and J Mönig. 2014. "Auswirkungen von Partitionierung und Transmutation auf Endlagerkonzepte und Langzeitsicherheit von Endlagern für Wärme entwickelnde radioaktive Abfälle." 318. GRS. https://www.grs.de/de/aktuelles/publikationen/grs-318-auswirkungen-von-partitionierung-und-transmutation-auf.

Böse, Fanny, Alexander Wimmers, Björn Steigerwald, and Christian von Hirschhausen. 2024. "Questioning Nuclear Scale-up Propositions: Availability and Economic Prospects of Light Water, Small Modular and Advanced Reactor Technologies." *Energy Research & Social Science* 110 (April 2024): 103448. https://doi.org/10.1016/j.erss.2024.103448.

Frieß, Friederike, Nikolaus Arnold, Wolfgang Liebert, and Nikolaus Müllner. 2021. "Sicherheitstechnische Analyse und Risikobewertung von Konzepten zu Partitionierungs- und Transmutationsanlagen für hochradioaktive Abfälle." Wissenschaftliches Gutachten im Auftrag des Bundesamtes für die Sicherheit der nuklearen Entsorgung (BASE) urn:nbn:de:0221-2021030826033. Berlin; Wien: Institut für Sicherheits- und Risikowissenschaften Universität für Bodenkultur (BOKU). https://www.base.bund.de/shareddocs/downloads/de/berichte/kt/gutachten-partitionierung-und-transmutation.pdf?__blob=publicationFile&v=1

Gompper, Klaus, Andreas Geist, and Horst Geckeis. 2010. "Actinoidenabtrennung Aus Hochradioaktiven Abfällen." *Nachrichten Aus Der Chemie* 58 (10): 1015–19. https://doi.org/10.1002/nadc.201075744.

Grambow, B. 2021. "Some Impacts of Advanced Fuel Cycle Options on Waste Management and Long Term Disposal Safety." https://www.nationalacademies.org/event/06-07-2021/merits-and-viability-of-different-nuclearfuel-cycles-and-technology-options-and-the-waste-aspects-of-advanced-nuclearreactorsmeeting-6-part-2-june-7-2021, zuletzt geprüft am 10.01.24.

Kirchner, Gerald, Matthias Englert, Christoph Pistner, Beate Kallenbach-Herbert, and Julia Mareike Neles. 2015. "Gutachten „Transmutation"." K-MAT 48. Kommission Lagerung hoch radioaktiver Abfallstoffe. Hamburg, Darmstadt: Öko-Institut, Universität Hamburg, ZNF. https://www.bundestag.de/resource/blob/400438/f54e3da4bbe76395bce2e40721212379/kmat_48-data.pdf.

Kuperman, Alan J. 2019. "Challenges of Plutonium Fuel Fabrication: Explaining the Decline of Spent Fuel Recycling." *International Journal of Nuclear Governance, Economy and Ecology* 4 (4): 302. https://doi.org/10.1504/IJNGEE.2019.106023.

Lensa, W., R. Nabbi, and M. Rossbach. 2007. "Impact of Partitioning, Transmutation and Waste Reduction Technologies on the Final Nuclear Waste Disposal, SYNTHESIS REPORT." 15. Schriften Des Forschungszentrums Jülich, Reihe Energie & Umwelt/Energy & Environment. https://inis.iaea.org/collection/NCLCollectionStore/_Public/39/117/39117966.pdf?r=1&r=1, last accessed on 15 Jan 2021.

NEA. 2002. "Accelerator-Driven Systems (ADS) and Fast Reactors (FR) in Advanced Nuclear Fuel Cycles." Paris: Nuclear Energy Agency, OECD.

NEA. 2011. *Potential Benefits and Impacts of Advanced Nuclear Fuel Cycles with Actinide Partitioning and Transmutation [Electronic Resource.* Paris: Nuclear Energy Agency, Organisation for Economic Co-operation and Development. https://www.oecd-nea.org/jcms/pl_14572/potential-benefits-and-impacts-of-advanced-nuclear-fuel-cycles-with-actinide-partitioning-and-transmutation?details=true.

NRC. 1996. *Nuclear Wastes: Technologies for Separations and Transmutation.* Washington, D.C.: National Academies Press. https://doi.org/10.17226/4912.

Pistner, Christoph, Matthias Englert, Christian von Hirschhausen, Fanny Böse, Björn Steigerwald, Lukas Gast, and Richard Donderer. 2024. "Analysis and Evaluation of the Development Status, Safety and Regulatory Framework for So-Called Novel Reactor Concepts." BASE – Research reports on the safety of nuclear waste disposal BASE-018/24. Berlin: Federal Office for the Saftey of Nuclear Waste Management. https://www.base.bund.de/SharedDocs/Downloads/BASE/DE/fachinfo/fa/Abschlussbericht_neuartige_Reaktorkonzepte_2024.pdf?__blob=publicationFile&v=5.

Salvatores, M., and G. Palmiotti. 2011. "Radioactive Waste Partitioning and Transmutation within Advanced Fuel Cycles: Achievements and Challenges." *Progress in Particle and Nuclear Physics* 66 (1): 144–66. https://doi.org/10.1016/j.ppnp.2010.10.001.

Scheer, Dirk, Frank Becker, Thomas Hassel, Peter Hocke, Thorsten Leusmann, and Volker Metz. 2024. "Trittsicherheit auf Zukunftspfaden? Ungewissheitsbewältigung bei der Entsorgung hochradioaktiver Abfälle." In *Entscheidungen in die weite Zukunft*, edited by Anne Eckhardt, Frank Becker, Volker Mintzlaff, Dirk Scheer, and Roman Seidl, 113–40. Energiepolitik und Klimaschutz. Energy Policy and Climate Protection. Wiesbaden: Springer Fachmedien Wiesbaden. https://doi.org/10.1007/978-3-658-42698-9_7.

Swift, P.N., and Sassani. 2019. "Choices on Permanent Disposal of High-Activity Radioactive Wastes." SAND2019-5941C. SNL - Sandia National Laboratories. https://www.osti.gov/biblio/1640197.

Correction to: Nuclear Power Plays: Complexities of Armament, Disarmament, and (Non-)Proliferation of Nuclear Weapons

Maren Vieluf, Sophie Kretzschmar, and Elisabeth Saar

Correction to:
Chapter 6 in: A. Wimmers et al. (eds.), *Nuclear Power*, Studies in Energy, Resource and Environmental Economics, https://doi.org/10.1007/978-3-031-99894-2_6

This chapter were inadvertently published with incorrect disclaimer. As per the request and confirmation from responsible editor, the disclaimer has been updated. The erratum book has been updated with the changes.

The updated version of this chapter can be found at
https://doi.org/10.1007/978-3-031-99894-2_6

A. Wimmers et al. (eds.), *Nuclear Power*, Studies in Energy, Resource and Environmental Economics, https://doi.org/10.1007/978-3-031-99894-2_19

MIX
Papier aus verantwortungsvollen Quellen
Paper from responsible sources
FSC® C105338

If you have any concerns about our products,
you can contact us on
ProductSafety@springernature.com

In case Publisher is established outside the EU,
the EU authorized representative is:
Springer Nature Customer Service Center GmbH
Europaplatz 3, 69115 Heidelberg, Germany

Printed by Libri Plureos GmbH
in Hamburg, Germany